Straight from the Book

Titu Andreescu
Gabriel Dospinescu

Straight from the Book

XYZ Press

Titu Andreescu
University of Texas at Dallas

Gabriel Dospinescu
École Normale Supérieure, Lyon

Library of Congress Control Number: 2012951362

ISBN-13: 978-0-9799269-3-8 ISBN-10: 0-9799269-3-9

9 8 7 6 5 4 3 2 1

www.awesomemath.org

Cover design by Iury Ulzutuev

The only way to learn mathematics is to do mathematics.

–Paul Halmos

Foreword

This book is a follow-on from the authors' earlier book, 'Problems from the Book'. However, it can certainly be read as a stand-alone book: it is not vital to have read the earlier book.

The previous book was based around a collection of problems. In contrast, this book is based around a collection of solutions. These are solutions to some of the (often extremely challenging) problems from the earlier book. The topics chosen reflect those from the first twelve chapters of the previous book: so we have Cauchy-Schwarz, Algebraic Number Theory, Formal Series, Lagrange Interpolation, to name but a few.

The book is one of the most remarkable mathematical texts I have ever seen. First of all, there is the richness of the problems, and the huge variety of solutions. The authors try to give several solutions to each problem, and moreover give insight about why each proof is the way it is, in what way the solutions differ from each other, and so on. The amount of work that has been put in, to compile and interrelate these solutions, is simply staggering. There is enough here to keep any devotee of problems going for years and years.

Secondly, the book is far more than a collection of solutions. The solutions are used as motivation for the introduction of some very clear expositions of mathematics. And this is modern, current, up-to-the-minute mathematics. For example, a discussion of Extremal Graph Theory leads to the celebrated Szemerédi-Trotter theorem on crossing numbers, and to the amazing applications of this by Székely and then on to the very recent sum-product estimates of Elekes, Bourgain, Katz, Tao and others. This is absolutely state-of-the-art material. It is presented very clearly: in fact, it is probably the best exposition of this that I have seen in print.

As another example, the Cauchy-Schwarz section leads on to the developements of sieve theory, like the Large Sieve of Linnik and the Turán-Kubilius inequality. And again, everything is incredibly clearly presented. The same applies to the very large sections on Algebraic Number Theory, on p-adic Analysis, and many others.

It is quite remarkable that the authors even know so much current mathematics – I do not think any of my colleagues would be so well-informed over so wide an area. It is also remarkable that, at least in the areas in which I am competent to judge, their explanations of these topics are polished and exceptionally well thought-out: they give just the right words to help someone understand what is going on.

Overall, this seems to me like an 'instant classic'. There is so much material, of such a high quality, wherever one turns. Indeed, if one opens the book at random (as I have done several times), one is pulled in immediately by the lovely exposition. Everyone who loves mathematics and mathematical thinking should acquire this book.

Imre Leader
Professor of Pure Mathematics
University of Cambridge

Preface

This book is a compilation of many suggestions, much advice, an even more hard work. Its main objective is to provide solutions to the problems which were originally proposed in the first 12 chapters of *Problems from the Book*. We were not able to provide full solutions in our first volume due to the lack of space. In addition, the statements of the proposed problems contained typos and some elementary mistakes which needed further editing. Finally, the problems were also considered to be quite difficult to tackle. With these points in mind, we came up with a two-part plan: to correct the identified errors and to publish comprehensive solutions to the problems.

The first task, editing the statements of the proposed problems, was simple and has already been completed in the second edition of *Problems from the Book*. Although we focused on changing several problems, we also introduced many new ones. The second task, providing full solutions, however, proved to be more challenging than expected, so we asked for help. We created a forum on www.mathlinks.ro (a familiar site for problem-solving enthusiasts) where solutions to the proposed problems were gathered. It was a great pleasure to witness the passion with which some of the best problem-solvers on *mathlinks* attacked these tough-nuts. This new book is the result of their common effort, and we thank them.

Providing solutions to every problem within the limited space of one volume turned out to be an optimistic plan. Only the solutions to problems from the first 12 chapters of the second edition of *Problems from the Book* are presented here. Furthermore, many of the problems are difficult and require a rather extensive mathematical background. We decided, therefore, to complement the problems and solutions with a series of addenda, using various problems as starting points for excursions into "real mathematics". Although we never underestimated the role of problem solving, we strongly believe that the reader will benefit more from embarking on a mathematical journey rather than navigating a huge list of scattered problems. This book tries to reconcile problem-solving with "professional mathematics".

Let us now delve into the structure of the book which consists of 12 chapters and presents the problems and solutions proposed in the corresponding chapters of *Problems from the Book*, second edition. Many of these problems are fairly difficult and different approaches are presented for a majority of them. At the end of each chapter, we acknowledge those who provided solutions. Some chapters are followed by one or two addenda, which present topics of more advanced mathematics stemming from the elementary topics discussed in the problems of that chapter.

The first two chapters focus on elementary algebraic inequalities (a notable caveat is that some of the problems are quite challenging) and there is not much to comment regarding these chapters except for the fact that algebraic inequalities have proven fairly popular at mathematics competitions. To provide relief from this rather dry landscape (the reader will notice that most of the problems in these two chapters start with "Let a, b, c be positive real numbers"), we included an addendum presenting deep applications of the Cauchy-Schwarz inequality in analytic number theory. For instance, we discuss Gallagher's sieve, Linnik's large sieve and its version due to Montgomery. We apply these results to the distribution of prime numbers (for instance Brun's famous theorem stating that the sum of the inverses of the twin primes converges). The note-worthy Turán-Kubilius inequality and its classical applications (the Hardy-Ramanujan theorem on the distribution of prime factors of n, Erdős' multiplication table problem, Wirsing's generalization of the prime number theorem) are also discussed. The reader will be exposed to the power of the Cauchy-Schwarz inequality in real mathematics and, hopefully, will understand that the gymnastics of three-variables algebraic inequalities is not the Holy Grail.

Chapter 3 discusses problems related to the unique factorization of integers and the p-adic valuation maps. Among the topics discussed, we cover the local-global principle (which is extremely helpful in proving divisibilities or arithmetic identities), Legendre's formula giving the p-adic valuation of $n!$, a beneficial elementary result called *lifting the exponent lemma*, as well as more advanced techniques from p-adic analysis. One of the most beautiful results presented in this chapter in the celebrated Skolem-Mahler-Lech theorem concerning the zeros of a linearly recursive sequence. Readers will perhaps appreciate the applications of p-adic analysis, which is covered extensively in a long addendum. This addendum discusses, from a foundational level, the arithmetic of p-adic numbers, a subject that plays a central role in modern number theory. Once the basic groundwork has been laid, we discuss the p-adic analogues of classical functions (exponential, logarithm, Gamma function) and their applications to difficult congruences (for instance, Kazandzidis' famous supercongruence). This serves as a good opportunity to explore the arithmetic of Bernoulli numbers, Volkenborn's theory of p-adic integration, Mahler and Amice's classical theorems characterizing continuous and locally analytic functions on the ring of p-adic integers or Morita's construction of the p-adic Gamma function. A second addendum to this chapter discusses various classical estimates on prime numbers, which are used throughout the book.

Chapter 4 discusses problems and elementary topics related to prime numbers of the form $4k+1$ and $4k+3$. The most intriguing problem discussed is, without any doubt, Cohn's renowned theorem characterizing the perfect squares in the Lucas sequence. Chapter 5 is dedicated again to the yoga of algebraic inequalities and is followed by an addendum discussing applications of Hölder's inequality.

Chapter 6 focuses on extremal graph theory. Most of the problems revolve around Turán's theorem, however the reader will also be exposed to topics such as chromatic numbers, bipartite graphs, etc. This chapter is followed by a relatively advanced addendum, discussing various topics related to the Szemerédi-Trotter theorem, which gives bounds for the number of incidences between a set of points and a set of curves. We discuss the theorem's classical probabilistic proof, its generalization to multi-graphs due to Székely and its application to the sum-product problem due to Elekes, as well as more recent developments due to Bourgain, Katz, and Tao. These results are then applied to natural and nontrivial geometric questions (for instance: what is the least number of distinct distances determined by n points in the plane? what is the maximal number of triangles of the same area?). Finally, another addendum completes this chapter and is dedicated to the powerful probabilistic method. After a short discussion on finite probability spaces, we provide many examples of combinatorial applications.

Chapter 7 involves combinatorial and number theoretic applications of finite Fourier analysis. The central principles of this chapter include the roots of unity and the fact that congruences between integers can be expressed in terms of sums of powers of roots of unity. To provide the reader with a broader view, we beefed-up this chapter with an addendum discussing Fourier analysis on finite abelian groups and applying it to Gauss sums of Dirichlet characters, additive problems, combinatorial or analytic number theory. For instance, the reader will find a discussion of the Pólya-Vinogradov inequality and Vinogradov's beautiful use of this inequality to deduce a rather strong bound on the least quadratic non-residue mod p. At the same time, we explore Dirichlet's L-functions, culminating in a proof of Dirichlet's theorem on arithmetic progressions. This section is structured to first present the usual analytic proof (since it is really a masterpiece from all points of view), up to some important simplification due to Monsky. At the same time, we also discuss how to turn this into an elementary proof that avoids complex analysis and dramatically uses Abel's summation formula.

Chapter 8 focuses on diverse applications of generating functions. This is an absolutely crucial tool in combinatorics, be it additive or enumerative. In this section, the reader will have the opportunity to explore enumerative problems (Catalan's problem, counting the number of solutions of linear diophantine equations or the number of irreducible polynomials mod p, of fixed degree). We also discuss exotic combinatorial identities or recursive sequences, which can be solved elegantly using generating functions, but also rather challenging congruences that appear so often in number theory. The chapter is followed by an addendum presenting a very classical topic in enumerative combinatorics, Lagrange's inversion formula. Among the applications, let us mention Abel's identity, other derived combinatorial identities, Cayley's theorem on labeled trees, and various related problems.

Chapter 9 is rather extensive, due to the vast nature of the topic covered, algebraic number theory. While we are able only to scratch the surface, nevertheless the reader will find a variety of intriguing techniques and ideas. For instance, we discuss arithmetic properties of cyclotomic polynomials (including Mann's beautiful theorem on linear equations in roots of unity), rationality problems, and various applications of the theorem of symmetric polynomials. In addition, we present techniques rooted in the theory of ideals in number fields, finite fields and p-adic methods. We also give an overview of the elementary algebraic number theory in the addendum following this chapter. After a brief review on ideals, field extensions, and algebraic numbers, we proceed with a discussion of the primitive element theorem and embeddings of number fields into $\mathbf{C}$. We also briefly survey Galois theory, and the fundamental theorem on the prime factorization of ideals in number fields, due to Kummer and Dedekind. Once the foundation has been set, we discuss the prime factorization in quadratic and cyclotomic fields and apply these techniques to basic problems that explore the aforementioned theories. Finally, we discuss various applications of Bauer's theorem and of Chebotarev's theorem. The next addendum is concerned with the fascinating topic of counting the number of solutions modulo p of systems of polynomial equations. We use this as an opportunity to state and prove the basic structural results on finite fields and introduce the Gauss and Jacobi sums. We go ahead and count the number of points over a finite field of a diagonal hypersurface and to compute its zeta function. This is a beautiful theorem of Weil, the very tip of a massive iceberg.

Chapter 10 focuses on the arithmetic of polynomials with integer coefficients. An essential aspect of the discussion concentrates on Mahler expansions, the theory of finite differences, and their applications. The techniques used in this chapter are rather diverse. Although the problems can be considered basic, they are challenging and require advanced problem-solving skills.

Chapter 11 provides respite from the difficult tasks mentioned above. It discusses Lagrange's interpolation formula, allowing a unified presentation of various estimates on polynomials.

The longest and certainly most challenging chapter is the last one. It explores several algebraic techniques in combinatorics. The methods are standard but powerful. The last part of the chapter deals with applications to geometry, presenting some of Dehn's wonderful ideas. The last problem presented in the book is the famous Freiling, Laczkovich, Rinne, Szekeres theorem, a stunningly beautiful application of algebraic combinatorics.

We would like to thank, again, the members of the mathlinks site for their invaluable contribution in providing solutions to many of the problems in this book. Special thanks are due to Richard Stong, who did a remarkable job by pointing out many inaccuracies and suggesting numerous alternative solutions. We would also like to thank Joshua Nichols-Barrer, Kathy Cordeiro and Radu Sorici, who gave the manuscript a readable form and corrected several infelicities. Many of the problems and results in this book were used by the authors in courses at the AwesomeMath Summer Program, and students' reactions guided us in the process of simplifying or adding more details to the discussed problems. We wish to thank them all, for their courage in taking and sticking with these courses, as well as for their valuable suggestions.

Titu Andreescu
titu.andreescu@utdallas.edu

Gabriel Dospinescu
gdospi2002@yahoo.com

Contents

Chapter 1

Some Useful Substitutions

Let us first recall the classical substitutions that will be used in the following problems. All of these are discussed in detail in [3], chapter 1 and the reader is invited to take a closer look there.

Consider three positive real numbers a, b, c. If $abc = 1$, a classical substitution is

$$a = \frac{x}{y}, \quad b = \frac{y}{z}, \quad c = \frac{z}{x}.$$

A less classical one is

$$a = \frac{x}{y+z}, \quad b = \frac{y}{z+x}, \quad c = \frac{z}{x+y}$$

(for some positive real numbers x, y, z) whenever $ab + bc + ca + 2abc = 1$, or its equivalent version

$$a = \frac{y+z}{x}, \quad b = \frac{z+x}{y}, \quad c = \frac{x+y}{z}$$

when $abc = a + b + c + 2$ (the equivalence between the two relations follows from the substitution $(a, b, c) \to \left(\frac{1}{a}, \frac{1}{b}, \frac{1}{c}\right)$). Two other very useful substitutions concern the relations $a^2 + b^2 + c^2 = abc + 4$ and $a^2 + b^2 + c^2 + 2abc = 1$. In the first case, with the extra assumption $\max(a, b, c) \geq 2$, one can find positive numbers x, y, z such that $xyz = 1$ and

$$a = x + \frac{1}{x}, \quad b = y + \frac{1}{y}, \quad c = z + \frac{1}{z}.$$

In the second case one can find an acute-angled triangle ABC such that

$$a = \cos(A), \quad b = \cos(B), \quad c = \cos(C).$$

Of course, in practice one often needs to use a mixture of these substitutions and to be rather familiar with classical identities and inequalities. But experience comes with practice, so let us delve into some exercises and problems to see how things really work.

1.1 The relation $a^2 + b^2 + c^2 = abc + 4$

We start with an easy exercise, based on the resolution of a quadratic equation.

1. Prove that if $a, b, c \geq 0$ satisfy $|a^2 + b^2 + c^2 - 4| = abc$, then

$$(a-2)(b-2) + (b-2)(c-2) + (c-2)(a-2) \geq 0.$$

Titu Andreescu, Gazeta Matematică

Proof. If $\max(a, b, c) < 2$, then everything is clear, so assume that

$$\max(a, b, c) \geq 2.$$

Then $a^2 + b^2 + c^2 - abc = 4$, so there exist positive numbers x, y, z such that $xyz = 1$ and

$$a = x + \frac{1}{x}, \quad b = y + \frac{1}{y}, \quad c = z + \frac{1}{z}.$$

But then $a, b, c \geq 2$ and we are done again. □

Proof. The most natural idea is to consider the hypothesis as a quadratic equation in a, for instance. It becomes $a^2 \pm abc + b^2 + c^2 - 4 = 0$, and solving the equation yields

$$a = \frac{\mp bc \pm \sqrt{(b^2-4)(c^2-4)}}{2}.$$

Thus $(b^2-4)(c^2-4)=(bc\pm 2a)^2$, which can also be written as

$$(b-2)(c-2)=\frac{(bc\pm 2a)^2}{(b+2)(c+2)}\geq 0.$$

Writing similar expressions for the other two variables, we are done. □

The following exercise is trickier and one needs some algebraic skills in order to solve it. We present two solutions, neither of which is really easy.

2. Find all triples x,y,z of positive real numbers such that

$$\begin{cases} x^2+y^2+z^2=xyz+4 \\ xy+yz+zx=2(x+y+z) \end{cases}$$

Proof. By the second equation we have $\max(x,y,z)\geq 2$ and so the first equation yields the existence of positive real numbers a,b,c such that

$$x=a+\frac{1}{a},\quad y=b+\frac{1}{b},\quad z=c+\frac{1}{c}$$

and $abc=1$.

$$\sum\left(ab+\frac{1}{ab}+\frac{a}{b}+\frac{b}{a}\right)=2\sum\left(a+\frac{1}{a}\right).$$

Since $abc=1$, we have

$$\sum\frac{1}{a}=\sum ab,\quad \sum\frac{1}{ab}=\sum a,$$

so the second equation can be written

$$\sum\left(\frac{a}{b}+\frac{b}{a}\right)=\sum a+\sum ab.$$

The left-hand side is also equal to

$$\sum c(a^2+b^2)=\left(\sum a\right)\left(\sum ab\right)-3,$$

because

$$\frac{a}{b} + \frac{b}{a} = \frac{a^2 + b^2}{ab} = c(a^2 + b^2).$$

We deduce that $\left(\sum a - 1\right)\left(\sum ab - 1\right) = 4$. Since $\sum a \geq 3$ and $\sum ab \geq 3$ (by the AM-GM inequality and the fact that $abc = 1$), this can only happen if $a = b = c = 1$ and thus when $x = y = z = 2$. □

Proof. If $x + y = 2$, the second equation yields $xy = 4$, so that $(x - y)^2 = -12$ which is a contradiction. Thus $x + y \neq 2$ and similarly $y + z \neq 2$, $z + x \neq 2$. The second equation yields

$$z = 2 + \frac{4 - xy}{x + y - 2},$$

and a rather brutal insertion of this expression in the first equation gives

$$(x - y)^2 + \left(\frac{4 - xy}{x + y - 2}\right)^2 = \frac{(4 - xy)^2}{2 - x - y}.$$

Unless $x = y = 2$, this implies the inequality $2 > x + y$. If two of the numbers x, y, z are equal to 2, then trivially so is the third one. If not, the previous argument shows that $2 > x + y, 2 > y + z$ and $2 > x + z$. But then the second equation yields

$$x + y + z > \sum_{cyc} x\left(\frac{y + z}{2}\right) = xy + yz + zx = 2(x + y + z),$$

a contradiction. Thus, the only solution is $x = y = z = 2$. □

The following problem hides under a clever algebraic manipulation a very simple AM-GM argument. The inequality is quite strong, as the reader can easily see by trying a brute-force approach.

3. Prove that if $a, b, c \geq 2$ satisfy $a^2 + b^2 + c^2 = abc + 4$, then

$$a + b + c + ab + ac + bc \geq 2\sqrt{(a + b + c + 3)(a^2 + b^2 + c^2 - 3)}.$$

Marian Tetiva

Proof. The hypothesis yields the existence of positive real numbers x, y, z such that

$$a=\frac{x}{y}+\frac{y}{x}, \quad b=\frac{y}{z}+\frac{z}{y}, \quad c=\frac{x}{z}+\frac{z}{x}.$$

The miracle is that both sides of the inequality have very nice factorizations. For the right-hand side, this is easy to observe, since

$$a+b+c+3=(xy+yz+zx)\left(\frac{1}{xy}+\frac{1}{yz}+\frac{1}{zx}\right)$$

and

$$a^2+b^2+c^2-3=\sum\left(\frac{x^2}{y^2}+\frac{y^2}{x^2}\right)+3=\left(\sum x^2\right)\left(\sum\frac{1}{x^2}\right).$$

For the left-hand side, things are more subtle, but one finally reaches the identity

$$a+b+c+ab+bc+ca=\left(\sum x^2\right)\left(\sum\frac{1}{xy}\right)+\left(\sum\frac{1}{x^2}\right)\left(\sum xy\right).$$

The desired inequality becomes simply the AM-GM inequality for two numbers! □

The following problems are rather tricky mixtures of algebraic manipulations and elementary number theory.

4. Find all triplets of positive integers (k, l, m) with sum 2002 and for which the system

$$\begin{cases} \dfrac{x}{y}+\dfrac{y}{x}=k \\[2ex] \dfrac{y}{z}+\dfrac{z}{y}=l \\[2ex] \dfrac{z}{x}+\dfrac{x}{z}=m \end{cases}$$

has real solutions.

Titu Andreescu, proposed for IMO 2002

Proof. The system has solutions if and only if

$$k^2 + l^2 + m^2 = lkm + 4.$$

An easy computation shows that this relation is equivalent to

$$(k+2)(l+2)(m+2) = (k+l+m+2)^2.$$

As $k + l + m = 2002$, we deduce that any solution of the problem satisfies $k + l + m = 2002$ and

$$(k+2)(l+2)(m+2) = (k+l+m+2)^2 = 2004^2 = 2^4 \cdot 3^2 \cdot 167^2.$$

A simple case analysis shows that the only solutions are $k = l = 1000, m = 2$ and its permutations. The result follows. □

5. Solve in positive integers the equation

$$(x+2)(y+2)(z+2) = (x+y+z+2)^2.$$

Titu Andreescu

Proof. A simple algebraic manipulation shows that the equation is equivalent to $x^2+y^2+z^2 = xyz+4$ and, seeing this as a quadratic equation in z, we obtain the equivalent form $(x^2-4)(y^2-4) = (xy-2z)^2$. If $x^2 < 4$, then $y^2 < 4$ as well and so $x = y = 1$, yielding $z = 2$. If $x = 2$, then $y = z$. In all other cases, we can find a positive square-free integer D (which is easily seen to be different from 1) and positive integers u, v such that $x^2 - 4 = Du^2$ and $y^2 - 4 = Dv^2$. Thus, solving the problem comes down to solving the generalized Pell equation $a^2 - Db^2 = 4$, which is a classical topic: this equation always has nontrivial integer solutions and if (a_0, b_0) is the smallest solution with $a_0, b_0 > 0$, then all solutions are given by

$$a_n = \left(\frac{a_0 + b_0\sqrt{D}}{2}\right)^n + \left(\frac{a_0 - b_0\sqrt{D}}{2}\right)^n,$$

$$b_n = \frac{1}{\sqrt{D}} \left[\left(\frac{a_0 + b_0\sqrt{D}}{2} \right)^n - \left(\frac{a_0 - b_0\sqrt{D}}{2} \right)^n \right]. \qquad \square$$

Part of the following problem can be dealt in a classical way, but we do not know how to solve it entirely without using the trick of substitutions.

6. The sequence $(a_n)_{n\geq 0}$ is defined by $a_0 = a_1 = 97$ and

$$a_{n+1} = a_n a_{n-1} + \sqrt{(a_n^2 - 1)(a_{n-1}^2 - 1)}$$

for all $n \geq 1$. Prove that $2 + \sqrt{2 + 2a_n}$ is a perfect square for all n.

Titu Andreescu

Proof. Writing

$$(a_{n+1} - a_n a_{n-1})^2 = (a_n^2 - 1)(a_{n-1}^2 - 1)$$

and simplifying this expression yields

$$a_{n-1}^2 + a_n^2 + a_{n+1}^2 - 2a_n a_{n-1} a_{n+1} = 1,$$

thus

$$(2a_{n-1})^2 + (2a_n)^2 + (2a_{n+1})^2 - (2a_{n-1})(2a_n)(2a_{n+1}) = 4.$$

Since we clearly have $a_n > 2$ for all n, this implies the existence of a sequence $x_n > 1$ such that $2a_n = x_n + x_n^{-1}$ and such that $x_{n+1} = x_n x_{n-1}$. Thus $\log x_n$ satisfies a Fibonacci-type recursive relation and so we can immediately find out the general term of the sequence $(a_n)_n$. Namely, a small computation shows that if we define $\alpha = 2 + \sqrt{3}$, then $x_n = \alpha^{4F_n}$, where F_n is the nth Fibonacci number. Thus

$$a_n = \frac{1}{2}\left(\alpha^{4F_n} + \frac{1}{\alpha^{4F_n}} \right)$$

and so

$$2 + \sqrt{2 + 2a_n} = 2 + \left(\alpha^{2F_n} + \frac{1}{\alpha^{2F_n}} \right) = \left(\alpha^{F_n} + \frac{1}{\alpha^{F_n}} \right)^2.$$

The result follows, since $\alpha^n + \alpha^{-n} \in \mathbb{Z}$ for all n, by the binomial formula. $\square$

Remark 1.1. Here is an alternative proof of the fact that all terms of the sequence are integers, without the use of substitutions. The method that we will use for this problem appears in many other problems. As we saw in the previous solution, the sequence satisfies the recursive relation

$$a_{n+1}^2 + a_n^2 + a_{n-1}^2 - 2a_{n+1}a_na_{n-1} = 1.$$

Writing the same relation for $n+1$ instead of n and subtracting the two yields the identity

$$a_{n+2}^2 - a_{n-1}^2 = 2a_na_{n+1}(a_{n+2} - a_{n-1}).$$

Note that $(a_n)_n$ is an increasing sequence (this follows trivially by induction from the recursive relation), so that we can divide by $a_{n+2} - a_{n-1} \neq 0$ in the previous relation and get $a_{n+2} = 2a_na_{n+1} - a_{n-1}$. The last relation clearly implies that all terms of the sequence are integers (since one can immediately check that this is the case with the first three terms of the sequence). Note however that it does not seem to follow easily that $2 + \sqrt{2 + 2a_n}$ is a perfect square using this method.

Remark 1.2. There are a lot of examples of very complicated recurrence relations that rather unexpectedly yield integers. For instance, the reader can try to prove the following result concerning Somos-5 sequences: let $a_1 = a_2 = \cdots = a_5 = 1$ and let

$$a_{n+5} = \frac{a_{n+1}a_{n+4} + a_{n+2}a_{n+3}}{a_n}$$

for $n \geq 0$. Then a_n is an integer for all n. Similarly one defines Somos-6, Somos-7, etc sequences by the formulas

$$a_0 = a_1 = \cdots = a_5 = 1, \quad a_{n+6} = \frac{a_{n+1}a_{n+5} + a_{n+2}a_{n+4} + a_{n+3}^2}{a_n},$$

$$a_0 = a_1 = \cdots = a_6 = 1, \quad a_{n+7} = \frac{a_{n+1}a_{n+6} + a_{n+2}a_{n+5} + a_{n+3}a_{n+4}}{a_n},$$

etc. One can prove (though this is not easy) that all terms of Somos-6 and Somos-7 sequences are integers. Surprisingly, this fails for Somos-8 sequences (in which case a_{17} is no longer an integer!).

1.2 The relations $abc = a + b + c + 2$ and $ab + bc + ca + 2abc = 1$

The first inequality in the following problem is very useful in practice and we will meet it very often in the following problems.

7. Prove that if $x, y, z > 0$ and $xyz = x + y + z + 2$, then

$$xy + yz + zx \geq 2(x + y + z) \text{ and } \sqrt{x} + \sqrt{y} + \sqrt{z} \leq \frac{3}{2}\sqrt{xyz}.$$

Proof. With the usual substitutions

$$x = \frac{b+c}{a}, \quad y = \frac{c+a}{b}, \quad z = \frac{a+b}{c},$$

the first inequality comes down (after clearing denominators and canceling out similar terms) to Schur's inequality

$$a(a-b)(a-c) + b(b-a)(b-c) + c(c-a)(c-b) \geq 0,$$

while the second one follows by adding up the inequalities

$$\sqrt{\frac{1}{xy}} = \sqrt{\frac{a}{a+c} \cdot \frac{b}{b+c}} \leq \frac{1}{2}\left(\frac{a}{c+a} + \frac{b}{b+c}\right). \qquad \square$$

The reader may find a bit strange the first method of proof of the following problem, but it is actually a quite powerful one. We will use again this kind of argument, see problem 11 for instance. Also, the third solution uses a very useful technique.

8. Let $x, y, z > 0$ be such that $xy + yz + zx = 2(x + y + z)$. Prove that

$$xyz \leq x + y + z + 2.$$

Gabriel Dospinescu, Mircea Lascu

Proof. We will argue by contradiction, assuming that $xyz > x+y+z+2$. We claim that we can find $0 < r < 1$ such that $X = rx, Y = ry, Z = rz$ satisfy $XYZ = X + Y + Z + 2$. Indeed, this comes down to the vanishing of

$$f(r) = r^3xyz - r(x + y + z) - 2$$

between 0 and 1, and this is clear, since $f(0) < 0$ and $f(1) > 0$. Next, the condition $xy + yz + zx = 2(x + y + z)$ yields

$$XY + YZ + ZX = 2r(X + Y + Z) < 2(X + Y + Z).$$

This contradicts the first inequality of problem 7. □

Proof. The condition can also be rewritten in the form

$$(x - 1)(y - 1) + (y - 1)(z - 1) + (z - 1)(x - 1) = 3$$

or in the form

$$xyz - x - y - z - 1 = (x - 1)(y - 1)(z - 1).$$

We will discuss several cases. If $x, y, z \geq 1$, then by the AM-GM inequality and the first identity, we get

$$1 \geq \sqrt[3]{(x - 1)^2(y - 1)^2(z - 1)^2},$$

which yields, thanks to the second identity, the desired estimate.

If $x, y, z \leq 1$ or if only one of the numbers x, y, z is smaller than or equal to 1, then $(x - 1)(y - 1)(z - 1) \leq 0$ and so $xyz \leq x + y + z + 1$ in this case. Finally, if two of the numbers are smaller than 1, say $x, y \leq 1$, the desired inequality can be written in the form $0 \leq x + y + z(1 - xy) + 2$, which is obvious. □

Proof. For three positive real numbers x, y, z consider fixing the first two elementary symmetric polynomials $\sigma_1 = x + y + z$ and $\sigma_2 = xy + yz + zx$ and letting $\sigma_3 = xyz$ vary. This amounts to varying only the constant term in the polynomial

$$p(t) = t^3 - \sigma_1 t^2 + \sigma_2 t - \sigma_3 = (t - x)(t - y)(t - z)$$

and defining x, y, z to be the three roots of this polynomial (in some order). Increasing σ_3, i.e. lowering the constant term, corresponds geometrically to lowering the graph. As we lower the graph, the smallest root increases, thus we maintain three positive real roots until the smallest root becomes a double root. If the double root is at $t = a$ and the larger root at $t = b$, then we have $\sigma_1 = 2a + b$, $\sigma_2 = a^2 + 2ab$ and $\sigma_3 = a^2 b$.

If we fix σ_1 and σ_2 with $\sigma_2 = 2\sigma_1$ as hypothesized, then we find $b = \frac{(4-a)a}{2(a-1)}$ and because $0 < a \leq b$, we see that $1 < a \leq 2$. By the discussion above $xyz = \sigma_3 \leq a^2 b$, so it suffices to show that

$$a^2 b = \frac{(4-a)a^3}{2(a-1)} \leq 2a + b + 2 = 2a + \frac{(4-a)a}{2(a-1)} + 2$$

for $1 < a \leq 2$. But this rearranges to $(a-2)^2(a^2-1) \geq 0$ and we are done. □

The technique used in the first solution of the next problem is rather versatile and the reader is invited to read the addendum 5.A for more examples.

9. Let $x, y, z > 0$ be such that $xy + yz + zx + xyz = 4$. Prove that

$$3\left(\frac{1}{\sqrt{x}} + \frac{1}{\sqrt{y}} + \frac{1}{\sqrt{z}}\right)^2 \geq (x+2)(y+2)(z+2).$$

Gabriel Dospinescu

Proof. Using the usual substitution

$$x = \frac{2a}{b+c}, \quad y = \frac{2b}{c+a}, \quad z = \frac{2c}{a+b},$$

the problem reduces to proving the inequality

$$3\left(\sum \sqrt{\frac{b+c}{a}}\right)^2 \geq 16\frac{(a+b+c)^3}{(a+b)(b+c)(c+a)}.$$

This is a quite strong inequality and it is easy to convince oneself that most applications of classical techniques fail. However, the following smart application of Hölder's inequality does the job:

$$\left(\sum \sqrt{\frac{b+c}{a}}\right)^2 \cdot \left(\sum a(b+c)^2\right) \geq \left(\sum (b+c)\right)^3,$$

so it is enough to prove that

$$3\prod(a+b) \geq 2\sum a(b+c)^2.$$

This reduces after expanding to $\sum a(b-c)^2 \geq 0$, which is clear. $\square$

Proof. First, we get rid of those nasty square roots, via the substitution

$$xy = 4c^2, \quad yz = 4a^2, \quad zx = 4b^2.$$

Then

$$x = \frac{2bc}{a}, \quad y = \frac{2ca}{b}, \quad z = \frac{2ab}{c}$$

and replacing these values in the inequality yields the equivalent form

$$3(a+b+c)^2 \geq 16(a+bc)(b+ca)(c+ab).$$

The hypothesis becomes $a^2+b^2+c^2+2abc = 1$, so that there exists an acute-angled triangle ABC such that $a = \cos A, \quad b = \cos B, \quad c = \cos C$. Next, observe that

$$c + ab = \cos C + \cos A \cdot \cos B = -\cos(A+B) + \cos A \cdot \cos B = \sin A \cdot \sin B.$$

Using this (and similar identities obtained by permuting the variables), the desired inequality becomes

$$\sqrt{3}\sum \cos A \geq 4\prod \sin A.$$

Using the well-known identities

$$s = 4R\prod \cos\frac{A}{2}, \quad r = 4R\prod \sin\frac{A}{2}, \quad \sum \cos A = 1 + \frac{r}{R},$$

the inequality becomes

$$\sqrt{3}\frac{r+R}{R} \geq \frac{2rs}{R^2},$$

or $(R+r)R\sqrt{3} \geq 2rs$. This splits trivially into $R+r \geq 3r$ (Euler's inequality) and $s \leq \frac{3\sqrt{3}}{2}R$, the last one being well-known and easy. □

Here is yet another easy application of problem 7.

10. Let $u, v, w > 0$ be positive real numbers such that

$$u + v + w + \sqrt{uvw} = 4.$$

Prove that

$$\sqrt{\frac{uv}{w}} + \sqrt{\frac{vw}{u}} + \sqrt{\frac{wu}{v}} \geq u + v + w.$$

China TST 2007

Proof. To get rid of those nasty square roots, let us perform the substitution

$$\sqrt{\frac{uv}{w}} = c, \quad \sqrt{\frac{vw}{u}} = a, \quad \sqrt{\frac{uw}{v}} = b.$$

Then $u = bc, v = ca$ and $w = ab$, so that the inequality becomes $a + b + c \geq ab + bc + ca$ and the hypothesis is $ab + bc + ca + abc = 4$. Another substitution

$$a = \frac{2}{x}, \quad b = \frac{2}{y}, \quad c = \frac{2}{z}$$

yields $xyz = x + y + z + 2$. We need to prove that $xy + yz + zx \geq 2(x + y + z)$, which is the first component of problem 7. □

The following problem has some common points with problem 7 and one can actually deduce it from that result. But the proof is not formal.

11. Prove that if $a, b, c > 0$ and $x = a + \frac{1}{b}$, $y = b + \frac{1}{c}$, $z = c + \frac{1}{a}$, then

$$xy + yz + zx \geq 2(x + y + z).$$

Vasile Cârtoaje

Proof. The method is the same as the one used in problem 8. The starting point is the observation that $xyz \geq 2 + x + y + z$. Indeed,

$$xyz = abc + \frac{1}{abc} + a + b + c + \frac{1}{a} + \frac{1}{b} + \frac{1}{c} \geq 2 + x + y + z.$$

Let us assume for sake of contradiction that $xy + yz + zx < 2(x + y + z)$ and let us choose $r \in (0, 1]$ such that $X = rx, Y = ry, Z = rz$ satisfy $XYZ = 2 + X + Y + Z$. This equality is equivalent to $r^3 xyz = 2 + r(x+y+z)$ and such r exists by continuity of the function $f(r) = r^3 xyz - 2 - r(x+y+z)$ and by the fact that $f(1) \geq 0$ and $f(0) < 0$. The hypothesis $xy + yz + zx < 2(x + y + z)$ can also be written as

$$XY + YZ + ZX < 2r(X + Y + Z) \leq 2(X + Y + Z).$$

This contradicts the first inequality in problem 7. □

Proof. As in the previous proof, we obtain that $xyz \geq x + y + z + 2$. Since we obviously have $\min(xy, yz, zx) > 1$, this implies that $z \geq \frac{2+x+y}{xy-1}$ and similar inequalities obtained by permuting the variables. Next, we have

$$x + y + z = a + \frac{1}{a} + b + \frac{1}{b} + c + \frac{1}{c} \geq 6.$$

In particular, there are two numbers, say x, y, such that $x + y \geq 4$. The inequality to be proved is equivalent to $z \geq \frac{2(x+y)-xy}{x+y-2}$, so we are done if we can prove that

$$\frac{2 + x + y}{xy - 1} \geq \frac{2(x + y) - xy}{x + y - 2}.$$

But this is equivalent, after an easy computation (in which it is convenient to denote $S = x + y$ and $P = xy$), to $(xy - y - x)^2 \geq (x - 2)(y - 2)$. If $(x - 2)(y - 2) \leq 0$, this is clear; otherwise $x, y \geq 2$ as $x + y \geq 4$. Let $u = x - 2, v = y - 2$, then the inequality becomes $(uv + u + v)^2 \geq uv$, with $u, v \geq 0$ and it is obvious. □

The form of the following inequality strongly suggests the use of the Cauchy-Schwarz inequality. It turns out that this approach works, but in a rather indirect and mysterious way, which makes the problem rather hard.

12. Prove that for all $a, b, c > 0$,

$$\frac{(b+c-a)^2}{(b+c)^2+a^2} + \frac{(c+a-b)^2}{(c+a)^2+b^2} + \frac{(a+b-c)^2}{(a+b)^2+c^2} \geq \frac{3}{5}.$$

Japan 1997

Proof. With the usual substitution

$$x = \frac{b+c}{a}, \quad y = \frac{a+c}{b}, \quad z = \frac{a+b}{c},$$

the problem asks to prove that for any positive numbers x, y, z such that $xyz = x + y + z + 2$ we have

$$\frac{(x-1)^2}{x^2+1} + \frac{(y-1)^2}{y^2+1} + \frac{(z-1)^2}{z^2+1} \geq \frac{3}{5}.$$

Applying the Cauchy-Schwarz inequality (or what is also called Titu's lemma), we obtain the bound

$$\frac{(x-1)^2}{x^2+1} + \frac{(y-1)^2}{y^2+1} + \frac{(z-1)^2}{z^2+1} \geq \frac{(x+y+z-3)^2}{x^2+y^2+z^2+3}.$$

It remains to prove that the last quantity is at least $\frac{3}{5}$. Let $S = x + y + z$ and $P = xy + yz + zx$, so that the inequality becomes

$$\frac{(S-3)^2}{S^2-2P+3} \geq \frac{3}{5}.$$

Notice that as $\frac{a}{b} + \frac{b}{a} \geq 2$ and cyclic permutations of this, we have $S \geq 6$. Also, by problem 7 we have $P \geq 2S$. Therefore,

$$\frac{(S-3)^2}{S^2-2P+3} \geq \frac{(S-3)^2}{S^2-4S+3} = \frac{S-3}{S-1} = 1 - \frac{2}{S-1} \geq \frac{3}{5}. \qquad \square$$

Remark 1.3. There are other solutions for this problem: some of them are shorter, but not easy to find. Here is a particularly elegant one, based on the linearization technique: the inequality is homogeneous, so we may assume that $a+b+c=1$. We need to prove that

$$\sum_{cyc} \frac{(1-2a)^2}{(1-a)^2+a^2} \geq \frac{3}{5}.$$

The point is to bound from below $\frac{(1-2a)^2}{(1-a)^2+a^2}$ by an affine function of a, suitably chosen. The best choice is the following

$$\frac{(1-2a)^2}{(1-a)^2+a^2} \geq \frac{69}{75} - \frac{54a}{25} \iff \left(a-\frac{1}{3}\right)^2\left(a+\frac{1}{6}\right) \geq 0$$

Of course, the question is how one came up with something like this. Well, it is actually very easy: we need constants A, B such that

$$\frac{(1-2a)^2}{(1-a)^2+a^2} \geq Aa+B$$

for all $a \in [0,1]$, with equality for $a=\frac{1}{3}$. Imposing also the vanishing of the derivative of the difference between the left and right-hand side at $\frac{1}{3}$ yields the desired constants A, B. Once we have the previous estimates, it is very easy to conclude by adding them and taking into account that $a+b+c=1$.

13. Find all real numbers k with the following property: for all positive numbers a, b, c the following inequality holds

$$\left(k+\frac{a}{b+c}\right)\left(k+\frac{b}{c+a}\right)\left(k+\frac{c}{a+b}\right) \geq \left(k+\frac{1}{2}\right)^3.$$

Vietnam TST 2009

Proof. Take first of all $a=b=1$ and arbitrary c to get that

$$\left(k+\frac{1}{c+1}\right)^2\left(k+\frac{c}{2}\right) \geq \left(k+\frac{1}{2}\right)^3.$$

Letting $c \to 0$, we deduce that any k as in the statement must satisfy

$$k(k+1)^2 \geq \left(k + \frac{1}{2}\right)^3,$$

which is equivalent to $4k^2 + 2k \geq 1$. We claim that any such k is a solution of the problem.

Pick such k and perform the usual substitution

$$x = \frac{a}{b+c}, \quad y = \frac{b}{c+a}, \quad z = \frac{c}{a+b}$$

to reduce the problem to

$$k^3 + k^2(x+y+z) + k(xy+yz+zx) + xyz \geq \left(k + \frac{1}{2}\right)^3.$$

Now, we know that $xy + yz + zx + 2xyz = 1$ and it is by now a classical fact (use problem 7 for $1/x, 1/y, 1/z$) that $x + y + z \geq 2(xy + yz + zx)$. Thus, it is enough to ensure that

$$k^3 + (2k^2 + k)(xy+yz+zx) + \frac{1 - (xy+yz+zx)}{2} \geq \left(k + \frac{1}{2}\right)^3.$$

This can be rewritten as

$$\left(2k^2 + k - \frac{1}{2}\right)\left(xy + yz + zx - \frac{3}{4}\right) \geq 0.$$

But the last inequality follows from the fact that $2k^2 + k - \frac{1}{2} \geq 0$ and

$$xy + yz + zx \geq \frac{3}{4},$$

which, by returning to the substitution

$$x = \frac{a}{b+c}, \quad y = \frac{b}{c+a}, \quad z = \frac{c}{a+b},$$

is equivalent to $\sum a(b-c)^2 \geq 0$.

In conclusion, the solutions of the problem are the real numbers k such that $4k^2 + 2k \geq 1$. □

We end this section with a very challenging problem, which answers the following natural question: what would be the analogue of the classical substitution $x = \frac{b+c}{a}, y = \frac{c+a}{b}, z = \frac{a+b}{c}$ when we have five variables? We warn the reader that the first solution is really not natural...

14. Let $a_1, a_2, ..., a_5$ be positive real numbers such that

$$a_1 a_2 ... a_5 = a_1(1 + a_2) + a_2(1 + a_3) + \cdots + a_5(1 + a_1) + 2.$$

What is the least possible value of $\frac{1}{a_1} + \frac{1}{a_2} + \cdots + \frac{1}{a_5}$?

Gabriel Dospinescu, Mathematical Reflections

Proof. Consider the linear system

$$x_2 + x_3 = a_1 x_1, \quad x_3 + x_4 = a_2 x_2, \quad x_4 + x_5 = a_3 x_3,$$

$$x_5 + x_1 = a_4 x_4, \quad x_1 + x_2 = a_5 x_5.$$

We can try to solve this system by expressing, for example, all variables in terms of x_1, x_5. Namely, we can use the fifth, first and second equation to express x_2, x_3, x_4 in terms of x_1, x_5. Replacing the obtained values in the other two equations and eliminating x_1, x_5 between them, we obtain that the system has a nontrivial solution if and only if

$$\prod_{i=1}^{5} a_i = 2 + \sum_{i=1}^{5} a_i + a_1 a_4 + a_4 a_2 + a_2 a_5 + a_5 a_3 + a_3 a_1.$$

All the previous (painful!) computations are left to the reader, since they are far from having anything conceptual. Note that the same result can be obtained by computing the determinant of the associated matrix. Now, the previous relation is almost the one given in the statement, up to a permutation of the variables. So, since the conclusion is symmetric in the five variables, we will assume from now on that the previous relation is satisfied instead of the one given in the statement.

The crucial claim is that under the previous hypothesis, the system has solutions whose unknowns x_i are positive. Probably the easiest way to prove this is to exhibit such a solution. Well, simply take $x_1 = 1$,

$$x_5 = \frac{1 + a_1 + a_2 + a_3 + a_1 a_3}{1 + a_5 + a_3 a_5 + a_2 a_5}$$

and the define

$$x_4 = \frac{1 + x_5}{a_4}, \quad x_3 = \frac{x_4 + x_5}{a_3}, \quad x_2 = \frac{x_3 + x_4}{a_2}.$$

The question is, of course, how on earth did we choose the value of x_5? Well, simply by solving the system, as indicated in the beginning of the solution. Note that we clearly have $x_i > 0$ and an easy computation shows that these are solutions of the system (we only have to check two equations, since three are satisfied by construction).

The conclusion is that if the a_i's satisfy the condition

$$\prod_{i=1}^{5} a_i = 2 + \sum_{i=1}^{5} a_i + a_1 a_4 + a_4 a_2 + a_2 a_5 + a_5 a_3 + a_3 a_1,$$

then we can find $x_i > 0$ such that

$$a_1 = \frac{x_2 + x_3}{x_1}, \quad a_2 = \frac{x_3 + x_4}{x_2}, \ldots, a_5 = \frac{x_1 + x_2}{x_5}.$$

So, the problem reduces to finding the minimal value of

$$\frac{x_1}{x_2 + x_3} + \frac{x_2}{x_3 + x_4} + \cdots + \frac{x_5}{x_1 + x_2}.$$

We claim that the previous expression is always at least $5/2$. By Titu's lemma, this reduces to proving that

$$(x_1 + x_2 + x_3 + x_4 + x_5)^2 \geq \frac{5}{2} \sum_{i<j} x_i x_j.$$

But since

$$2 \sum x_i x_j = \left(\sum x_i \right)^2 - \sum x_i^2,$$

this reduces to $\left(\sum x_i\right)^2 \leq 5\sum x_i^2$, which is Cauchy-Schwarz.

Putting everything together, we obtain that the minimal value of

$$\frac{1}{a_1}+\frac{1}{a_2}+\cdots+\frac{1}{a_5}$$

is $5/2$, attained when all a_i's are equal to 2. Who wants to play now the same game with 7 variables? □

Proof. Here is another proof, also far from being evident... We will use the following nontrivial

Lemma 1.4. *For all nonnegative real numbers $x_1, x_2, \ldots, x_5$ we have*

$$(x_1+x_2+\cdots+x_5)^3 \geq 25(x_1x_2x_3+x_2x_3x_4+\cdots+x_5x_1x_2).$$

Proof. This is not easy at all, but here is a very elegant (but unnatural) proof: consider the identity

$$\begin{aligned} &x_1x_2x_3+x_2x_3x_4+x_3x_4x_5+x_4x_5x_1+x_5x_1x_2\\ &\qquad = x_5(x_1+x_3)(x_2+x_4)+x_2x_3(x_1+x_4-x_5).\end{aligned}$$

We may assume that $x_5 = \min x_i$ (since the inequality is cyclic), so that $x_1+x_4-x_5 \geq 0$. Denoting $x_1+x_2+x_3+x_4 = 4t$ and using the AM-GM inequality, we deduce that

$$x_5(x_1+x_3)(x_2+x_4)+x_2x_3(x_1+x_4-x_5) \leq 4t^2x_5+\left(\frac{4t-x_5}{3}\right)^3.$$

Thus, it remains to prove that

$$4t^2x_5+\left(\frac{4t-x_5}{3}\right)^3 \leq \frac{(x_5+4t)^3}{25}.$$

By homogeneity, we may assume that $x_5 = 1$ and then expanding everything the inequality becomes, with the substitution $4t-1 = 3x$, $(x-1)^2(8x+7) \geq 0$, which is clear. □

Using this lemma for the inverses of the a_i's and denoting

$$S=\frac{1}{a_1}+\frac{1}{a_2}+\cdots+\frac{1}{a_5},$$

we deduce that

$$S^3\geq 25\sum\frac{1}{a_ia_{i+1}a_{i+2}}=25\frac{\sum a_ia_{i+1}}{a_1a_2\cdots a_5}.$$

On the other hand, by Mac-Laurin's and AM-GM inequalities we also have

$$S^5\geq\frac{5^5}{a_1a_2\cdots a_5},\quad S^4\geq 125\frac{a_1+a_2+\cdots+a_5}{a_1a_2\cdots a_5}.$$

Taking into account these inequalities and the hypothesis, we deduce that

$$\frac{S^4}{5^3}+\frac{S^3}{5^2}+\frac{2S^5}{5^5}\geq 1,$$

which immediately implies that $S\geq\frac{5}{2}$. □

1.3 The relation $a^2+b^2+c^2+2abc=1$

The following problem is a rather tricky application of the AM-GM inequality.

15. Prove that in all acute-angled triangles the following inequality holds

$$\left(\frac{\cos A}{\cos B}\right)^2+\left(\frac{\cos B}{\cos C}\right)^2+\left(\frac{\cos C}{\cos A}\right)^2+8\cos A\cos B\cos C\geq 4.$$

Titu Andreescu, MOSP 2000

Proof. Since

$$\cos^2 A+\cos^2 B+\cos^2 C+2\cos A\cos B\cos C=1,$$

the inequality can be written

$$\left(\frac{\cos A}{\cos B}\right)^2 + \left(\frac{\cos B}{\cos C}\right)^2 + \left(\frac{\cos C}{\cos A}\right)^2 \geq 4(\cos^2 A + \cos^2 B + \cos^2 C).$$

Let

$$x = \cos^2 A, \quad y = \cos^2 B, \quad z = \cos^2 C,$$

so we need to prove that

$$\frac{x}{y} + \frac{y}{z} + \frac{z}{x} \geq 4(x + y + z).$$

The point is that we actually have

$$\frac{x}{y} + \frac{y}{z} + \frac{z}{x} \geq \frac{x + y + z}{\sqrt[3]{xyz}}$$

for any positive real numbers x, y, z, as follows by adding up the AM-GM inequalities

$$\frac{2x}{y} + \frac{y}{z} \geq 3\sqrt[3]{\frac{x^2}{yz}}.$$

Thus, it is enough to prove that $xyz \leq \frac{1}{64}$. But this is well-known and easy to prove. □

The following problem is a preparation for a hard problem to come, but has also an independent interest.

16. Prove that in every acute-angled triangle ABC,

$$(\cos A + \cos B)^2 + (\cos B + \cos C)^2 + (\cos C + \cos A)^2 \leq 3.$$

Proof. Letting

$$\cos A = \sqrt{\frac{1}{yz}}, \quad \cos B = \sqrt{\frac{1}{zx}}, \quad \cos C = \sqrt{\frac{1}{xy}}$$

for some positive real x, y, z, we obtain the relation $xyz = x + y + z + 2$ from the classical identity $\sum \cos^2 A + 2\prod \cos A = 1$. Thus we can find positive real numbers a, b, c such that

$$x = \frac{b+c}{a}, \quad y = \frac{c+a}{b}, \quad z = \frac{a+b}{c}.$$

The desired inequality can be written

$$\sum \frac{1}{xy} + \sum \frac{1}{z\sqrt{xy}} \le \frac{3}{2} \Longleftrightarrow \sum x + \sum \sqrt{xy} \le \frac{3}{2} xyz$$

$$\Longleftrightarrow \sum x + \sum \sqrt{xy} \le \frac{3}{2}(x+y+z+2) \Longleftrightarrow \left(\sum \sqrt{x}\right)^2 \le 2\left(\sum x + 3\right).$$

This follows from Cauchy-Schwarz

$$\left(\sum \sqrt{x}\right)^2 \le \left(\sum (b+c)\right) \cdot \left(\sum \frac{1}{a}\right) = 2\sum x + 6. \qquad \square$$

Proof. Since the triangle is acute, there are positive numbers x, y, z such that

$$a^2 = y+z, \quad b^2 = z+x, \quad c^2 = x+y.$$

Then

$$\cos A = \frac{x}{\sqrt{(x+y)(x+z)}}$$

and similar identities for $\cos B, \cos C$. The desired inequality can be written as

$$\sum \frac{x^2}{(x+y)(x+z)} + \sum \frac{yz}{(y+z)\sqrt{(x+y)(x+z)}} \le \frac{3}{2},$$

which (after clearing denominators) is equivalent to

$$\sum \left(x^2(y+z) + yz\sqrt{(x+z)(x+y)}\right) \le \frac{3}{2}(x+y)(y+z)(z+x)$$

and then to

$$\sum_{cyc} \sqrt{yz(x+y) \cdot yz(x+z)} \le 3xyz + \frac{1}{2}\sum_{cyc} yz(y+z).$$

However, this follows immediately from the AM-GM inequality:

$$\sqrt{yz(x+y)\cdot yz(x+z)} \le xyz + \frac{1}{2}yz(y+z)$$

and two similar inequalities. $\square$

Even though the following problem seems classical, it is actually rather hard. Fortunately, we did the difficult job in the previous problem. We also present an independent and very elegant approach due to Oaz Nair and Richard Stong.

17. Prove that if $a, b, c \ge 0$ satisfy $a^2 + b^2 + c^2 + abc = 4$, then

$$0 \le ab + bc + ca - abc \le 2.$$

Titu Andreescu, USAMO 2001

Proof. The inequality on the left is easy: the hypothesis and the AM-GM inequality imply that $abc \le 1$, so that $\min(a, b, c) \le 1$. We may assume that $c = \min(a, b, c)$, so that

$$ab + bc + ca - abc = c(a+b) + ab(1-c) \ge 0.$$

The hard point is proving that $ab+bc+ca-abc \le 2$. Taking into account the hypothesis, this can also be written as

$$\left(\frac{a+b}{2}\right)^2 + \left(\frac{b+c}{2}\right)^2 + \left(\frac{a+c}{2}\right)^2 \le 3.$$

If we denote $a = 2x, b = 2y, c = 2z$ then $x^2 + y^2 + z^2 + 2xyz = 1$ and so there exists a triangle ABC such that $x = \cos(A), y = \cos(B), z = \cos(C)$. Thus, the problem reduces to the previous one (since a, b, c are nonnegative, the triangle ABC is acute angled). $\square$

Proof. Here is a very elegant proof for the hard part of the inequality. Among the three numbers $a-1, b-1, c-1$, two have the same sign, say $b-1$ and $c-1$. Thus $(1-b)(1-c) \ge 0$, implying that $b + c \le bc + 1$ and $ab + ac - abc \le a$.

Thus, it is enough to prove that $a + bc \leq 2$. But the given condition and the fact that a is nonnegative imply that

$$a = \frac{-bc + \sqrt{(b^2 - 4)(c^2 - 4)}}{2}.$$

Thus, it is enough to prove that $\sqrt{(b^2 - 4)(c^2 - 4)} \leq 4 - bc$. Note that $bc \leq 4$, since $b^2 \leq 4$ and $c^2 \leq 4$. Squaring the previous inequality thus yields an equivalent one

$$b^2c^2 - 4b^2 - 4c^2 + 16 \leq b^2c^2 - 8bc + 16,$$

which is obvious. □

We end this chapter with another challenging problem, which reduces after some tricky algebraic manipulations to the infamous "Iran 1996 inequality."

18. Prove that in all acute-angled triangles the following inequality holds

$$\frac{a^2b^2}{c^2} + \frac{a^2c^2}{b^2} + \frac{b^2c^2}{a^2} \geq 9R^2.$$

Nguyen Son Ha

Proof. If A, B, C are the angles of the triangle, the sine law and the identity $\sin^2 x + \cos^2 x = 1$ yield the equivalent form of the inequality

$$\sum \frac{(1 - \cos^2 A)(1 - \cos^2 B)}{1 - \cos^2 C} \geq \frac{9}{4}.$$

Write

$$\cos^2 A = yz, \quad \cos^2 B = zx, \quad \cos^2 C = xy.$$

Then $xy + yz + zx + 2xyz = 1$ and so there exist positive numbers X, Y, Z such that

$$x = \frac{X}{Y + Z}, \quad y = \frac{Y}{X + Z}, \quad z = \frac{Z}{X + Y}.$$

We want to prove that

$$\sum \frac{(1 - yz)(1 - zx)}{1 - xy} \geq \frac{9}{4}.$$

On the other hand, we have

$$1 - yz = \frac{X(X+Y+Z)}{(X+Y)(X+Z)},$$

so that the inequality is equivalent to

$$\sum \frac{XY(X+Y+Z)}{Z(X+Y)^2} \geq \frac{9}{4}.$$

This can also be written in the form

$$XYZ(X+Y+Z) \cdot \sum \frac{1}{(ZX+ZY)^2} \geq \frac{9}{4}$$

and it is a consequence of the following famous inequality

Lemma 1.5. *For all positive numbers a, b, c we have*

$$(ab+bc+ca)\left(\frac{1}{(b+c)^2} + \frac{1}{(c+a)^2} + \frac{1}{(a+b)^2}\right) \geq \frac{9}{4}.$$

Proof. We may assume that $a \geq b \geq c$. First, we show that

$$\frac{1}{(a+b)^2} + \frac{1}{(a+c)^2} + \frac{1}{(b+c)^2} \geq \frac{1}{4ab} + \frac{2}{(a+c)(b+c)}.$$

This can be rewritten

$$\left(\frac{1}{a+c} - \frac{1}{b+c}\right)^2 \geq \frac{(a-b)^2}{4ab(a+b)^2},$$

or equivalently $4ab(a+b)^2 \geq (a+c)^2(b+c)^2$. This is clear, as $(a+b)^2 \geq (a+c)^2$ and $4ab \geq (b+c)^2$.

Thus, it remains to prove that

$$(ab+bc+ca)\left[\frac{1}{4ab} + \frac{2}{(a+c)(b+c)}\right] \geq \frac{9}{4}.$$

Using the identities

$$\frac{ab+bc+ca}{4ab} = \frac{1}{4} + \frac{c(a+b)}{4ab}, \quad \frac{2(ab+bc+ca)}{(a+c)(b+c)} = 2 - \frac{2c^2}{(a+c)(b+c)},$$

this becomes

$$\frac{c(a+b)}{4ab} \geq \frac{2c^2}{(a+c)(b+c)}.$$

But this is simply the standard inequality $(a+b)(a+c)(b+c) \geq 8abc$ and we are done. □

This completes the solution to the problem. □

1.4 Notes

We would like to thank the following people for their solutions: Vo Quoc Ba Can (problems 13, 18), Xiangyi Huang (problems 4, 5), Logeswaran Lajanugen (problem 1, 9), Oaz Nair (problem 17), Dusan Sobot (problems 2, 9, 16), Richard Stong (problems 8, 17), Gjergji Zaimi (problems 2, 3, 6, 7, 8, 11, 12, 15, 16, 17).

Chapter 2

Always Cauchy-Schwarz...

As the title suggests, all problems in this chapter can be solved using the Cauchy-Schwarz inequality, even though sometimes this will require quite a lot of work. Let us recall the statement of the Cauchy-Schwarz inequality: if $a_1, a_2, \ldots, a_n$ and $b_1, b_2, \ldots, b_n$ are real numbers, then

$$(a_1^2 + a_2^2 + \cdots + a_n^2) \cdot (b_1^2 + b_2^2 + \cdots + b_n^2) \geq (a_1 b_1 + a_2 b_2 + \cdots + a_n b_n)^2.$$

This follows easily from the fact that $\sum_{i=1}^{n} (a_i x + b_i)^2 \geq 0$ for all real numbers x. Indeed, the left-hand side is a quadratic function of x with nonnegative values, so its discriminant is negative or 0. But this is precisely the content of the Cauchy-Schwarz inequality. Another proof is based on Lagrange's identity

$$\left(\sum_{i=1}^{n} a_i^2\right) \cdot \left(\sum_{i=1}^{n} b_i^2\right) - \left(\sum_{i=1}^{n} a_i b_i\right)^2 = \sum_{1 \leq i < j \leq n} (a_i b_j - a_j b_i)^2.$$

This useful identity will be used several times in this chapter.

As the best way to get familiar with this inequality is via a lot of examples of all levels of difficulty, we will not insist on any theoretical aspects and go directly to battle. We start with two easy examples, destined to give the reader some confidence. He will surely need it for the more difficult problems to come...

1. Let a, b, c be nonnegative real numbers. Prove that

$$(ax^2 + bx + c)(cx^2 + bx + a) \geq (a + b + c)^2 x^2$$

for all nonnegative real numbers x.

Titu Andreescu, Gazeta Matematică

Proof. This is just a matter of re-arranging terms and applying Cauchy-Schwarz:

$$\begin{aligned}(ax^2 + bx + c)(cx^2 + bx + a) &= (ax^2 + bx + c)(a + bx + cx^2) \\ &\geq (ax + bx + cx)^2 \\ &= (a + b + c)^2 x^2.\end{aligned}$$

□

2. Let p be a polynomial with positive real coefficients. Prove that if $p\left(\frac{1}{x}\right) \geq \frac{1}{p(x)}$ is true for $x = 1$, then it is true for all $x > 0$.

Titu Andreescu, Revista Matematică Timişoara

Proof. Write $p(X) = a_0 + a_1 X + \cdots + a_n X^n$ and observe that

$$\begin{aligned}p(x)p\left(\frac{1}{x}\right) &= (a_0 + a_1 x + \cdots + a_n x^n)\left(a_0 + \frac{a_1}{x} + \cdots + \frac{a_n}{x^n}\right) \\ &\geq (a_0 + a_1 + \cdots + a_n)^2 \\ &= p(1)^2.\end{aligned}$$

The result follows. □

The following exercise is already a bit trickier, due to the lack of symmetry.

3. Prove that for all real numbers $a, b, c \geq 1$ the following inequality holds:

$$\sqrt{a-1} + \sqrt{b-1} + \sqrt{c-1} \leq \sqrt{a(bc+1)}.$$

Proof. The inequality being symmetric in b, c, but not in a, it is natural to deal first with $\sqrt{b-1}+\sqrt{c-1}$. This is easy to bound using Cauchy-Schwarz:

$$\sqrt{b-1}+\sqrt{c-1} \leq \sqrt{(b-1+1)(1+c-1)} = \sqrt{bc}.$$

So, it is enough to prove that $\sqrt{bc}+\sqrt{a-1} \leq \sqrt{a(bc+1)}$. But this is once more the Cauchy-Schwarz inequality. □

Another easy, but a bit exotic application of the Cauchy-Schwarz inequality is the following Chinese olympiad problem.

4. Let n be a positive integer. Find the number of ordered n-tuples of integers $(a_1, a_2, \ldots, a_n)$ such that

$$a_1 + a_2 + \cdots + a_n \geq n^2 \text{ and } a_1^2 + a_2^2 + \cdots + a_n^2 \leq n^3 + 1.$$

China 2002

proof By the Cauchy-Schwarz inequality we have

$$a_1 + a_2 + \cdots + a_n \leq \sqrt{n(n^3+1)} < n^2 + 1.$$

Since a_i are integers and $a_1 + a_2 + \cdots + a_n \geq n^2$, we must have

$$a_1 + a_2 + \cdots + a_n = n^2.$$

But then (again by Cauchy-Schwarz) we have $a_1^2 + a_2^2 + \cdots + a_n^2 \geq n^3$, forcing $a_1^2 + a_2^2 + \cdots + a_n^2 \in \{n^3, n^3+1\}$. If $a_1^2 + a_2^2 + \cdots + a_n^2 = n^3$, then we must have equality in Cauchy-Schwarz, implying that all a_i's are equal to n. Assume now that $a_1^2 + a_2^2 + \cdots + a_n^2 = n^3 + 1$ and let $b_i = a_i - n$. Then

$$b_1^2 + b_2^2 + \cdots + b_n^2 = n^3 + 1 - 2n \cdot n^2 + n^3 = 1,$$

forcing all but one b_i vanish. This is however impossible, as $b_1 + b_2 + \cdots + b_n = 0$. Therefore, this second case will not occur and the only solution is

$$a_1 = a_2 = \cdots = a_n = n.$$ □

We continue the series of easy exercises:

5. Let $x_1, x_2, \ldots, x_{10}$ be real numbers between 0 and $\frac{\pi}{2}$ such that

$$\sin^2 x_1 + \sin^2 x_2 + \cdots + \sin^2 x_{10} = 1.$$

Prove that

$$3(\sin x_1 + \sin x_2 + \cdots + \sin x_{10}) \leq \cos x_1 + \cos x_2 + \cdots + \cos x_{10}.$$

Saint Petersburg 2001

Proof. The Cauchy-Schwarz inequality and the hypothesis yield

$$\begin{aligned} \cos x_1 &= \sqrt{\sin^2 x_2 + \sin^2 x_3 + \cdots + \sin^2 x_{10}} \\ &\geq \frac{1}{3}(\sin x_2 + \sin x_3 + \cdots + \sin x_{10}) \end{aligned}$$

and similarly for the other variables. The result follows by adding up these inequalities. □

The following exercise combines an easy application of the Cauchy-Schwarz inequality with some classical formulae from geometry. Recall that r is the inradius and that s is the semi-perimeter of a triangle.

6. The triangle ABC satisfies

$$\left(\cot \frac{A}{2}\right)^2 + \left(2 \cot \frac{B}{2}\right)^2 + \left(3 \cot \frac{C}{2}\right)^2 = \left(\frac{6s}{7r}\right)^2.$$

Show that ABC is similar to a triangle whose sides are integers, and find the smallest set of such integers.

Titu Andreescu, USAMO 2002

Proof. Let the incircle be tangent to the sides of the triangle at points which split the sides into segments of length x and y, x and z, and y and z. Thus, the sides of the triangle have lengths $x + y$, $x + z$, $y + z$.

The equation

$$\left(\cot\frac{A}{2}\right)^2+\left(2\cot\frac{B}{2}\right)^2+\left(3\cot\frac{C}{2}\right)^2=\left(\frac{6s}{7r}\right)^2$$

can also be rewritten as

$$\frac{z^2}{r^2}+4\frac{y^2}{r^2}+9\frac{x^2}{r^2}=\left(\frac{6(x+y+z)}{7r}\right)^2.$$

This simplifies to

$$49(z^2+4y^2+9x^2)=36(x+y+z)^2.$$

Now, by Cauchy-Schwarz we have

$$\left(1+\frac{1}{4}+\frac{1}{9}\right)(z^2+4y^2+9x^2)\geq(x+y+z)^2,$$

yielding

$$49(z^2+4y^2+9x^2)\geq 36(x+y+z)^2.$$

Thus, we are in the equality case of the Cauchy-Schwarz inequality, which is precisely the case when there is a $k>0$ such that $x=4k, z=36k, y=9k$. Then the sides of the triangle are $13k, 40k, 45k$. Thus the triangle is similar to the triangle with sidelengths $13, 40, 45$, and the problem is solved. □

The statement of the following problem looks rather classical. There are however some technical problems which make the problem more difficult than expected.

7. Let $n\geq 2$ be an even integer. We consider all polynomials of the form $x^n+a_{n-1}x^{n-1}+\cdots+a_1x+1$, with real coefficients and having at least one real zero. Determine the least possible value of $a_1^2+a_2^2+\cdots+a_{n-1}^2$.

Czech-Polish-Slovak Competition 2002

Proof. Suppose that the corresponding polynomial has a real zero x. Using Cauchy-Schwarz, we obtain

$$\begin{aligned}(x^n+1)^2 &= (a_1x+a_2x^2+\cdots+a_{n-1}x^{n-1})^2\\ &\le (a_1^2+a_2^2+\cdots+a_{n-1}^2)(x^2+x^4+\cdots+x^{2(n-1)}).\end{aligned}$$

Thus

$$a_1^2+a_2^2+\cdots+a_{n-1}^2 \ge \frac{(x^n+1)^2}{x^2+x^4+\cdots+x^{2(n-1)}} = f(x)$$

and so we need to find first the minimal value of f. Now, looking at the zeros of the derivative of f suggests that the minimal value might be taken at $x=1$, so that it equals $\frac{4}{n-1}$. This is not easy to prove using derivatives, as the computations are a bit nasty. Instead, we will prove in an elementary way that

$$\frac{n-1}{4}(x^n+1)^2 \ge x^2+x^4+\cdots+x^{2(n-1)}.$$

In order to prove this, note that we have the inequalities

$$x^n+1\ge x^2+x^{n-2},\quad x^n+1\ge x^4+x^{n-4},\ldots,\quad x^n+1\ge x^{n-2}+x^2.$$

Adding them shows that

$$\frac{n-2}{4}(x^n+1)\ge x^2+x^4+\cdots+x^{n-2}.$$

On the other hand, multiplying the last inequality by x^n and adding the result to the previous inequality gives the inequality

$$\frac{n-2}{4}(x^n+1)^2\ge x^2+\cdots+x^{n-2}+x^{n+2}+\cdots+x^{2(n-1)}.$$

Thus it remains to prove that $(x^n+1)^2\ge 4x^n$, which is clear.

The previous paragraph shows that

$$a_1^2+a_2^2+\cdots+a_{n-1}^2\ge\frac{4}{n-1}$$

if $x^n+a_{n-1}x^{n-1}+\cdots+a_1x+1$ has at least one real zero. On the other hand, choosing $a_1=a_2=\cdots=a_{n-1}=-\frac{2}{n-1}$ shows that $\frac{4}{n-1}$ is optimal. $\square$

The denominators in the following inequality look awful, but a clever application of the Cauchy-Schwarz inequality can make all of them equal. The method of proof is worth remembering, since it appears quite often.

8. Prove that for any positive real numbers x, y, z such that $xyz \geq 1$ the following inequality holds

$$\frac{x}{x^3+y^2+z}+\frac{y}{y^3+z^2+x}+\frac{z}{z^3+x^2+y} \leq 1.$$

Tuan Le, KöMaL magazine

Proof. Using Cauchy-Schwarz, we obtain

$$(x^3+y^2+z)\left(\frac{1}{x}+1+z\right) \geq (x+y+z)^2.$$

Thus

$$\frac{x}{x^3+y^2+z} \leq \frac{x\left(\frac{1}{x}+1+z\right)}{(x+y+z)^2}.$$

Writing down similar inequalities for the second and third terms of the left-hand side, we reduce the problem to proving that

$$\sum(1+x+xz) \leq (x+y+z)^2 \iff \sum x^2 + \sum xy \geq 3 + \sum x.$$

This is immediate from

$$\sum xy \geq 3\sqrt[3]{(xyz)^2} \geq 3$$

and

$$\sum x^2 \geq \frac{(x+y+z)^2}{3} \geq \sqrt[3]{xyz}(x+y+z) \geq x+y+z. \qquad \square$$

Here is a rather nice-looking inequality taken from a Romanian Team Selection Test. There are plenty of ways to prove it, but what follows is particularly elegant and natural.

9. Let $n \geq 2$ and let $a_1, a_2, \ldots, a_n$ and $b_1, b_2, \ldots, b_n$ be real numbers such that
$$a_1^2 + a_2^2 + \cdots + a_n^2 = b_1^2 + b_2^2 + \cdots + b_n^2 = 1$$
and $a_1b_1 + a_2b_2 + \cdots + a_nb_n = 0$. Prove that
$$(a_1 + a_2 + \cdots + a_n)^2 + (b_1 + b_2 + \cdots + b_n)^2 \leq n.$$

Cezar and Tudorel Lupu, Romanian TST 2007

Proof. The proof mimics the proof of Cauchy-Schwarz: take any real number x and apply the Cauchy-Schwarz inequality to obtain
$$\sum_{i=1}^{n}(a_i + xb_i)^2 \geq \frac{\left(\sum_{i=1}^{n} a_i + x\sum_{i=1}^{n} b_i\right)^2}{n}.$$
By hypothesis, the left-hand side equals $1 + x^2$. Therefore
$$\left(\sum_{i=1}^{n} a_i + x\sum_{i=1}^{n} b_i\right)^2 \leq n(x^2 + 1)$$
for any real number x. The difference between the right-hand side and the left-hand side being a quadratic polynomial which takes nonnegative values on the whole real line, its discriminant has to be negative or zero, thus
$$4A^2B^2 \leq 4(n - A^2)(n - B^2),$$
where $A = \sum_{i=1}^{n} a_i$ and $B = \sum_{i=1}^{n} b_i$. It is immediate to check that this is equivalent to the desired inequality. $\square$

The following problem is easy, but the lack of symmetry might make it appear more difficult than it really is.

10. Find the largest real number T with the following property: if a, b, c, d, e are nonnegative real numbers such that $a + b = c + d + e$, then
$$\sqrt{a^2 + b^2 + c^2 + d^2 + e^2} \geq T(\sqrt{a} + \sqrt{b} + \sqrt{c} + \sqrt{d} + \sqrt{e})^2.$$

Proof. This is a simple application of the Cauchy-Schwarz inequality. Namely, we have

$$a^2 + b^2 \geq \frac{(a+b)^2}{2}, \quad c^2 + d^2 + e^2 \geq \frac{(c+d+e)^2}{3} = \frac{(a+b)^2}{3},$$

so that

$$\sqrt{a^2 + b^2 + c^2 + d^2 + e^2} \geq (a+b)\sqrt{\frac{5}{6}}.$$

On the other hand,

$$\sqrt{a} + \sqrt{b} \leq \sqrt{2(a+b)}, \quad \sqrt{c} + \sqrt{d} + \sqrt{e} \leq \sqrt{3(c+d+e)} = \sqrt{3(a+b)},$$

therefore

$$(\sqrt{a} + \sqrt{b} + \sqrt{c} + \sqrt{d} + \sqrt{e})^2 \leq (\sqrt{2} + \sqrt{3})^2(a+b).$$

Combining these two inequalities yields the estimate

$$\sqrt{a^2 + b^2 + c^2 + d^2 + e^2} \geq \frac{\sqrt{30}}{6(\sqrt{2}+\sqrt{3})^2}(\sqrt{a} + \sqrt{b} + \sqrt{c} + \sqrt{d} + \sqrt{e})^2.$$

To see that this is optimal, it suffices to keep track of the equalities in the previous inequalities. For instance, we can take $a = b = 3$ and $c = d = e = 2$. Thus the answer is $T = \frac{\sqrt{30}}{6(\sqrt{3}+\sqrt{2})^2}$. □

The next problem requires a whole series of applications of Cauchy-Schwarz.

11. Let $x_1, x_2, \ldots, x_n$ be positive real numbers such that

$$\frac{1}{1+x_1} + \frac{1}{1+x_2} + \cdots + \frac{1}{1+x_n} = 1.$$

Prove the inequality

$$\sqrt{x_1} + \sqrt{x_2} + \cdots + \sqrt{x_n} \geq (n-1)\left(\frac{1}{\sqrt{x_1}} + \frac{1}{\sqrt{x_2}} + \cdots + \frac{1}{\sqrt{x_n}}\right).$$

Vojtech Jarnik Competition 2002

Proof. Let $a_i = \frac{1}{1+x_i}$, so that $\sum a_i = 1$ and $x_i = \frac{\sum_{j\neq i} a_j}{a_i}$. Then, using Cauchy-Schwarz we can write:

$$\sum \sqrt{x_i} = \sum \sqrt{\frac{a_2 + a_3 + \cdots + a_n}{a_1}} \geq \sum \frac{\sqrt{a_2} + \sqrt{a_3} + \cdots + \sqrt{a_n}}{\sqrt{(n-1)a_1}}.$$

Rearranging terms in the previous sum gives

$$\frac{1}{\sqrt{n-1}} \cdot \sum \sqrt{a_1}\left(\frac{1}{\sqrt{a_2}} + \frac{1}{\sqrt{a_3}} + \cdots + \frac{1}{\sqrt{a_n}}\right)$$

and using again Cauchy-Schwarz we can bound this from below by

$$\frac{1}{\sqrt{n-1}} \cdot \sum \frac{(n-1)^2\sqrt{a_1}}{\sqrt{a_2} + \sqrt{a_3} + \cdots + \sqrt{a_n}}.$$

Using once more Cauchy-Schwarz for the denominators of each fraction

$$\sqrt{a_2} + \sqrt{a_3} + \cdots + \sqrt{a_n} \leq \sqrt{(n-1)(a_2 + a_3 + \cdots + a_n)}$$

yields the desired inequality. □

The idea of the following example is worth keeping in mind, since it turns out to be useful in a wide range of problems.

12. For $n \geq 2$ let $a_1, a_2, \ldots a_n$ be positive real numbers such that

$$(a_1 + a_2 + \cdots + a_n)\left(\frac{1}{a_1} + \frac{1}{a_2} + \cdots + \frac{1}{a_n}\right) \leq \left(n + \frac{1}{2}\right)^2.$$

Prove that $\max(a_1, a_2, \ldots, a_n) \leq 4\min(a_1, a_2, \ldots, a_n)$.

Titu Andreescu, USAMO 2009

Proof. The idea is to fix two of the variables, say a_1, a_2 and apply the Cauchy-Schwarz inequality to get rid of the remaining variables. Explicitly, this can be written in the form

$$\left(n + \frac{1}{2}\right)^2 \geq (a_1 + a_2 + \cdots + a_n)\left(\frac{1}{a_1} + \frac{1}{a_2} + \cdots + \frac{1}{a_n}\right)$$
$$\geq \left(\sqrt{(a_1 + a_2)\left(\frac{1}{a_1} + \frac{1}{a_2}\right)} + n - 2\right)^2,$$

where the first inequality is simply the hypothesis, while the second one is Cauchy-Schwarz applied to $n-1$ terms, grouping the terms indexed 1 and 2 in each multiplicand. We deduce that $\frac{(a_1+a_2)^2}{a_1a_2} \le \frac{25}{4}$, which immediately implies that

$$\max(a_1, a_2) \le 4\min(a_1, a_2).$$

Since everything is symmetric in $a_1, a_2, \dots, a_n$, the result follows. □

A bit more difficult, but with a similar flavor is the following problem.

13. Let $n > 2$ and let $x_1, x_2, \dots, x_n$ be positive real numbers such that

$$(x_1 + x_2 + \dots + x_n)\left(\frac{1}{x_1} + \frac{1}{x_2} + \dots + \frac{1}{x_n}\right) = n^2 + 1.$$

Prove that

$$(x_1^2 + x_2^2 + \dots + x_n^2)\left(\frac{1}{x_1^2} + \frac{1}{x_2^2} + \dots + \frac{1}{x_n^2}\right) > n^2 + 4 + \frac{2}{n(n-1)}.$$

Gabriel Dospinescu

Proof. The crucial idea is to write the hypothesis in a different way: expanding the product and rearranging terms shows that the hypothesis can also be written

$$\sum_{1\le i<j\le n}\left(\sqrt{\frac{x_i}{x_j}} - \sqrt{\frac{x_j}{x_i}}\right)^2 = 1.$$

Let us write

$$t_{i,j} = \left(\sqrt{\frac{x_i}{x_j}} - \sqrt{\frac{x_j}{x_i}}\right)^2.$$

Now, expanding again gives us

$$\begin{aligned}\left(\sum_{i=1}^n x_i^2\right)\cdot\left(\sum_{i=1}^n \frac{1}{x_i^2}\right) - n^2 &= \sum_{1\le i<j\le n}\left(\frac{x_i}{x_j} - \frac{x_j}{x_i}\right)^2 \\ &= \sum_{1\le i<j\le n} t_{i,j}\cdot\left(\sqrt{\frac{x_i}{x_j}} + \sqrt{\frac{x_j}{x_i}}\right)^2.\end{aligned}$$

Since

$$\left(\sqrt{\frac{x_i}{x_j}}+\sqrt{\frac{x_j}{x_i}}\right)^2=4+t_{i,j},$$

the inequality to be proved becomes

$$\sum_{1\le i<j\le n}(4+t_{i,j})t_{i,j}>4+\frac{2}{n(n-1)},$$

which is equivalent (taking into account the hypothesis that the sum of all $t_{i,j}$ is 1) to

$$\sum_{1\le i<j\le n}t_{i,j}^2>\frac{2}{n(n-1)}.$$

This follows from Cauchy-Schwarz inequality, but there is still a detail to be explained: we have to show that we cannot have equality. But if we had equality, all $t_{i,j}$ would be equal to $\frac{2}{n(n-1)}$ and so all numbers $\frac{x_i}{x_j}+\frac{x_j}{x_i}$ would be equal. Since $n>2$, this would force an equality $x_j=x_k$ for some $j\neq k$. But then $t_{j,k}=0$, a contradiction. □

We continue with an inequality which combines a rather direct application of the Cauchy-Schwarz inequality with a nice telescopic identity. We also present a beautiful alternate solution, due to Richard Stong.

14. Prove that for any real numbers $x_1,x_2,\ldots,x_n$ the following inequality holds:

$$\frac{x_1}{1+x_1^2}+\frac{x_2}{1+x_1^2+x_2^2}+\cdots+\frac{x_n}{1+x_1^2+\cdots+x_n^2}<\sqrt{n}.$$

Bogdan Enescu, IMO Shortlist 2001

Proof. The solution is very short, but far from obvious. The idea is to use the Cauchy-Schwarz inequality to reduce the problem to

$$\frac{x_1^2}{(1+x_1^2)^2}+\frac{x_2^2}{(1+x_1^2+x_2^2)^2}+\cdots+\frac{x_n^2}{(1+x_1^2+\cdots+x_n^2)^2}<1.$$

Thus, we need to prove that for any $a_1,a_2,\ldots,a_n \geq 0$ we have

$$\sum_{i=1}^{n} \frac{a_i}{(1+a_1+\cdots+a_i)^2} < 1.$$

Define $S_i = a_1+a_2+\cdots+a_i$, with the convention $S_0 = 0$. This is an increasing sequence and so

$$\sum \frac{a_i}{(1+a_1+\cdots+a_i)^2} = \sum \frac{S_i - S_{i-1}}{(1+S_i)^2} \leq$$

$$\sum \frac{S_i - S_{i-1}}{(1+S_i)(1+S_{i-1})} = \sum \left(\frac{1}{1+S_{i-1}} - \frac{1}{1+S_i}\right) = 1 - \frac{1}{1+S_n} < 1. \quad \square$$

Proof. First note that for any nonnegative real numbers c and A we have

$$\begin{aligned}\frac{x}{1+A^2+x^2} + \sqrt{\frac{c}{1+A^2+x^2}} &\leq \frac{x+\sqrt{c}\cdot\sqrt{1+A^2}}{\sqrt{(1+A^2)(1+A^2+x^2)}} \\ &\leq \frac{\sqrt{1+c}\cdot\sqrt{1+A^2+x^2}}{\sqrt{(1+A^2)(1+A^2+x^2)}} \\ &= \sqrt{\frac{1+c}{1+A^2}}.\end{aligned}$$

Here the first inequality is just $1+A^2+x^2 \geq 1+A^2$ with equality if and only if $x=0$; the second is Cauchy-Schwarz with equality if and only if $x\sqrt{c} = \sqrt{1+A^2}$. Thus we cannot have equality in both cases and the final inequality is actually strict. Applying this inequality repeatedly, it is easy to see by downwards induction on k that

$$\begin{aligned}&\frac{x_1}{1+x_1^2} + \cdots + \frac{x_n}{1+x_1^2+\cdots+x_n^2} \\ &\quad < \frac{x_1}{1+x_1^2} + \cdots + \frac{x_k}{1+x_1^2+\cdots+x_k^2} + \sqrt{\frac{n-k}{1+x_1^2+\cdots+x_k^2}}.\end{aligned}$$

The case $k=0$ is the desired result. $\square$

The trick of making the denominators equal thanks to a smart application of Cauchy-Schwarz or AM-GM inequality is also used in the following problem.

15. Prove that if a, b, c, d are positive real numbers, then

$$\frac{a}{b^2+c^2+d^2}+\frac{b}{a^2+c^2+d^2}+\frac{c}{a^2+b^2+d^2}+\frac{d}{a^2+b^2+c^2}>\frac{4}{a+b+c+d}.$$

P.K. Hung

Proof. Using the AM-GM inequality in the form $(x+y)^2 \geq 4xy$, we obtain

$$\frac{a}{b^2+c^2+d^2} \geq \frac{4a^3}{(a^2+b^2+c^2+d^2)^2}.$$

Adding these inequalities yields

$$\sum \frac{a}{b^2+c^2+d^2} \geq 4\frac{\sum a^3}{(\sum a^2)^2}.$$

By Cauchy-Schwarz,

$$\sum a^3 \geq \frac{\left(\sum a^2\right)^2}{\sum a}.$$

Inserting this in the previous inequality yields the desired result (note that the inequality is strict, since otherwise the equality in the Cauchy-Schwarz inequality would yield $a = b = c = d$, for which the inequality is strict). Note that even though the inequalities we used were very rough, the constant 4 is optimal: simply take $a = b$ and c, d close to 0. □

One needs rather good gymnastics with Cauchy-Schwarz to deal with the following problem.

16. Let $n \geq 2$ be an integer and let $x_1, x_2, \ldots, x_n$ be real numbers satisfying

$$x_1^2 + x_2^2 + \cdots + x_n^2 + x_1x_2 + x_2x_3 + \cdots + x_{n-1}x_n = 1.$$

For a fixed $1 \leq k \leq n$, find the maximum value that $|x_k|$ can take.

China 1998

Proof. Let us write the condition in the form

$$x_1^2 + (x_1 + x_2)^2 + (x_2 + x_3)^2 + \cdots + (x_{n-1} + x_n)^2 + x_n^2 = 2.$$

If we fix $1 \le k \le n$ and apply the Cauchy-Schwarz inequality twice, we obtain

$$\begin{aligned} &x_1^2 + (x_1 + x_2)^2 + \cdots + (x_{k-1} + x_k)^2 \\ &\quad \ge \frac{(x_1 - (x_1 + x_2) + (x_2 + x_3) - \cdots + (-1)^{k-1}(x_{k-1} + x_k))^2}{k} = \frac{x_k^2}{k} \end{aligned}$$

and

$$\begin{aligned} &(x_k + x_{k+1})^2 + \cdots + (x_{n-1} + x_n)^2 + x_n^2 \\ &\qquad \ge \frac{((x_k + x_{k+1}) - (x_{k+1} + x_{k+2}) + \cdots - x_n)^2}{n - k + 1} = \frac{x_k^2}{n - k + 1}. \end{aligned}$$

Taking into account the hypothesis and these two inequalities, we deduce that $2 \ge \frac{n+1}{k(n+1-k)} x_k^2$ for all k, so that

$$|x_k| \le \sqrt{\frac{2k(n+1-k)}{n+1}}.$$

By studying the equality case in the previous inequalities, one immediately sees that this value is attainable for each fixed k. Specifically, if we define $a_k = \sqrt{\frac{2k(n+1-k)}{n+1}}$ and take $x_0 = x_{n+1} = 0$ by convention, then equality occurs if $(-1)^{k-j} x_j$ interpolates linearly and evenly between these endpoints and $x_k = a_k$. The explicit formula is

$$x_j = \begin{cases} (-1)^{k-j} \frac{j a_k}{k} & j \le k \\ (-1)^{j-k} \frac{(n+1-j) a_k}{(n+1-k)} & j \ge k \end{cases} \qquad \square$$

Another ingenious application of the Cauchy-Schwarz inequality can be found in the following problem.

17. Let a, b, c be positive real numbers. Prove that

$$\frac{1}{a^2} + \frac{1}{b^2} + \frac{1}{c^2} + \frac{1}{(a+b+c)^2} \ge \frac{7}{25} \left(\frac{1}{a} + \frac{1}{b} + \frac{1}{c} + \frac{1}{a+b+c} \right)^2$$

Iran 2010

Proof. We see that we have equality when $a = b = c$, so we have to apply the Cauchy-Schwarz inequality in a smart way for the left-hand side of the inequality. Namely, start with

$$\left(1+1+1+\frac{1}{9}\right)\left(\frac{1}{a^2}+\frac{1}{b^2}+\frac{1}{c^2}+\frac{1}{(a+b+c)^2}\right) \geq \left(\frac{1}{a}+\frac{1}{b}+\frac{1}{c}+\frac{1}{3(a+b+c)}\right)^2.$$

This reduces the problem to showing that

$$\frac{1}{a}+\frac{1}{b}+\frac{1}{c}+\frac{1}{3(a+b+c)} \geq \frac{14}{15}\left(\frac{1}{a}+\frac{1}{b}+\frac{1}{c}+\frac{1}{a+b+c}\right).$$

Fortunately, this is equivalent to the classical inequality

$$\frac{1}{a}+\frac{1}{b}+\frac{1}{c} \geq \frac{9}{a+b+c}$$

and we are done. □

The form of the following problem strongly suggests using Cauchy-Schwarz. However, it is rather easy to check that many attempts fail...

18. Let x, y, z be real numbers and let A, B, C be the angles of a triangle. Prove that

$$x\sin A + y\sin B + z\sin C \leq \sqrt{(1+x^2)(1+y^2)(1+z^2)}.$$

Proof. Using that $C = \pi - A - B$, we obtain the identity

$$x\sin A + y\sin B + z\sin C = (x + z\cos B)\sin A + z\sin B\cos A + y\sin B.$$

Using Cauchy-Schwarz and the fact that $\sin^2 A + \cos^2 A = 1$, we obtain

$$(x + z\cos B)\sin A + z\sin B\cos A \leq \sqrt{(x+z\cos B)^2 + z^2\sin^2 B} = \sqrt{x^2+z^2+2xz\cos B}.$$

Another application of Cauchy-Schwarz thus yields

$$x\sin A + y\sin B + z\sin C \le \sqrt{(1+y^2)(x^2+z^2+2xz\cos B+\sin^2 B)}.$$

Thus, it remains to prove that

$$x^2+z^2+2xz\cos B+1-\cos^2 B \le 1+x^2+z^2+x^2z^2,$$

which is equivalent to $(xz-\cos B)^2 \ge 0$. □

We now enter the zone of challenging problems, with an unusual one. The first solution seems to come from nowhere (well, it actually came from the author's imagination...), but we also present a beautiful geometric proof of Richard Stong, which makes things rather clear.

19. Let a, b, c, x, y, z be real numbers and let

$$A = ax+by+cz, \quad B = ay+bz+cx, \quad C = az+bx+cy.$$

Assuming that $\min(|A-B|, |B-C|, |C-A|) \ge 1$, find the smallest possible value of $(a^2+b^2+c^2)(x^2+y^2+z^2)$.

Adrian Zahariuc, Mathematical Reflections

Proof. Note that by Lagrange's identity and the Cauchy-Schwarz inequality we can write

$$\begin{aligned}
&(a^2+b^2+c^2)(x^2+y^2+z^2)\\
&\quad = (ax+by+cz)^2+(ay-bx)^2+(bz-cy)^2+(cx-az)^2\\
&\quad \ge A^2+\frac{(ay-bx+bz-cy+cx-az)^2}{3}\\
&\quad = A^2+\frac{|B-C|^2}{3}.
\end{aligned}$$

Since everything is symmetric, we obtain

$$\begin{aligned}
&(a^2+b^2+c^2)(x^2+y^2+z^2)\\
&\qquad \ge \max\left(A^2+\frac{|B-C|^2}{3}, B^2+\frac{|C-A|^2}{3}, C^2+\frac{|A-B|^2}{3}\right).
\end{aligned}$$

Using the hypothesis, it is immediate to check that the last quantity is at least $\frac{4}{3}$.

To see that the answer of the problem is $\frac{4}{3}$, it remains to find a 6-tuple (a, b, c, x, y, z) satisfying the conditions of the problem and for which

$$(a^2 + b^2 + c^2)(x^2 + y^2 + z^2) = \frac{4}{3}.$$

Taking $A = 1, B = 0, C = -1$ (this is a triple which minimizes

$$\max\left(A^2 + \frac{|B - C|^2}{3}, B^2 + \frac{|C - A|^2}{3}, C^2 + \frac{|A - B|^2}{3}\right)$$

under the restrictions of the problem), we must ensure that we have equality in the Cauchy-Schwarz inequality. Solving the corresponding system yields, after some tedious but easy work, suitable values for a, b, c, x, y, z, namely

$$a = \frac{1}{3}, \quad b = 0, \quad c = -\frac{1}{3}, \quad x = y = 1, \quad z = -2. \qquad \square$$

Proof. Let u be the column vector with entries (a, b, c), v the column vector with entries (x, y, z) and let R be the linear map

$$R\begin{pmatrix} x \\ y \\ z \end{pmatrix} = \begin{pmatrix} 0 & 1 & 0 \\ 0 & 0 & 1 \\ 1 & 0 & 0 \end{pmatrix}\begin{pmatrix} x \\ y \\ z \end{pmatrix} = \begin{pmatrix} y \\ z \\ x \end{pmatrix}.$$

Then $A = u^T v$, $B = u^T Rv$, $C = u^T R^2 v$ and we want to minimize $\|u\|^2 \cdot \|v\|^2$. Note that R acts on $\mathbb{R}^3$ by fixing the line $x = y = z$ and rotating $120°$ in the orthogonal plane $x + y + z = 0$.

Write $u = u_1 + u_2$, where u_1 lies on the line $x = y = z$ and u_2 lies in the plane $x + y + z = 0$ and similarly for v. Then

$$\|u\|^2 = \|u_1\|^2 + \|u_2\|^2 \text{ and } u^T R^k v = u_1^T v_1 + u_2^T R^k v_2.$$

Thus the contributions of u_1 and v_1 to A, B, C are all the same (hence cancel in $|A - B|$, etc.). Thus clearly the minimum occurs when $u_1 = v_1 = 0$. In this case, R acts as a $120°$ rotation on v, hence $v + Rv + R^2 v = 0$ and thus

$A+B+C=0$. Recall that $\cos\alpha+\cos(\alpha+2\pi/3)+\cos(\alpha+4\pi/3)=0$ for any α. If we now take ω to be the angle between u and v, then we compute

$$\begin{aligned}A^2+B^2+C^2 &= \|u\|^2\cdot\|v\|^2(\cos^2\omega+\cos^2(\omega+2\pi/3)+\cos^2(\omega+4\pi/3))\\ &= \frac{1}{2}\|u\|^2\cdot\|v\|^2(3+\cos 2\omega+\cos 2(\omega+2\pi/3)+\cos 2(\omega+4\pi/3))\\ &= \frac{3}{2}\|u\|^2\cdot\|v\|^2.\end{aligned}$$

The conditions $A+B+C=0$ and $\min(|A-B|,|B-C|,|C-A|)\geq 1$ imply that $A^2+B^2+C^2\geq 2$ (without loss of generality, assume that A and B are nonnegative. As $|A-B|\geq 1$, we have $\max(A,B)\geq 1$, so $A^2+B^2+C^2=2A^2+2B^2+2AB\geq 2\max(A,B)^2\geq 2$) and so

$$\|u\|^2\cdot\|v\|^2=(a^2+b^2+c^2)(x^2+y^2+z^2)\geq\frac{4}{3}.$$

It is easy to see from the proof above that equality is attained. Simply choose any (x,y,z) with $x+y+z=0$ and choose (a,b,c) with $a+b+c=0$, orthogonal to (x,y,z) so that $A=0$, and scaled so that $B=1$. For example, taking $(x,y,z)=(1,-2,1)$ we see that $(a,b,c)=(-1,0,1)/3$ suffices. □

We present three short solutions for the following problem. However, none of them is really easy or natural.

20. Let a,b,c,d be real numbers such that

$$(a^2+1)(b^2+1)(c^2+1)(d^2+1)=16.$$

Prove that

$$-3\leq ab+bc+cd+da+ac+bd-abcd\leq 5.$$

Titu Andreescu, Gabriel Dospinescu

Proof. Write the inequality in the form

$$(ab+bc+cd+da+ac+bd-abcd-1)^2\leq 16.$$

Next, apply Cauchy-Schwarz in the form

$$\begin{aligned}&(a(b+c+d-bcd)+bc+bd+cd-1)^2\\ &\quad\le (a^2+1)\left[(b+c+d-bcd)^2+(bc+bd+cd-1)^2\right].\end{aligned}$$

The miracle is that we actually have

$$(b+c+d-bcd)^2+(bc+bd+cd-1)^2=(b^2+1)(c^2+1)(d^2+1).$$

Of course, this can be checked by brute force, but perhaps nicer is determining it through two applications of Lagrange's identity. □

Proof. We will use Lagrange's identity and Cauchy-Schwarz in the form

$$\begin{aligned}\prod(a^2+1)&=\left[(a+b)^2+(1-ab)^2\right]\left[(c+d)^2+(1-cd)^2\right]\\ &\ge\left[(a+b)(c+d)-(1-ab)(1-cd)\right]^2.\end{aligned}$$

Note that

$$(a+b)(c+d)-(1-ab)(1-cd)=-1+ab+bc+cd+da+ac+bd-abcd.$$

A moment of thought shows that what we have just proved is exactly the desired inequality. □

Proof. Here is a rather unnatural, but very elegant proof. Consider the polynomial

$$P(X)=(X-a)(X-b)(X-c)(X-d)$$

and observe that the hypothesis becomes $P(i)P(-i)=16$. This can also be written as $|P(i)|^2=16$. On the other hand, we have

$$P(i)=1-\sum ab+abcd+i\left(\sum a-\sum abc\right).$$

We deduce that

$$(1-\sum ab+abcd)^2+\left(\sum a-\sum abc\right)^2=16,$$

from where the conclusion follows, as the desired inequality can be written as

$$|1-\sum ab+abcd|\leq 4.$$

Straight from the Book, isn't it? □

We continue with a really nice, but rather technical problem. It requires some delicate algebraic computations and a rather exotic application of Cauchy-Schwarz. We also present a more conceptual proof, due to Richard Stong, which uses more advanced tools, but proves much more.

21. Let a, b, c, d, e be nonnegative real numbers such that $a^2+b^2+c^2=d^2+e^2$ and $a^4+b^4+c^4=d^4+e^4$. Prove that $a^3+b^3+c^3\leq d^3+e^3$.

IMC 2006

Proof. Since we only deal with even exponents in the hypothesis, let us square the desired inequality and write it in the form

$$\sum a^6+2\sum a^3b^3\leq d^6+e^6+2d^3e^3.$$

Next, the identity

$$\sum a^6-3a^2b^2c^2=\frac{1}{2}\left(\sum a^2\right)\left(3\sum a^4-\left(\sum a^2\right)^2\right)$$

combined with the hypothesis easily yield

$$a^6+b^6+c^6=3a^2b^2c^2+d^6+e^6.$$

Thus, we need to prove that

$$2\sum a^3b^3+3a^2b^2c^2\leq 2(de)^3=2(a^2b^2+b^2c^2+c^2a^2)^{3/2}.$$

With the obvious substitutions, this is equivalent to the inequality

$$2\sum x^3+3xyz\leq 2\left(\sum x^2\right)^{3/2}.$$

Using Cauchy-Schwarz

$$(2\sum x^3 + 3xyz)^2 = \left(\sum x(2x^2 + yz)\right)^2 \le \left(\sum x^2\right)\left(\sum (2x^2 + yz)^2\right),$$

it remains to prove that

$$\sum (2x^2 + yz)^2 \le 4\left(\sum x^2\right)^2,$$

which follows immediately from $x^2y^2 + y^2z^2 + z^2x^2 \ge xyz(x + y + z)$, itself equivalent to $\sum (xy - xz)^2 \ge 0$. □

Proof. Consider the polynomial $p(t) = t^3 - \sigma_1 t^2 + \sigma_2 t - \sigma_3$ with σ_1 and σ_2 fixed and σ_3 allowed to vary. Regard the roots x, y, z of $p(t)$ as functions of σ_3. Suppose $f : \mathbb{R} \to \mathbb{R}$ is any smooth function (at least three times differentiable for the discussion below). Define a function

$$g(\sigma_3) = f(x) + f(y) + f(z).$$

Then g is a differentiable function of σ_3 and

$$g'(\sigma_3) = \frac{f'(x)}{(x-y)(x-z)} + \frac{f'(y)}{(y-z)(y-x)} + \frac{f'(z)}{(z-x)(z-y)} = \frac{1}{2} f'''(\zeta),$$

for some ζ with $\min(x, y, z) < \zeta < \max(x, y, z)$. To prove the first equality one merely checks that the curve

$$c(s) = \left(x + \frac{s}{(x-y)(x-z)}, y + \frac{s}{(y-z)(y-x)}, z + \frac{s}{(z-x)(z-y)}\right)$$

preserves σ_1, has $\left.\frac{d\sigma_2}{ds}\right|_{s=0} = 0$, and has $\left.\frac{d\sigma_3}{ds}\right|_{s=0} = 1$. For the second equality note that

$$f'(t) - \frac{f'(x)(t-y)(t-z)}{(x-y)(x-z)} - \frac{f'(y)(t-z)(t-x)}{(y-z)(y-x)} - \frac{f'(z)(t-x)(t-y)}{(z-x)(z-y)}$$

vanishes at $t = x, y, z$. Hence by two applications of Rolle's Theorem, its second derivative vanishes at some ζ in $(\min(x, y, z), \max(x, y, z))$ and this is the required ζ.

Apply this to $x = a^2$, $y = b^2$, $z = c^2$, and $f(u) = u^{3/2}$. This amounts to choosing x, y, z to be the roots of $p(t) = t^3 - (d^2+e^2)t^2 + d^2e^2t - \sigma_3$. Since $f'''(\zeta) < 0$ for all $\zeta > 0$, we see that g is a decreasing function of σ_3 for $\sigma_3 \geq 0$. Thus $g(0) \geq g(\sigma_3)$ which, unwinding definitions, is the desired inequality.

This argument shows that the same inequality holds for any exponent strictly between 2 and 4 and gives similar (sometimes reversed) inequalities for other exponents. □

The reader will probably appreciate the beauty of the following inequality. It is more difficult than it appears at first sight, as the obvious application of Cauchy-Schwarz fails rather badly.

22. Prove that for any real numbers $x_1, x_2, \ldots, x_n$ the following inequality holds

$$\left(\sum_{i=1}^{n}\sum_{j=1}^{n}|x_i - x_j|\right)^2 \leq \frac{2(n^2-1)}{3}\left(\sum_{i=1}^{n}\sum_{j=1}^{n}|x_i - x_j|^2\right).$$

IMO 2003

Proof. The first step is to order the x_i's, say $x_1 \leq x_2 \leq \cdots \leq x_n$. Then

$$\sum_{i,j=1}^{n}|x_i - x_j| = 2\sum_{i<j}(x_j - x_i) = 2(-(n-1)x_1 - (n-3)x_2 + \cdots + (n-1)x_n).$$

Thus, applying Cauchy-Schwarz yields

$$\left(\sum_{i,j}|x_i - x_j|\right)^2 \leq 4 \cdot \left[\sum_{i=1}^{n}(n-2i+1)^2\right] \cdot \sum_{i=1}^{n} x_i^2.$$

It is easy to compute the last expression and the final estimate that we obtain is

$$\left(\sum_{i,j}|x_i - x_j|\right)^2 \leq \frac{4n(n^2-1)}{3}\sum_{i=1}^{n} x_i^2.$$

On the other hand, Legendre's identity shows that

$$\sum_{i,j=1}^{n}(x_i - x_j)^2 = 2n\sum_{i=1}^{n} x_i^2 - 2\left(\sum_{i=1}^{n} x_i\right)^2.$$

Well, unfortunately, when we combine all this we see that we are not done, because of the bad term $2\left(\sum_{i=1}^{n} x_i\right)^2$. Fortunately, it is easy to repair the argument: indeed, we can always add the same number to all x_i'swithout changing the hypothesis or the conclusion of the problem. Thus, we may assume that $x_1 + x_2 + \cdots + x_n = 0$. But then the previous inequalities allow us to conclude the proof. □

The following problem is a very tricky application of the Cauchy-Schwarz inequality. The technique used in the proof is worth remembering, since it is quite useful. It is also a standard tool in analysis and probability (it is actually likely that the following problem is inspired by probability theory).

23. Let $a_1, a_2, \ldots, a_n$ be positive real numbers which add up to 1. Let n_i be the number of integers k such that $2^{1-i} \geq a_k > 2^{-i}$. Prove that

$$\sum_{i\geq 1}\sqrt{\frac{n_i}{2^i}} \leq 4 + \sqrt{\log_2(n)}.$$

L. Leindler, Miklos Schweitzer Competition

Proof. Choose a positive integer N and split the sum in two parts: the one for $i \leq N$ and the one for $i > N$. Apply the Cauchy-Schwarz inequality for each of them to obtain

$$\sum_{i>N}\sqrt{\frac{n_i}{2^i}} \leq \sqrt{\sum_{i>N} n_i}\cdot\sqrt{\sum_{i>N}\frac{1}{2^i}} \quad \text{and} \quad \sum_{i=1}^{N}\sqrt{\frac{n_i}{2^i}} \leq \sqrt{N}\cdot\sqrt{\sum_{i=1}^{N}\frac{n_i}{2^i}}.$$

On the other hand, we have

$$\sum_{i>N}\frac{1}{2^i} = \frac{1}{2^N} \quad \text{and} \quad \sum_{i>N} n_i \leq \sum_{i\geq 1} n_i = n,$$

the last relation being obvious by definition of the n_i's. Finally, and most importantly we have

$$\sum_{i=1}^{N} \frac{n_i}{2^i} \le \sum_{i\ge 1} 2^{-i} \cdot \left(\sum_{2^{-i}<a_k\le 2^{1-i}} 1 \right) < \sum_{k=1}^{n} a_k = 1.$$

Putting these inequalities together, we deduce that

$$\sum_{i\ge 1} \sqrt{\frac{n_i}{2^i}} < \sqrt{\frac{n}{2^N}} + \sqrt{N}.$$

Taking $N = \log_2(n)$, we obtain an even stronger (and strict!) inequality, in which 4 is replaced by 1. $\square$

We end the series of moderately difficult problems with a very nice-looking improvement of an IMO 2004 problem.

24. Let $n > 2$ be an integer. Find the greatest real number k with the following property: if the positive real numbers $x_1, x_2, \dots, x_n$ satisfy

$$k > (x_1 + x_2 + \cdots + x_n)\left(\frac{1}{x_1} + \frac{1}{x_2} + \cdots + \frac{1}{x_n}\right),$$

then any three of them are sides of a triangle.

Adapted after IMO 2004

Proof. The main point is to solve this problem for $n = 3$. If b, c are positive real numbers, let us look at the possible values of

$$f(x) = (x + b + c)\left(\frac{1}{x} + \frac{1}{b} + \frac{1}{c}\right)$$

when $x \ge b + c$. It is not difficult to check (either directly or by computing the derivative) that f is increasing in this domain of x, so that

$$f(x) \ge f(b+c) = 2(b+c)\left(\frac{1}{b} + \frac{1}{c} + \frac{1}{b+c}\right) = 2 + 2\frac{(b+c)^2}{bc} \ge 10.$$

We deduce that if x, b, c satisfy $f(x) < 10$, then x, b, c are the sides of a triangle (since everything is symmetric in x, b, c). Thus, for $n = 3$ the answer is 10.

To reduce the general problem to the case $n = 3$, we use Cauchy-Schwarz, which allows us to obtain information about x_1, x_2, x_3 knowing that

$$k > (x_1 + x_2 + \cdots + x_n)\left(\frac{1}{x_1} + \frac{1}{x_2} + \cdots + \frac{1}{x_n}\right).$$

Indeed, using Cauchy-Schwarz and the inequality

$$(x_4 + \cdots + x_n) \cdot \left(\frac{1}{x_4} + \cdots + \frac{1}{x_n}\right) \geq (n-3)^2,$$

we obtain

$$\sqrt{k} > n - 3 + \sqrt{(x_1 + x_2 + x_3)\left(\frac{1}{x_1} + \frac{1}{x_2} + \frac{1}{x_3}\right)},$$

so that

$$(x_1 + x_2 + x_3)\left(\frac{1}{x_1} + \frac{1}{x_2} + \frac{1}{x_3}\right) < (\sqrt{k} - n + 3)^2.$$

Using this and the previous case ($n = 3$), we deduce that if

$$(n - 3 + \sqrt{10})^2 > (x_1 + x_2 + \cdots + x_n)\left(\frac{1}{x_1} + \frac{1}{x_2} + \cdots + \frac{1}{x_n}\right),$$

then any three among the numbers $x_1, x_2, \ldots, x_n$ are the sides of a triangle.

It remains to check that this is indeed optimal. To see this, choose

$$x_4 = x_5 = \cdots = x_n = 1, \quad x_1 = x_2 = \frac{\sqrt{10}}{4}, \quad x_3 = \frac{\sqrt{10}}{2}.$$

Then x_1, x_2, x_3 are not the sides of a triangle and

$$(x_1 + x_2 + \cdots + x_n)\left(\frac{1}{x_1} + \frac{1}{x_2} + \cdots + \frac{1}{x_n}\right) = (n - 3 + \sqrt{10})^2.$$

Thus, the answer is $k_n = (n - 3 + \sqrt{10})^2$. □

We give two proofs for the next difficult problem. The first is based on a very tricky application of Cauchy-Schwarz combined with a mixing-variables argument, while the second one is a pure, but very technical, mixing-variables argument.

25. If a, b, c, d, e are real numbers such that $a+b+c+d+e=0$, then

$$(a^2+b^2+c^2+d^2+e^2)^2 \leq \frac{30}{7}(a^4+b^4+c^4+d^4+e^4).$$

Vasile Cârtoaje

Proof. First of all, we may assume that among a, b, c, d, e at least three of the numbers are nonnegative, say a, b, c. This follows immediately from the pigeonhole principle, possibly after changing the sign of all numbers. The key step is the following very tricky application of Cauchy-Schwarz:

$$\begin{aligned} 9(a^4+b^4+c^4+d^4+e^4) &= [9(a^4+b^4+c^4)+2(d^4+e^4)]+(7d^4)+(7e^4) \\ &\geq \frac{(2\sqrt{21(9(a^4+b^4+c^4)+2(d^4+e^4))}+21(d^2+e^2))^2}{84+63+63}. \end{aligned}$$

A small computation yields the equivalent form of this inequality

$$\begin{aligned} &\frac{30}{7}(a^4+b^4+c^4+d^4+e^4) \\ &\quad \geq \left(2\sqrt{\frac{9(a^4+b^4+c^4)+2(d^4+e^4)}{21}}+d^2+e^2\right)^2. \end{aligned}$$

This reduces the problem to showing that

$$2\sqrt{9(a^4+b^4+c^4)+2(d^4+e^4)} \geq \sqrt{21}(a^2+b^2+c^2).$$

To exploit the relationship between a, b, c, d, e, we use the fact that

$$d^4+e^4 \geq \frac{(d+e)^4}{8} = \frac{(a+b+c)^4}{8}.$$

Thus, it remains to prove that

$$36(a^4+b^4+c^4)+(a+b+c)^4 \geq 21(a^2+b^2+c^2)^2$$

for all nonnegative numbers a, b, c.

This inequality does not seem to follow easily from well-known results, so we will employ the powerful technique of mixing variables to prove it. Let

$$\begin{aligned} f(a,b,c) &= 36(a^4+b^4+c^4)+(a+b+c)^4-21(a^2+b^2+c^2)^2 \\ &= 15(a^4+b^4+c^4)-42(a^2b^2+b^2c^2+c^2a^2)+(a+b+c)^4. \end{aligned}$$

We will first show that for $a=\min(a,b,c)$ we have

$$f(a,b,c) \geq f\left(a, \frac{b+c}{2}, \frac{b+c}{2}\right).$$

For this we compute

$$\begin{aligned} f(a,b,c) - f\left(a, \frac{b+c}{2}, \frac{b+c}{2}\right) &= 15\left(b^4+c^4-\frac{(b+c)^4}{8}\right) \\ &- 42a^2\left(b^2+c^2-\frac{(b+c)^2}{2}\right) - 42\left(b^2c^2-\frac{(b+c)^4}{16}\right) \\ &= \frac{3}{8}(b-c)^2\left[42b^2+92bc+42c^2-56a^2\right] \geq 0, \end{aligned}$$

where the final inequality follows since $28b^2+28c^2-56a^2 \geq 0$. Thus we reduce to the case where $a \leq b = c$. In this case we compute

$$\begin{aligned} f(a,b,b) &= 4(b^4+8b^3-15b^2a^2+2ba^3+4a^4) \\ &= 4(b-a)^2(b^2+10ab+4a^2) \\ &\geq 0. \end{aligned}$$

The result follows.

Note that we have equality for $a=b=c=2$ and $d=e=-3$. $\square$

Proof. We will use a mixing variables argument. Let

$$F(a,b,c,d,e) = 30(a^4+b^4+c^4+d^4+e^4) - 7(a^2+b^2+c^2+d^2+e^2)^2.$$

We want to prove that for $a+b+c+d+e=0$ we have $F(a,b,c,d,e) \geq 0$. The basic formula we need is that

$$\begin{aligned} F(a,b,c,d,e) - F\left(a,b,c,\frac{d+e}{2},\frac{d+e}{2}\right) \\ = (d-e)^2(21d^2+34de+21e^2-7(a^2+b^2+c^2)), \end{aligned}$$

which can be checked by tedious computation.

By the pigeonhole principle three of a,b,c,d,e must have the same sign (counting zero as having either sign). By symmetry, we may assume $a,b,c \geq 0$. Then

$$\begin{aligned} 7(a^2+b^2+c^2) &\leq 7(a+b+c)^2 = 7(d+e)^2 \\ &\leq 17(d+e)^2 + 4(d^2+e^2) = 21d^2+34de+21e^2. \end{aligned}$$

Therefore by the formula above $F(a,b,c,d,e) \geq F\left(a,b,c,\frac{d+e}{2},\frac{d+e}{2}\right)$. Thus we may assume three of a,b,c,d,e are nonnegative and the other two are equal. To keep the same basic formula, we invoke symmetry and switch our assumption to $c,d,e \geq 0$ and $a=b=-(c+d+e)/2$. Now suppose $c \leq e$. Then we have

$$\begin{aligned} 7(a^2+b^2+c^2) &= \frac{7}{2}(c+d+e)^2 + 7c^2 \\ &\leq \frac{7}{2}(d+2e)^2 + 7e^2 = \frac{7}{2}d^2 + 14de + 21e^2 \\ &\leq 21d^2+34de+21e^2. \end{aligned}$$

Therefore we again have $F(a,b,c,d,e) \geq F\left(a,b,c,\frac{d+e}{2},\frac{d+e}{2}\right)$. We conclude that we can repeatedly average the largest of c,d,e with either of the other two and we will only lower F. By continuity of F, we therefore reduce to the case where $a=b$ and $c=d=e$. In this case it is easy to check that $F(a,b,c,d,e)=0$ and we are done. □

We end this chapter with a challenging inequality, which combines some clever uses of Cauchy-Schwarz with a tricky homogeneity argument. This result also appears in [35] and it is a generalization of a problem discussed in [3], chapter 2, example 11.

26. Prove that for any positive real numbers $a_1, a_2, \ldots, a_n$, x_1, $x_2, \ldots, x_n$ such that

$$\sum_{i \le i < j \le n} x_i x_j = \binom{n}{2},$$

the following inequality holds

$$\frac{a_1}{a_2 + \cdots + a_n}(x_2 + \cdots + x_n) + \cdots + \frac{a_n}{a_1 + \cdots + a_{n-1}}(x_1 + \cdots + x_{n-1}) \ge n.$$

Vasile Cârtoaje, Gabriel Dospinescu

Proof. First of all, it is enough to prove that for any $x_1, x_2, \ldots, x_n > 0$ and any $a_1, a_2, \ldots, a_n > 0$ we have

$$\sum \frac{a_1}{a_2 + \cdots + a_n}(x_2 + \cdots + x_n) \ge n\sqrt{\frac{\sum_{1 \le i < j \le n} x_i x_j}{\binom{n}{2}}}.$$

Since this is homogeneous, it is enough to prove it when $x_1 + x_2 + \cdots + x_n = 1$. In this case, it is equivalent to

$$\sum \frac{a_1}{a_2 + \cdots + a_n} \ge \sum \frac{a_1 x_1}{a_2 + \cdots + a_n} + n\sqrt{\frac{\sum_{1 \le i < j \le n} x_i x_j}{\binom{n}{2}}}.$$

But Cauchy-Schwarz shows that

$$\sum \frac{a_1}{a_2 + \cdots + a_n} x_1 \le \sqrt{\sum \left(\frac{a_1}{a_2 + \cdots + a_n}\right)^2} \cdot \sqrt{\sum_{i=1}^{n} x_i^2}$$

and a second application of Cauchy-Schwarz yields

$$\sqrt{\sum\left(\frac{a_1}{a_2+\cdots+a_n}\right)^2}\cdot\sqrt{\sum_{i=1}^{n}x_i^2+n\sqrt{\frac{\sum_{1\le i<j\le n}x_ix_j}{\binom{n}{2}}}}$$
$$\le\sqrt{\frac{n}{n-1}+\sum\left(\frac{a_1}{a_2+\cdots+a_n}\right)^2}\cdot\sqrt{\sum_{i=1}^{n}x_i^2+2\sum_{1\le i<j\le n}x_ix_j}.$$

So, it remains to prove that

$$\left(\sum\frac{a_1}{a_2+\cdots+a_n}\right)^2\ge\frac{n}{n-1}+\sum\left(\frac{a_1}{a_2+\cdots+a_n}\right)^2.$$

A very elegant approach for this inequality was proposed by Darij Grinberg in [35], where a more general result is proved. We may assume that

$$a_1+a_2+\cdots+a_n=1.$$

We need to prove that

$$\sum_{i<j}\frac{a_ia_j}{(1-a_i)(1-a_j)}\ge\frac{n}{2n-2}.$$

Using $T2$'s lemma (a form of Cauchy-Schwarz), we reduce this to proving that

$$\left(\sum_{i<j}a_ia_j\right)^2\ge\frac{n}{2n-2}\sum_{i<j}a_ia_j(1-a_i)(1-a_j).$$

Note that

$$\sum_{i<j}a_ia_j=\frac{1-\sum_i a_i^2}{2}=\frac{1}{2}\sum_i a_i(1-a_i).$$

Thus, if we let $b_i=a_i(1-a_i)$, it remains to prove that

$$(b_1+b_2+\cdots+b_n)^2\ge\frac{2n}{n-1}\sum_{i<j}b_ib_j.$$

This follows directly from Cauchy-Schwarz and the identity

$$2\sum_{i<j} b_i b_j = (b_1 + b_2 + \cdots + b_n)^2 - (b_1^2 + b_2^2 + \cdots + b_n^2). \qquad \square$$

Remark 2.1. We can also prove the inequality

$$\sum_{i<j} \frac{a_i a_j}{(1-a_i)(1-a_j)} \ge \frac{n}{2n-2}$$

by mixing variables: consider the map

$$F(a_1, a_2, \ldots, a_n) = 2\sum_{i<j} \frac{a_i a_j}{(1-a_i)(1-a_j)}$$

and set $x = \frac{a_1+a_2}{2}$. We claim that $F(a_1, a_2, \ldots, a_n) \ge F(x, x, a_3, \ldots, a_n)$. A small computation shows that this is equivalent to

$$\left(\frac{a_1}{1-a_1} + \frac{a_2}{1-a_2} - 2\frac{x}{1-x}\right) \cdot 2\sum_{i\ge 3} \frac{a_i}{1-a_i} + \frac{2a_1a_2}{(1-a_1)(1-a_2)} - \frac{2x^2}{(1-x)^2} \ge 0.$$

Another computation shows that

$$\frac{a_1}{1-a_1} + \frac{a_2}{1-a_2} - \frac{2x}{1-x} = \frac{(a_1-a_2)^2}{2(1-a_1)(1-a_2)(1-x)}$$

and

$$\frac{2x^2}{(1-x)^2} - \frac{2a_1a_2}{(1-a_1)(1-a_2)} = \frac{(a_1-a_2)^2}{2} \cdot \frac{1-2x}{(1-x)^2(1-a_1)(1-a_2)}.$$

Thus, the inequality $F(a_1, a_2, \ldots, a_n) \ge F(x, x, a_3, \ldots, a_n)$ is equivalent to

$$2\sum_{i\ge 3} \frac{a_i}{1-a_i} \ge \frac{1-2x}{1-x}.$$

But this is easy, since the left-hand side is at least $2\sum_{i\geq 3} a_i = 2(1-2x)$ and $2(1-2x) \geq \frac{1-2x}{1-x}$ is equivalent to $(1-2x)^2 \geq 0$. Thus, we have $F(a_1, a_2, \ldots, a_n) \geq F(x, x, a_3, \ldots, a_n)$. Continuing to mix variables in this way and using the continuity of F implies that $F(a_1, a_2, \ldots, a_n)$ is at least $F(m, m, \ldots, m)$, where m is the arithmetic mean of the a_i's, namely $1/n$. The result follows.

2.1 Notes

The following people provided solutions to the problems discussed in this chapter: Vo Quoc Ba Can (problem 20), Ta Minh Hoang (problem 17), Mitchell Lee (problem 6), Dung Tran Nam (problem 25), Dusan Sobot (problems 1, 2, 5, 7), Richard Stong (problems 14, 19, 21, 25), Gjergji Zaimi (problems 3, 4, 10, 13).

Addendum 2.A Cauchy-Schwarz in Number Theory

This addendum shows some very beautiful applications of the Cauchy-Schwarz inequality to number theory problems. We present Gallagher's sieve and a beautiful arithmetic application; we discuss the large sieve, an amazing tool invented by Linnik and extensively developed by a series of brilliant mathematicians; finally we discuss another famous result, the Turán-Kubilius inequality. The Cauchy-Schwarz inequality plays an important role in the proofs of all these theorems, which are elementary but quite powerful: this ought to show the reader how Cauchy-Schwarz appears in "real mathematics" and not only in olympiad-type problems. Analytic number theory has the reputation of being rather technical and this addendum is no exception. We hope that the results presented here (especially their applications) will compensate for this nonetheless.

The following two sections try to give satisfactory answers to the following natural problem: suppose that A is a set of integers such that

$$A \pmod p = \{x \pmod p | x \in A\}$$

is relatively small for all primes p in a finite set $\mathbb{P}$. Are there nontrivial bounds on the size of A?

2.A.1 Gallagher's sieve and a nice application

Recall that the von Mangoldt function Λ is defined by $\Lambda(p^n) = \log p$ if p is a prime and $n \geq 1$ and $\Lambda(x) = 0$ for any other integer x. The crucial property of Λ is that

$$\sum_{d|n} \Lambda(d) = \log n$$

for all n. The following theorem is a pretty tricky application of the Cauchy-Schwarz inequality, but the application given below reveals its usefulness and power.

Theorem 2.A.1. *(Gallagher's larger sieve) Let S be a finite nonempty set of integers and let P be a finite set of prime powers. Assume that for each $p \in P$ we can find a real number $u(p) \geq |\{s \pmod p | s \in S\}|$ such that*

$$\sum_{p \in P} \frac{\Lambda(p)}{u(p)} > 2 \log X,$$

where $X = \max_{s \in S} |s|$. Then

$$|S| \leq \frac{\sum_{p \in P} \Lambda(p) - \log 2X}{\sum_{p \in P} \frac{\Lambda(p)}{u(p)} - \log 2X}.$$

Proof. Let $p \in P$ and let $s(r,p)$ be the number of elements of S that are congruent to r modulo p. Then by Cauchy-Schwarz and the fact that

$$u(p) \geq |\{s \pmod p | s \in S\}|$$

we have

$$|S|^2 = \left(\sum_{r=0}^{p-1} s(r,p) \right)^2 \leq u(p) \sum_{r=0}^{p-1} s(r,p)^2,$$

thus

$$\frac{|S|^2}{u(p)} \leq \sum_{r=0}^{p-1} \sum_{\substack{s_1, s_2 \in S \\ s_1 \equiv s_2 \equiv r \pmod p}} 1 = |S| + \sum_{\substack{s_1 \neq s_2 \in S \\ p | s_1 - s_2}} 1.$$

Multiplying this by $\Lambda(p)$ and summing over all p yields

$$|S|^2 \sum_{p \in P} \frac{\Lambda(p)}{u(p)} \leq |S| \sum_{p \in P} \Lambda(p) + \sum_{s_1 \neq s_2} \sum_{p | s_1 - s_2} \Lambda(p).$$

As

$$\sum_{p | s_1 - s_2} \Lambda(p) \leq \log(|s_1 - s_2|) \leq \log 2X,$$

we deduce that

$$|S|^2 \sum_{p \in P} \frac{\Lambda(p)}{u(p)} \leq |S| \sum_{p \in P} \Lambda(p) + (|S|^2 - |S|) \log 2X,$$

from which the result follows immediately. □

The promised application (taken from [19]) requires some preliminaries. The following result is very classical, being related to the following natural question: given two positive integers, what is the probability that they are relatively prime?

Proposition 2.A.2. *As $x \to \infty$, we have*

$$\sum_{k \le x} \varphi(k) = \frac{3}{\pi^2} x^2 + O(x \ln x).$$

Proof. The key point is the equality

$$\frac{\varphi(k)}{k} = \sum_{d|k} \frac{\mu(d)}{d},$$

which follows easily from the classical formula

$$\varphi(k) = k \cdot \prod_{p|k} \left(1 - \frac{1}{p}\right).$$

This relation yields

$$\begin{aligned}
\sum_{k \le x} \varphi(k) &= \sum_{k \le x} k \cdot \sum_{d|k} \frac{\mu(d)}{d} \\
&= \sum_{d \le x} \frac{\mu(d)}{d} \cdot \sum_{j \le \frac{x}{d}} jd \\
&= \sum_{d \le x} \mu(d) \cdot \frac{\left[\frac{x}{d}\right]\left(1 + \left[\frac{x}{d}\right]\right)}{2} \\
&= \frac{x^2}{2} \cdot \sum_{d \le x} \frac{\mu(d)}{d^2} + O\left(\sum_{d \le x} \frac{x}{d}\right).
\end{aligned}$$

Note that $\sum_{d \le x} \frac{x}{d} = O(x \ln x)$. Next, we claim that

$$\sum_{d \ge 1} \frac{\mu(d)}{d^2} = \frac{6}{\pi^2}.$$

This follows from Euler's celebrated formula $\sum_{n\geq 1}\frac{1}{n^2}=\frac{\pi^2}{6}$ and the following computation[1]

$$\sum_{d\geq 1}\frac{\mu(d)}{d^2}\cdot\sum_{n\geq 1}\frac{1}{n^2}=\sum_{n,d\geq 1}\frac{\mu(d)}{(nd)^2}=\sum_{k\geq 1}\frac{1}{k^2}\cdot\sum_{d|k}\mu(d)=1.$$

Finally, as

$$\left|\sum_{d>x}\frac{\mu(d)}{d^2}\right|<\sum_{d>x}\frac{1}{d(d-1)}=O\left(\frac{1}{x}\right),$$

we obtain the desired result by combining the previous observations. □

Theorem 2.A.3. *Let $a,b>1$ be integers such that for any prime power p there exists $k\geq 1$ (depending on p) such that $b\equiv a^k \pmod p$. Then b is an integral power of a.*

Proof. We may assume that $a\geq 3$ (as we may work with a^2 and b^2 instead of a and b). The most difficult step is to establish that $\ln a$ and $\ln b$ are linearly dependent over $\mathbb{Q}$. Let us assume that this is not true and consider a large number x. Let S_x be the set of numbers smaller than x and of the form $a^i\cdot b^j$, with i,j nonnegative integers. As $\ln a$ and $\ln b$ are linearly independent over $\mathbb{Q}$, the set S_x has the same number of elements as the number of pairs (i,j) for which $i\ln a+j\ln b<\ln x$. So, there is an absolute constant $c>0$ depending only on a and b such that $|S_x|>c(\ln x)^2$. We will bound $|S_x|$ from above using Gallagher's sieve.

For any positive integer y let P_y be the set of prime powers dividing at least one of the numbers $a-1,a^2-1,\ldots,a^y-1$. For each $p\in P_y$, let $u(p)$ be the order of a mod p. Then $u(p)\leq y$ for all $p\in P_y$, and, since b is a power of a modulo p, we have $|S_x \pmod p|\leq u(p)$ for all $p\in P_y$. To continue, we need a technical lemma.

---

[1] Which uses the absolute convergence of the double series, as well as the fact that $\sum_{d|k}\mu(d)$ equals 0 for all $k>1$ and 1 for $k=1$.

Lemma 2.A.4. *There exist constants $c_1, c_2 > 0$, depending only on $a \geq 3$, such that for all $y \geq 1$ we have*

$$c_1 y^2 \leq \sum_{p \in P_y} \Lambda(p) \leq c_2 y^2.$$

Proof. One estimate is very easy, since

$$\sum_{p \in P_y} \Lambda(p) \leq \sum_{j=1}^{y} \sum_{d | a^j - 1} \Lambda(d) = \sum_{j=1}^{y} \ln(a^j - 1) < \frac{y(y+1)}{2} \cdot \ln a.$$

The other estimate is more delicate and crucially uses properties of cyclotomic polynomials[2] and proposition 2.A.2. Let ϕ_n be the nth cyclotomic polynomial. Remark 9.5 implies that

$$\sum_{u(p)=d} \Lambda(p) \geq \sum_{p | \phi_d(a)} \Lambda(p) - \sum_{p | d} \Lambda(p) = \ln \phi_d(a) - \ln d.$$

By the very definition of ϕ_d we have $\ln \phi_d(a) \geq \varphi(d) \cdot \ln(a-1)$. Hence

$$\sum_{p \in P_y} \Lambda(p) = \sum_{d \leq y} \sum_{u(p)=d} \Lambda(p) \geq \left(\sum_{d \leq y} \varphi(d) \right) \cdot \ln(a-1) - \sum_{d \leq y} \ln d.$$

Since $\sum_{d \leq y} \ln d = O(y \ln y)$, it suffices to use proposition 2.A.2 to conclude. $\square$

The previous lemma shows that by choosing c correctly and y about $c \ln x$, we can ensure that

$$\sum_{p \in P_y} \frac{\Lambda(p)}{u(p)} \geq \frac{1}{y} \sum_{p \in P_y} \Lambda(p) > 2 \ln(2x)$$

and so by Gallagher's sieve

$$|S_x| \leq \frac{1}{\ln(2x)} \left(\sum_{p \in P_y} \Lambda(p) - \ln(2x) \right) \leq c_3 \ln x$$

[2]For more details the reader is referred to section 9.2.

for an absolute constant c_3. This contradicts the first paragraph for large x.

Hence $\ln a$ and $\ln b$ are linearly dependent over $\mathbb{Q}$ and so there exists an integer $c > 1$ and positive integers i, j, relatively prime, such that $a = c^i$ and $b = c^j$. If p is a prime power divisor of $c^i - 1$, there is k_p such that p divides $c^{j-ik_p} - 1$ and so p divides $c^j - 1$. We deduce that $c^i - 1$ divides $c^j - 1$ and so i divides j. The result follows. □

Remark 2.A.5. The result does not hold if we consider only primes instead of prime powers. For instance, using properties of quadratic residues (not more than the multiplicativity of Legendre's symbol) one can easily prove that 16 is an 8th power modulo any prime. Of course, 16 is not an eighth power of an integer. In the beautiful papers [4] and [32],[3] the following general result is proved fully using techniques of algebraic number theory:

Theorem 2.A.6. *Let $n > 1$ be an integer and let a be an integer such that a is an nth power modulo any sufficiently large[4] prime. Then either a is the nth power of an integer or $8|n$ and $a = 2^{\frac{n}{2}} b^n$ for some integer b.*

2.A.2 The large sieve

This rather long section presents a very deep result in analytic number theory, known as the large sieve. Introduced by Y. Linnik and extensively developed by Renyi, Bombieri, Davenport, Montgomery, Vaughan, Gallagher (see [8], [9], [14], [25], [37], [44], [46], [57], [55], [56] to cite only a few references), this became a basic tool in modern analytic number theory, with rather spectacular results. We start by presenting a vast generalization of a famous inequality of Hilbert, due to Montgomery and Vaughan, which, combined with some very clever tricks, yields the analytic form of the large sieve inequality. Combined with standard results in finite Fourier analysis (for which the reader is invited to read addendum 7.A) this yields arithmetic forms of the large sieve. This has some amazing applications to the distribution of prime numbers, twin

[3] Also see theorem 9.B.60.

[4] Actually, the proofs show that it is enough to assume that this holds for a set of primes of Dirichlet density 1.

primes, least quadratic residue, prime numbers in arithmetic progressions, etc. We follow rather closely the amazingly well-written article [56].

The analytic form of the large sieve inequality

We will spend some time proving the following deep theorem, known as the analytic form of the large sieve inequality. Let $||x|| = \min_{n\in\mathbb{Z}} |x - n|$ be the distance from the real number x to the discrete set $\mathbb{Z}$.

Theorem 2.A.7. *Let $x_1, x_2, \ldots, x_n$ be real numbers such that*

$$||x_i - x_j|| \geq \varepsilon > 0$$

for all $i \neq j$. Let

$$T(x) = \sum_{M<k\leq M+N} z_k \cdot e^{2i\pi kx}$$

be a trigonometric polynomial, where $M, N \in \mathbb{N}$ and $z_{M+1}, \ldots, z_{M+N} \in \mathbb{C}$. Then

$$\sum_{j=1}^{n} |T(x_j)|^2 \leq \left(N + \frac{1}{\varepsilon}\right) \sum_{M<k\leq M+N} |z_k|^2.$$

Theorem 2.A.7 has a long history and many mathematicians contributed to it: Davenport-Halberstam, Gallagher (who gave a very simple proof of the inequality with $\pi N + \frac{1}{\varepsilon}$ instead of $N + \frac{1}{\varepsilon}$), Montgomery and Vaughan, Selberg. One can prove (this is due to Selberg) that the factor $N + \frac{1}{\varepsilon}$ can be improved to $N - 1 + \frac{1}{\varepsilon}$ and that this is sharp. The proof of theorem 2.A.7 will span over the next sections.

Tools from linear algebra

In this section we recall a few standard facts about inequalities concerning matrix norms and we prove a duality principle. Recall that the standard hermitian product on $\mathbb{C}^n$ is defined by $\langle x, y\rangle = \sum_{i=1}^{n} x_i \overline{y_i}$, where x_i (respectively y_i) are the coordinates of x (respectively y). If $v \in \mathbb{C}^n$, we denote $|v|^2 = \langle v, v\rangle$.

Definition 2.A.8. A matrix $A = (a_{ij}) \in M_n(\mathbb{C})$ is called hermitian if for all $x, y \in \mathbb{C}^n$ we have $\langle Ax, y\rangle = \langle x, Ay\rangle$.

The reader can easily check that $A = (a_{ij}) \in M_n(\mathbb{C})$ is hermitian if and only if $a_{ij} = \overline{a_{ji}}$ for all i, j. A fundamental result of linear algebra is that for any such matrix A we can find real numbers $\lambda_1, \lambda_2, \ldots, \lambda_n$ and an orthonormal basis $v_1, v_2, \ldots, v_n$ of $\mathbb{C}^n$ (i.e. $\langle v_i, v_j\rangle = 0$ if $i \neq j$ and 1 otherwise) such that $Av_i = \lambda_i \cdot v_i$ for all i. This can also be stated as: all eigenvalues of a hermitian matrix are real numbers and there exists an orthonormal basis consisting of eigenvectors.[5] The following result will be very useful in the next sections.

Proposition 2.A.9. *Let A be a hermitian matrix and let $C > 0$. If the inequality $|\langle Av, v\rangle| \le C|v|^2$ holds for any eigenvector v of A, then it holds for any $v \in \mathbb{C}^n$.*

Proof. Let $v_1, v_2, \ldots, v_n$ be an orthonormal basis of eigenvectors, with corresponding eigenvalues $\lambda_1, \lambda_2, \ldots, \lambda_n$. Consider any $v \in \mathbb{C}^n$ and write $v = \sum_{i=1}^n x_i v_i$ for some $x_i \in \mathbb{C}$. Then

$$\langle Av, v\rangle = \sum_{i,j} x_i \overline{x_j} \lambda_i \cdot \langle v_i, v_j\rangle = \sum_{i=1}^n \lambda_i \cdot |x_i|^2.$$

By hypothesis we have $|\langle Av_i, v_i\rangle| \le C|v_i|^2$ for all i, and since $Av_i = \lambda_i \cdot v_i$, we must have $|\lambda_i| \le C$. Hence

$$|\langle Av, v\rangle| \le C \sum_{i=1}^n |x_i|^2 = C|v|^2. \qquad \square$$

We end this section with a very useful technique. Though elementary, it will play a key role in the proof of theorem 2.A.7.

Proposition 2.A.10. *(Duality principle) Let $(a_{ij})_{1\le i\le m, 1\le j\le n}$ be complex numbers and let $C > 0$. The following are equivalent:*

1) For all $z_i \in \mathbb{C}$ we have

$$\sum_{j=1}^n \left|\sum_{i=1}^m a_{ij} z_i\right|^2 \le C \sum_{i=1}^m |z_i|^2.$$

[5] Recall that if $A = (a_{ij}) \in M_n(\mathbb{C})$ is any matrix and if $\lambda \in \mathbb{C}$ and $v \in \mathbb{C}^n - \{0\}$ satisfy $Av = \lambda \cdot v$, then we say that v is an eigenvector of A associated to the eigenvalue λ.

2) For all $y_i \in \mathbb{C}$ we have

$$\sum_{i=1}^{m} \left| \sum_{j=1}^{n} a_{ij} y_j \right|^2 \leq C \sum_{i=1}^{n} |y_i|^2.$$

Proof. Assume for instance that 1) holds. Then for all $z_i, y_i \in \mathbb{C}$ we have, by Cauchy-Schwarz

$$\left| \sum_{i=1}^{m} \sum_{j=1}^{n} a_{ij} z_i y_j \right|^2 \leq \sum_{j=1}^{n} |y_j|^2 \cdot \sum_j \left| \sum_i a_{ij} z_i \right|^2 \leq C \sum_i |z_i|^2 \cdot \sum_j |y_j|^2.$$

By choosing for z_i the complex conjugate of $\sum_j a_{ij} y_j$, we obtain the inequality in 2). The converse is proved in exactly the same way. □

Montgomery and Vaughan's theorem

The key technical ingredient in the proof of theorem 2.A.7 is the following delicate result [57].

Theorem 2.A.11. *(Montgomery and Vaughan) Let $x_1, x_2, \ldots, x_n$ be real numbers such that $||x_i - x_j|| \geq \varepsilon > 0$ for all $i \neq j$.*

Then for all $z_1, z_2, \ldots, z_n \in \mathbb{C}$

$$\left| \sum_{i \neq j} \frac{z_i \overline{z_j}}{\sin \pi (x_i - x_j)} \right| \leq \frac{1}{\varepsilon} \sum_{i=1}^{n} |z_i|^2.$$

The proof of theorem 2.A.11 is a very nice mixture of elementary, analytic and algebraic arguments. A first crucial ingredient is Euler's famous identity (that we will take for granted)

$$\frac{\pi}{\sin \pi x} = \lim_{N \to \infty} \sum_{|k| \leq N} \frac{(-1)^k}{x+k}.$$

Since

$$\sum_{|k|\leq N} \frac{(-1)^k |k|}{x+k} = \sum_{k=1}^{N} \frac{2x(-1)^k k}{x^2-k^2} = o(N)$$

(the last equality is immediate, since $\frac{2x(-1)^n n}{x^2-n^2} \to 0$ as $n \to \infty$), we can also write

$$\frac{\pi}{\sin \pi x} = \lim_{N\to\infty} \sum_{|k|\leq N} \left(1 - \frac{|k|}{N}\right) \frac{(-1)^k}{x+k}.$$

Thus, it is enough to prove that for all $N \geq n$ we have

$$\left| \sum_{i\neq j} z_i \overline{z_j} \sum_{|k|\leq N} \left(1 - \frac{|k|}{N}\right) \frac{(-1)^k}{k + x_i - x_j} \right| \leq \frac{\pi}{\varepsilon} \sum_{i=1}^{n} |z_i|^2.$$

Now, since there are $N - |k|$ solutions of the equation $j_1 - j_2 = k$ with $j_1, j_2 \in \{1, 2, \ldots, N\}$, it is not difficult to see that the inequality is equivalent to

$$\left| \sum_{i_1\neq i_2} \sum_{1\leq j_1, j_2\leq N} \frac{(-1)^{j_1+j_2} z_{i_1} \overline{z_{i_2}}}{x_{i_1} - x_{i_2} + j_1 - j_2} \right| \leq \frac{\pi N}{\varepsilon} \sum |z_i|^2.$$

This follows from the following vast generalization of a famous inequality due to Hilbert, applied to the family of real numbers $(x_i + j)_{1\leq i\leq n, 1\leq j\leq N}$ and to the family of complex numbers $((-1)^j z_i)_{1\leq i\leq n, 1\leq j\leq N}$.

Theorem 2.A.12. *(Montgomery-Vaughan) Let $\varepsilon > 0$ and let $x_1, x_2, \ldots, x_n \in \mathbb{R}$ be such that $|x_i - x_j| \geq \varepsilon$ for all $i \neq j$. Then for any complex numbers $z_1, z_2, \ldots, z_n$*

$$\left| \sum_{i\neq j} \frac{z_i \cdot \overline{z_j}}{x_i - x_j} \right| \leq \frac{\pi}{\varepsilon} \sum_{i=1}^{n} |z_i|^2.$$

Proof. By homogeneity and symmetry we may assume that $\varepsilon = 1$ and that $x_1 < x_2 < \cdots < x_n$. Then the hypothesis implies that $x_j - x_i \geq j - i$ for $i < j$. Consider the matrix A, where $a_{ij} = 1_{i\neq j} \cdot \frac{1}{x_i - x_j}$. Note that A is antisymmetric, thus $i \cdot A$ is a hermitian matrix, to which proposition 2.A.9 can be applied with

$C = \pi$. Thus, we may assume that $z = (z_1, \dots, z_n)$ is an eigenvector of iA, so also of A, say $Az = i \cdot \lambda z$ for some $\lambda \in \mathbb{R}$.

By Cauchy-Schwarz, the square of the left-hand side of the desired inequality is bounded by $\sum_i |z_i|^2 \cdot \sum_i \left|\sum_{j \neq i} a_{ij}\overline{z_j}\right|^2$, so it is enough to prove the inequality

$$\sum_i \left|\sum_{j \neq i} a_{ij}\overline{z_j}\right|^2 \leq \pi^2 \sum_{i=1}^n |z_i|^2.$$

Expanding brutally the left-hand side and using the crucial observation that $a_{ij}a_{ik} = a_{jk}(a_{ij} - a_{ik})$ yields

$$\begin{aligned} \text{LHS} &= \sum_{j,k} \overline{z_j} z_k \cdot \sum_{i \neq j,k} a_{ij}a_{ik} \\ &= \sum_j |z_j|^2 \cdot \sum_{i \neq j} a_{ij}^2 + \sum_{j \neq k} \overline{z_j} z_k a_{jk} \left(\sum_{i \neq j} a_{ij} - \sum_{i \neq k} a_{ik} + 2a_{jk} \right). \end{aligned}$$

It is now time to use the fact that z is an eigenvector:

$$\begin{aligned} \sum_{j \neq k} \overline{z_j} z_k a_{jk} \left(\sum_{i \neq j} a_{ij} \right) &= \sum_j \overline{z_j} \cdot \left(\sum_{k \neq j} a_{jk} z_k \right) \left(\sum_{i \neq j} a_{ij} \right) \\ &= i\lambda \sum_j |z_j|^2 \left(\sum_{i \neq j} a_{ij} \right). \end{aligned}$$

Doing the same with the other sum, we finally obtain that the term

$$\sum_{j \neq k} \overline{z_j} z_k a_{jk} \left(\sum_{i \neq j} a_{ij} \right)$$

cancels with the other similar term, so that

$$\text{LHS} = \sum_j |z_j|^2 \cdot \sum_{i \neq j} a_{ij}^2 + \sum_{j \neq k} 2\overline{z_j} z_k a_{jk}^2.$$

Now, using the AM-GM inequality $|2\overline{z_j}z_k| \le |z_j|^2 + |z_k|^2$ and rearranging terms, we obtain

$$\text{LHS} \le 3\sum_j |z_j|^2 \cdot \sum_{i\ne j} a_{ij}^2.$$

As $|x_i - x_j| \ge |i-j|$, we have for any fixed j that[6]

$$\sum_{i\ne j} a_{ij}^2 \le \sum_{i\ne j} \frac{1}{(i-j)^2} \le 2\sum_{n\ge 1}\frac{1}{n^2} = \frac{\pi^2}{3}.$$

Combining the last two inequalities yields the desired result. □

Proof of theorem 2.A.7

Let $a_{kj} = e^{2i\pi k x_j}$ for $1 \le j \le n$ and $M < k \le M+N$. We need to prove that for all $(z_k)_{M<k\le M+N}$ we have

$$\sum_{j=1}^{n}\left|\sum_{M<k\le M+N} a_{kj}z_k\right|^2 \le \left(N+\frac{1}{\varepsilon}\right)\sum_k |z_k|^2$$

and by the duality principle (proposition 2.A.10) it is enough to prove that for all $y_1, \dots, y_n$ we have

$$\sum_{M<k\le M+N}\left|\sum_{j=1}^{n} a_{kj}y_j\right|^2 \le \left(N+\frac{1}{\varepsilon}\right)\sum_j |y_j|^2.$$

Expanding the left-hand side, we obtain the equivalent inequality

$$\sum_{j_1\ne j_2} y_{j_1}\overline{y_{j_2}} \cdot \sum_{M<k\le M+N} e^{2i\pi k(x_{j_1}-x_{j_2})} \le \frac{1}{\varepsilon}\sum_j |y_j|^2.$$

---

[6]We use here Euler's famous identity

$$\sum_{n\ge 1}\frac{1}{n^2} = \frac{\pi^2}{6}.$$

Using the formula for the sum of a geometric series, a small computation shows that for all u

$$\sum_{M<k\leq M+N} e^{2i\pi ku} = \frac{1}{2i\sin(\pi u)}\left(e^{i\pi u(2M+2N+1)} - e^{i\pi u(2M+1)}\right).$$

By the triangle inequality, it is thus enough to prove that for any

$$s \in \{2M+2N+1, 2M+1\}$$

and for any $y_1, \ldots, y_n$ we have

$$\left|\sum_{j_1\neq j_2} \frac{y_{j_1}\overline{y_{j_2}}}{\sin\pi(x_{j_1}-x_{j_2})} e^{i\pi(x_{j_1}-x_{j_2})s}\right| \leq \frac{1}{\varepsilon}\sum_j |y_j|^2.$$

But this follows from theorem 2.A.11 with $z_j = y_j \cdot e^{i\pi s\cdot x_j}$.

A quick proof of a weaker form of the sieve inequality

In this section we present Gallagher's short and beautiful proof of the following weaker form of the large sieve inequality. It is much simpler than the proof presented in the previous sections and the result it establishes is good enough in most applications of the large sieve.

Theorem 2.A.13. *Let $x_1, x_2, \ldots, x_n$ be real numbers such that*

$$||x_i - x_j|| \geq \varepsilon > 0$$

for all $i \neq j$. Let

$$T(x) = \sum_{M<k\leq M+N} z_k \cdot e^{2i\pi kx}$$

be a trigonometric polynomial, where $M, N \in \mathbb{N}$ and $z_{M+1}, \ldots, z_{M+N} \in \mathbb{C}$. Then

$$\sum_{j=1}^{n} |T(x_j)|^2 \leq \left(\pi N + \frac{1}{\varepsilon}\right) \sum_{M<k\leq M+N} |z_k|^2.$$

Note that the only difference between this theorem and theorem 2.A.7 is the factor πN, which is N in theorem 2.A.7.

Proof. Assume that f is a continuously differentiable complex-valued map on $\mathbb{R}$. Integrating by parts, it is easy to establish the equality

$$\varepsilon \cdot f(x_j) = \int_{x_j-\frac{\varepsilon}{2}}^{x_j+\frac{\varepsilon}{2}} f(t)dt + \int_{x_j-\frac{\varepsilon}{2}}^{x_j} \left(t - x_j + \frac{\varepsilon}{2}\right) f'(t)dt + \int_{x_j}^{x_j+\frac{\varepsilon}{2}} \left(t - x_j - \frac{\varepsilon}{2}\right) f'(t)dt.$$

Using the triangle inequality and the fact that

$$\left|t - x_j + \frac{\varepsilon}{2}\right| \leq \frac{\varepsilon}{2}$$

for $t \in \left[x_j - \frac{\varepsilon}{2}, x_j\right]$ (and a similar inequality for $t \in \left[x_j, x_j + \frac{\varepsilon}{2}\right]$), we obtain

$$|f(x_j)| \leq \frac{1}{\varepsilon} \cdot \int_{x_j-\frac{\varepsilon}{2}}^{x_j+\frac{\varepsilon}{2}} |f(t)|dt + \frac{1}{2} \int_{x_j-\frac{\varepsilon}{2}}^{x_j+\frac{\varepsilon}{2}} |f'(t)|dt.$$

Take $f(x) = T(x)^2$ and add the corresponding inequalities. The hypothesis $||x_i - x_j|| \geq \varepsilon$ ensures that the intervals $\left(x_j - \frac{\varepsilon}{2}, x_j + \frac{\varepsilon}{2}\right)$ do not overlap mod 1, so using the 1-periodicity of f we obtain

$$\sum_{j=1}^{n} |T(x_j)|^2 \leq \frac{1}{\varepsilon} \int_0^1 |T(x)|^2 dx + \int_0^1 |T(x)| \cdot |T'(x)|dx.$$

Using Parseval's equality and the Cauchy-Schwarz inequality, we obtain

$$\sum_{j=1}^{n} |T(x_j)|^2 \leq \frac{1}{\varepsilon} \cdot \sum_{k=M+1}^{M+N} |z_k|^2 + \sqrt{\sum_{k=M+1}^{M+N} |z_k|^2} \cdot \sqrt{\sum_{k=M+1}^{M+N} |2\pi k z_k|^2}$$
$$\leq \left(\frac{1}{\varepsilon} + 2\pi \max_{M<k\leq M+N} |k|\right) \sum_{k=M+1}^{M+N} |z_k|^2.$$

So we are done if we can ensure that $\max_{M<k\leq M+N}|k|\leq\frac{N}{2}$. Of course, for arbitrary M and N it is unreasonable to hope for such an inequality, but a moment of thought shows that M plays no role in the theorem we are trying to prove, so we can simply choose $M=-\left[\frac{N+1}{2}\right]$ to finish the proof. □

Arithmetic forms of the large sieve inequality

We will apply theorem 2.A.7 for a special family of well-spaced numbers x_i: pick an integer $Q>1$ and consider all numbers of the form $\frac{a}{q}$, with $\gcd(a,q)=1$ and $1\leq a\leq q\leq Q$. It is clear that the difference between two such numbers is (in absolute value) at least $\frac{1}{Q^2}$. Applying the large sieve inequality to this collection, we deduce the following

Theorem 2.A.14. *Let*

$$T(x)=\sum_{M<k\leq M+N}a_k e^{2i\pi kx}$$

be a trigonometric polynomial. Then for all $Q>1$ we have

$$\sum_{q=1}^{Q}\sum_{\substack{1\leq a\leq q\\ \gcd(a,q)=1}}\left|T\left(\frac{a}{q}\right)\right|^2\leq(N+Q^2)\sum_{M<k\leq M+N}|a_k|^2.$$

Next, let us observe that

$$T\left(\frac{a}{q}\right)=\sum_k a_k e^{\frac{2i\pi ka}{q}}=\sum_{h=1}^{q}\left(\sum_{k\equiv h\,(\mathrm{mod}\ q)}a_k\right)e^{\frac{2i\pi ah}{q}}.$$

Therefore, using the techniques of addendum 7.A, more precisely Plancherel's formula (theorem 7.A.5, but this can also be done by expanding everything), we obtain

$$\sum_{a=1}^{q}\left|T\left(\frac{a}{q}\right)\right|^2=q\sum_{h=1}^{q}\left|\sum_{k\equiv h\,(\mathrm{mod}\ q)}a_k\right|^2.$$

Now, if the a_n were uniformly distributed, one would expect that $\sum_{k\equiv h \pmod q} a_k$ behaves as $E_q = \frac{1}{q}\sum_k a_k$. Actually, a small computation reveals that

$$\sum_{h=1}^{q}\left|\sum_{k\equiv h\,(\mathrm{mod}\ q)} a_k - E_q\right|^2 = \sum_{h=1}^{q}\left|\sum_{k\equiv h\,(\mathrm{mod}\ q)} a_k\right|^2 - q|E_q|^2,$$

which combined with the previous formula yields

$$\sum_{a=1}^{q-1}\left|T\left(\frac{a}{q}\right)\right|^2 = q\sum_{h=1}^{q}\left|\sum_{k\equiv h\,(\mathrm{mod}\ q)} a_k - E_q\right|^2.$$

Unfortunately, $\sum_{a=1}^{q-1}\left|T\left(\frac{a}{q}\right)\right|^2$ is usually larger than

$$\sum_{1\le a\le q, \gcd(a,q)=1}\left|T\left(\frac{a}{q}\right)\right|^2,$$

but if q is a prime number, they are actually equal! Using these remarks and the previous theorem, we obtain the following strong inequality:

Theorem 2.A.15. *Let $(a_k)_{M<k\le M+N}$ be a sequence of complex numbers. Then for all integers $Q>1$ we have*

$$\sum_{p\le Q} p\cdot\sum_{h=1}^{p}\left|\sum_{k\equiv h\,(\mathrm{mod}\ p)} a_k - \frac{1}{p}\sum_k a_k\right|^2 \le (N+Q^2)\sum_k |a_k|^2,$$

the sum being taken over all primes $p\le Q$.

By specializing $a_k = 1_{k\in A}$ for a subset $A\subset[M+1, M+N]$, we obtain the following equidistribution result:

Corollary 2.A.16. *Let $A\subset[M+1,M+N]$ be a set of integers. Then for all integers $Q>1$ we have*

$$\sum_{p\le Q}\sum_{h=1}^{p} p\cdot\left||A\cap(h+p\mathbb{Z})| - \frac{1}{p}|A|\right|^2 \le (N+Q^2)|A|.$$

This finally yields a "sieve inequality."

Theorem 2.A.17. *Let $A \subset [M+1, M+N]$ be a set of integers and suppose that for each prime $p \leq Q$ at least $\omega(p)$ residue classes mod p contain no element of A. Then*

$$|A| \leq \frac{N+Q^2}{\sum_{p\leq Q}\frac{\omega(p)}{p}}.$$

Proof. Use the previous corollary and the observation that for each p we have

$$\sum_{h=1}^{p}\left|A\cap(h+p\mathbb{Z})-\frac{1}{p}|A|\right|^2 \geq \omega(p)\frac{|A|^2}{p^2},$$

as at least $\omega(p)$ terms in the sum are equal to $\frac{|A|^2}{p^2}$. □

The previous theorem helps understanding the name "large sieve." Indeed, in typical applications A (mod p) will miss a good part of the residue classes modulo p (i.e. the integers $\omega(p)$ will be quite large) and the sieve inequality will yield nontrivial bounds on $|A|$. The next result, an amazing theorem of Linnik, is at the very origin of the large sieve and one of its best applications. It concerns the distribution of the least quadratic non residue modulo p. Vinogradov conjectured that it is smaller than $c(\varepsilon)p^{\varepsilon}$ for any $\varepsilon > 0$ and if one assumes the Generalized Riemann Hypothesis, then one can actually prove that it is smaller than $c(\log p)^2$. The following deep theorem due to Linnik [47] is a first step in the direction of Vinogradov's conjecture (which is still largely open):

Theorem 2.A.18. *(Linnik) Let $n(p)$ be the least positive integer which is not a quadratic residue mod p. Then for all $\varepsilon > 0$ there is a constant $c(\varepsilon)$ such that for all N we have*

$$|\{p \leq N | n(p) > N^{\varepsilon}\}| \leq c(\varepsilon).$$

Proof. Fix $\varepsilon > 0$, which for simplicity (but without loss of generality) we take of the form $\frac{1}{d}$ for some integer d greater than 2. For a positive integer N, let

$$P_N = \{p \leq \sqrt{N} | n(p) > N^{\varepsilon}\}$$

and

$$A_N = \left\{ n \le N | \left(\frac{n}{p}\right) = 1 \quad \text{for all} \quad p \in P_N \right\}.$$

We will prove the following technical result:

Lemma 2.A.19. *There exist positive constants $c(\varepsilon)$ and $c_1(\varepsilon)$ such that for all $N > c_1(\varepsilon)$ we have $|A_N| > c(\varepsilon)N$.*

Let us assume for a moment that this holds. Since $A_N \pmod p$ misses all quadratic non-residues mod p, we can choose $\omega(p) = \frac{p-1}{2} \ge \frac{p}{4}$ in the previous theorem and obtain $|A_N| \le \frac{8N}{|P_N|}$. Hence for all $N > c_1(\varepsilon)$ we have $|P_N| < \frac{8}{c(\varepsilon)}$, which is enough to conclude the theorem.

Let us prove the lemma. Note that A_N contains all numbers $n \le N$ all of whose prime factors are smaller than N^ε, as Legendre's symbol is multiplicative. We will consider all numbers $n = kp_1 \cdots p_d$ smaller than N, with $p_i \in [N^{\varepsilon - \frac{\varepsilon^2}{2}}, N^\varepsilon]$ in nondecreasing order and some integer k. Note that $k \le N^{\frac{1}{2}\varepsilon}$, as $n \le N$, so k is relatively prime to any prime greater than $N^{\varepsilon - \frac{\varepsilon^2}{2}}$. This easily implies that all such numbers are different and they clearly belong to A_N. Since there are more than $\frac{N}{p_1 p_2 \cdots p_d} - 1$ possible values for k, it follows that we have at least

$$\sum_{p_i \in [N^{\varepsilon - \frac{\varepsilon^2}{2}}, N^\varepsilon]} \frac{N}{p_1 \cdots p_d} - \pi(N^\varepsilon)^d \ge \frac{1}{d!} N \left(\sum_{p \in [N^{\varepsilon - \frac{\varepsilon^2}{2}}, N^\varepsilon]} \frac{1}{p} \right)^d - \pi(N^\varepsilon)^d$$

such numbers. Taking into account the fact[7] that $\pi(x) = o(x)$ and

$$\sum_{p \in [a,b]} \frac{1}{p} = \log \frac{\log b}{\log a} + O(1)$$

as $a, b \to \infty$, the last quantity is easily seen to be greater than $c(\varepsilon)N$ for some positive constant $c(\varepsilon)$ and all sufficiently large (depending on ε) N. The desired result follows. $\square$

---

[7]Both these results are proved in addendum 3.A.

Though theorem 2.A.17 is already a very strong result, in most applications one needs more refined estimates, in which one takes into account all $q \leq Q$, not only prime numbers. In order to do this, we need another nice application of finite Fourier analysis and Cauchy-Schwarz, but before doing that recall that μ is Möbius's function, defined by $\mu(n) = (-1)^k$ if n is a product of $k \geq 0$ distinct prime numbers and $\mu(n) = 0$ otherwise.

Theorem 2.A.20. *Let $A \subset [M+1, M+N]$ be a set of integers and suppose that for each prime $p \leq Q$ we have*

$$|\{a \pmod p | a \in A\}| \leq p - \omega(p).$$

Then for all trigonometric polynomials $T(x) = \sum_{k\in A} a_k e^{2i\pi kx}$ and all positive integers q

$$\sum_{\substack{\gcd(a,q)=1\\ a\leq q}} \left|T\left(\frac{a}{q}\right)\right|^2 \geq \mu(q)^2 \prod_{p|q} \frac{\omega(p)}{p-\omega(p)} |T(0)|^2.$$

Proof. Let

$$g(q) = \mu(q)^2 \prod_{p|q} \frac{\omega(p)}{p-\omega(p)},$$

a multiplicative function. We will prove the theorem in two steps: first we will prove it when q is prime and then we will show that if the theorem holds for two relatively prime numbers q, q', then it also holds for qq'. First, assume that q is prime. Let

$$x_h = \sum_{k\in A, k\equiv h \pmod q} a_k.$$

Then at least $\omega(q)$ of the numbers x_h are zero (by hypothesis), so by Cauchy-Schwarz and Plancherel's identity we get

$$|T(0)|^2 = \left|\sum_{h=1}^{q} x_k\right|^2 \leq (q-\omega(q)) \sum_h |x_h|^2 = \left(1 - \frac{\omega(q)}{q}\right) \sum_{a=1}^{q} \left|T\left(\frac{a}{q}\right)\right|^2,$$

from where the result follows easily. Next, assume that the result holds for q and q', with $\gcd(q, q') = 1$. Using the Chinese Remainder Theorem and the fact that the theorem holds for q (with $T\left(\frac{b}{q'}+x\right)$ instead of $T(x)$) and then for q', we have

$$\begin{aligned}\sum_{a\in(\mathbb{Z}/qq'\mathbb{Z})^*}\left|T\left(\frac{a}{qq'}\right)\right|^2 &= \sum_{b\in(\mathbb{Z}/q'\mathbb{Z})^*}\sum_{a\in(\mathbb{Z}/q\mathbb{Z})^*}\left|T\left(\frac{a}{q}+\frac{b}{q'}\right)\right|^2\\ &\geq \sum_{b\in(\mathbb{Z}/q'\mathbb{Z})^*} g(q)\left|T\left(\frac{b}{q'}\right)\right|^2\\ &\geq g(q)g(q')|T(0)|^2 = g(qq')|T(0)|^2.\end{aligned}$$

□

Combining the special case $T(x)=\sum_{k\in A}e^{2i\pi kx}$ of the previous theorem with the large sieve inequality (theorem 2.A.14), we obtain the following strong form of theorem 2.A.17:

Theorem 2.A.21. *(Montgomery) Let $A\subset[M+1,M+N]$ be a set of integers and suppose that for each prime $p\leq Q$ we have*

$$|\{a \pmod p)|a\in A\}|\leq p-\omega(p).$$

Then $|A|\leq\frac{N+Q^2}{L}$, where

$$L=\sum_{q\leq Q}\mu(q)^2\prod_{p|q}\frac{\omega(p)}{p-\omega(p)}.$$

Some applications

We present here two classical applications of the large sieve: a uniform upper bound on $\pi(m+n)-\pi(n)$ and an upper bound for the number of twin primes smaller than n.

Theorem 2.A.22. *If $m\geq\sqrt{n}$, then there are at most $\frac{4n}{\log n}$ primes between $m+1$ and $m+n$.*

Proof. Let A be the set of prime numbers between $m+1$ and $m+n$. No element of A is divisible by some prime smaller than or equal to $[\sqrt{n}]$, so by theorem 2.A.21 we obtain

$$|A| \leq \frac{n+n}{L}, \quad L = \sum_{q \leq \sqrt{n}} \mu(q)^2 \prod_{p|q} \frac{1}{p-1}.$$

Using the fact that $\frac{1}{p-1} = \sum_{j \geq 1} \left(\frac{1}{p}\right)^j$, we obtain

$$L \geq \sum_{p_1^{j_1} \cdots p_s^{j_s} \leq \sqrt{n}} \frac{1}{p_1^{j_1} \cdots p_s^{j_s}} = \sum_{j \leq \sqrt{n}} \frac{1}{j} > \log \sqrt{n} = \frac{1}{2} \log n.$$

Inserting this result in the previous inequality yields the desired result. □

Theorem 2.A.23. *There exists an absolute constant c such that for all n there are at most $\frac{cn}{(\log n)^2}$ primes $p \leq n$ such that $p+2$ is also prime.*

Proof. Let $f(n)$ be the number of twin primes smaller than n and let A be the set of twin primes in $([\sqrt{n}], n]$. If $p \in (2, [\sqrt{n}]]$ is a prime number, then clearly $0, -2 \notin \{a \pmod{p} | a \in A\}$, so we can take $\omega(p) = 2$ for such primes in theorem 2.A.21 (and $\omega(2) = 0$), deducing that

$$f(n) - f([\sqrt{n}]) = |A| \leq \frac{2n}{L}, \quad L = \sum_{\substack{q \leq \sqrt{n} \\ \gcd(2,q)=1}} \mu(q)^2 \prod_{p|q} \frac{2}{p-2}.$$

It remains to obtain a lower bound on L. Since $\frac{2}{p-2} = \sum_{j \geq 1} \left(\frac{2}{p}\right)^j$, we obtain

$$L \geq \sum_{\substack{p_1^{i_1} \cdots p_s^{i_s} \leq \sqrt{n} \\ \min p_i > 2}} \frac{2^{i_1 + \cdots + i_s}}{p_1^{i_1} \cdots p_s^{i_s}} \geq \sum_{\substack{p_1^{i_1} \cdots p_s^{i_s} \leq \sqrt{n} \\ \min p_i > 2}} \frac{(i_1+1) \cdots (i_s+1)}{p_1^{i_1} \cdots p_s^{i_s}}$$

$$= \sum_{\substack{k \leq \sqrt{n} \\ \gcd(2,k)=1}} \frac{d(k)}{k} \geq \left(\sum_{\substack{k \leq \sqrt[4]{n} \\ \gcd(2,k)=1}} \frac{1}{k} \right)^2 \geq c(\log n)^2,$$

where $d(k)$ is the number of divisors of k.

Combining the previous estimates, we obtain

$$f(n) - f([\sqrt{n}]) \leq \frac{c_1 n}{(\log n)^2}$$

for some constant $c_1 > 0$. Using the trivial bound $f([\sqrt{n}]) \leq \sqrt{n} = O\left(\frac{n}{\log^2 n}\right)$, the result follows. □

Using the previous theorem, we can easily prove the following famous result, which is probably the origin of all modern sieve theories:

Theorem 2.A.24. *(V.Brun) The sum of the inverses of all twin prime numbers converges.*

Proof. By the previous theorem, there are at most $\frac{c \cdot 2^j}{j^2}$ twin primes between 2^j and 2^{j+1}, for some absolute constant c. The sum of the inverses of these twin primes is smaller than $\frac{1}{2^j} \cdot \frac{c \cdot 2^j}{j^2} = \frac{c}{j^2}$. As $\sum_{j \geq 1} \frac{1}{j^2}$ converges, the result follows. □

These are far from being the most spectacular applications of the large sieve (though they are already deep theorems!). We refer the reader to [9], [19], [37], [44], for many other applications.

2.A.3 The Turán-Kubilius inequality

In this section, we use again the Cauchy-Schwarz inequality to prove a famous inequality in analytic number theory, the Turán-Kubilius inequality. We then apply it to establish a well-known result of Hardy and Ramanujan (which was actually the origin of this inequality) and then a very amusing consequence due to Erdős. We also present a dual form of the Turán-Kubilius inequality, due to Elliott, which has a similar form to the inequality established in theorem 2.A.15 and which turned out to have some pretty far-reaching consequences (for instance, a proof of the prime number theorem!).

A function $f : \{1, 2, \dots\} \to \mathbb{C}$ is called additive if $f(ab) = f(a) + f(b)$ for all relatively prime positive integers a, b. It is called strongly additive if moreover $f(p^k) = f(p)$ for all primes p and all positive integers k. Thus

$$f(n) = \sum_{p|n} f(p^{v_p(n)}), \text{ respectively } f(n) = \sum_{p|n} f(p)$$

when f is additive, respectively strongly additive. In particular, if f is additive we have

$$\frac{1}{n}\sum_{i=1}^{n} f(i) = \frac{1}{n}\sum_{p^k \le n}\left(\left[\frac{n}{p^k}\right] - \left[\frac{n}{p^{k+1}}\right]\right) f(p^k),$$

while if f is strongly additive, then

$$\frac{1}{n}\sum_{i=1}^{n} f(i) = \frac{1}{n}\sum_{p \le n}\left[\frac{n}{p}\right] f(p).$$

Thus

$$E_f(n) = \sum_{p^k \le n}\left(1 - \frac{1}{p}\right)\frac{f(p^k)}{p^k}, \text{ respectively } E_f^{sa}(n) = \sum_{p \le n}\frac{f(p)}{p}$$

is a good approximation of the average value of $f(1), \dots, f(n)$ if f is additive, respectively strongly additive. Our fundamental result will allow us to see

$$B_f(n) = \sum_{p^k \le n}\frac{|f(p^k)|^2}{p^k}, \text{ respectively } B_f^{sa}(n) = \sum_{p \le n}\frac{|f(p)|^2}{p}$$

as an upper bound for the variance of f as a random variable on $\{1, 2, \dots, n\}$ (with the uniform distribution). Without further ado, we follow [24], which contains a very easy proof of this theorem.

Theorem 2.A.25. *(Turán-Kubilius inequality) There exists an absolute constant $C > 0$ with the following property:*

1) For all strongly additive maps $f : \{1, 2, \dots\} \to \mathbb{C}$ and all n

$$\frac{1}{n}\sum_{k=1}^{n}|f(k) - E_f^{sa}(n)|^2 \le C \cdot B_f^{sa}(n).$$

2) For all additive maps $f : \{1, 2, \dots\} \to \mathbb{C}$ and all n

$$\frac{1}{n}\sum_{k=1}^{n}|f(k) - E_f(n)|^2 \le C \cdot B_f(n).$$

Proof. We will prove only 1), as 2) is proved in exactly the same way. Suppose first that f takes nonnegative values. Denote $E = E_f^{sa}(n)$, $B = \sqrt{B_f^{sa}(n)}$ and observe that

$$S = \sum_{k=1}^{n}|f(k) - E|^2 = \sum_{k\le n} f(k)^2 - 2E\sum_{k\le n} f(k) + nE^2.$$

As f is strongly additive, we have

$$\sum_{k\le n} f(k)^2 = \sum_{k\le n}\sum_{p,q|k} f(p)f(q) = \sum_{p\le n}\left[\frac{n}{p}\right] f(p)^2 + \sum_{\substack{pq\le n\\ p\ne q}}\left[\frac{n}{pq}\right] f(p)f(q)$$

and the last quantity does not exceed $nB^2 + nE^2$, as f takes nonnegative values and $[x] \le x$ for all x. On the other hand, since $[x] > x - 1$ for all x, we can write

$$-2E\sum_{k\le n} f(k) = -2E\sum_{p\le n}\left[\frac{n}{p}\right] f(p) \le -2nE^2 + 2E\sum_{p\le n} f(p).$$

All in all, we obtain

$$S \le nB^2 + 2E\sum_{p\le n} f(p).$$

Next, two applications of Cauchy-Schwarz show that

$$E \cdot \sum_{p\le n} f(p) \le \sqrt{\sum_{p\le n}\frac{1}{p}\cdot B^2}\cdot\sqrt{\sum_{p\le n} p\cdot B^2} = B^2\cdot\sqrt{\sum_{p\le n} p\cdot\sum_{p\le n}\frac{1}{p}}.$$

Using the classical estimates[8]

$$\sum_{p\le n}\frac{1}{p}=O(\log\log n),\quad \sum_{p\le n}p=O\left(\frac{n^2}{\log n}\right),$$

we deduce that

$$S\le nB^2\left(1+O\left(\sqrt{\frac{\log\log n}{\log n}}\right)\right)$$

and the result follows.

Suppose now that f is real-valued and define two strongly additive functions g_1, g_2 by

$$g_1(p)=1_{f(p)\ge 0}\cdot f(p),\quad g_2(p)=-1_{f(p)<0}\cdot f(p).$$

Using Cauchy-Schwarz, we get (with $E=E_f(n)$, $E_j=E_{g_j}(n)$)

$$\sum_{k\le n}|f(k)-E|^2\le 2\sum_{j=1}^{2}\sum_{k=1}^{n}|g_j(k)-E_j|^2$$

and the result follows easily by applying the result of the previous paragraph to g_1, g_2. Finally, if f takes on complex values, set $g_1=\mathrm{Re}(f)$ and $g_2=\mathrm{Im}(f)$ and use the same argument. □

Corollary 2.A.26. *(Turán) If $\omega(n)$, $\Omega(n)$ is the number of prime factors of n without (respectively with) multiplicity, then*

$$\sum_{k\le n}(\omega(k)-\log\log n)^2=O(n\log\log n)$$

and

$$\sum_{k\le n}(\Omega(k)-\log\log n)^2=O(n\log\log n).$$

[8]Which follow easily from the results of the addendum 3.A.

Proof. Use the Turán-Kubilius inequality with $f = \omega$, respectively $f = \Omega$. Note that if $f = \omega$, then

$$E_f^{sa}(n) = B_f^{sa}(n) = \sum_{p \leq n} \frac{1}{p} = \log\log n + O(1),$$

as follows from Mertens' theorem 3.A.5.
Similar arguments apply for $f = \Omega$. □

The following theorem is an immediate consequence of the previous result. Its original proof was much more intricate (see [39]).

Theorem 2.A.27. *(Hardy-Ramanujan)*
The functions ω and Ω have normal order $\log\log n$*, i.e. if $f \in \{\omega, \Omega\}$, then for all $\varepsilon > 0$ we have*

$$\lim_{x\to\infty} \frac{1}{x} \left| \left\{ n \leq x | 1 - \varepsilon < \frac{f(n)}{\log\log x} < 1 + \varepsilon \right\} \right| = 1.$$

Proof. Let $f \in \{\omega, \Omega\}$ and let

$$A_x = \left\{ n \leq x \middle| \left| \frac{f(n)}{\log\log x} - 1 \right| \geq \varepsilon \right\}.$$

Since for any $n \in A_x$ we have

$$(f(n) - \log\log x)^2 \geq \varepsilon^2 \cdot (\log\log x)^2,$$

we deduce that

$$\varepsilon^2 \cdot (\log\log x)^2 \cdot |A_x| \leq \sum_{n \in A_x} (f(n) - \log\log x)^2 \leq \sum_{n \leq x} (f(n) - \log\log x)^2.$$

Using the previous corollary, the result follows. □

Here is a very beautiful consequence of the Hardy-Ramanujan theorem, which is surprisingly difficult to prove by other means:

Theorem 2.A.28. *(Erdős) We have*

$$\lim_{n\to\infty} \frac{|\{a \cdot b | 1 \le a, b \le n\}|}{n^2} = 0.$$

Proof. By the previous theorem, there are $n^2 + o(n^2)$ pairs (a, b) with $1 \le a, b \le n$ and such that $\Omega(ab)$ is about $2 \log \log n$. The same theorem shows that $n^2 + o(n^2)$ numbers $k \in \{1, 2, \ldots, n^2\}$ have $\Omega(k)$ about

$$\log \log n^2 = \log \log n + \log 2 << 2 \log \log n,$$

so the number of numbers of the form ab (with $1 \le a, b \le n$) must be $o(n^2)$. $\square$

The following difficult theorem (see [28]) refines considerably Hardy-Ramanujan's result. It is also known as the fundamental theorem of probabilistic number theory.

Theorem 2.A.29. *(Erdős-Kac) Let $f : \{1, 2, \ldots\} \to \mathbb{R}$ be a strongly additive function such that the sequence $(f(p))_p$ is bounded (p runs over the prime numbers) and such that $\lim_{n\to\infty} B_f(n) = \infty$. Then for all real numbers a*

$$\lim_{x\to\infty} \frac{1}{x} |\{n \le x | f(n) - E_f(x) < a\sqrt{B_f(x)}\}| = \frac{1}{\sqrt{2\pi}} \int_{-\infty}^{a} e^{-u^2/2} du.$$

Let us end this section and this addendum with another beautiful application of the Turán-Kubilius inequality, due to Elliott. We let $v_p(n)$ denote the exponent of the prime number p in the factorization of n.

Theorem 2.A.30. *There exists an absolute constant $C > 0$ such that:*

1) For all x and all $a_n \in \mathbb{C}$

$$\sum_{p^k \le x} p^k \left| \sum_{\substack{n \le x \\ v_p(n) = k}} a_n - \frac{1}{p^k}\left(1 - \frac{1}{p}\right) \sum_{n \le x} a_n \right|^2 \le Cx \sum_{n \le x} |a_n|^2.$$

2) For all x and all complex numbers a_n

$$\sum_{p\le x} p\left|\sum_{\substack{n\le x\\ p|n}} a_n - \frac{1}{p}\sum_{n\le x} a_n\right|^2 \le Cx\cdot\sum_{n\le x}|a_n|^2.$$

Before looking at the proof, the reader is strongly advised to compare this theorem and theorem 2.A.15.

Proof. We will prove only the first part, as the same argument yields the second part. Observe that Turán-Kubilius's inequality can also be written in the form

$$\frac{1}{n}\sum_{k=1}^{n}\left|\sum_{p^u\le n} f(p^u)\left(1_{u=v_p(k)} - \left(1-\frac{1}{p}\right)\frac{1}{p^u}\right)\right|^2 \le C\sum_{p^u\le n}\frac{|f(p^u)|^2}{p^u}.$$

This suggests defining for $j,k\in[1,n]$ the quantity

$$a_{kj} = \left(1_{u=v_p(k)} - \left(1-\frac{1}{p}\right)\frac{1}{j}\right)\sqrt{j}$$

if $j=p^u$ for some u and some prime p and $a_{jk}=0$ otherwise. Given any sequence $y_1,\dots,y_n$ of complex numbers, we can certainly define an additive map f such that $f(p^u)=\sqrt{p^u}y_{p^u}$ for all u,p such that $p^u\le n$. Then the previous inequality can be written

$$\sum_{k=1}^{n}\left|\sum_{j=1}^{n} a_{kj}y_j\right|^2 \le nC\sum_{p^u\le n}|y_{p^u}|^2 \le nC\sum_{j=1}^{n}|y_j|^2.$$

Applying the duality principle 2.A.10 and unwinding definitions, we obtain the desired inequality. □

The previous theorem plays a key role in Hildebrand's "elementary" proof (see [40]) of an extremely difficult theorem of Wirsing [85], [86], that confirmed a long-standing conjecture of Erdős and Wintner.

Theorem 2.A.31. *(Wirsing) Let $f : \{1, 2, \ldots\} \to [-1, 1]$ be a multiplicative function (i.e. $f(ab) = f(a)f(b)$ whenever $\gcd(a, b) = 1$). Then*

$$M(f) = \lim_{x\to\infty} \frac{1}{x} \sum_{n\leq x} f(n)$$

exists. Moreover, $M(f) = 0$ if

$$\sum_p \frac{1 - f(p)}{p} = \infty.$$

To see how subtle this theorem is, take Möbius' function for f: it is fairly elementary that $\sum_p \frac{1}{p} = \infty$, so the previous theorem yields $\sum_{n\leq x} \mu(n) = o(x)$. But it is elementary to prove that this is equivalent to the prime number theorem! Even though much easier than Wirsing's original proof, Hildebrand's proof is still rather technical and we refer the reader to his article [40] for details.

Chapter 3

Look at the Exponent

3.1 Introduction

If p is a prime, we define the p-adic valuation map $v_p : \mathbb{Z} \to \mathbb{N} \cup \{\infty\}$ by: $v_p(0) = \infty$ and for $n \neq 0$ we have $v_p(n) = k$ if p^k divides n, but p^{k+1} does not divide n. More concretely, for $n \neq 0$, $v_p(n)$ is the exponent of the prime number p in the prime factorization of n. The unique factorization of integers into powers of prime numbers easily yields

$$v_p(ab) = v_p(a) + v_p(b), \quad v_p(a+b) \geq \min(v_p(a), v_p(b)),$$

with equality if $v_p(a) \neq v_p(b)$. The first property allows us to extend this map v_p to $\mathbb{Q}$ by $v_p\left(\frac{a}{b}\right) = v_p(a) - v_p(b)$ for all nonzero integers a, b. It is an easy exercise to check that this is well-defined (i.e. if $\frac{a}{b} = \frac{c}{d}$, then we get the same value for $v_p\left(\frac{a}{b}\right)$ and $v_p\left(\frac{c}{d}\right)$).

We will frequently use the following easy properties of the p-adic valuation:

$$v_p(\gcd(x_1, x_2, \ldots, x_n)) = \min_{1 \leq i \leq n} v_p(x_i),$$
$$v_p(\operatorname{lcm}(x_1, x_2, \ldots, x_n)) = \max_{1 \leq i \leq n} v_p(x_i).$$

3.2 Local-global principle

A very useful idea when dealing with divisibilities (or even equalities concerning arithmetic objects) is the local-global principle: if a and b are nonzero integers, then a divides b if and only if for all primes p we have $v_p(a) \le v_p(b)$. This "local-global principle" is the simplest of a series of such statements, concerning the way in which the arithmetic of integers is governed by the "behavior at each prime." Some of these statements are very deep and most of them are an active area of research. Here are some other such examples: a positive integer is a perfect square if and only if it is a perfect square mod p for all primes p (this is already a quite serious result, using the quadratic reciprocity law). Another famous example is Hasse-Minkowski's local-global principle: if a_i are nonzero rational numbers, the equation $a_1x_1^2 + a_2x_2^2 + \cdots + a_nx_n^2 = 0$ has nontrivial rational solutions if and only if it has nontrivial real solutions and nontrivial p-adic solutions for all primes p. This is a deep theorem and we refer the reader to the addendum concerning p-adic numbers for more details.

We start with some easy applications of the local-global principle.

1. Prove the identity

$$\frac{\operatorname{lcm}(a,b,c)^2}{\operatorname{lcm}(a,b)\cdot\operatorname{lcm}(b,c)\cdot\operatorname{lcm}(c,a)} = \frac{\gcd(a,b,c)^2}{\gcd(a,b)\cdot\gcd(b,c)\cdot\gcd(c,a)}$$

for all positive integers a, b, c.

USAMO 1972

Proof. Fix any prime p and let $x = v_p(a), y = v_p(b)$ and $z = v_p(c)$. The p-adic valuation of the left-hand side is

$$2\max(x,y,z) - \max(x,y) - \max(y,z) - \max(z,x),$$

while that of the right-hand side is

$$2\min(x,y,z) - \min(x,y) - \min(y,z) - \min(z,x).$$

We claim that these two quantities are equal. Since the two expressions are symmetric in x, y, z, we may very well assume that $x \ge y \ge z$. Then we need to prove that $2x - x - y - x = 2z - y - z - z$, which is clear. □

Remark 3.1. Another natural proof uses the identity

$$\operatorname{lcm}(x,y) = \frac{xy}{\gcd(x,y)}$$

and its analogue

$$\operatorname{lcm}(x,y,z) = \frac{xyz\gcd(x,y,z)}{\gcd(x,y)\gcd(y,z)\gcd(z,x)}.$$

Plugging in these values immediately yields the result.

2. Let $a_1, a_2, \ldots, a_k$ and $b_1, b_2, \ldots, b_k$ be positive integers such that a_i and b_i are relatively prime for all $i \in \{1, 2, \ldots, k\}$. Prove that

$$\gcd\left(\frac{a_1 m}{b_1}, \frac{a_2 m}{b_2}, \ldots, \frac{a_k m}{b_k}\right) = \gcd(a_1, a_2, \ldots, a_k),$$

where $m = \operatorname{lcm}(b_1, b_2, \ldots, b_k)$.

IMO 1974 Shortlist

Proof. Fix a prime p and let $x_i = v_p(a_i)$ and $y_i = v_p(b_i)$. By hypothesis, we have $\min(x_i, y_i) = 0$ for all i and we need to prove that if

$$z = \max(y_1, y_2, \ldots, y_k),$$

then

$$\min(x_1 - y_1 + z, x_2 - y_2 + z, \ldots, x_k - y_k + z) = \min(x_1, x_2, \ldots, x_k).$$

Note that $x_i - y_i + z \geq x_i$ for all i, so certainly the left-hand side of the previous inequality is at least the right-hand side. On the other hand, if $z = 0$, then this forces all y_i to vanish and in this case the equality is clear. So, we may assume that there is some j such that $y_j = z > 0$. Then $x_j = 0$ and $x_j - y_j + z = 0$, making the equality clear again. □

The next problem is quite challenging and requires some preliminaries. A very common situation in analytic number theory is the following one: we have two maps f and g and we assume that $g(n) = \sum_{d|n} f(d)$ for all n. Möbius found a very nice inversion formula, which expresses f in terms of g. Define the Möbius function μ by $\mu(1) = 1$, $\mu(n) = 0$ if n is not squarefree and $\mu(p_1p_2\cdots p_k) = (-1)^k$ if $p_1, p_2, \dots, p_k$ are distinct prime numbers. Then the Möbius' inversion formula reads $f(n) = \sum_{d|n} \mu\left(\frac{n}{d}\right) g(d)$. The main ingredient in the proof is the fact that $\sum_{d|n} \mu(d) = 0$ for all $n > 1$ and it equals 1 for $n = 1$. This is immediate, since if $p_1, p_2, \dots, p_k$ are the different primes dividing n, then for $n > 1$

$$\begin{aligned}\sum_{d|n} \mu(d) &= 1 - \sum_{i=1}^{k} \mu(p_i) + \sum_{i<j} \mu(p_ip_j) + \cdots \\ &= 1 - \binom{k}{1} + \binom{k}{2} - \cdots \\ &= (1-1)^k = 0,\end{aligned}$$

by the binomial formula. Using this observation, we can write

$$\begin{aligned}\sum_{d|n} \mu\left(\frac{n}{d}\right) g(d) &= \sum_{d|n} \mu(d) g\left(\frac{n}{d}\right) \\ &= \sum_{d|n} \mu(d) \cdot \sum_{d_1 d|n} f(d_1) \\ &= \sum_{d_1|n} f(d_1) \sum_{d|\frac{n}{d_1}} \mu(d) \\ &= f(n).\end{aligned}$$

The next problem requires the multiplicative version of Möbius's formula, namely if $g(n) = \prod_{d|n} f(d)$, then $f(n) = \prod_{d|n} g(d)^{\mu\left(\frac{n}{d}\right)}$. The proof being exactly the same as above, we leave it to the reader to fill in the details.

3. Let $(a_n)_{n\geq 1}$ be a sequence of positive integers such that

$$\gcd(a_m, a_n) = a_{\gcd(m,n)}$$

for all positive integers m, n. Prove that there exists a unique sequence of positive integers $(b_n)_{n\geq 1}$ such that $a_n = \prod_{d|n} b_d$ for all n.

Marcel Ţena, Romanian TST

Proof. In the light of the previous discussion, we are forced to define

$$b_n = \prod_{d|n} a_{\frac{n}{d}}^{\mu(d)}.$$

Of course, the hard point is to prove that b_n is an integer. In order to prove this, we will first transform the expression defining b_n, using the Mersenne property[1] of the sequence $(a_n)_n$. Namely, let $n = p_1^{\alpha_1} p_2^{\alpha_2} \cdots p_k^{\alpha_k}$, then

$$b_n = \frac{a_n}{\prod a_{\frac{n}{p_i}}} \cdot \frac{\prod a_{\frac{n}{p_i p_j}}}{\prod a_{\frac{n}{p_i p_j p_k}}} \cdots$$

The key remark is that the Mersenne property yields the equality

$$\gcd(a_{\frac{n}{p_i}})_{i\in I} = a_{\frac{n}{\prod_{i\in I} p_i}}$$

for all subsets I of $\{1, 2, \ldots, k\}$. Therefore we can write

$$b_n = \frac{a_n}{\prod a_{\frac{n}{p_i}}} \cdot \frac{\prod \gcd(a_{\frac{n}{p_i}}, a_{\frac{n}{p_j}})}{\prod \gcd(a_{\frac{n}{p_i}}, a_{\frac{n}{p_j}}, a_{\frac{n}{p_k}})} \cdots.$$

Finally, the inclusion-exclusion principle coupled with the local-global principle easily yields the formula

$$\frac{\prod x_i}{\prod_{i<j} \gcd(x_i, x_j)} \cdot \frac{\prod \gcd(x_i, x_j, x_k)}{\prod \gcd(x_i, x_j, x_k, x_l)} \cdots = \operatorname{lcm}(x_1, x_2, \ldots, x_n)$$

for all nonzero integers $x_1, x_2, \ldots, x_n$. Using this observation, we deduce that

$$b_n = \frac{a_n}{\operatorname{lcm}\left[a_{\frac{n}{p_1}}, a_{\frac{n}{p_2}}, \ldots, a_{\frac{n}{p_k}}\right]},$$

[1]See chapter 10, the introduction to problem 10.

which is clearly an integer (by the Mersenne property, if m divides n then a_m divides a_n). The result follows. $\square$

Proof. Let p be a prime. Note that if a sequence $(a_n)_{n\geq 1}$ has the hypothesized property, then so does $a'_m = p^{v_p(a_m)}$. Further, if the desired conclusion holds for all such a'_m, then it clearly holds for a_m. Thus we may assume $a_m = p^{u_m}$ and we see that the sequence $(u_n)_{n\geq 1}$ consists of nonnegative integers and satisfies $\min(u_m, u_n) = u_{\gcd(m,n)}$.

Let $0 \leq r_1 < r_2 < \cdots$ be the values actually taken on by the sequence $(u_n)_{n\geq 1}$. For any two elements m and m' of the set $S_k = \{m : u_m = r_k\}$ we have $u_{\gcd(m,m')} = \min(u_m, u_{m'}) = r_k$ and hence $\gcd(m, m') \in S_k$. Therefore S_k has a unique minimal element m_k and all elements of S_k are multiples of m_k. Note that if $j > k$, then $\min(u_{m_j}, u_{m_k}) = \min(r_j, r_k) = r_k$, so $\gcd(m_j, m_k) = m_k$ or $m_k | m_j$. Thus in the sequence $1 = m_1, m_2, m_3, \ldots$ each term divides the later terms. Now suppose $m_k | m$ and $m_{k+1} \nmid m$. Then $\gcd(m_k, m) = m_k$ so $\min(u_{m_k}, u_m) = r_k$ or $u_m \geq r_k$. However $\gcd(m_{k+1}, m) < m_{k+1}$ so $\min(u_{m_{k+1}}, u_m) < r_{k+1}$ and hence $u_m < r_{k+1}$. Since the r_i are all the values taken on by (u_n), we conclude that $u_m = r_k$.

To recap, the sequences $r_1 < r_2 < \cdots$ and $m_1, m_2, \ldots$ determine u_m uniquely. If k is the largest index for which $m_k | m$, then $u_k = r_k$. Define $b_1 = p^{r_1}$, $b_{m_k} = p^{r_k - r_{k-1}}$ for $k \geq 2$, and $b_m = 1$ if m is not one of the m_k. Then we immediately see that if k is the largest index for which $m_k | m$, then

$$\prod_{d|m} b_d = \prod_{i=1}^{k} b_{m_i} = p^{r_1} \prod_{i=2}^{k} p^{r_i - r_{i-1}} = p^{r_k} = p^{u_m} = a_m. \qquad \square$$

3.3 Legendre's formula

Legendre discovered a very beautiful and useful formula for $v_p(n!)$:

$$v_p(n!) = \sum_{j\geq 1} \left[\frac{n}{p^j}\right] = \frac{n - s_p(n)}{p-1},$$

where $s_p(n)$ is the sum of digits of n when written in base p. The proof of the first equality is easy, since there are $\left[\frac{n}{p^j}\right] - \left[\frac{n}{p^{j+1}}\right]$ positive integers $x \leq n$

such that $v_p(x) = j$. The second equality is a direct computation left to the reader. The purpose of this section is to present some nice applications of this formula.

We present two solutions for the next problem: one uses the local-global principle combined with Legendre's formula, while the second one is more exotic, but very powerful.

4. Show that if n is a positive integer and a and b are integers, then

$$\frac{1}{n!}a(a+b)(a+2b)\cdots(a+(n-1)b)b^{n-1} \in \mathbb{Z}.$$

IMO 1985 Shortlist

Proof. We will prove that for any prime $p \leq n$ we have

$$v_p(n!) \leq v_p(a(a+b)\cdots(a+(n-1)b)b^{n-1}),$$

which is enough to solve the problem. If p divides b, things are rather clear, as $v_p(n!) \leq n-1$ while $v_p(b^{n-1}) \geq n-1$. So assume that p does not divide b. But then there are at least $[n/p]$ multiples of p among $a, a+b, \ldots, a+(n-1)b$ (since among $a, a+b, \ldots, a+(p-1)b$ there is at least one multiple of p, the same for $a+pb, \ldots, a+(2p-1)b$,...), at least $[n/p^2]$ multiples of p^2 and so on. So, the p-adic valuation of $a(a+b)\cdots(a+(n-1)b)$ is at least $[n/p]+[n/p^2]+\cdots$ and this is exactly the p-adic valuation of $n!$, by Legendre's formula. $\square$

Proof. Consider the matrix $A = \{a_{i,j}\}_{1\leq i,j\leq n}$ where $a_{i,j} = \binom{a+(i-1)b}{j}$. We claim that we have

$$\det(A) = \frac{b^{\frac{n(n-1)}{2}}}{n!} \cdot a(a+b)\cdots(a+(n-1)b)$$

To prove this, note that one can evaluate any determinant of the form

$$\begin{vmatrix} P_1(x_1) & P_1(x_2) & \cdots & P_1(x_n) \\ P_2(x_1) & P_2(x_2) & \cdots & P_2(x_n) \\ \vdots & & & \vdots \\ P_n(x_1) & P_n(x_2) & \cdots & P_n(x_n) \end{vmatrix},$$

where P_i are polynomials of degree at most $n-1$, by multiplying the matrix of the coefficients of the P_i's by the Vandermonde matrix $(x_j^{i-1})_{i,j}$ and then taking the determinant. Using this observation, it is easy to prove the above formula.

Now, since $\det(A)$ is an integer (as it is a polynomial expression with integer coefficients in the entries of A), it follows that for any p relatively prime to b, the p-adic valuation of $a(a+b)\cdots(a+(n-1)b)$ is at least that of $n!$. As for primes dividing b the problem is easy (since $v_p(n!) \le n-1$), the result follows. □

5. Prove that for all integers a, b with $b \neq 0$ there exists a positive integer n such that $v_2(n!) \equiv a \pmod{b}$.

KöMaL

Proof. If $s_2(n)$ is the sum of digits of n when written in base 2, we need to find n such that $n - s_2(n) \equiv a \pmod{b}$. Choose $n = 2^{x_1} + 2^{x_2} + \cdots + 2^{x_k}$ with $x_1 < x_2 < \cdots < x_k$. Then $s_2(n) = k$ and we need to ensure that

$$2^{x_1} - 1 + \cdots + 2^{x_k} - 1 \equiv a \pmod{b}.$$

Write $b = 2^r c$ with odd c and choose $r < x_1 < x_2 < \cdots < x_k$ such that $x_i \equiv 1 \pmod{\varphi(c)}$. Then $2^{x_i} - 1 \equiv 1 \pmod{c}$ and so

$$2^{x_1} - 1 + \cdots + 2^{x_k} - 1 \equiv k \pmod{c}.$$

Also, we have

$$2^{x_1} - 1 + \cdots + 2^{x_k} - 1 \equiv -k \pmod{2^r}.$$

Thus, it is enough to choose k such that $k \equiv a \pmod{c}$ and $k \equiv -a \pmod{2^r}$, which is possible by the Chinese Remainder Theorem. □

Remark 3.2. This problem admits a vast generalization, that appeared on the IMO Shortlist 2007. Namely, we can prove that for any positive integer d and positive integers $b_1, b_2, \ldots, b_m$ there exist infinitely many positive integers n such that $d | n - s_{b_i}(n)$ for all i. Here $s_b(n)$ is the sum of digits of n when written in base b.

Proof. Let $c_b(n)$ be the number of digits of n in base b and consider the sequence defined by

$$a_0 = db_1b_2\cdots b_m, \quad a_{j+1} = a_0^{1+\max_{1\le i\le m} c_{b_i}(a_j)}.$$

It is trivial to check that d divides a_i for all i and that

$$s_{b_i}(a_{i_1} + a_{i_2} + \cdots + a_{i_l}) = s_{b_i}(a_{i_1}) + s_{b_i}(a_{i_2}) + \cdots + s_{b_i}(a_{i_l})$$

for any distinct numbers $i_1, i_2, \ldots, i_l$.

Since the m-tuples

$$S_i = (s_{b_1}(a_i) \pmod d, s_{b_2}(a_i) \pmod d, \ldots, s_{b_m}(a_i) \pmod d)$$

take only finitely many values, the pigeonhole principle yields the existence of an m-tuple S that repeats infinitely often, say $S = S_{i_0} = S_{i_1} = \cdots$ for an infinite increasing sequence $i_0, i_1, \ldots$.

Let

$$c_k = a_{i_{dk}} + a_{i_{dk+1}} + \cdots + a_{i_{dk+d-1}}.$$

We will prove that for all k we have $d|c_k - s_{b_j}(c_k)$ for all j, which will end the proof. Since clearly $d|c_k$, it remains to check that d divides $s_{b_j}(c_k)$, which follows from

$$s_{b_j}(c_k) = s_{b_j}(a_{i_{dk}}) + s_{b_j}(a_{i_{dk+1}}) + \cdots + s_{b_j}(a_{i_{dk+d-1}})$$

and

$$s_{b_j}(a_{i_{dk}}) \equiv s_{b_j}(a_{i_{dk+1}}) \equiv \cdots \equiv s_{b_j}(a_{i_{dk+d-1}}) \pmod d. \qquad \square$$

A trickier combination of the local-global principle and Legendre's formula can be found in the following problem, which implies the classical inequality $\mathrm{lcm}(1, 2, \ldots, n) \ge 2^{n-1}$. This nice observation as well as the problem appeared in [31]. Amusingly, the same result appeared much before [31] in the same journal, in the form of a proposed problem! There is also an elementary, but more difficult proof of the fact that $\mathrm{lcm}(1, 2, \ldots, n) \le 3^n$ for all n, see for instance [38]. Using the prime number theorem, one can prove that $\mathrm{lcm}(1, 2, \ldots, n)$ behaves like e^n.

6. Prove the identity

$$(n+1)\operatorname{lcm}\left(\binom{n}{0},\binom{n}{1},\dots,\binom{n}{n}\right)=\operatorname{lcm}(1,2,\dots,n+1)$$

for any positive integer n.

Peter L. Montgomery, AMM E 2686

Proof. We will prove that for any prime p the expressions in the two sides have the same p-adic valuation. Note that

$$(n+1)\binom{n}{i}=(i+1)\binom{n+1}{i+1}.$$

Let k denote the number of times p divides the right-hand side of the equality from the problem statement, so $p^k \le n+1 < p^{k+1}$. Taking $i=p^k-1$, we obtain that

$$v_p\left((n+1)\binom{n}{i}\right)\ge v_p(i+1)=k,$$

which shows that the p-adic valuation of the left-hand side is at least k. The opposite inequality is more delicate. Fix $0\le i\le n$ and note that Legendre's formula gives

$$v_p\left(\binom{n+1}{i+1}\right)=\sum_{r\ge 1}x_r,\quad x_r=\left[\frac{n+1}{p^r}\right]-\left[\frac{i+1}{p^r}\right]-\left[\frac{n-i}{p^r}\right].\tag{3.1}$$

Now, since

$$[a+b]-[a]-[b]=[\{a\}+\{b\}]\in\{0,1\}$$

for any real numbers a,b, we have $x_r\in\{0,1\}$ for all r. Moreover, we have $x_r=0$ if $r>k$, since in this case $p^r>n+1$. The crucial point is that for all $r\le v_p(i+1)$ we also have $x_r=0$. Indeed, writing $i+1=p^r u$ for some integer u, we have

$$x_r=\left[\frac{n+1}{p^r}\right]-u-\left[\frac{n+1}{p^r}-u\right]=0.$$

Putting these remarks together yields the inequality

$$\sum_{r\geq 1} x_r \leq k - v_p(i+1),$$

which, combined with (3.1), yields the estimate

$$v_p\left((i+1)\binom{n+1}{i+1}\right) \leq k,$$

establishing the opposite inequality. □

We continue with a rather tricky diophantine equation.

7. Solve in positive integers $x^{2007} - y^{2007} = x! - y!$.

Romania TST 2007

Proof. We claim that there are only trivial solutions $x = y$. Suppose that $x > y$ is a solution of the problem. We will distinguish two cases.

In the first case, assume that $y \leq 2007$. If $y = 1$, then $x^{2007} = x!$ and trivially $x = 1$ (otherwise $x - 1$ would divide x^{2007}, so $x - 1 = 1$, which is clearly not possible). Thus $y > 1$ and we may choose a prime $p|y$. Then p divides $y^{2007}, x!, y!$, so that $p|x$. But then

$$2007 \leq v_p(x^{2007} - y^{2007}) = v_p(y!(x!/y! - 1)) = v_p(y!) < y,$$

a contradiction.

Now, assume that $y > 2007$. Then $x-y$ is a multiple of any prime p smaller than 2007 such that $(2007, p-1) = 1$. Indeed, if p is such a prime, then p divides $x^{2007} - y^{2007} = x! - y!$. If p divides x, then clearly it also divides y and so it divides $x - y$. If not, since p divides $x^{p-1} - y^{p-1}$ and $\gcd(2007, p-1) = 1$, it follows that p divides $x - y$. We deduce that $x > y + 2007$ in this case. But then

$$x! - y! = y!(x!/y! - 1) > 2007! \cdot x(x-1)\cdots(x-2006) > x^{2007},$$

again a contradiction. □

The next problem is also tricky.

8. Prove that for all positive integers n different from 3 and 5, $n!$ is divisible by the number of its positive divisors.

Paul Erdős, Miklos Schweitzer Competition

Proof. Since the number of positive divisors of $n!$ is precisely $\prod_{p\le n}(1+v_p(n!))$, it suffices to prove the following

Lemma 3.3. *Let P_n be the set of prime numbers less than or equal to n. For all $n \neq 3,5,7$ there exists an injective map $f : P_n \to \{1,2,\ldots,n\}$ such that $1+v_p(n!)$ divides $f(p)$ for all $p \in P_n$.*

Indeed, if this is true, then $\prod_{p\le n}(1+v_p(n!))$ is a divisor of $\prod_{p\in P_n} f(p)$, which divides $n!$, because f is injective. Thus the problem is solved for $n \neq 3,5,7$. For $n=7$, we can check by hand that the result holds.

It remains to construct f. For $p \le \sqrt{n}$ simply choose $f(p) = 1+v_p(n!)$. Note that $f(p) \le n$ follows from Legendre's formula and for $p < q \le \sqrt{n}$ we have $\left[\frac{n}{p^j}\right] \ge \left[\frac{n}{q^j}\right]$ for all j, the inequality being strict for $j=1$ (because $\frac{n}{p}-\frac{n}{q} \ge \frac{n}{pq} > 1$). For the other primes, we will define f by induction. Assume we defined f for all primes $q<p$, where $p \in P_n, p>\sqrt{n}$ is given. There are

$$\left[\frac{n}{1+v_p(n!)}\right] = \left[\frac{n}{1+\left[\frac{n}{p}\right]}\right]$$

multiples of $1+v_p(n!)$ (we used the fact that $p>\sqrt{n}$ to obtain $v_p(n!) = \left[\frac{n}{p}\right]$). If we manage to prove that this number is greater than $\pi(p)-1$, which is the number of occupied values for $f(p)$, we are done, since we can take for $f(p)$ any of the remaining values. However, note that $\pi(p) \le \frac{p+1}{2}$, with equality only for $p=3,5,7$. Let us evaluate $\left[\frac{n}{1+\left[\frac{n}{p}\right]}\right]$. Write $n=kp+r$ for some $0 \le r < p$, so that

$$\left[\frac{n}{1+\left[\frac{n}{p}\right]}\right] = \left[\frac{kp+r}{k+1}\right] = p + \left[\frac{r-p}{1+k}\right] > p-1+\frac{r-p}{1+k} \ge p-1-\frac{p}{1+k}.$$

So, for $k \geq 2$ we have

$$\left[\frac{n}{1+v_p(n!)}\right] > \pi(p) - 1$$

and we are done. For $k = 1$ it remains to see if we can ensure that $\left[\frac{n}{2}\right] > \frac{p-1}{2}$. But this trivially holds for $n > p$. So, the only obstruction is $n = p$ and $\pi(p) = \frac{p+1}{2}$. As we observed, this only happens for $n = 3, 5, 7$, which is excluded by the hypothesis. □

For the next problem, we will need some basic estimates about prime numbers. We refer the reader to the addendum 3.A for proofs and more details.

9. Let n, k be positive integers such that $n > 9^k$. Prove that $\binom{n}{k}$ has at least k distinct prime factors.

Paul Erdős, Miklos Schweitzer Competition

Proof. We will prove that for n large enough, $\binom{n}{k}$ is a multiple of a product of k numbers that are pairwise relatively prime and greater than 1. This will clearly imply that it has at least k prime factors. Define $L_k = \text{lcm}(1, 2, \ldots, k)$. The key point of the proof is the following

Lemma 3.4. *For $n \geq k + L_k$, $\binom{n}{k}$ is a multiple of $\prod_{i=0}^{k-1} \frac{n-i}{\gcd(n-i, L_k)}$.*

If this happens, we are done, because the numbers $\frac{n-i}{\gcd(n-i,L_k)}, \frac{n-j}{\gcd(n-j,L_k)}$ are pairwise relatively prime and greater than 1.

To prove the lemma, we will compare p-adic valuations for all primes p. Note that the lemma is equivalent to

$$k!| \prod_{i=n-k+1}^{n} \gcd(i, L_k).$$

Thus, we need to prove that for all primes p we have

$$v_p(k!) \leq \sum_{i=n-k+1}^{n} v_p(\gcd(i, L_k)). \tag{3.2}$$

Let $r = v_p(L_k) = [\log_p k]$. For all $i \le r$ we can find at least $\left[\frac{k}{p^i}\right]$ multiples of p^i among the k consecutive numbers $n-k+1, n-k+2, \dots, n$. Also, if u is a multiple of p^i with $i \le r$, then so is $\gcd(L_k, u)$. We conclude that (3.2) follows from Legendre's formula, as the only nonzero terms on the left-hand side are of the form $\left[\frac{k}{p^i}\right]$ with $i \le r$.

Finally, it remains to prove that $k + L_k < 9^k$. Note that

$$L_k = \prod_{p \le k} p^{[\log_p k]} = \prod_{p > \sqrt{k}} p \cdot \prod_{p \le \sqrt{k}} p^{[\log_p k]} \le 4^k \cdot \left(\prod_{p \le \sqrt{k}} k \right),$$

where here we used theorem 3.A.3. Thus

$$L_k < k^{\sqrt{k}} \cdot 4^k < 9^k,$$

the last inequality being easy to prove. □

3.4 Problems with combinatorial and valuation-theoretic aspects

The problems in this section are fairly elementary, but none of them is easy.

10. Let $n \ge 2$ and let $a_1, a_2, \dots, a_n$ be positive integers, not all of them equal. Prove that there are infinitely many prime numbers p for which there exists a positive integer k such that

$$p \mid a_1^k + a_2^k + \cdots + a_n^k.$$

Iranian Olympiad 2004

Proof. By dividing all a_i's by their greatest common divisor, we may assume that they are relatively prime. Suppose that there are only finitely many primes $p_1, p_2, \dots, p_N$ such that all prime factors of $a_1^k + a_2^k + \cdots + a_n^k$ (where k varies over the positive integers) are among $p_1, p_2, \dots, p_N$.

Assume that among $a_1, a_2, \ldots, a_n$ there are b_i numbers not divisible by p_i. Since $a_1, a_2, \ldots, a_n$ are relatively prime, we have $b_i \geq 1$. Consider

$$k = 2\prod_{i=1}^{N} \varphi\left(p_i^{v_{p_i}(b_i)+1}\right).$$

Since $k > v_{p_i}(b_i) + 1$, we have $p_i^{v_{p_i}(b_i)+1} | a_j^k$ whenever $p_i | a_j$. Using this and Euler's theorem, we infer that

$$a_1^k + a_2^k + \cdots + a_n^k \equiv b_i \pmod{p_i^{v_{p_i}(b_i)+1}}.$$

Hence

$$v_{p_i}(a_1^k + a_2^k + \cdots + a_n^k) = v_{p_i}(b_i)$$

for all i. Since all prime factors of $a_1^k + a_2^k + \cdots + a_n^k$ are among $p_1, p_2, \ldots, p_N$, we deduce that

$$a_1^k + a_2^k + \cdots + a_n^k = p_1^{v_{p_1}(b_1)} p_2^{v_{p_2}(b_2)} \cdots p_N^{v_{p_N}(b_N)}.$$

Now, at least one of the a_i's is greater than 1, thus

$$a_1^k + a_2^k + \cdots + a_n^k \geq 2^k > k > \prod_{i=1}^{N} p_i^{v_{p_i}(b_i)}.$$

The two relations are clearly contradictory and the problem is solved. □

A classical problem of Erdős is the following: if $a_1, a_2, \ldots, a_{n+1}$ are different positive integers not exceeding $2n$, then one can find $i \neq j$ such that a_i divides a_j. The idea is very simple and beautiful: the largest odd divisors of the a_i's form a sequence of $n+1$ odd numbers between 1 and $2n-1$, so there must be two equal terms in this sequence. But then the corresponding a_i and a_j have a quotient which is a power of 2. The next problem is a variation on this classical gem.

11. Let $f(n)$ be the maximum size of a subset of $\{1,2,\ldots,n\}$ which does not contain two distinct elements i,j such that $i|2j$. Prove that there exists a constant $C>0$ such that for all n we have

$$\left|f(n)-\frac{4n}{9}\right|\leq C\ln n.$$

Paul Erdős, AMM E 3403

Proof. We will actually exhibit the optimal set with the property that it does not contain two distinct elements i,j with $i|2j$. Defining

$$A_n=\{a\in\mathbb{Z}|\quad n/3<a\leq n,\quad v_2(a)\equiv 0\pmod 2\},$$

it is easy to see that A_n satisfies the conditions of the problem. Indeed, if $i,j\in A_n$ and $i|2j$, we must have $i|j$, since we cannot have $v_2(i)=v_2(j)+1$ as both $v_2(i)$ and $v_2(j)$ are even. So j/i is either 1 or 2. It cannot be 2, because in this case one of $v_2(i)$, $v_2(j)$ would be odd, so it has to be 1 and $i=j$.

Next, we prove that this set is optimal, in the sense that it has the maximal number of elements among all sets satisfying the conditions of the problem. Take any such set A and fix k such that 3 does not divide k and $v_2(k)$ is even. Look at all the numbers $k,3k,9k,\ldots$ and $2k,6k,18k,\ldots$. There is at most one element of A among the union of these numbers and by definition there is exactly one element of A_n among them. On the other hand, if one varies k, the numbers $k,3k,\ldots,2k,6k,\ldots$ form a partition of the positive integers. This clearly implies that A has at most as many elements as A_n.

Finally, we have to estimate the size of A_n. The elements of A_n are exactly the numbers of the form $4^j b$ with b odd, $j\geq 0$ such that $\frac{n}{3\cdot 4^j}<b\leq\frac{n}{4^j}$. There are approximately $\frac{n}{3\cdot 4^j}$ odd numbers b such that $\frac{n}{3\cdot 4^j}<b\leq\frac{n}{4^j}$, with an error at most 2 if $4^j<n$ and error at most $\frac{n}{3\cdot 4^j}$ for $4^j>n$. The total error is thus logarithmic in n and since

$$\sum_{j\geq 0}\frac{n}{3\cdot 4^j}=\frac{4n}{9},$$

we have

$$|A_n|=\frac{4n}{9}+O(\log n),$$

finishing the proof. □

12. Find all positive integers n with the following property: there exist natural numbers $b_1, b_2, \ldots, b_n$, not all equal and such that the number $(b_1+k)(b_2+k)\cdots(b_n+k)$ is a power of an integer for each natural number k. Here, a power means a number of the form x^y with $x, y > 1$.

Russia 2008

Proof. There are some obvious solutions: for instance, if n is composite, say $n = ab$ with $a, b > 1$, then we can choose $b_1 = b_2 = \cdots = b_a = 2$ and all the other b_i's equal to 1. Then for any k we have

$$(b_1+k)(b_2+k)\cdots(b_n+k) = (k+2)^a(k+1)^{a(b-1)},$$

which is certainly a power.

So, the real question is to decide whether numbers $b_1, b_2, \ldots, b_n$ can exist when n is a prime. It turns out that the answer is negative. Assume that such numbers existed and let $c_1, c_2, \ldots, c_N$ be the set of distinct numbers among $b_1, b_2, \ldots, b_n$, with multiplicities $m_1, m_2, \ldots, m_N$. By assumption, we have $N > 1$ and clearly $n = m_1 + m_2 + \cdots + m_N$. Moreover, we know that for any k, the number $(c_1+k)^{m_1}(c_2+k)^{m_2}\cdots(c_N+k)^{m_N}$ is a perfect power. The key point is to choose numbers k for which we can find distinct primes $p_1, p_2, \ldots, p_N$ such that $v_{p_i}(c_j + k) = 1$ if $i = j$ and 0 otherwise. In this case, if

$$(c_1+k)^{m_1}(c_2+k)^{m_2}\cdots(c_N+k)^{m_N} = x^y$$

for some $x, y > 1$, we have $yv_{p_i}(x) = m_i$, so that y divides all m_i. But then y divides their sum, which is n and since n is a prime, it follows that $n = y$. Thus $n = y$ will divide all m_i and this obviously contradicts the fact that $N > 1$ and $m_1 + m_2 + \cdots + m_N = n$.

Thus, we are done if we can find distinct primes $p_1, p_2, \ldots, p_N$ and k such that $v_{p_i}(c_j + k) = 1$ if $i = j$ and 0 otherwise. This is very simple: first, we choose some distinct prime numbers $p_1, p_2, \ldots, p_N$ sufficiently large, say not dividing any of the numbers $c_i - c_j$ with $i \neq j$, and then choose k such that $k + c_i \equiv p_i \pmod{p_i^2}$ for all i. Such k exists by the Chinese Remainder Theorem. Of course, $v_{p_i}(k + c_i) = 1$ and for $j \neq i$ we cannot have $p_i | c_j + k$, since otherwise p_i would divide $c_i - c_j$, contradicting the choice of p_i.Thus, such k satisfies all desired conditions and the answer to the problem is that n must be composite. $\square$

A nice mixture of valuation-theoretic arguments and pigeonhole principle can be found in the following problem.

13. Let a be a positive integer. Prove that the set of prime divisors of $2^{2^n}+a$ for $n = 1, 2, \ldots$ is infinite.

Iranian TST 2009

Proof. Assuming the contrary, let $p_1, p_2, \ldots, p_N$ be such that all prime factors of $2^{2^n}+a$ are among $p_1, p_2, \ldots, p_N$ for all n. Pick a large number r such that $2^r > a^{2^{N+1}}+a$ and n_0 such that $2^{2^{n_0}}+a > (p_1p_2\cdots p_N)^r$. Then for all $n \geq n_0$ we have

$$(p_1p_2\cdots p_N)^r < 2^{2^n}+a = \prod_{i=1}^{N} p_i^{v_{p_i}(2^{2^n}+a)},$$

so that we can find $1 \leq i \leq N$ with $v_{p_i}(2^{2^n}+a) > r$. This p_i depends on the choice of n, but among the indices i associated to the numbers $n = n_0+1, n_0+2, \ldots, n_0+N+1$ there will two identical ones. Thus we can write $p_i^r|2^{2^n}+a$ and $p_i^r|2^{2^{n+m}}+a$ for some $n \geq n_0$, some $1 \leq m \leq N+1$ and some $1 \leq i \leq N$. But then $2^{2^n} \equiv -a \pmod{p_i^r}$, so that $2^{2^{m+n}} \equiv a^{2^m} \pmod{p_i^r}$ and$p_i^r|a^{2^m}+a$. In particular,

$$a^{2^{N+1}}+a \geq a^{2^m}+a \geq p_i^r \geq 2^r,$$

contradicting the choice of r. The result follows. $\square$

We continue with two more unusual problems.

14. Let $p_1, p_2, \ldots, p_k$ be distinct prime numbers and let S be the set of positive integers all of whose prime factors are among $p_1, p_2, \ldots, p_k$. If A is a finite set of integers, let $G(A)$ be the graph whose set of vertices is A, two vertices $a, b \in A$ being connected if $a - b \in S$. Is it true that for all $m \geq 3$ we can find A with m elements such that

a) the graph $G(A)$ is complete?

b) the graph $G(A)$ is connected with all vertices of degree at most 2?

Miklos Schweitzer, Competition 2009

Proof. The answer to the first question is negative. Let p be the smallest prime different from $p_1, p_2, \dots, p_k$ and suppose that $G(A)$ is complete for some finite set A with $|A| > p$. Then there exist $a, b \in A$ different such that p divides $a - b$, so that a and b arenot connected.

On the other hand, the answer to the second question is positive! We will construct a set A with m elements $a_1, a_2, \dots, a_m$ such that $G(A)$ is a simple path. It is enough to ensure that for all $1 \leq n \leq m$, $a_{n+1} - a_n \in S$ and $a_{n+2} - a_n, a_{n+3} - a_n, \dots$ are not in S. But we can choose

$$a_n = p_1 p_2 \cdots p_k + (p_1 p_2 \cdots p_k)^2 + \cdots + (p_1 p_2 \cdots p_k)^n.$$

Then clearly $a_{n+1} - a_n \in S$. On the other hand, for any $i \geq 2$ we have $v_{p_j}(a_{n+i} - a_n) = n + 1$ for all $1 \leq j \leq k$. Since $a_{n+i} - a_n > (p_1 p_2 \cdots p_k)^{n+1}$, it follows that $a_{n+i} - a_n$ is not in S and the result follows. □

15. Let m and n be positive integers such that $m + 1, m + 2, \cdots, m + n$ are composite numbers and $m > n^{n-1}$. Prove that we can find pairwise distinct prime numbers $p_1, p_2, \dots, p_n$ such that p_i divides $m + i$ for all $1 \leq i \leq n$.

Tuymaada Olympiad 2004

Proof. First, we will deal with those i such that $m + i$ has at most $n - 1$ prime factors. For such i choose a prime p_i for which $p_i^{v_{p_i}(m+i)}$ is maximal (note that p_i is unique with this property). Since $m + i > n^{n-1}$ and since $m + i$ has at most $n - 1$ prime factors, we have $p_i^{v_{p_i}(m+i)} > n$. We claim that if $i \neq j$ and $m + i, m + j$ have at most $n - 1$ prime factors, then $p_i \neq p_j$. Indeed, assume that $p_i = p_j = p$. Then $\min(p^{v_p(m+i)}, p^{v_p(m+j)})$ divides $m + i$ and $m + j$ and moreover is greater than n. But any common divisor of $m + i$ and $m + j$ divides $j - i$ and so it is smaller than n. This shows that $i \mapsto p_i$ is injective.

It is now easy to conclude: make a list with those numbers $m+i$ having at most $n-1$ prime factors. For each of them, the previous paragraph yields a prime factor p_i and the associated p_i's are distinct. If all $m+i$ are in the list, we are done, otherwise successively pick remaining numbers and choose one of their prime factors which is not among the p_i's or among primes previously selected. This is possible at every step, since any $m+i$ not in the list has at least n prime factors. $\square$

3.5 Lifting exponent lemma

This section is devoted to some applications of the following useful result:

Theorem 3.5. *(Lifting exponent lemma) Let p be an* **odd** *prime and let a and b be integers relatively prime to p, such that $p|a-b$. Then for all positive integers n*

$$v_p(a^n - b^n) = v_p(n) + v_p(a-b).$$

Proof. Consider first the case $v_p(n) = 0$. We need to prove that p does not divide $\frac{a^n-b^n}{a-b}$, which is clear, as by hypothesis

$$\frac{a^n - b^n}{a-b} = a^{n-1} + a^{n-2}b + \cdots + b^{n-1} \equiv na^{n-1} \pmod{p}$$

and p does not divide na^{n-1}. Next, we prove the result for $n = p$; we need to check that p divides exactly once into $a^{p-1} + \cdots + b^{p-1}$. Write $b = a + pk$ for some integer k. Then by the binomial formula we have $b^i \equiv a^i + ia^{i-1}pk \pmod{p^2}$, so that

$$\frac{a^p - b^p}{a-b} = \sum_{i=0}^{p-1} a^{p-1-i}b^i \equiv \sum_{i=0}^{p-1}(a^{p-1} + ipka^{p-2})$$

$$\equiv pa^{p-1} + p^2k\frac{p-1}{2}a^{p-2} \equiv pa^{p-1} \pmod{p^2}.$$

Let us apply the case $n = p$ to $a^{n/p}$ and $b^{n/p}$ (note that they still satisfy the hypotheses of the problem) to get $v_p(a^n - b^n) = 1 + v_p(a^{n/p} - b^{n/p})$. The result follows now by an immediate induction on $v_p(n)$. $\square$

The reader might wonder what happens for $p = 2$. There is of course a version of the theorem for $p = 2$, but the formula is more complicated. The proof is however much easier.

Theorem 3.6. *Let x, y be odd integers and let n be an even positive integer. Then*

$$v_2(x^n - y^n) = v_2\left(\frac{x^2 - y^2}{2}\right) + v_2(n).$$

Proof. Write $n = 2^k a$ for some odd number a. Then

$$x^n - y^n = (x^a - y^a)(x^a + y^a)(x^{2a} + y^{2a}) \cdots (x^{2^{k-1}a} + y^{2^{k-1}a}).$$

Now observe that if u, v are odd numbers, then $u^2 + v^2 \equiv 2 \pmod 4$. Thus

$$v_2(x^n - y^n) = v_2(x^{2a} - y^{2a}) + k - 1.$$

Finally, since a, x, y are odd, it is easy to see that $\frac{x^{2a} - y^{2a}}{x^2 - y^2}$ is odd. The result follows. □

An easy consequence of these results is the following estimate. It is much weaker than the previous theorems and can be proved directly by induction, too.

Corollary 3.7. *If a, b are integers and p is any prime dividing $a - b$, then for all n we have*

$$v_p(a^n - b^n) \geq v_p(a - b) + v_p(n).$$

The next problem is an immediate consequence of this corollary.

16. Let a, b, c be positive integers such that $c | a^c - b^c$. Prove that $c | \frac{a^c - b^c}{a - b}$.

I. Niven, AMM E 564

Proof. First of all, note that $\frac{a^c - b^c}{a - b}$ is an integer, so the statement makes sense. Let p be a prime dividing c. We will prove that $v_p(c) \leq v_p\left(\frac{a^c - b^c}{a - b}\right)$. If p does not divide $a - b$, this follows from the hypothesis that c divides $a^c - b^c$, so we may assume that p divides $a - b$. But then everything follows from corollary 3.7. □

The next problem appeared as a Chinese TST 2009 problem, but the result is much older (see [61]).

17. Let n be a positive integer and let $a > b > 1$ be integers such that b is odd and $b^n | a^n - 1$. Prove that $a^b > \frac{3^n}{n}$.

M.B. Nathanson

Proof. Take any prime factor p of b. Since b is odd, we have $p > 2$. Note that

$$n \leq v_p(b^n) \leq v_p(a^n - 1) \leq v_p(a^{(p-1)n} - 1) \leq v_p(a^{p-1} - 1) + v_p(n).$$

We deduce that

$$a^b > a^{p-1} - 1 \geq p^{n - v_p(n)} \geq \frac{p^n}{n} \geq \frac{3^n}{n},$$

and the result follows. □

Remark 3.8. Using a generalization of the deep Thue-Siegel theorem, Mahler proved in [48] the following result: if a, b, u, v are nonzero integers with $u > v > 1$, then there are only finitely many positive integers n such that

$$au^n \equiv b \pmod{v^n}.$$

It is quite rare to see two very similar problems at the IMO; nevertheless the following problem appeared in weaker forms at the IMO in 1990 and 1999.

18. Find all primes p and all positive integers n such that n^{p-1} divides $(p-1)^n + 1$.

After IMO 1990 and 1999

Proof. Let p and n be as in the statement. Note that if $p = 2$, then $n = 1$ or $n = 2$. From now on, we assume that $p > 2$. If n is even, then 4 divides n^{p-1}, but it does not divide $(p-1)^n + 1$, a contradiction. So, n is odd. Let q be the smallest prime factor of n. Since

$$(p-1)^n \equiv -1 \text{ and } (p-1)^{q-1} \equiv 1 \pmod{q},$$

the facts that n is odd and $\gcd(n, q-1) = 1$ imply that $p - 1 \equiv -1 \pmod{q}$, or $p = q$.

By the lifting exponent lemma (using that n is odd) we have

$$(p-1)v_p(n) = v_p(n^{p-1}) \leq v_p((p-1)^n + 1) = 1 + v_p(n).$$

Thus $(p-2)v_p(n) \leq 1$. In particular, $p = 3$ and $v_p(n) = 1$. Write $n = 3a$ with $\gcd(a, 3) = 1$ and observe that a^2 divides $8^a + 1$. We claim that $a = 1$. Otherwise, let r be the smallest prime factor of a, so that

$$8^a \equiv -1 \text{ and } 8^{r-1} \equiv 1 \pmod{r},$$

whence $8 \equiv -1 \pmod{r}$ as $\gcd(a, r-1) = 1$. But then $r = 3$, which is impossible. It follows that $a = 1$ and $n = 3$. □

Remark 3.9. Another interesting problem that can be solved using the same ideas is the following one: find all positive integers a and b such that a^b divides $b^a - 1$.

19. Find all positive integers a, b, c such that

$$(2^a - 1)(3^b - 1) = c!.$$

Gabriel Dospinescu, Mathematical Reflections

Proof. We will only give the key ideas, since the computations are a bit tedious. Take a solution (a, b, c) with $c > 3$ and note that a is even, since 3 divides $2^a - 1$. Also, as 4 divides $c!$, 4 must divide $3^b - 1$ and so b is even. Then, using the lifting exponent lemma, we deduce that

$$\frac{c - s_3(c)}{2} = v_3(c!) = v_3(2^a - 1) = v_3(4^{a/2} - 1) = 1 + v_3(a).$$

Similarly, by writing $b = 2^k r$ with r odd, we have

$$\begin{aligned} v_2(3^b - 1) &= v_2\left(\frac{1}{2}(9^r - 1)\right) + v_2(b) \\ &= v_2(4(9^{r-1} + 9^{r-2} + \cdots + 1)) + v_2(b) \\ &= 2 + v_2(b). \end{aligned}$$

Thus

$$c - s_2(c) = v_2(c!) = v_2(3^b - 1) = 2 + v_2(b).$$

We deduce that

$$a \geq 3^{v_3(a)} \geq 3^{\frac{c}{2}-1-\log_3(c)-1} = \frac{3^{\frac{c}{2}-2}}{c}$$

and

$$b \geq 2^{v_2(b)} \geq 2^{c-1-\log_2 c-2} = \frac{2^{c-3}}{c}.$$

From here it follows that

$$c! \geq (2^{\frac{3^{c/2-2}}{c}} - 1)(3^{\frac{2^{c-3}}{c}} - 1).$$

It is not difficult (even if it is quite tedious) to check that this does not hold for any $c \geq 12$. We deduce that in any solution we have $c \leq 11$. We can easily exclude the cases $c \in \{8, 9, 10, 11\}$, since in this case we obtain $v_2(b) \geq 5$, forcing $b \geq 32$, which is too large. For $c \in \{4, 5, 6, 7\}$, we have $c! \geq 3^b - 1$ for only one value of b with $v_2(b) = c - s_2(c) - 2$, so we simply check to see if there is a compatible value of a. Combining these with the obvious solutions for $c \leq 3$, we end up with $(a, b, c) \in \{(1, 1, 2), (2, 1, 3), (2, 2, 4), (4, 2, 5), (6, 4, 7)\}$. □

20. Let a be a fixed positive integer. Prove that the equation $n! = a^b - a^c$ has only finitely many solutions (n, b, c) in positive integers.

Chinese TST 2004

Proof. Let p be an odd prime not dividing a. Then by the lifting exponent lemma we have

$$v_p(a^n - 1) \leq v_p(a^{p-1} - 1) + v_p(n).$$

Taking $n = b - c$ and noting that $v_p(n!) > \frac{n}{p} - 1$, we conclude that

$$v_p(b - c) \geq v_p(n!) - v_p(a^{p-1} - 1) \geq \frac{n}{p} - K$$

for some constant K, independent of n. Letting $\epsilon = p^{-K} > 0$, we conclude that $b - c \geq \epsilon p^{n/p}$ for all n. Thus

$$n^n > n! = a^b - a^c > a^{b-c} \geq a^{\epsilon p^{n/p}}.$$

Taking logarithms, we deduce that n is bounded in terms of a. Since $c, b - c < n!$, the conclusion follows. $\square$

We continue with a very beautiful, but hard problem.

21. Let x, y be relatively prime integers greater than 1. Prove that for infinitely many primes p, the exponent of p in $x^{p-1} - y^{p-1}$ is odd.

Barry Powell, AMM E 2948

Proof. Choose any integer $k > 2$. It is a well-known result of Fermat that the equations $a^4 + b^4 = c^2$ and $a^4 + b^4 = 2c^2$ have only trivial solutions. Therefore $x^{2^{k-1}} + y^{2^{k-1}}$ is neither a perfect square nor twice a perfect square, whence there is some odd prime p such that $v_p(x^{2^{k-1}} + y^{2^{k-1}})$ is odd. Since $\gcd(x, y) = 1$, p cannot divide xy. Since p divides $x^{2^k} - y^{2^k}$ and does not divide $x^{2^{k-1}} - y^{2^{k-1}}$, the order of x/y mod p is 2^k and 2^k divides $p - 1$. Hence the lifting exponent lemma gives

$$v_p(x^{p-1} - y^{p-1}) = v_p(x^{2^k} - y^{2^k}) = v_p(x^{2^{k-1}} + y^{2^{k-1}}) \equiv 1 \pmod 2.$$

The result follows by taking successively $k = 3, 4, \ldots$ and observing that the prime p associated to k in the previous discussion satisfies $p \geq 1 + 2^k$. $\square$

Proof. We will argue by contradiction: assume that there exists $N > |x - y|$ such that for all primes $p > N$ we have

$$v_p(x^{p-1} - y^{p-1}) \equiv 0 \pmod 2.$$

We claim that for all $p > N$ the number $\frac{x^p - y^p}{x - y}$ is a perfect square. Take a prime factor q of $\frac{x^p - y^p}{x - y}$. Since q does not divide x (otherwise q divides y, contradicting the fact that $\gcd(x, y) = 1$) and since q divides $x^p - y^p$, the order of x/y mod q is 1 or p. If q divides $x - y$, then q divides px^{p-1} and so $q = p$, contradicting the fact that $p > N > |x - y|$. So the order of x/y mod q is p and so p divides $q - 1$. Then the lifting exponent lemma implies that $v_q(x^{q-1} - y^{q-1}) = v_q(x^p - y^p)$ is an even number, finishing the proof of the claim.

Next, take a prime $r \equiv 3 \pmod 4$ dividing $4xy - 1$ and (using Dirichlet's theorem) take a prime $p \equiv -1 \pmod{r-1}$ larger than N. Thus there exists z such that

$$z^2 = \frac{x^p - y^p}{x - y} \equiv \frac{\frac{1}{x} - \frac{1}{y}}{x - y} \equiv -\frac{1}{xy} \equiv -4 \pmod r,$$

i.e. -1 is a quadratic residue mod r, a contradiction. □

3.6 p-adic techniques

For more details on the results and techniques used in this section, the reader is (strongly) advised to read the addendum of this chapter, which is a modest introduction to p-adic numbers. Before passing to the next problem, let us recall the notion of congruence for rational numbers. Let p be a prime. The localization of $\mathbb{Z}$ with respect to the ideal $p\mathbb{Z}$ is the set

$$\mathbb{Z}_{(p)} = \{\frac{a}{b} \in \mathbb{Q} | \gcd(a,b) = 1, \quad \gcd(b,p) = 1\}.$$

It is easy to check that $\mathbb{Z}_{(p)}$ is a subring of $\mathbb{Q}$. If $a, b \in \mathbb{Z}_{(p)}$, we write $a \equiv b \pmod{p^j}$ if $a - b \in p^j \cdot \mathbb{Z}_{(p)}$. This is equivalent to the fact that the numerator of the fraction $a - b$, when written in lowest terms, is a multiple of p^j. It is easy to see that this is an equivalence relation, satisfying the usual properties of congruences over $\mathbb{Z}$. Moreover, we have a natural morphism f of rings $\mathbb{Z}_{(p)} \to \mathbb{Z}/p\mathbb{Z}$, sending $\frac{a}{b}$ to $\overline{a} \cdot \overline{b}^{-1}$, ($\overline{b}^{-1}$ being the inverse of b mod p). Then for $x \in \mathbb{Z}_{(p)}$ we have $x \equiv 0 \pmod p$ if and only if $f(x) = 0$. This observation is very useful when trying to prove congruences for rational numbers. For instance, let p be a prime greater than 3 and consider the element

$$x = \sum_{k=1}^{p-1} \frac{1}{k^2}$$

of $\mathbb{Z}_{(p)}$. We have

$$f(x) = \sum_{k=1}^{p-1} \overline{k^2}^{-1} = \overline{\sum_{k=1}^{p-1} k^2} = 0,$$

since the map $x \to x^{-1}$ is a bijection of $(\mathbb{Z}/p\mathbb{Z})^*$ and

$$\sum_{k=1}^{p-1} k^2 = \frac{(p-1)p(2p-1)}{6}$$

is a multiple of p. Thus the numerator of the fraction $\sum_{k=1}^{p-1} \frac{1}{k^2}$, when written in lowest terms, is a multiple of p. The next problem is a variation on this idea combined with some ideas from p-adic analysis.

22. Let $p > 5$ be a prime. Prove that p^4 divides the numerator of the fraction

$$2 \cdot \sum_{k=1}^{p-1} \frac{1}{k} + p \cdot \sum_{k=1}^{p-1} \frac{1}{k^2}$$

when written in lowest terms.

Gabriel Dospinescu

Proof. The first step is to note that

$$2\sum_{k=1}^{p-1} \frac{1}{k} = \sum_{k=1}^{p-1} \left(\frac{1}{k} + \frac{1}{p-k} \right) = \sum_{k=1}^{p-1} \frac{p}{k(p-k)}.$$

Thus, it is enough to prove that

$$\sum_{k=1}^{p-1} \left(\frac{1}{k(p-k)} + \frac{1}{k^2} \right) \equiv 0 \pmod{p^3}.$$

Now, the crucial remark is that in the field of p-adic numbers we have the convergent expansion

$$\frac{1}{k(p-k)} = -\frac{1}{k^2} \frac{1}{1-\frac{p}{k}} = -\frac{1}{k^2} \left(1 + \frac{p}{k} + \frac{p^2}{k^2} + \cdots \right).$$

By truncating at level p^3 we obtain the congruence

$$\frac{1}{k(p-k)} \equiv -\frac{1}{k^2} - \frac{p}{k^3} - \frac{p^2}{k^4} \pmod{p^3}. \tag{3.3}$$

Of course, one does not need p-adic numbers to find or check the previous congruence, since obtaining the appropriate polynomials in p and k and validating that they work are each formal algebraic matters. The p-adic approach simply gives a straightforward route to both, here.

Using (3.3), it remains to prove that

$$\sum_{k=1}^{p-1} \frac{1}{k^3} + p \sum_{k=1}^{p-1} \frac{1}{k^4} \equiv 0 \pmod{p^2}.$$

We will actually prove that

$$\sum_{k=1}^{p-1} \frac{1}{k^3} \equiv 0 \pmod{p^2}, \quad \sum_{k=1}^{p-1} \frac{1}{k^4} \equiv 0 \pmod{p}.$$

The same argument as in the preliminary discussion yields

$$\sum_{k=1}^{p-1} \frac{1}{k^4} \equiv \sum_{k=1}^{p-1} k^4 \equiv 0 \pmod{p},$$

the last congruence being established either by using the existence of primitive roots mod p (which makes the corresponding sum the sum of a geometric progression with ratio g^4, where g is a primitive root mod p) or simply by using explicit formulae for this kind of sum. In order to prove the other congruence, note that

$$\frac{1}{k^3} + \frac{1}{(p-k)^3} = p \cdot \frac{k^2 - k(p-k) + (p-k)^2}{k^3(p-k)^3} \equiv p \frac{-3}{k^4} \pmod{p^2},$$

so

$$2 \sum_{k=1}^{p-1} \frac{1}{k^3} \equiv -3p \sum_{k=1}^{p-1} \frac{1}{k^4} \equiv 0 \pmod{p^2}.$$

The result follows. □

The following problem is a generalization of a classical congruence due to Fleck. For the reader not familiar with p-adic numbers, it is worth reading the addendum 3.B before attacking the proof.

23. Let p be a prime and n, s positive integers with $n > s+1$. Prove that p^d divides

$$\sum_{\substack{0\le k\le n\\ p|k}} (-1)^k k^s \binom{n}{k},$$

where $d = \left[\frac{n-s-1}{p-1}\right]$.

Gabriel Dospinescu, Mathematical Reflections

Proof. Fix a primitive root of unity $z \in \mathbb{C}$ of order p. The field $K = \mathbb{Q}(z)$ is an extension of degree $p-1$ of $\mathbb{Q}$, as the minimal polynomial of z is

$$X^{p-1} + X^{p-2} + \cdots + X + 1.$$

Choosing a prime (ideal) β above p in the ring of algebraic integers O_K of K and completing β-adically, we obtain a valuation v_p on K (seen inside its β-adic completion), which extends the usual p-adic valuation on $\mathbb{Q}$ and such that $v_p(1-z) = \frac{1}{p-1}$. To prove the last equality, note that $\frac{1-z^i}{1-z}$ is a unit in O_K for all i relatively prime to p, so $v_p(1-z) = v_p(1-z^i)$. Since we also have

$$\prod_{i=1}^{p-1}(1-z^i) = p,$$

the result follows.

Note that

$$\frac{1}{p}\sum_{j=0}^{p-1} z^{kj} = 0$$

if k is not a multiple of p and 1 otherwise. We deduce that

$$\sum_{\substack{0\le k\le n\\ p|k}} (-1)^k \cdot k^s \binom{n}{k} = \frac{1}{p}\sum_{j=0}^{p-1}\sum_{k=0}^{n}(-z^j)^k k^s \binom{n}{k}.$$

Now, let $n-s-1=d(p-1)+r$ for some $0\leq r<p-1$. We will prove that

$$v_p\left(\sum_{k=0}^{n}(-z^j)^k k^s\binom{n}{k}\right)>d$$

for all $0\leq j\leq p-1$. This will imply that

$$v_p\left(\sum_{\substack{0\leq k\leq n\\ p\mid k}}(-1)^k\cdot k^s\binom{n}{k}\right)>d-1$$

and since this p-adic valuation is an integer, the result will follow.

Now, to prove the claim, we will use the following:

Lemma 3.10. *The polynomial* $\sum_{k=0}^{n}k^s\binom{n}{k}X^k$ *is a multiple of* $(1+X)^{n-s}$ *for all* $s<n$.

Proof. This is very easy: for $s=0$ it is clear, and if

$$\sum_{k=0}^{n}k^s\binom{n}{k}X^k=(1+X)^{n-s}f(X)$$

then we get the inductive step by differentiating and multiplying both sides by X. □

Coming back to the proof, write

$$\sum_{k=0}^{n}k^s\binom{n}{k}X^k=(1+X)^{n-s}f(X)$$

for some $f\in\mathbb{Z}[X]$ (note that we necessarily have $f\in\mathbb{Z}[X]$, as $(1+X)^{n-s}$ and $\sum_{k=0}^{n}k^s\binom{n}{k}X^k$ have integer coefficients and $(1+X)^{n-s}$ is monic). Then for[2] $1\leq j<p$ we have

$$\sum_{k=0}^{n}(-z^j)k^s\binom{n}{k}=(1-z^j)^{n-s}f(-z^j)$$

[2]Note that by taking $X=-1$ in the previous relation we obtain $\sum_{k=0}^{n}(-1)^k k^s\binom{n}{k}=0$, so we only have to deal with $j\geq 1$.

and so

$$v_p\left(\sum_{k=0}^{n}(-z^j)k^s\binom{n}{k}\right) \geq \frac{n-s}{p-1} = d + \frac{r+1}{p-1} > d.$$

Thus, the claim is proved and the result follows. □

We end this chapter with a wonderful result due to Skolem, Mahler and Lech. It truly shows what a versatile tool p-adic analysis can be. Even though the result still holds for $p = 2$, we will assume that p is odd in order to avoid some technical problems.

24. a) Let p be an odd prime and let $a_0, a_1, \ldots$ be integers such that

$$\sum_{k=0}^{n} p^k \binom{n}{k} a_k = 0$$

for infinitely many positive integers n. Prove that for all n we have that $a_n = 0$.

b) A sequence $(a_n)_n$ of integers satisfies

$$a_{n+d} = x_1 a_{n+d-1} + x_2 a_{n+d-2} + \cdots + x_d a_n$$

for all $n \geq 0$, where $d \geq 1$ and $x_1, x_2, \ldots, x_d$ are integers. Prove that there exists a finite set S and integers $c_1, c_2, \ldots, c_N, d_1, d_2, \ldots, d_N$ such that

$$\{n \geq 0 | a_n = 0\} = S \cup (c_1 + d_1\mathbb{N}) \cup \cdots \cup (c_N + d_N\mathbb{N}).$$

Skolem-Mahler-Lech theorem

Proof. a) Consider the following function

$$f(x) = \sum_{k\geq 0} p^k a_k \binom{x}{k}.$$

Of course, such a series cannot converge as a series of real numbers, but it does converge as a series of p-adic numbers, since $v_p\left(p^k a_k \binom{x}{k}\right) \geq k$ for all $x \in \mathbb{Z}_p$

(note that $\binom{x}{n} \in \mathbb{Z}_p$ because the map $x \to \binom{x}{n}$ is continuous and takes values in $\mathbb{Z}_p$ when x is in the dense subset $\mathbb{Z}$ of $\mathbb{Z}_p$). Thus f defines a map $f : \mathbb{Z}_p \to \mathbb{Z}_p$. The crucial property of f is that it interpolates the sequence $\sum_{k=0}^{n} p^k \binom{n}{k} a_k$ and that it converges to its Taylor series (has good analytic behavior). The first claim is obvious, since by definition we have

$$f(n) = \sum_{k=0}^{n} p^k \binom{n}{k} a_k.$$

The second claim is more delicate. Let us write

$$\binom{x}{k} = \frac{1}{k!} \sum_{j=0}^{k} b_{j,k} x^j$$

for some integers $b_{j,k}$. Then we can write

$$f(x) = \sum_{k \geq 0} \frac{p^k a_k}{k!} \left(\sum_{j=0}^{k} b_{j,k} x^j \right) = \sum_{j \geq 0} c_j x^j,$$

where $c_j = \sum_{k \geq j} \frac{p^k a_k b_{j,k}}{k!}$. Note that the series defining c_j converges, since

$$v_p \left(\frac{p^k b_{j,k}}{k!} \right) \geq k \cdot \frac{p-2}{p-1}$$

tends to ∞ as $k \to \infty$. The same estimate shows that

$$v_p(c_j) \geq \inf_{k \geq j} k \cdot \frac{p-2}{p-1} = j \frac{p-2}{p-1}.$$

(this is why we assumed that $p > 2$). The conclusion is that f converges to its Taylor series and so behaves as a holomorphic function, i.e. has good analytic properties.

Now, by hypothesis we know that $f(n) = 0$ for infinitely many integers n. We want to prove that $f = 0$. This will imply that $f(n) = 0$ for all n and it is easy to see that this forces $a_n = 0$ for all n. Since $\mathbb{Z}_p$ is compact, we deduce that there exists $a \in \mathbb{Z}_p$ and an infinite sequence of integers n_j such that $f(n_j) = 0$ and n_j converges p-adically to a. Now, for all $x \in \mathbb{Z}_p$ we can write

$$\begin{aligned} f(x) &= \sum_{j\geq 0} c_j((x-a)+a)^j \\ &= \sum_{j\geq 0} c_j \left(\sum_{k=0}^{j} \binom{j}{k} (x-a)^k a^{j-k} \right) \\ &= \sum_{k\geq 0} \left(\sum_{j\geq k} c_j \binom{j}{k} a^{j-k} \right) (x-a)^k. \end{aligned}$$

Again, the series defining $d_k = \sum_{j\geq k} c_j \binom{j}{k} a^{j-k}$ converges because $v_p(c_j) \to \infty$ and we also have

$$v_p(d_k) \geq \inf_{j\geq k} v_p(c_j) \geq k\frac{p-2}{p-1}.$$

Recall that $f(n_j) = 0$ for all j. On the other hand

$$v_p\left(\sum_{k\geq 1} d_k (n_j - a)^k \right) \geq v_p(n_j - a) \to \infty,$$

so that $\lim_{j\to\infty} f(n_j) - d_0 = 0$. We deduce that $d_0 = 0$. Dividing the equality $f(n_j) = 0$ by $a - n_j$ and repeating the argument yields $d_1 = 0$, then $d_2 = 0$ and so on. We deduce that all d_j's are zero and so $f = 0$. The result follows.

b) We may assume that $x_d \neq 0$. Consider the matrix M defined by

$$m_{ij} = 1_{j=i+1}$$

for $i < n$ and whose last row is $x_d, x_{d-1}, \ldots, x_1$. This is the companion matrix associated to the characteristic polynomial $X^d - x_1 X^{d-1} - \cdots - x_d$ of the recursive relation. Let V_n be the column vector whose coordinates are $a_n, a_{n+1}, \ldots, a_{n+d-1}$. Then the recursive relation becomes $V_{n+1} = MV_n$, thus $V_n = M^n V_0$. If e_1 is the column vector whose coordinates are $1, 0, 0, \ldots, 0$ and if $\langle , \rangle$ is the standard inner product in $\mathbb{R}^d$, we deduce that $a_n = \langle M^n V_0, e_1 \rangle$. It is easy to check that $\det M$ equals x_d up to a sign. Pick a prime $p > 2 + |x_d|$, so M is invertible mod p. Using either Lagrange's theorem or the pigeonhole principle, we can find $r \geq 1$ such that $M^r \equiv I_d \pmod{p}$. Thus we can write $M^r = I_d + pN$ for some matrix N with integral coefficients. It is then enough to prove that for any $0 \leq j \leq r-1$, the set $A_j = \{n \geq 0 | a_{nd+j} = 0\}$ is either finite or contains all nonnegative integers. In order to prove this, the point is to write

$$a_{nd+j} = \langle (I_d + pN)^n M^j V_0, e_1 \rangle = \sum_{k=0}^{n} \binom{n}{k} p^k b_k,$$

where $b_k = \langle N^k M^j V_0, e_1 \rangle$ is a sequence of integers. If A_j is infinite, then part a) of the problem shows that A_j contains all nonnegative integers and the result follows. □

Remark 3.11. Under the hypothesis $p > 2$, the map f is analytic on the whole $\mathbb{Z}_p$. If $p = 2$, one can still prove (using for instance Amice's theorem 3.B.30) that f is locally analytic on $\mathbb{Z}_p$. The proof is then exactly the same as in the case $p > 2$.

Remark 3.12. The result holds for sequences with values in any field of characteristic 0, as Lech proved. The key point is that we have a p-adic version of the Lefschetz principle (the proof is not easy, but elementary, see [13] or [45]): if S is a finite subset of a field K which is finitely generated over $\mathbb{Q}$, then for infinitely many primes p there is an embedding of K into $\mathbb{Q}_p$ sending all elements of S to $\mathbb{Z}_p$. Applied to the roots of the characteristic polynomial of the recurrence relation, this reduces the proof to the p-adic case, which has already been discussed.

Remark 3.13. The result does not hold for fields of positive characteristic. For instance, the sequence $a_n = (1+t)^n - 1 - t^n$ is linearly recursive with values in $\mathbb{F}_p((t))$, but the reader can easily check that it vanishes precisely at $\{p^n | n \geq 0\}$, which is not the union of a finite set and finitely many arithmetic progressions.

3.7 Miscellaneous problems

Before attacking the following problem, let us recall aclassical property of the Fermat number $F_n = 2^{2^n} + 1$, namely that any prime factor p of F_n satisfies $p \equiv 1 \pmod{2^{n+1}}$. Indeed, if p divides F_n, then p divides $2^{2^{n+1}} - 1$, but does not divide $2^{2^n} - 1$. Thus the order of 2 mod p is 2^{n+1} and since this order divides $p-1$, the result follows. The next problem is a generalization of some very classical problems, such as the following one: find all integers n such that the congruence $xy \equiv 1 \pmod{n}$ implies the congruence $x \equiv y \pmod{n}$. It is easy to see that this is equivalent to $n|x^2 - 1$ for any x relatively prime to n and that this happens if and only if n divides 24. The next problem is a bit trickier.

25. Let m be an integer greater than 1. Suppose that a positive integer n satisfies $n|a^m - 1$ for all integers a relatively prime to n. Prove that $n \leq 4m(2^m - 1)$ and find all cases of equality.

Gabriel Dospinescu, Marian Andronache, Romanian TST 2004

Proof. Write $n = 2^k r$ with r odd. By the Chinese Remainder Theorem, there is an a with $a \equiv 3 \pmod{2^k}$ and $a \equiv 2 \pmod{r}$. Such an a is clearly relatively prime to n, so $n|a^m - 1$. Hence $2^m \equiv a^m \equiv 1 \pmod{r}$ and $3^m \equiv a^m \equiv 1 \pmod{2^k}$. Thus $r|2^m - 1$ and $2^k|3^m - 1$. One easily checks that $v_2(3^m - 1) = 2 + v_2(m)$, hence $2^k|4m$. We conclude that $n|4m(2^m - 1)$ and in particular $n \leq 4m(2^m - 1)$.

Now suppose equality holds. The above argument actually shows that $n \leq 4 \cdot 2^{v_2(m)} \cdot (2^m - 1)$ so for equality to hold we must have $m = 2^{v_2(m)}$, that is, m must be a power of 2. Suppose $m = 2^s$. Then

$$n = 2^{s+2}\left(2^{2^s} - 1\right) = 2^{s+2} \cdot 3 \cdot 5 \cdots \left(2^{2^{s-1}} + 1\right),$$

that is, n is a power of 2 times some number of consecutive Fermat numbers. The cases $s = 1$ and $s = 2$ give two equality cases $(m, n) = (2, 24)$ and $(m, n) = (4, 240)$. Suppose now that $s \geq 3$, and let p be a prime divisor of $2^{2^{s-1}} + 1$. The preliminary discussion shows that $p \equiv 1 \pmod{2^s}$, in particular $p \equiv 1 \pmod{8}$, so that by quadratic reciprocity, 2 is a square mod p. The Chinese

Remainder Theorem gives an a relatively prime to n with $a^2 \equiv 2 \pmod{p}$. Then $n | a^m - 1$, so $2^{2^{s-1}} \equiv a^{2^s} = a^m \equiv 1 \pmod{p}$. But this contradicts the fact that $2^{2^{s-1}} \equiv -1 \pmod{p}$. Thus there are no solutions with $s \geq 3$. $\square$

Here is a beautiful problem, for which we present two elementary proofs, both ingenious, neither quite natural. In the addendum to this chapter, we will present a more natural proof which uses p-adic analysis.

26. Let x_n be the exponent of 2 in the prime factorization of the numerator of $\frac{2}{1} + \frac{2^2}{2} + \cdots + \frac{2^n}{n}$, when written in lowest terms. Prove that

$$\lim_{n \to \infty} x_n = \infty$$

and that $x_{2^n} \geq 2^n - n + 1$.

Adapted from a Kvant problem

Proof. The first proof is based on the following nice identity known as Staver's identity, which also appeared as a problem in the USA TST 2000.

Lemma 3.14. *For any positive integer n we have*

$$\binom{n}{0}^{-1} + \binom{n}{1}^{-1} + \cdots + \binom{n}{n}^{-1} = \frac{n+1}{2^{n+1}} \left(\frac{2}{1} + \frac{2^2}{2} + \cdots + \frac{2^{n+1}}{n+1} \right).$$

Let us accept this lemma for a moment and see how we can conclude. Using the extension of v_2 to all rational numbers, we can write

$$v_2 \left(\sum_{k=1}^{n+1} \frac{2^k}{k} \right) \geq n + 1 - v_2(n+1) - \max_{0 \leq j \leq n} v_2 \left(\binom{n}{j} \right).$$

Since

$$v_2 \left(\binom{n}{j} \right) = s_2(j) + s_2(n-j) - s_2(n) \leq 2(1 + \log_2(n)) - 1$$

and since $v_2(n+1) \leq \log_2(n+1)$, this trivially implies that $x_n \to \infty$. Moreover, since each of the numbers $\binom{2^N - 1}{k}$ is odd and since there are an even number of terms in the sum $\sum_{k=0}^{2^N-1} \frac{1}{\binom{2^N-1}{k}}$, it follows that $x_{2^N} \geq 2^N - N + 1$, ending the proof of the second part of the problem.

Now, let us prove the lemma. We will present two completely different approaches, the first one very down-to-earth, the second one more exotic, but more powerful also.

Denote the left-hand side by a_n and the right-hand side by b_n. Clearly, $a_0 = b_0 = 1$. Note that

$$b_n = \frac{n+1}{2n}\frac{n}{2^n}\left(\frac{2}{1}+\frac{2^2}{2}+\cdots+\frac{2^n}{n}\right)+1 = \frac{n+1}{2n}b_{n-1}+1.$$

On the other hand,

$$\begin{aligned}
\frac{n+1}{2n}a_{n-1}+1 &= 1+\frac{n+1}{2n}\sum_{i=0}^{n-1}\frac{i!(n-1-i)!}{(n-1)!}\\
&= 1+\frac{1}{2}\sum_{i=0}^{n-1}\frac{((i+1)+(n-i))\cdot i!(n-1-i)!}{n\cdot(n-1)!}\\
&= 1+\frac{1}{2}\sum_{i=0}^{n-1}\frac{(i+1)!(n-1-i)!+i!(n-i)!}{n!}\\
&= 1+\frac{1}{2}\left(\sum_{i=1}^{n}\binom{n}{i}^{-1}+\sum_{i=0}^{n-1}\binom{n}{i}^{-1}\right)\\
&= a_n.
\end{aligned}$$

Taking into account these two relations, it follows by induction that $a_n = b_n$ for all n and the lemma is proved.

Now, let us turn to the second approach. We will use here the classical formula

$$\int_0^1 t^a(1-t)^b dt = \frac{a!b!}{(a+b+1)!},$$

which can be proved separately for each value of $a+b$ and then by induction on say a and integration by parts. We deduce that

$$\sum_{k=0}^{n}\frac{1}{\binom{n}{k}}=\frac{1}{n!}\cdot\sum_{k=0}^{n}(n+1)!\int_0^1 t^k(1-t)^{n-k}dt$$
$$=(n+1)\int_0^1\frac{t^{n+1}-(1-t)^{n+1}}{2t-1}dt.$$

Making the substitution $2t-1=s$ yields

$$\int_0^1\frac{t^{n+1}-(1-t)^{n+1}}{2t-1}dt=\frac{1}{2^{n+2}}\int_{-1}^1\frac{(1+s)^{n+1}-(1-s)^{n+1}}{s}ds$$

and noting that the integrand is an even function of s, the last expression is also equal to

$$\frac{1}{2^{n+1}}\int_0^1\left(\frac{(1+s)^{n+1}-1}{s}+\frac{1-(1-s)^{n+1}}{s}\right)ds$$
$$=\frac{1}{2^{n+1}}\cdot\sum_{i=0}^{n}\int_0^1\left((1+s)^i+(1-s)^i\right)ds$$
$$=\frac{1}{2^{n+1}}\sum_{i=0}^{n}\left(\frac{2^{i+1}-1}{i+1}+\frac{1}{i+1}\right),$$

from where the result follows trivially. □

Proof. This second elementary solution actually mimics the p-adic analytic proof, without mentioning p-adic numbers. Define the sequence of polynomials

$$L_n(X)=\frac{X}{1}+\frac{X^2}{2}+\cdots+\frac{X^n}{n}$$

and $F_n(X)=L_n(2X-X^2)-2L_n(X)$. Since

$$F_n'(X)=2(1-X)L_n'(2X-X^2)-2L_n'(X)$$
$$=\frac{2X^n(1-(2-X)^n)}{1-X}$$
$$=-2X^n(1+(2-X)+\cdots+(2-X)^{n-1})$$

is of the form $X^n(a_0+a_1X+\cdots+a_{n-1}X^{n-1})$ for some integers a_i and moreover $F_n(0)=0$, we deduce that we can write

$$F_n(X)=X^{n+1}\left(\frac{a_0}{n+1}+\frac{a_1}{n+2}X+\cdots+\frac{a_{n-1}}{2n}X^{n-1}\right).$$

Therefore, since $L_n(2)=-\frac{1}{2}F_n(2)$, we can write

$$\frac{2}{1}+\frac{2^2}{2}+\cdots+\frac{2^n}{n}=-\sum_{k=0}^{n-1}\frac{2^{n+k}a_k}{n+k+1},$$

immediately implying the estimate

$$v_2\left(\frac{2}{1}+\frac{2^2}{2}+\cdots+\frac{2^n}{n}\right)\geq\min_{0\leq k\leq n-1}(n+k-v_2(n+k+1)).$$

As $v_2(n+k+1)\leq\log_2(n+k+1)\leq\log_2(2n)$, the fact that $x_n\to\infty$ is now clear. The other inequality is also trivial using the previous estimates. □

We continue with a rather technical problem. The solution we present is long and complicated, but nevertheless rather natural. This problem also appeared on the IMO 2007 Shortlist.

27. Find the exponent of 2 in the prime factorization of the number

$$\binom{2^{n+1}}{2^n}-\binom{2^n}{2^{n-1}}.$$

J. Desmong, W.R. Hastings, AMM E 2640

Proof. Note that

$$\frac{\binom{2^{n+1}}{2^n}}{\binom{2^n}{2^{n-1}}}=\frac{1\cdot 3\cdot\dots\cdot(2^{n+1}-1)}{(1\cdot 3\cdot\dots\cdot(2^n-1))^2}=\frac{(2^n+1)(2^n+3)\cdots(2^n+2^n-1)}{1\cdot 3\cdot\dots\cdot(2^n-1)},$$

so that we need to compute

$$v_2\left(\binom{2^n}{2^{n-1}}\left(\frac{(2^n+1)(2^n+3)\cdots(2^n+2^n-1)}{1\cdot 3\cdot\cdots\cdot(2^n-1)}-1\right)\right)$$
$$=1+v_2((2^n+1)(2^n+3)\cdots(2^n+2^n-1)-1\cdot 3\cdot\cdots\cdot(2^n-1)).$$

Looking at small values of n, we conjecture that for $n\geq 2$

$$v_2((2^n+1)(2^n+3)\cdots(2^n+2^n-1)-1\cdot 3\cdot\cdots\cdot(2^n-1))=3n-1.$$

To prove this, we start by expanding the product

$$(2^n+1)(2^n+3)\cdots(2^n+2^n-1)$$

as if it were a polynomial in 2^n. If we work mod 2^{3n}, the only terms that count are the first three: if $a=1\cdot 3\cdot\cdots\cdot(2^n-1)$, then

$$(2^n+1)(2^n+3)\cdots(2^n+2^n-1)\equiv a+a\cdot 2^n\sum_{i=0}^{2^{n-1}-1}\frac{1}{2i+1}$$
$$+2^{2n}\cdot a\sum_{0\leq i<j\leq 2^{n-1}-1}\frac{1}{(2i+1)(2j+1)}\pmod{2^{3n}}.$$

For simplicity, let $x_i=\frac{1}{2i+1}$ and $y_i=\frac{1}{2^n-2i-1}$. Then

$$2\sum x_i=2^n\sum x_iy_i,$$

which combined with the previous observation yields

$$v_2((2^n+1)(2^n+3)\cdots(2^n+2^n-1)-1\cdot 3\cdot\cdots\cdot(2^n-1))$$
$$=2n-1+v_2\left(\sum_{i=0}^{2^{n-1}-1}x_iy_i+2\sum_{0\leq i<j\leq 2^{n-1}-1}x_ix_j\right).$$

We will assume that $n>2$. Then

$$v_2\left(\sum_{i=0}^{2^{n-1}-1}x_i\right)=v_2\left(2^{n-1}\sum_{i=0}^{2^{n-1}-1}x_iy_i\right)\geq n-1,$$

so

$$2\sum_{0\le i<j\le 2^{n-1}-1} x_ix_j = \left(\sum x_i\right)^2 - \sum x_i^2 \equiv -\sum_{i=0}^{2^{n-1}-1} x_i^2 \pmod{2^{n+1}}.$$

Since

$$x_iy_i = -x_i^2\frac{1}{1-2^nx_i} \equiv -x_i^2 - 2^nx_i^3 \pmod{2^{2n}},$$

we obtain

$$\sum_{i=0}^{2^{n-1}-1} x_iy_i + 2\sum_{0\le i<j\le 2^{n-1}-1} x_ix_j$$
$$\equiv -2\sum_{i=0}^{2^{n-1}-1} x_i^2 - 2^n\sum_{i=0}^{2^{n-1}-1} x_i^3 \pmod{2^{n+1}}.$$

Finally, we clearly have

$$2^n\sum_{i=0}^{2^{n-1}-1} x_i^3 \equiv 0 \pmod{2^{n+1}}$$

and

$$\sum_{i=0}^{2^{n-1}-1} x_i^2 \equiv 1^2+3^2+\cdots+(2^n-1)^2 \equiv \frac{2^n(4^n-1)}{6} \pmod{2^n},$$

because the remainders mod 2^n of the numbers $\frac{1}{2i+1}$ are the same as the remainders of the numbers $2i+1$. Putting everything together, the result follows. □

Remark 3.15. There are incredibly many congruences concerning binomial coefficients. A very good exercise for the reader is to prove the following theorem of Ljunggren from 1952: for any $p \ge 5$ and any integers a and b, the number $\binom{pa}{pb} - \binom{a}{b}$ is a multiple of p^3. A more difficult result, due to Zieve (1999) is the fact that $\binom{ap^{n-1}}{bp^{n-1}} - \binom{a}{b}$ is a multiple of p^{3n} for any $n \ge 1$ and any integers a, b, if p is a prime greater than 3.

The last problem of the chapter is quite technical, but also very beautiful. A much weaker result was proposed as problem 3 at the IMO 2008. This kind of result is actually very old and classical.

28. Prove that for any $c > 0$ there are infinitely many n such that the largest prime divisor of $n^2 + 1$ is greater than cn.

Chebyshev, Nagell

Proof. Let $f(n)$ be the largest prime factor of $P_n = (1^2+1)(2^2+1)\cdots(n^2+1)$. We will prove that $\lim_{n\to\infty} \frac{f(n)}{n} = \infty$. This is enough to deduce the desired result: indeed, choose any $c > 0$, then for sufficiently large n we have $f(n) > cn$. By definition there exists $k_n \le n$ such that $f(n)|k_n^2 + 1$. Then the largest prime factor of $k_n^2 + 1$ is greater than ck_n and we are done (note that $k_n \to \infty$ as $n \to \infty$, since $f(n)$ divides $k_n^2 + 1$).

To prove the result on $f(n)$, we will bound the p-adic valuation of prime factors of P_n. Let A_n be the set of prime numbers smaller than or equal to n and such that $p \equiv 1 \pmod 4$. Any prime factor of P_n is in the set A_{n^2+1}. The following lemma is the first crucial ingredient:

Lemma 3.16. *a) For any odd prime p we have*

$$v_p(P_n) \le \log_p(n^2 + 1) + \frac{2n}{p-1}.$$

b) We have that $v_2(P_n) = \left\lfloor \frac{n+1}{2} \right\rfloor$.

Proof. The second part follows from the fact that the factors of P_n alternate 2 (mod 8) and 1 (mod 4). To prove a), let n_i be the number of $j \in \{1, 2, \dots, n\}$ such that $p^i|j^2 + 1$. Clearly

$$v_p(P_n) = \sum_{k=1}^{n} v_p(k^2 + 1) = \sum_{i=1}^{[\log_p(n^2+1)]} n_i.$$

Note that if $p > 2$ and p^i divides both j^2+1 and k^2+1, then p^i divides $j-k$ or $j+k$. Indeed, p^i divides $(j-k)(j+k)$ and p cannot divide both $j-k, j+k$

as p does not divide j, k and $p > 2$. This being said, if j_1 is the smallest index with $p^i | j_1^2 + 1$, then all k such that $p^i | k^2 + 1$ are of the form $j_1 + sp^i$ for some $0 \le s \le \frac{n-j_1}{p^i}$ or of the form $-j_1 + tp^i$ for some $\frac{1+j_1}{p^i} \le t \le \frac{n+j_1}{p^i}$. We easily deduce that $n_i \le 1 + \frac{2n}{p^i}$ when p is odd and the result follows by adding up these estimates. □

Using the previous observations, we can write ($c_1 > 0$ is an absolute constant and $\pi(n)$ is the number of prime numbers not exceeding n)

$$\begin{aligned} 2\log n! &< \log P_n \\ &= \sum_{\substack{p \le f(n) \\ p \in A_{n^2+1}}} v_p(P_n)\log p + v_2(P_n)\log 2 \\ &\le \sum_{\substack{p \le f(n) \\ p \in A_{n^2+1}}} \left(\log_p(n^2+1) + \frac{2n}{p-1}\right)\log p + c_1 n \\ &\le 3\log n \cdot \pi(f(n)) + 2n \sum_{\substack{p \le f(n) \\ p \in A_{n^2+1}}} \frac{\log p}{p-1} + c_1 n. \end{aligned}$$

Now, corollary 3.A.2 shows that $\pi(n) \le c_2 \frac{n}{\log n}$ for all n and some absolute constant $c_2 > 0$. The crucial fact is that

$$\sum_{\substack{4|p-1 \\ p \le f(n)}} \frac{\log p}{p-1} = \frac{\log f(n)}{2} + O(1),$$

a nontrivial result that follows from our proof of Dirichlet's theorem (see addendum 7.A for more details). Taking this into account and dividing the previous inequality by $n \log n$ yields

$$2\frac{\log n!}{n\log n} \le \frac{3c_2 f(n)}{n \log f(n)} + \frac{\log f(n)}{\log n} + O\left(\frac{1}{\log n}\right).$$

Combining this with the fact that $\lim_{n\to\infty} \frac{\log n!}{n \log n} = 1$ easily yields what we want here. □

Remark 3.17. Nagell [60] actually proved the following general result: if $f \in \mathbb{Z}[X]$ is not a product of linear factors with integer coefficients and if $P(x)$ is the largest prime factor of $f(1)f(2)\cdots f(x)$, then there exists $c > 0$ (depending on f) such that $P(x) > cx \log x$. The proof consists in a refinement of the previous method, combined with rather deep estimates in analytic and algebraic number theory (more precisely, the prime ideal theorem). Erdős [27] proved that there exists $c > 0$ depending on f such that $P(x) > x(\log x)^{c \log \log \log x}$. It is a very deep theorem of Hooley [41] that there exists $c > 0$ such that for all $x > 2$ the largest prime factor of $\prod_{n \le x}(n^2+1)$ is greater than $cx^{\frac{11}{10}}$. The proofs of these theorems are very involved.

3.8 Notes

We would like to thank the following people for providing solutions to some of the problems in this chapter: Xiangyi Huang (problems 8,26), Ofir Nachum (problem 10), Fedja Nazarov (problem 8), Richard Stong (problems 3, 11, 20, 25, 27), Victor Wang (problems 8, 21), Gjergji Zaimi (problems 3, 4, 25), Alex Zhu (problem 27).

Addendum 3.A Classical Estimates on Prime Numbers

Since we are using quite a lot of estimates about prime numbers in various places of this book, gathering these results in one addendum seemed more than appropriate. All results here are absolutely classical and go back to the beautiful ideas of Chebyshev, who was probably the first person to put some order in the chaotic world of prime numbers. These ideas were revisited by Erdős and their exposition is heavily influenced by this "update." Basically, everything follows from some very smart applications of Legendre's formula to middle binomial coefficients.

For the reader's convenience, let us recall some standard notation. We let $\pi(n)$ be the number of primes less than or equal to n. If g is a map taking positive values when the argument is large enough and if f is any complex-valued map, we say that $f = O(g)$ (respectively $f = o(g)$) if $\frac{f(x)}{g(x)}$ is bounded (respectively tends to 0) when $x \to \infty$. The crucial estimate that we will use when dealing with the behavior of $\pi(n)$ is the following easy consequence of Legendre's formula.

Theorem 3.A.1. *For any $n \geq 2$, $\binom{n}{[\frac{n}{2}]}$ divides $\prod_{p \leq n} p^{[\log_p n]}$ and is a multiple of $\prod_{[\frac{n+1}{2}] < p \leq n} p$.*

Proof. The second part follows immediately from the identity (note that $n = \left[\frac{n}{2}\right] + \left[\frac{n+1}{2}\right]$)

$$\left[\frac{n}{2}\right]! \binom{n}{\left[\frac{n}{2}\right]} = \prod_{\left[\frac{n+1}{2}\right] < j \leq n} j$$

and the fact that $\prod_{[\frac{n+1}{2}] < p \leq n} p$ divides the right-hand side and is relatively prime to $\left[\frac{n}{2}\right]!$. For the first part, Legendre's theorem yields

$$v_p\left(\binom{n}{\left[\frac{n}{2}\right]}\right) = \sum_{j \geq 1} \left(\left[\frac{n}{p^j}\right] - \left[\frac{[n/2]}{p^j}\right] - \left[\frac{[(n+1)/2]}{p^j}\right]\right).$$

As for all $a, b \in \mathbb{R}$ we have $[a+b]-[a]-[b] \in \{0,1\}$, all terms in the sum are equal to 0 or 1 and all terms for $j > \log_p n$ are zero. Thus $v_p\left(\binom{n}{[\frac{n}{2}]}\right) \leq [\log_p n]$ and the result follows. $\square$

The famous (and deep) prime number theorem asserts that $\frac{\pi(n)\cdot\log n}{n}$ converges to 1 as $n \to \infty$. The following result gives a uniform lower bound estimate. Of course, it is weaker than the prime number theorem, but it is rather amazing that with so few tools it already gives the "correct" lower bound. Note that $\log 2$ is approximately 0.69.

Corollary 3.A.2. *For all $n \geq 2$ we have*

$$6\log 2\frac{n}{\log n} > \pi(n) \geq \frac{\log 2}{2}\frac{n}{\log n}.$$

Proof. Using theorem 3.A.1 for $N = \binom{2n}{n}$, we obtain

$$\log N = \sum_{p\leq 2n} v_p(N)\log p \leq \sum_{p\leq 2n}[\log_p(2n)]\log p \leq \pi(2n)\log(2n).$$

Next, N is the largest among the $\binom{2n}{k}$ and $\sum_k \binom{2n}{k} = 4^n$, hence $N \geq \frac{4^n}{2n+1}$. We obtain

$$\pi(2n) \geq \frac{\log N}{\log(2n)} \geq \frac{2n\log 2 - \log(2n+1)}{\log(2n)}.$$

Using this inequality as well as $\pi(2n+1) \geq \pi(2n)$, a small computation yields the lower bound for $n \geq 6$ as in the original statement; it is easy to check directly the cases $2 \leq n \leq 5$.

Using theorem 3.A.1 again yields

$$n^{\pi(2n)-\pi(n)} < \prod_{n<p\leq 2n} p \leq \binom{2n}{n} \leq 4^n,$$

which applied to $n = 2^k$ gives

$$\pi(2^{k+1}) - \pi(2^k) \leq \frac{2^{k+1}}{k}.$$

Combined with the obvious inequality $\pi(2^{k+1}) \leq 2^k$, this implies the inequality

$$(k+1)\pi(2^{k+1}) - k\pi(2^k) \leq \pi(2^{k+1}) + 2^{k+1} \leq 3 \cdot 2^k,$$

which easily implies that $\pi(2^n) < \frac{3 \cdot 2^n}{n}$. Finally, we have

$$\pi(n) \leq \pi(2^{1+[\log_2 n]}) < 3 \cdot \frac{2^{1+[\log_2 n]}}{1+[\log_2 n]} < \frac{6n}{\log_2 n}. \qquad \square$$

Before tackling the next result, we will prove a very elementary, yet powerful inequality due to Erdős:

Theorem 3.A.3. *For any $n \geq 2$ we have $\prod_{p \leq n} p < 4^{n-1}$.*

Proof. The proof is by strong induction, the inequality being clear for $n = 2$. Suppose that it holds for all numbers smaller than or equal to n and let us prove that $\prod_{p \leq n+1} p < 4^n$. If $n+1$ is even, this is clear, so suppose that $n = 2k$. Note that $\prod_{k+2 \leq p \leq 2k+1} p$ divides $\binom{2k+1}{k}$, so it is certainly at most equal to $\binom{2k+1}{k}$. Combining this with the induction hypothesis for k gives

$$\prod_{p \leq n+1} p = \prod_{p \leq k+1} p \cdot \prod_{k+2 \leq p \leq 2k+1} p < 4^k \cdot \binom{2k+1}{k} \leq 4^n,$$

the last inequality being a consequence of

$$2 \cdot 4^k = (1+1)^{2k+1} = \cdots + \binom{2k+1}{k} + \binom{2k+1}{k+1} + \cdots \geq 2\binom{2k+1}{k}. \qquad \square$$

A slightly trickier argument yields the following beautiful theorem of Mertens.

Theorem 3.A.4. *We have*

$$\sum_{p \leq n} \frac{\log p}{p} = \log n + O(1).$$

Proof. We will use the prime factorization of $n!$. Legendre's formula yields

$$\frac{n}{p-1}-\left(1+\frac{\log n}{\log p}\right)<v_p(n!)<\frac{n}{p-1}.$$

Multiplying this by $\log p$ and summing over $p\leq n$ yields

$$-\log\prod_{p\leq n}p-\pi(n)\cdot\log n<\log n!-n\sum_{p\leq n}\frac{\log p}{p-1}<0.$$

Using Erdős' inequality $\prod_{p\leq n}p<4^n$, the previous estimates on $\pi(n)$, and the inequalities $n\log n>\log n!>n(\log n-1)$ (the first one is obvious, the second one follows easily by induction using the inequality $\log\left(1+\frac{1}{n}\right)<\frac{1}{n}$) yields

$$8\log 2>\sum_{p\leq n}\frac{\log p}{p-1}-\log n>-1.$$

The theorem follows from this estimate and the fact that the series $\sum_p\frac{\log p}{p(p-1)}$ converges (since $\frac{\log p}{p(p-1)}<\frac{1}{p\sqrt{p}}$ if p is large enough). □

The next result is a simple application of Abel's summation and of the previous theorem.

Theorem 3.A.5. *We have*

$$\sum_{p\leq n}\frac{1}{p}=\log\log n+O(1).$$

Proof. Define $a_n=\frac{\log n}{n}$ if n is a prime and 0 otherwise. By the previous theorem there is a bounded sequence r_n such that

$$S_n=a_2+\cdots+a_n=\log n+r_n.$$

Thus

$$\sum_{p\leq n}\frac{1}{p}=\sum_{k=2}^{n}\frac{S_k-S_{k-1}}{\log k}=\frac{S_n}{\log n}+\sum_{k=2}^{n-1}S_k\cdot\left(\frac{1}{\log k}-\frac{1}{\log(k+1)}\right).$$

The triangle inequality yields

$$\sum_{k=2}^{n-1} r_k \cdot \left(\frac{1}{\log k} - \frac{1}{\log(k+1)}\right) = O(1),$$

so it remains to prove that

$$\sum_{k=2}^{n-1} \left(1 - \frac{\log k}{\log(k+1)}\right) = \log\log n + O(1).$$

Note that for $k \geq 2$

$$1 - \frac{\log k}{\log(k+1)} = \frac{1}{\log(k+1)} \int_k^{k+1} \frac{dt}{t}.$$

Hence

$$\begin{aligned}0 &\leq \int_k^{k+1} \frac{dt}{t\log t} - \left(1 - \frac{\log k}{\log(k+1)}\right)\\ &= \int_k^{k+1} \frac{(\log(k+1) - \log t)\,dt}{t\log t\log(k+1)}\\ &\leq \frac{1}{k^2(\log 2)^2},\end{aligned}$$

where we have used

$$\log(k+1) - \log t \leq \log(k+1) - \log(k) = \log(1+1/k) \leq 1/k.$$

Since the sum of the upper bounds converges, we have

$$\sum_{k=2}^{n-1} \left(1 - \frac{\log k}{\log(k+1)}\right) = \int_2^n \frac{dt}{t\log t} + O(1) = \log\log n + O(1). \qquad \square$$

Remark 3.A.6. One can actually say much more and the true content of Mertens' theorems is the following: there is a constant c such that

$$\sum_{p\leq n} \frac{1}{p} = \log\log n + c + O\left(\frac{1}{\log n}\right)$$

and if $\gamma = \lim_{n\to\infty}\left(1+\frac{1}{2}+\cdots+\frac{1}{n}-\log n\right)$, then

$$\prod_{p\le n}\left(1-\frac{1}{p}\right)=\frac{e^{-\gamma}}{\log n}+O\left(\frac{1}{\log^2 n}\right).$$

The first estimate is not difficult and uses again Abel's summation formula combined with basic integral calculus. The second estimate is much trickier.

Addendum 3.B An Introduction to p-adic Numbers

This rather long addendum is an introduction to the wonderful theory of p-adic numbers and their applications. This is a vast subject and the literature concerning it is huge, so that we cannot even properly scratch its surface. However, even a glimpse into the subject reveals amazing things...

For a variety of reasons, reduction mod p is awfully insufficient. The best way to reduce modulo arbitrary powers of a prime p while still working in a reasonable algebraic (and especially analytic) context is using p-adic numbers. Very roughly, a p-adic number is a kind of "analytic function of p" with coefficients taken mod p. So, p-adic numbers will be infinite expansions $\sum_{k>>-\infty} a_k p^k$, where a_k are integers between 0 and $p-1$. This is a mixture of the classical idea of decimal expansion and the more analytic Taylor expansion of a nicely behaved complex-valued function around 0. The idea of seeing integers as analytic functions of primes is incredibly powerful and appears all the time in modern number theory.

Though the best way to define the field of p-adic numbers is by completing $\mathbb{Q}$ with respect to the p-adic absolute value, we will introduce p-adic numbers algebraically and then develop their analytic properties. We think that this is a bit easier to digest. Next, we briefly discuss what happens when one takes a finite extension of the field of p-adic numbers, which gives us the opportunity to discuss valuations and absolute values from a more abstract (and useful) point of view. This discussion reveals a huge complete extension of $\mathbb{Q}_p$, denoted $\mathbb{C}_p$ and called the field of complex p-adic numbers. Its mere existence has some amazing consequences, for instance a beautiful geometric result of Monsky, concerning tiling of squares by triangles of the same area.

After discussing classical analogues of the exponential and logarithm maps, we focus on the p-adic analogue of the complex Gamma function. This is the most technical part of the addendum, but also the most rewarding. It requires a preliminary discussion of Mahler expansions and p-adic integration, which are rather complicated, but once the machine is sufficiently developed one can prove deep congruences in a quite straightforward way. We highly recommend the wonderful books [13], [49] and [67] for a much more thorough treatment and many good examples.

3.B.1 Arithmetic of p-adic integers and p-adic numbers

A nice way (though maybe not the most intuitive) to see a p-adic integer is to understand it as a compatible system of residue classes mod p^n, for all n. That is, following Hensel, a p-adic integer is a sequence $(\overline{x}_n)_{n\geq 1}$, where each $\overline{x}_n$ is a residue class mod p^n, say the class of an integer a_n and where $a_{n+1} \equiv a_n \pmod{p^n}$ for all n. This simply says that $\overline{x}_{n+1}$ maps to $\overline{x}_n$ under the natural map $\mathbb{Z}/p^{n+1}\mathbb{Z} \to \mathbb{Z}/p^n\mathbb{Z}$. With this description, it is fairly clear that the set of p-adic integers becomes a ring, if we define the addition and multiplication component-wise (i.e. the sum of the sequences $(\overline{x}_n)_n$ and $(\overline{y}_n)_n$ is declared to be the sequence $(\overline{x_n + y_n})_n$, similarly for multiplication). To avoid useless repetitions, let us give a name to such sequences:

Definition 3.B.1. A sequence $(\bar{x}_n)_{n\geq 1}$, $\bar{x}_n \in \mathbb{Z}/p^n\mathbb{Z}$ is called *compatible* if $x_{n+1} \equiv x_n \pmod{p^n}$ for all n, where $x_n \in \mathbb{Z}$ is any lifting of $\overline{x}_n$. Let $\mathbb{Z}_p$ be the ring (for the previously defined operations) of all compatible sequences and call it the ring of p-adic integers.

Note that $\mathbb{Z}$ lives inside $\mathbb{Z}_p$, as any integer n can be thought of as the compatible sequence $(n \pmod{p^k})_k$. The map sending n to this sequence gives an injective ring morphism from $\mathbb{Z}$ to $\mathbb{Z}_p$. We always identify an integer and its image in $\mathbb{Z}_p$.

It turns out that our new ring $\mathbb{Z}_p$ has very nice properties, both algebraically and topologically, making it by far easier to handle than $\mathbb{Z}$ (you might have not noticed, but $\mathbb{Z}$ is a very complicated object, actually...). We discussed quite a lot the p-adic valuation for integers and rational numbers and we will see that it naturally extends to $\mathbb{Z}_p$, making $\mathbb{Z}_p$ a beautiful place to do analysis. However, we need some preliminaries in order to make this dream reality. The following result is crucial for the arithmetic of $\mathbb{Z}_p$ and shows the big difference between $\mathbb{Z}$ and $\mathbb{Z}_p$ as far as arithmetic is concerned. Recall that a unit in a ring is an element that has a multiplicative inverse in that ring.

Theorem 3.B.2. *Any nonzero p-adic integer x can be uniquely written $x = p^k u$ for some nonnegative integer k and some unit u.*

Proof. Before going on to the proof, let us characterize the units of $\mathbb{Z}_p$. This will also play an important role in the proof of the theorem:

Lemma 3.B.3. *A compatible sequence (x_n) defines a unit in $\mathbb{Z}_p$ if and only if its first component is nonzero.*

Proof. One direction being obvious, let us assume that the first component is nonzero. By compatibility, all x_n are relatively prime to p, thus their classes mod p^n are invertible. Simply choose y_n to be the inverse of x_n mod p^n and check that it forms a compatible sequence, which is the inverse of x by construction. □

To prove uniqueness of the representation given in the theorem, we need the following easy

Lemma 3.B.4. *If $x \in \mathbb{Z}_p$ and $p^m x = 0$ then $x = 0$.*

Proof. By induction on m we may assume that $m = 1$. Next, write x as a compatible sequence $(\bar{x}_n)_n$ and observe that the condition $px = 0$ simply says that $p\bar{x}_n = 0$ in $\mathbb{Z}/p^n\mathbb{Z}$. This means that $p^{n-1}|x_n$ for each $n \geq 1$. But since p^n divides $x_{n+1} - x_n$, we see necessarily that $p^n|x_n$, so in fact all components of x are zero. □

Assume now that $x = p^k u = p^l v$ for some u, v units and some nonnegative integers k, l. If $k > l$, lemma 3.B.4 yields $p^{k-l}u = v$. As v is invertible (in other words a unit), we have a contradiction with lemma 3.B.3. Similarly, we cannot have $k < l$, so $k = l$. Applying lemma 3.B.4 once more, we get $u = v$, which proves the uniqueness part of the theorem.

To prove the existence, write x as a compatible sequence and let m be the largest integer j such that $x_j \equiv 0 \pmod{p^j}$. Then $y_n = \frac{x_{n+m}}{p^m}$ are integers, since by compatibility $x_{n+m} \equiv x_m \equiv 0 \pmod{p^m}$. Moreover, since x_n is compatible, so is y_n. Then by construction (and the compatibility of $(\bar{x}_n)_n$), the sequence y_n defines a p-adic integer y such that $p^m y = x$. We claim that y is a unit, which will finish the proof of the first part of the theorem. But the first component of y_n does not vanish, so the result follows from lemma 3.B.3. □

Here is an important consequence of the theorem:

Corollary 3.B.5. *$\mathbb{Z}_p$ is an integral domain. In other words, if $ab = 0$ then $a = 0$ or $b = 0$.*

Proof. This follows immediately from the previous theorem and the second lemma in its proof. □

If R is an integral domain, one can define its field of fractions: formally, it is the set of all symbols $\frac{a}{b}$ with $a, b \in R$ and $b \neq 0$, it being understood that we identify $\frac{a}{b}$ and $\frac{c}{d}$ if $ad = bc$. Addition and multiplication of fractions being done as in elementary school, this yields a field. Applying this to $\mathbb{Z}_p$, we obtain the field of p-adic numbers.

$$\mathbb{Q}_p = \left\{\frac{a}{b} | a, b \in \mathbb{Z}_p, b \neq 0\right\} = \{\frac{a}{p^n} | a \in \mathbb{Z}_p, n \in \mathbf{N}\},$$

the second equality being a consequence of theorem 3.B.2.

3.B.2 The p-adic valuation revisited

We will give a more analytic flavor to $\mathbb{Q}_p$, by endowing it with an absolute value, which plays the same role as the usual absolute value on real numbers.

Definition 3.B.6. Let $x \in \mathbb{Q}_p - \{0\}$ and write (according to theorem 3.B.2) $x = p^k u$ for a unique unit u and a unique integer k. Call $k = v_p(x)$ the p-adic valuation of x and $|x|_p = p^{-v_p(x)}$ the p-adic absolute value of x. Define $|0|_p = 0$.

The following is an immediate consequence of the definition:

Proposition 3.B.7. *For all $x, y \in \mathbb{Q}_p$ we have*

$$|xy|_p = |x|_p \cdot |y|_p \text{ and } |x + y|_p \leq \max(|x|_p, |y|_p),$$

with equality if $|x|_p \neq |y|_p$. Moreover, $|\cdot|_p$ extends the p-adic absolute value on $\mathbb{Q} \subset \mathbb{Q}_p$.

Note that the inequality $|x + y|_p \le \max(|x|_p, |y|_p)$ satisfied by the p-adic absolute value is stronger than the usual triangle inequality for real or complex numbers. This has a whole variety of consequences, which make p-adic numbers a rather exotic object from a geometric point of view. However, this will not affect us, since we will deal with analytic aspects and for that it is enough to introduce a distance on $\mathbb{Q}_p$, a measure of how close numbers are to one another. The distance between $x, y \in \mathbb{Q}_p$ is defined by $d(x, y) = |x - y|_p$. We can then define analytic objects as in the "real world" (i.e. in the world of real numbers, but we will leave it to you to decide whether that is really the real world...). For instance, there are obvious notions of convergent sequences, continuous functions, etc. Basically, any real-analytic object has a p-adic counterpart. Just to see how it works, let us give the definition of a convergent sequence:

Definition 3.B.8. Say a sequence of p-adic numbers x_n converges to a p-adic number a if $|x_n - a|_p$ converges to 0 in the usual sense, that is for all $N > 1$ there is n_0 such that $|x_n - a|_p < 1/N$ for all $n > n_0$.

Intuitively, the sequence x_n converges to a if the difference $x_n - a$ is more and more divisible by p when n is large, that is if $v_p(x_n - a)$ goes to infinity as $n \to \infty$. If you think of a p-adic number as a compatible sequence, this means that for any k, if n is large enough (depending on k) then the first k components of $x_n - a$ are zero. The following result is absolutely fundamental:

Theorem 3.B.9. *If $x_n \in \mathbb{Q}_p$ converges to 0 then the series $\sum_{n \ge 0} x_n$ converges in $\mathbb{Q}_p$. That is, the sequence whose general term is $x_0 + x_1 + \cdots + x_n$ converges in $\mathbb{Q}_p$.*

Note that this is NOT true for real numbers (think about the harmonic series!). Also, note the following important consequence: a sequence $x_n \in \mathbb{Q}_p$ converges if and only if $x_n - x_{n-1}$ tends to 0 in $\mathbb{Q}_p$, a fact that will be used a lot in subsequent sections.

Proof. Write $s_n = x_0 + x_1 + \cdots + x_n$, so that $s_n - s_{n-1}$ goes to 0. Note that we may assume that all x_n are p-adic integers: indeed, since x_n goes to 0, x_n is a p-adic integer for n large enough. Multiplying all x_n by the same large power of p so that the first terms also become p-adic integers does not affect the hypothesis or the conclusion. Next, write $s_i = (\bar{s}_{i1}, \bar{s}_{i2}, \ldots)$ as a compatible sequence. Thinking of these sequences as infinite rows of some infinite matrix, the crucial fact is the following:

Lemma 3.B.10. *For any n there exists k_n such that $\bar{s}_{in} = \bar{s}_{jn}$ for all $i, j \geq k_n$. That is, every column of this infinite matrix eventually becomes constant.*

Proof. Indeed, note that for $i > j$ we have

$$v_p(s_i - s_j) = v_p(x_{j+1} + \cdots + x_i) \geq \inf_{k \geq j+1} v_p(x_k)$$

and the last quantity goes to infinity as $j \to \infty$. Thus for $i > j$ large enough we have $v_p(s_i - s_j) > n$, which implies that $\bar{s}_{in} = \bar{s}_{jn}$. □

This lemma gives us a candidate for the limit of the sequence s_n: define the sequence $a = (\bar{a}_1, \bar{a}_2, \ldots)$, where $\bar{a}_n$ is the common value of the elements $\bar{s}_{in}$ for i large enough (using the notations of the lemma we have $\bar{a}_n = \bar{s}_{k_n n}$). It is then easy to check that this sequence is compatible and defines a p-adic integer which is the limit of the sequence s_n. □

The following basic result will be used frequently below.

Proposition 3.B.11. *$\mathbb{Z}_p$ is a compact subset of $\mathbb{Q}_p$.*

Proof. Consider any sequence x_n of elements of $\mathbb{Z}_p$. Since the first component of x_n (seen as a compatible sequence) takes only finitely many values, there exists a subsequence $x_{\varphi_1(n)}$ and an integer a_1 such that $x_{\varphi_1(n)} \equiv a_1 \pmod{p}$ for all n. The same argument yields a subsequence $x_{\varphi_1(\varphi_2(n))}$ and an integer a_2 such that $x_{\varphi_1(\varphi_2(n))} \equiv a_2 \pmod{p^2}$ for all n, etc. Considering

$$\varphi(n) = \varphi_1(\varphi_2(\cdots \varphi_n(n)) \cdots),$$

we obtain a subsequence such that $x_{\varphi(n)} \equiv a_k \pmod{p^k}$ for all $n \geq k$. It follows that $(a_k \pmod{p^k})_k$ is a compatible sequence, defining a p-adic integer a. By construction, we have $\lim_{n \to \infty} x_{\varphi(n)} = a$ and the result follows. □

Finally, let us give another fundamental property of p-adic integers, which shows that they are basically "formal power series in p" or "infinite base-p expansions."

Theorem 3.B.12. *For any p-adic integer x there exists a unique sequence $a_n \in \{0, 1, \ldots, p-1\}$ such that*

$$x = \sum_{n=0}^{\infty} a_n p^n.$$

By definition, this equality means that the sequence whose general term is $a_0 + a_1 p + \cdots + a_n p^n$ converges to x. Moreover, if a_n is the first nonzero term of this sequence, then $v_p(x) = n$.

Proof. If x is a p-adic integer, there exists a unique $a_0 \in \{0, 1, \ldots, p-1\}$ such that $x - a_0 \in p\mathbb{Z}_p$. Indeed, it is clear that a_0 has to be (the lifting to $\{0, 1, \ldots, p-1\}$ of) the first term of the compatible sequence x. Using this remark, we deduce by induction that for any n there are unique $a_0, a_1, \ldots, a_n \in \{0, 1, \ldots, p-1\}$ such that $x - (a_0 + a_1 p + \cdots + a_n p^n) \in p^{n+1}\mathbb{Z}_p$. But this implies that $x = \lim_{n\to\infty}(a_0 + a_1 p + \cdots + a_n p^n)$. The rest is essentially immediate using lemma 3.B.4 and theorem 3.B.3. $\square$

So any p-adic number x can be uniquely written as a Laurent series

$$x = \sum_{k>-N} p^k a_k$$

for some N and some $a_k \in \{0, 1, \ldots, p-1\}$. Moreover, we have the following nice criterion to establish when $x \in \mathbb{Q}$. The proof is a bit tricky.

Proposition 3.B.13. *The p-adic number*

$$x = \sum_{k>-N} p^k a_k$$

is a rational number if and only if the sequence $(a_k)_k$ becomes periodic from a certain point.

Proof. It is immediate to check that if $(a_k)_k$ is eventually periodic, then x is rational (simply because $p^a + p^{2a} + \cdots = \frac{p^a}{1-p^a}$ in $\mathbb{Q}_p$ for any $a > 0$). The amusing point is proving the converse. Multiplying x by a power of p, we may assume that $x \in \mathbb{Z}_p$, say $x = \sum_{k\geq 0} a_k p^k$. Write $x = \frac{u}{v}$ for some relatively prime integers u and v and consider the sequence $x_k = \sum_{j\geq k} a_j p^{j-k}$. Then clearly $x_k = a_k + px_{k+1}$. As $x_0 = x$ is rational, it is clear that all x_k are rational. But much more is true: we claim that we can find $y_k \in \mathbb{Z}$ such that $|y_k| \leq \max(|u|, |v|)$ (using the ordinary absolute value!) and $x_k = \frac{y_k}{v}$. Indeed, if this holds for x_k, then we can take $y_{k+1} = \frac{y_k - va_k}{p}$ (clearly $|y_{k+1}| \leq \max(|u|, |v|)$; to see that $y_{k+1} \in \mathbb{Z}$, note that $x_k - a_k \in p\mathbb{Z}_p$, so p must divide $y_k - va_k$). Now, the sequence $(y_k)_k$ is a bounded sequence of integers, so we can find $i < j$ such that $y_i = y_j$. Then $x_i = x_j$ and by uniqueness (proved in the previous theorem) we must have $a_{i+1} = a_{j+1}, a_{i+2} = a_{j+2}, \ldots$. This finishes the proof. □

The following is also absolutely crucial. It basically says that in many cases solving a polynomial equation in p-adic numbers is the same as solving it mod p, since any solution mod p will automatically lift to a compatible sequence of solutions mod p^n.

Theorem 3.B.14. *(Hensel's lemma) Let $f \in \mathbb{Z}_p[X]$ and let $a \in \mathbb{Z}_p$ be such that $|f(a)|_p < 1$ and $|f'(a)|_p = 1$. Then there exists unique $b \in \mathbb{Z}_p$ such that $f(b) = 0$ and $|b - a|_p < 1$.*

Proof. We will prove by induction that one can find a sequence of p-adic integers a_n with $a_0 = a, a_{n+1} \equiv a_n \pmod{p^{n+1}}$ and $v_p(f(a_n)) \geq n + 1$. By the previous theorem, the sequence a_n will converge to a p-adic integer b and since $v_p(f(a_n)) \geq n+1$ and $f(a_n)$ converges to $f(b)$, then $f(b) = 0$. To prove the existence of a sequence a_n, assume we constructed $a_0, \ldots, a_n$ and search for $a_{n+1} = a_n + p^{n+1}b_n$ for some p-adic integer b_n. We need to ensure that $f(a_n + p^{n+1}b_n) \equiv 0 \pmod{p^{n+2}}$. Since

$$f(a_n + p^{n+1}b_n) \equiv f(a_n) + p^{n+1}b_n f'(a_n) \pmod{p^{n+2}},$$

it is enough to take b_n such that $f(a_n) + p^{n+1}b_n f'(a_n) \equiv 0 \pmod{p^{n+2}}$, which is possible as $f'(a_n)$ is a unit (because $a_n \equiv a_0 \pmod{p}$ and $f'(a_0)$ is a unit). □

3.B.3 Absolute values and their extensions

Definitions and Ostrowski's theorem

Let us start with an easy observation: $\mathbb{Q}_p$ is not algebraically closed. Indeed, the equation $x^2 = p$ has no solution in $\mathbb{Q}_p$, since if $x \in \mathbb{Q}_p$ satisfies $x^2 = p$, then $p^{-1} = |p|_p = |x^2|_p = |x|_p^2$ and $|x|_p$ is of the form p^{-a} for an integer a, a contradiction. Thus, it is meaningful to study finite extensions of $\mathbb{Q}_p$, as one is often interested in solving polynomial equations over $\mathbb{Q}_p$. It turns out that all finite extensions of $\mathbb{Q}_p$ also have natural absolute values that extend the absolute value of $\mathbb{Q}_p$, though this is not trivial at all. It is thus better to abstract the situation, using the following

Definition 3.B.15. 1. An absolute value on a field K is a map $|\cdot| : K \to \mathbb{R}^+$ such that $|x| = 0$ if and only if $x = 0$, $|xy| = |x| \cdot |y|$ and $|x+y| \leq |x| + |y|$. The absolute value is called non-archimedean if $|x+y| \leq \max(|x|, |y|)$. The absolute value is called trivial if $|x| = 1$ whenever $x \neq 0$.

2. Two absolute values are called equivalent if there exists $c > 0$ such that $|x|_2 = |x|_1^c$.

3. A valuation on a field K is a map $v : K \to \mathbb{R} \cup \{\infty\}$ such that $v(x) = \infty$ if and only if $x = 0$, $v(xy) = v(x) + v(y)$ and $v(x+y) \geq \min(v(x), v(y))$.

It is clear that if $|\cdot|$ is an absolute value, then $v(x) = -\log|x|$ defines a valuation. If $|\cdot|$ is a non-archimedean absolute value on K, the ring of integers of K is by definition $O_K = \{x \in K \| x| \leq 1\}$. It is easy to check that this is a ring and that $m_K = \{x \in K \| x| < 1\}$ is a maximal ideal of O_K. Thus $k_K = O_K/m_K$ is a field, called the residue field of K.

It is clear that any non-archimedean absolute value is bounded by 1 on $\mathbb{Z}$, but the nice and somewhat tricky fact is that the converse holds. Indeed, if $|n| \le 1$ for all n, then for all x, y and all n we can write

$$|x+y|^n = |(x+y)^n| = \left|\sum_{k=0}^{n} \binom{n}{k} x^{n-k} y^k\right|$$
$$\le \sum_{k=0}^{n} |x|^k |y|^{n-k} \le (n+1)\max(|x|,|y|)^n.$$

Taking the nth root of this inequality and letting $n \to \infty$ yields $|x+y| \le \max(|x|,|y|)$, proving the claim. With these remarks being made, we are ready to prove the following beautiful result:

Theorem 3.B.16. *(Ostrowski) Any nontrivial norm on $\mathbb{Q}$ is equivalent to the p-adic absolute value for some prime p or to the usual absolute value.*

Proof. Suppose first that the absolute value $|\cdot|$ is non-archimedean. We know that $|m/n| = |m|/|n| \neq 1$ for some nonzero $m, n \in \mathbb{Z}$. Without loss of generality, then, let $|m| < 1$, so that for some prime $p|m$ we have $|p| < 1$. If q is another prime with $|q| < 1$, then we may find integers a and b with $ap+bq = 1$, and then

$$1 = |1| = |ap+bq| \le \max(|ap|,|bq|) \le \max(|p|,|q|) < 1,$$

a contradiction. We conclude that $|n| = |p|^{v_p(n)}$, and so $|\cdot|$ is equivalent to the p-adic absolute value.

The difficult case is when $|\cdot|$ is archimedean. We saw that in this case there is an integer $n > 1$ such that $|n| > 1$. Pick any such n and write for all $x > 1$ the number x in base n, say $x = x_0 + x_1 n + \cdots + x_k n^k$. Note that $k \le \log_n x$ and that if $C_n = \max_{1 \le j \le n-1} |j|$, then

$$|x| \le |x_0| + |x_1||n| + \cdots + |x_k||n|^k < C_n \frac{|n|^{k+1}}{|n|-1} < A x^{\log_n |n|}$$

for some constant A, independent of x. Applying this to x^N for N large enough yields $|x| \le x^{\log_n |n|}$.

Now, we claim that for any integer $x > 1$ we have $|x| > 1$. Indeed, if $|x| \leq 1$, by writing n^j in base x and using the same argument as before, we deduce that

$$|n|^j = |n^j| \leq C(1 + \log_x n^j).$$

As $|n| > 1$, this is certainly not true for j large enough, proving the claim.

We have, therefore, that for all $x, n > 1$ both $|x| \leq x^{\log_n |n|}$ and $|n| \leq n^{\log_x |x|}$, so that (e.g.) the first inequality is in fact an equality. This implies that $\log_n |n|$ is a constant function of $n > 1$. Thus, there is d such that $|n| = n^d$ for all integers $n > 1$ and the result follows. □

Extensions of absolute values

We prove now the following fundamental and nontrivial theorem. The proof is pretty acrobatic and uses a nice mixture of analytic and algebraic arguments.

Theorem 3.B.17. *Let K be a finite extension of $\mathbb{Q}_p$. There is a unique extension of the absolute value on $\mathbb{Q}_p$ to an absolute value on K. This absolute value is non archimedean and it is given by*

$$|x| = \sqrt[n]{|N_{K/\mathbb{Q}_p}(x)|_p}$$

if $x \in K$, where $n = [K : \mathbb{Q}_p]$ and $N_{K/\mathbb{Q}_p}$ is the norm.

Proof. We prove uniqueness first. We claim that for any two absolute values $|\cdot|_1, |\cdot|_2$ on K that extend $|\cdot|_p$ on $\mathbb{Q}_p$, one can find $c_1, c_2 > 0$ such that $c_1|x|_1 \leq |x|_2 \leq c_2|x|_1$ for all x. Assume that this happens for a moment. Then

$$c_1|x|_1^n = c_1|x^n|_1 \leq |x|_2^n \leq c_2|x|_1^n,$$

and taking nth roots and letting $n \to \infty$ we get $|x|_1 = |x|_2$, proving the uniqueness part. Now, to prove the claim, it is enough to prove the following: if $e_1, e_2, \ldots, e_n$ is a basis of K over $\mathbb{Q}_p$ and if we define

$$\left| \sum_{i=1}^{n} x_i e_i \right|_\infty = \max_{1 \leq i \leq n} |x_i|_p,$$

then there are $c_1, c_2 > 0$ such that $c_1|x|_\infty \le |x|_1 \le c_2|x|_\infty$. But clearly for

$$x = \sum_{i=1}^{n} x_i e_i$$

we have

$$|x|_1 \le \sum_{i=1}^{n} |x_i|_p \cdot |e_i|_1 \le \left(\sum_{i=1}^{n} |e_i|_1\right) |x|_\infty,$$

so we can take $c_2 = \sum |e_i|_1$. Obtaining c_1 is more subtle. Let

$$S = \{x \in K \,|\, |x|_\infty = 1\}.$$

If we equip K with the product topology induced from its vector-space structure over $\mathbb{Q}_p$, then S is easily seen to be a bounded, closed subset of K, whence compact. Moreover the map $x \to |x|_1$ is continuous on S, as say

$$||x|_1 - |y|_1| \le |x - y|_1 \le c_2|x - y|_\infty.$$

Because this map does not vanish on S, there is $c_1 > 0$ such that $|x|_1 \ge c_1$ for $x \in S$. As any $x \in K$ can be scaled by an element of $\mathbb{Q}_p$ to become an element of S, the claim is proved.

Existence is harder. Defining $|x| = \sqrt[n]{|N_{K/\mathbb{Q}_p}(x)|_p}$, standard properties of the norm yield $|xy| = |x| \cdot |y|$, $|x| = 0 \Leftrightarrow x = 0$ and $|x| = |x|_p$ for $x \in \mathbb{Q}_p$. The difficult point is proving that $|x + y| \le \max(|x|, |y|)$. By multiplicativity, it is enough to prove that $|x + 1| \le \max(1, |x|)$, which reduces to

$$\text{if } |x| \le 1, \text{ then } |x + 1| \le 1.$$

This is however quite subtle. We will actually prove the following result: there exists $c > 0$ such that $|x + 1| \le c$ whenever $|x| \le 1$. Assume that we proved this for a moment. Applying it to x/y or y/x (according to whether $|y| \ge |x|$ or not), we deduce that for all x, y we have $|x + y| \le c \max(|x|, |y|)$. But then for all d we have

$$|x + y|^d = |(x + y)^d| \le c \max_{0 \le i \le d} \left| \binom{d}{i} x^{d-i} y^i \right| \le c(\max(|x|, |y|))^d.$$

Taking dth roots and letting $d \to \infty$, we obtain $|x+y| \le \max(|x|, |y|)$, finishing the proof.

It remains to prove the existence of c. This is similar to the first part of the proof. Namely, let $e_1, \dots, e_n$ be a basis of K over $\mathbb{Q}_p$ and let $|\cdot|_\infty$ be as above. As the norm of an element of $\mathbb{Q}_p$ is a polynomial expression of the coordinates of that element in the basis $e_1, \dots, e_n$, it follows that $x \to |x|$ is a continuous map. Since it does not vanish on the compact set $\{x||x|_\infty = 1\}$, there are positive numbers a, b such that $a \le |x| \le b$ whenever $|x|_\infty = 1$. An obvious scaling argument implies that $a|x|_\infty \le |x| \le b|x|_\infty$ for all x, from where

$$|1+x| \le b|1+x|_\infty \le b\left(1 + \frac{|x|}{a}\right) \le b + \frac{b}{a}$$

whenever $|x| \le 1$. The existence of c is thus proved and we are done. □

The uniqueness property in the previous theorem ensures that if $K \subset L$ are two finite extensions of $\mathbb{Q}_p$, then the restriction to K of the unique absolute value on L extending that on $\mathbb{Q}_p$ is the unique absolute value on K extending that on $\mathbb{Q}_p$. This implies the following very useful

Corollary 3.B.18. *Fix an algebraic closure $\overline{\mathbb{Q}}_p$ of $\mathbb{Q}_p$. There is a unique extension of $|\cdot|_p$ to a non archimedean absolute value on $\overline{\mathbb{Q}}_p$.*

From $\overline{\mathbb{Q}}_p$ to $\mathbb{C}_p$

We have a bad news: after all the hard work in the previous section, we have to tell you that $\overline{\mathbb{Q}}_p$ is not a very good object. When dealing with p-adic numbers, analysis is intensively used and finite extensions of $\mathbb{Q}_p$ are very good places to do analysis because they are complete. This means that all Cauchy sequences converge in such a finite extension (this also happens in $\mathbb{R}$ or $\mathbb{C}$ or in a compact interval, but not in $(0,1)$ for instance: the sequence $1/n$ is Cauchy, but does not converge to an element of $(0,1)$). On the other hand, $\overline{\mathbb{Q}}_p$ is not complete (this is not really easy to prove, actually, but those with a good analytic background will observe that it follows immediately from Baire's lemma), so one cannot do reasonable analysis on this field.

Let us explain why finite extensions of $\mathbb{Q}_p$ are complete, since this is very important. The fact that $\mathbb{Q}_p$ is complete was essentially proved while proving theorem 3.B.9. To see that a finite extension K is complete for the unique absolute value $|\cdot|$ extending $|\cdot|_p$, choose a basis $e_1, \ldots, e_n$ of K over $\mathbb{Q}_p$ and define $|x|_\infty = \max_i |x_i|$ if $x = \sum_i x_i e_i$ and $x_i \in \mathbb{Q}_p$. This is a norm on K (but not an absolute value) and the same argument as in the first paragraph of the proof theorem 3.B.17 shows that there exist $c_1, c_2 > 0$ such that $c_1|x| \leq |x|_\infty \leq c_2|x|$ for all x. Thus, the notions of Cauchy sequence and convergent sequence are the same for $|\cdot|$ and $|\cdot|_\infty$. But it is clear (from the fact that $\mathbb{Q}_p$ is complete) that K is complete for $|\cdot|_\infty$, thus K is complete for $|\cdot|$, too.

Now, we would like to have a field that contains $\overline{\mathbb{Q}_p}$ and which is still complete. It turns out that there exists such a field which is moreover minimal. The situation is very similar to that of $\mathbb{Q}$ endowed with the p-adic absolute value: it is not complete for this absolute value, but adding all possible limits of all Cauchy sequences in $\mathbb{Q}$ one ends up with a much bigger field, $\mathbb{Q}_p$. One can play the same game starting with $\overline{\mathbb{Q}_p}$ and one ends up with a huge field $\mathbb{C}_p$, endowed with an absolute value $|\cdot|_p$ extending that on $\mathbb{Q}_p$ and having the properties:

1) The field $\mathbb{C}_p$ is complete with respect to $|\cdot|_p$, that is any Cauchy sequence in $\mathbb{C}_p$ (with respect to $|\cdot|_p$) converges in $\mathbb{C}_p$. Just as in theorem 3.B.9, we deduce that if $a_n \in \mathbb{C}_p$ is a sequence converging to 0, then $\sum_{n\geq 1} a_n$ converges in $\mathbb{C}_p$ (the notion of convergence is defined just as for $\mathbb{Q}_p$).

2) The field $\mathbb{C}_p$ is algebraically closed, in particular it contains $\overline{\mathbb{Q}_p}$. Moreover, $\overline{\mathbb{Q}_p}$ is dense in $\mathbb{C}_p$.

3) The residue field of $\mathbb{C}_p$ is an algebraic closure of $\mathbb{F}_p$.

Here is the way one constructs $\mathbb{C}_p$: consider the set C of all Cauchy sequences in $\overline{\mathbb{Q}_p}$. It is easy to check that this is actually a ring, addition and multiplication being defined component-wise. Next, one checks that the subset m of C consisting of sequences that converge to 0 is a maximal ideal of C. One defines $\mathbb{C}_p = C/m$. By definition, this is a field. It has a natural absolute value $|\cdot|_p$, defined by: if $x \in \mathbb{C}_p$ is the class of a Cauchy sequence $(a_n)_n$, then

$|x|_p = \lim_{n\to\infty} |a_n|_p$. One checks that this is well-defined (i.e. independent of the choice of the sequence $(a_n)_n$) and it is an absolute value that extends the one on $\mathbb{Q}_p$. It is not difficult to prove that $\mathbb{C}_p$ is complete for this absolute value and essentially by definition $\overline{\mathbb{Q}}_p$ is dense in $\mathbb{C}_p$. It is more difficult to prove that $\mathbb{C}_p$ is an algebraically closed field.

The previous construction does not use any special property of $\overline{\mathbb{Q}}_p$ except the fact that it has an absolute value. In general, for any field K endowed with an absolute value, one can construct (in exactly the same way as above) a field $\hat{K}$ that contains K and has an absolute value $|\cdot|$ extending that of K, such that K is dense in $\hat{K}$. This field is called the completion of K. If K is algebraically closed, then $\hat{K}$ is also algebraically closed, though this is not so easy to prove.

A summary

The upshot of this technical section is that finite extensions of $\mathbb{Q}_p$ behave as $\mathbb{Q}_p$ both algebraically and analytically, that $\overline{\mathbb{Q}}_p$ is a pretty bad field from the analytic point of view and that if one wants to do analysis, then one has to work with its completion $\mathbb{C}_p$. Moreover, in $\mathbb{C}_p$ a series $\sum_n a_n$ converges if and only if a_n converges to 0 (which means that $\lim_{n\to\infty} |a_n|_p = 0$) and then one can permute its terms as one wishes and still get the same value of the sum (this is definitely wrong in real or complex analysis!). Finally, one can deal with double sums in a rather leisurely way, since if $\lim_{\max(m,n)\to\infty} a_{m,n} = 0$, then

$$\sum_m \left(\sum_n a_{m,n}\right) = \sum_n \left(\sum_m a_{m,n}\right)$$

and all series converge. Note that we did not prove the last two assertions, but since they are rather easy, we leave them as good exercises for the reader.

3.B.4 p-adic analogues of classical functions

Recall that for any complex number x, the series $\sum_{n\geq 0} \frac{x^n}{n!}$ converges to a complex number called e^x and $x \to e^x$ is a surjective group morphism $\mathbb{C} \to \mathbb{C}^*$. Let us study the p-adic analogue of this construction: the problem is that

$v_p(n!)$ is quite large, so we cannot expect that the previous series converges for all x. Actually, by theorem 3.B.9 the previous series converges for some $x \in \mathbb{C}_p$ if and only if $v_p\left(\frac{x^n}{n!}\right) \to \infty$. Using Legendre's formula $v_p(n!) = \frac{n-s_p(n)}{p-1}$, where $s_p(n) = O(\log n)$ is the sum of digits of n when written in base p, we deduce that the series converges iff

$$\lim_{n\to\infty} n\left(v_p(x) - \frac{1}{p-1}\right) + \frac{s_p(n)}{p-1} = \infty,$$

which happens if and only if $v_p(x) > \frac{1}{p-1}$, i.e. $|x| < p^{-\frac{1}{p-1}}$. Moreover, one can easily check (using the remark on double sums made in the previous section) that if x, y satisfy these conditions, then so does $x + y$ and $e^x \cdot e^y = e^{x+y}$.

It turns out that one can construct an inverse to the exponential map, which is however defined on all $\mathbb{C}_p$. More precisely, we have the following nontrivial

Theorem 3.B.19. *There exists a unique continuous homomorphism* $\log_p$: $\mathbb{C}_p^* \to \mathbb{C}_p$ *such that* $\log_p(p) = 0$ *and*

$$\log_p(x) = \sum_{n\geq 1}(-1)^{n-1}\frac{(x-1)^n}{n}$$

for $|x-1|_p < 1$.

Proof. (sketch) The proof is pretty long, so we only give the main steps. The crucial point is the following

Lemma 3.B.20. *Any* $x \in \mathbb{C}_p^*$ *can be uniquely written* $x = p^r \cdot \zeta \cdot v$ *for some* $r \in \mathbb{Q}$, ζ *a root of unity of order prime to* p *and* $v \in \mathbb{C}_p$ *such that* $|v - 1| < 1$.

Proof. Let us prove the existence part. By construction, $v_p(\mathbb{C}_p^*) = \mathbb{Q}$, so that given any $x \in \mathbb{C}_p^*$ there is $r \in \mathbb{Q}$ and $u \in \mathbb{C}_p^*$ such that $x = p^r \cdot u$ and $v_p(u) = 0$. Consider the image of u in the residue field $\overline{\mathbb{F}_p}$ of $\mathbb{C}_p$. It is a nonzero element of some $\mathbb{F}_q^*$ for some power q of p. Thus $v_p(u^{q-1} - 1) > 0$ and then easily $u^{(q-1)q^n} \to 1$ as $n \to \infty$. We see similarly that $\zeta = \lim_{n\to\infty} u^{q^n}$ converges and then $\zeta^{q-1} = 1$ with $v_p(u - \zeta) > 0$. So one can take $v = u/\zeta$.

For uniqueness, it is clear that $r = v_p(x)$ is uniquely determined. It is thus enough to check that no root of unity $\zeta \neq 1$ of order prime to p satisfies $|1 - \zeta| < 1$. If ζ has order $n > 1$, then the norm (from $\mathbb{Q}_p(\zeta)$ to $\mathbb{Q}_p$) of $1 - \zeta$ is a divisor of n, and so cannot be a multiple of p. $\square$

Now, let us study $\log_p$. Let $x \in \mathbb{C}_p^*$ and write $x = p^r \cdot \zeta \cdot v$ as in the lemma. Note that if we admit that $\log_p$ exists, then necessarily $N \log_p(\zeta) = \log_p(\zeta^N) = 0$ if $\zeta^N = 1$, so necessarily $\log_p(\zeta) = 0$. As $\log_p(p) = 0$, we must have

$$\log_p x = \log_p(v) = \sum_{n \geq 1} (-1)^{n-1} \frac{(v-1)^n}{n}.$$

This shows that if $\log_p$ exists, then it is unique.

It is harder to prove existence. First, by the previous paragraph we must define

$$\log_p x = \log_p(v) = \sum_{n \geq 1} (-1)^{n-1} \frac{(v-1)^n}{n}$$

if $x = p^r \cdot \zeta \cdot v$. Note that the series converges, as

$$v_p\left(\frac{(v-1)^n}{n}\right) \geq n v_p(v-1) - \log_p(n) \to \infty.$$

Moreover, since the series converges uniformly, it is easy to see that $v \to \log_p(v)$ is continuous for $|v - 1| < 1$. From here it is not difficult to check that $x \to \log_p(x)$ is continuous on $\mathbb{C}_p^*$. It remains to check that it is additive. This easily reduces to

$$\log_p(1+u) + \log_p(1+v) = \log_p(1 + (u + v + uv))$$

for $|u| < 1$ and $|v| < 1$, which is the tricky point. First, one checks that as formal series in X, Y we have

$$\log(1+X) + \log(1+Y) = \log(1 + (X + Y + XY)),$$

for instance by differentiating both sides in X, respectively Y. Next, the series defining $\log_p(1+u)$, $\log_p(1+v)$ and $\log_p(1+u+v+uv)$ converge absolutely and one can permute their terms as one wants, without changing the value of the series. This implies that we can substitute $X = u$ and $Y = v$ in the formal series equality and finishes the proof of the theorem. □

With the same arguments as in the proof of the previous theorem we obtain $\log_p(e^x) = x$ if $|x| < p^{-\frac{1}{p-1}}$ (it is easy to check that $|e^x - 1|_p < 1$ for such x) and $e^{\log_p(x)} = x$ if x is close enough to 1 so that $v_p(\log_p(x)) > \frac{1}{p-1}$.

We end this section with another useful p-adic analogue, the binomial functions and power functions. Define, for $x \in \mathbb{Q}_p$ and $n \geq 0$

$$\binom{x}{n} = \frac{x(x-1)\cdots(x-n+1)}{n!}.$$

Proposition 3.B.21. *1) (Vandermonde's identity) If $x, y \in \mathbb{Q}_p$, then*

$$\binom{x+y}{n} = \sum_{i=0}^{n} \binom{x}{i} \cdot \binom{y}{n-i}.$$

2) If $x \in \mathbb{Z}_p$, then $\binom{x}{n} \in \mathbb{Z}_p$ for all n.

3) If $a \in \mathbb{C}_p$ satisfies $|a|_p < 1$ and $x \in \mathbb{Z}_p$, define

$$(1+a)^x = \sum_{n \geq 0} \binom{x}{n} a^n.$$

Then the series converges and $x \to (1+a)^x$ is a continuous additive homomorphism from $\mathbb{Z}_p$ to $\mathbb{C}_p^*$.

Proof. 1) If x, y are positive integers, simply compare coefficients in

$$(1+T)^{x+y} = (1+T)^x \cdot (1+T)^y.$$

The result then follows by density and continuity. The same argument works for 2). The convergence of the series in 3) follows immediately from 2) and theorem 3.B.9. The continuity follows from the uniform convergence of the series, while the equality $(1+a)^x \cdot (1+a)^y = (1+a)^{x+y}$ follows either by a simple computation using 1) or from the case $x, y \in \{1, 2, \ldots\}$ by continuity and density. □

Some applications

We discuss here some immediate applications of the preceding theoretical results. The reader will probably appreciate better the power of p-adic techniques, since none of the following problems are easy to solve by other means.

Example 3.B.22. (Kiran Kedlaya, USA TST) Let $p > 5$ and let

$$f_p(x) = \sum_{k=1}^{p-1} \frac{1}{(px+k)^2}.$$

Prove that for any integers x, y, p^3 divides the numerator of $f_p(x) - f_p(y)$ when written in lowest terms.

Proof. Using the tools previously introduced, this is very simple: working in $\mathbb{Q}_p$, we can write

$$\begin{aligned} f_p(x) &= \sum_{k=1}^{p-1} \frac{1}{k^2}\left(1+\frac{px}{k}\right)^{-2} \\ &= \sum_{k=1}^{p-1} \frac{1}{k^2} \sum_{j\geq 0} \binom{-2}{j} \frac{p^j}{k^j} x^j \\ &\equiv \sum_{k=1}^{p-1} \frac{1}{k^2}\left(1 - \frac{2px}{k} + 3\frac{p^2x^2}{k^2}\right) \\ &= \sum_{k=1}^{p-1} \frac{1}{k^2} - 2px \sum_{k=1}^{p-1} \frac{1}{k^3} + p^2x^2 \sum_{k=1}^{p-1} \frac{1}{k^4} \pmod{p^3}. \end{aligned}$$

It suffices thus to show that

$$p^2 \mid \sum_{k=1}^{p-1} \frac{1}{k^3} \quad \text{and} \quad p \mid \sum_{k=1}^{p-1} \frac{1}{k^4},$$

but these congruences have already been discussed in the solution of problem 22, chapter 3. $\square$

Example 3.B.23. (how not to prove Fermat's last theorem) Let p be a prime and let $k, N \geq 1$. There exist integers x, y, z, not all of them divisible by p and such that $x^N + y^N \equiv z^N \pmod{p^k}$.

Proof. It is enough to show the existence of $x, z \in \mathbb{Z}_p$ such that $x^N + 1 = z^N$, since then $x \pmod{p^k}, 1$ and $z \pmod{p^k}$ is a solution. We would like to take $z = (1 + x^N)^{1/N}$. Using the results of the previous section, we are tempted to take $z = \sum_{n \geq 0} \binom{\frac{1}{N}}{n} x^{nN}$. Unfortunately, N is not necessarily prime to p, so we cannot apply directly those results. However,

$$v_p\left(\binom{\frac{1}{N}}{n} x^{nN}\right) \geq Nnv_p(x) - \frac{n}{p-1} - nv_p(N)$$

and this tends to ∞ as $n \to \infty$ if $Nv_p(x) > \frac{1}{p-1} + v_p(N)$. We thus choose such x and define z by the previous series. Then $z \in \mathbb{Z}_p$ (by the previous estimate) and the usual argument with formal series shows that $z^N = 1 + x^N$. □

The next problem has already been discussed in chapter 3, problem 26, where two rather difficult solutions were given. Using 2-adic numbers, it becomes almost obvious.

Example 3.B.24. Write

$$\frac{2}{1} + \frac{2^2}{2} + \cdots + \frac{2^n}{n} = \frac{a_n}{b_n}$$

for relatively prime integers a_n, b_n. Then $v_2(a_n) > n - \log_2(n)$ (this is the ordinary logarithm here!) for $n \geq 2$.

Proof. Let us work in $\mathbb{Q}_2$. The series $\sum_n \frac{2^n}{n}$ suggests considering $\log_2(-1)$. Indeed, the series defining this is exactly $-\sum_n \frac{2^n}{n}$. On the other hand, since $\log_2$ is additive and since $(-1)^2 = 1$ and $\log_2(1) = 0$, we must have $\log_2(-1) = 0$, that is in $\mathbb{Q}_2$ we have the equality $\sum_{n \geq 1} \frac{2^n}{n} = 0$. But then

$$v_2\left(\sum_{k=1}^{n} \frac{2^k}{k}\right) = v_2\left(-\sum_{k>n} \frac{2^k}{k}\right) \geq \inf_{k>n}(k - \log_2 k) > n - \log_2(n). \qquad \square$$

3.B.5 A geometric application

In this section we reward the reader with a mathematical gem, due to Paul Monsky. This uses the existence of an absolute value on $\mathbb{C}_p$ extending the one on $\mathbb{Q}_p$, a result which was explained in previous sections. It is a nontrivial fact from field theory that $\mathbb{C}_p$ is isomorphic as a field with $\mathbb{C}$. The choice of an isomorphism allows us to transfer the absolute value on $\mathbb{C}_p$ to one on $\mathbb{C}$, that still extends the p-adic absolute value on $\mathbb{Q}$. The reader who finds this construction very indirect will probably spend some time trying to construct directly such an absolute value on $\mathbb{C}$. Inevitable failure will probably convince him of the power of the arguments in previous sections.

Theorem 3.B.25. *(Monsky) One cannot dissect a square into an odd number of triangles of the same area.*

It is absolutely remarkable that no geometric proof is known for this pretty innocent-looking problem. Monsky's proof (see [53]) is a stunning combination of arithmetic and combinatorics.

Proof. Consider the square with vertices $(0,0),(1,0),(0,1),(1,1)$. Using the extension of the 2-adic valuation to $\mathbb{R}$, color the point $(x,y) \in \mathbb{R}^2$ in red if $\max(|x|_2, |y|_2) < 1$, in blue if $|x|_2 \geq \max(1, |y|_2)$ and in green if $|y|_2 > |x|_2$ and $|y|_2 \geq 1$. We will repeatedly use the easy observation that translation by a red point is color-preserving.

Here is the crucial point:

Lemma 3.B.26. *If T is a triangle whose vertices have three different colors, then $|A(T)|_2 > 1$, where $A(T)$ is the area of T.*

Proof. By the remark on translations by red points, we may assume that one of the vertices of T is $(0,0)$. Let $b = (b_1, b_2)$ and $c = (c_1, c_2)$ be the other vertices, say b is blue and c is green. Then

$$|A(T)|_2 = \left|\frac{b_1c_2 - b_2c_1}{2}\right|_2 = 2|b_1|_2 \cdot |c_2|_2 \cdot \left|1 - \frac{c_1}{c_2} \cdot \frac{b_2}{b_1}\right|_2 > 1,$$

as $|b_1|_2, |c_2|_2 \geq 1$ and $\left|\frac{c_1}{c_2} \cdot \frac{b_2}{b_1}\right|_2 < 1$. □

Consider now a dissection of the square into n triangles of the same area, which is necessarily $1/n$. Color only the vertices of the triangles, as above. If we can prove that there is a triangle with vertices of different colors, we deduce from the previous lemma that $|n|_2 < 1$ and so n is even. The existence of such a triangle is a trivial consequence of Sperner's lemma, but it is perhaps useful to recall how things work in this easy two-dimensional case: consider segments on the boundary of the square whose endpoints are red and blue (i.e. one endpoint is red and the other one blue). All vertices on the edge $[0,1] \times \{0\}$ are either red or blue. All vertices on the edge $\{0\} \times [0,1]$ are either red or green. All vertices on the edges $[0,1] \times \{1\}$ or $\{1\} \times [0,1]$ are either blue or green. Therefore all segments on the sides with one endpoint red and the other blue are on the side $[0,1] \times \{0\}$. As $(0,0)$ is red and $(1,0)$ is blue, there must be an odd number of such segments. On the other hand, assume that no triangle has vertices of different colors. It is easy to check that all triangles have an even number of sides whose endpoints are red and blue. As the triangles partition the square, we deduce that the number of red-blue segments on the border of the square is even, a contradiction. Thus, there must be a "colorful" triangle and the theorem is proved. □

3.B.6 Mahler expansions

One of the miracles of p-adic analysis is that one has a fairly explicit description of all continuous functions on $\mathbb{Z}_p$. Of course, this is far from being true in real or complex analysis, so the following theorem is surprising to say the least. It is however absolutely crucial when dealing with more delicate aspects of p-adic numbers and we will use it constantly in the following sections.

Theorem 3.B.27. *(Mahler) For any continuous function $f : \mathbb{Z}_p \to \mathbb{Q}_p$ there is a unique sequence $(a_n(f))_{n\geq 0}$ of p-adic numbers such that $\lim_{n\to\infty} a_n = 0$ and for all $x \in \mathbb{Z}_p$*

$$f(x) = \sum_{n\geq 0} a_n(f) \binom{x}{n}.$$

Moreover, we have $\min_{x\in\mathbb{Z}_p} v_p(f(x)) = \min_{n\geq 0} v_p(a_n(f))$.

Proof. Note that if the equality

$$f(x) = \sum_{n\geq 0} a_n(f)\binom{x}{n}$$

holds for all $x \in \mathbb{Z}_p$, then for all n

$$f(n) = \sum_{k=0}^{n} a_k(f)\binom{n}{k}.$$

Either by considering the exponential generating function of $(f(n))_n$ and $(a_n(f))_n$ or by using the theory of finite differences (see chapter 10, section 10.3), we deduce that

$$a_n(f) = \sum_{k=0}^{n}(-1)^{n-k}\binom{n}{k}f(k).$$

Assume for a moment that we proved that $\lim_{n\to\infty} a_n(f) = 0$, which is the difficult point of the theorem. Then, since $\binom{x}{n} \in \mathbb{Z}_p$ for $x \in \mathbb{Z}_p$, we deduce that $g(x) = \sum_{n\geq 0} a_n(f)\binom{x}{n}$ converges uniformly for $x \in \mathbb{Z}_p$ and so g is a continuous function. Moreover, by construction $g(n) = f(n)$ for all $n \geq 1$, so by density of $\{1, 2, \ldots\}$ in $\mathbb{Z}_p$ we obtain $f = g$ and the first part of the theorem follows. Finally, from the previous relations between the values of f at positive integers and the $a_n(f)$ we obtain

$$v_p(f(n)) \geq \min_{0\leq i\leq n} v_p(a_i(f)), \quad v_p(a_n(f)) \geq \min_{0\leq i\leq n} v_p(f(i)),$$

so another density argument yields $\min_{x\in\mathbb{Z}_p} v_p(f(x)) = \min_{n\geq 0} v_p(a_n(f))$. Note that those minima exist, as $v_p(a_n(f))$ diverges to ∞ and as f is continuous on the compact set $\mathbb{Z}_p$.

It remains to prove that $v_p(a_n(f)) \to \infty$. As f is bounded (because it is continuous and $\mathbb{Z}_p$ is compact), by multiplying f by some power of p we may assume that $f(\mathbb{Z}_p) \subset \mathbb{Z}_p$. As $\mathbb{Z}_p$ is compact, f is uniformly continuous on $\mathbb{Z}_p$, so there is n_0 such that $v_p(f(x+p^{n_0}) - f(x)) \geq 1$ for all $x \in \mathbb{Z}_p$. Let

$$\Delta f(x) = f(x+1) - f(x),$$

then

$$\Delta^n f(x) = \sum_{k=0}^{n}(-1)^{n-k}\binom{n}{k}f(x+k)$$

and $a_n(f) = \Delta^n f(0)$. As p divides $\binom{p^{n_0}}{k}$ for all $1 \le k < p^{n_0}$, it follows that $v_p(\Delta^{p^{n_0}} f(x)) \ge 1$ for all $x \in \mathbb{Z}_p$ and so $v_p(\Delta^n f(x)) \ge 1$ for all $n \ge p^{n_0}$ and all x. The map $g = \frac{1}{p}\Delta^{p^{n_0}} f$ is continuous and $g(\mathbb{Z}_p) \subset \mathbb{Z}_p$. Applying the same argument to g, we find n_1 such that $v_p(\Delta^{p^{n_1}} g(x)) \ge 1$ for all x. Then $v_p(\Delta^n f(x)) \ge 2$ for all $n \ge p^{n_0} + p^{n_1}$. Continuing like this, we find integers n_i such that $v_p(\Delta^n f(x)) \ge d$ for all $n \ge p^{n_0} + \cdots + p^{n_{d-1}}$ and all $x \in \mathbb{Z}_p$. Taking $x = 0$ shows that $v_p(a_n(f)) \to \infty$ and finishes the proof. □

Remark 3.B.28. The numbers $a_n(f)$ are called the Mahler coefficients of the function f. We discussed only the case of $\mathbb{Q}_p$-valued functions, but the result holds for K-valued functions, where K is any complete subfield of $\mathbb{C}_p$, with basically the same proof.

Here is a nice application of the previous theorem, proposed at the USA TST 2011 by Josh Nichols-Barrer.

Example 3.B.29. Let p be a prime. We say that a sequence of integers $\{z_n\}_{n=0}^{\infty}$ is a p-pod if for each $e \ge 0$, there is an $N \ge 0$ such that whenever $m \ge N$, p^e divides the sum

$$\sum_{k=0}^{m}(-1)^k\binom{m}{k}z_k.$$

Prove that if both sequences $\{x_n\}_{n=0}^{\infty}$ and $\{y_n\}_{n=0}^{\infty}$ are p-pods, then the sequence $\{x_n y_n\}_{n=0}^{\infty}$ is a p-pod.

Proof. See the sequence z_n as a map on $\mathbb{N}$ in the obvious way. The Mahler coefficients of this map are precisely the numbers

$$a_m(z) = \sum_{k=0}^{m}(-1)^k\binom{m}{k}z_k.$$

By hypothesis, z is a p-pod if and only if $a_m(z)$ tends to 0 in $\mathbb{Q}_p$ and by Mahler's theorem this happens if and only if z_n extends to a continuous function on $\mathbb{Z}_p$ (namely $x \to \sum_{n\ge 0} a_n(z)\cdot\binom{x}{n}$). But the pointwise product of two continuous functions on $\mathbb{Z}_p$ is clearly a continuous function, from which the result follows. □

One also has a characterization of locally analytic functions $f : \mathbb{Z}_p \to \mathbb{Q}_p$ in terms of their Mahler coefficients, though the proof is much more difficult. Recall that a function $f : \mathbb{Z}_p \to \mathbb{Q}_p$ is called locally analytic if for any $a \in \mathbb{Z}_p$ there exists n_a and p-adic numbers $f^{(n)}(a)$ such that whenever $v_p(x-a) \geq n_a$ we have

$$f(x) = \sum_{n \geq 0} \frac{f^{(n)}(a)}{n!}(x-a)^n.$$

Theorem 3.B.30. *(Amice) A continuous function $f : \mathbb{Z}_p \to \mathbb{Q}_p$ is locally analytic on $\mathbb{Z}_p$ if and only if its Mahler coefficients $a_n(f)$ satisfy*[3]

$$\limsup_{n\to\infty} \sqrt[n]{|a_n(f)|_p} < 1.$$

3.B.7 The p-adic Gamma function: preliminaries

Recall that the Gamma function is defined for $\mathrm{Re}(s) > 0$ by

$$\Gamma(s) = \int_0^\infty e^{-t} t^s \frac{dt}{t}.$$

It has the nice property that it interpolates the numbers $n!$, as $\Gamma(n) = (n-1)!$ for any n. One would like to have a p-adic analogue of the Gamma function, as this function plays an amazingly important role in real and complex analysis. Unfortunately, it is not difficult to see that there is no continuous function $f : \mathbb{Z}_p \to \mathbb{Q}_p$ such that $f(n) = (n-1)!$ for all positive integers n. On the other hand, we have the following beautiful result, due to Morita. To simplify the exposition, we will assume from now on that $p > 2$. There are natural extensions of all the following results to the case when $p = 2$, but that would force us discuss two cases in both the statement and proof of the following results.

[3] This means that there is $r > 0$ such that $v_p(a_n(f)) \geq rn$ for all sufficiently large n.

Theorem 3.B.31. *(Morita) There is a unique continuous map $\Gamma_p : \mathbb{Z}_p \to \mathbb{Q}_p$ such that for all $n \geq 2$ we have*

$$\Gamma_p(n) = (-1)^n \prod_{\substack{j=1 \\ \gcd(p,j)=1}}^{n-1} j.$$

Proof. Defining

$$g(n) = (-1)^n \prod_{\substack{j=1 \\ \gcd(p,j)=1}}^{n-1} j$$

for $n \geq 2$, let us prove the following

Lemma 3.B.32. $g(n+p^k) \equiv g(n) \pmod{p^k}$ *for all n and all $k \geq 1$.*

We have

$$g(n) - g(n+p^k) = (-1)^n \prod_{\substack{j=1 \\ \gcd(j,p)=1}}^{n-1} j \cdot \left(1 + \prod_{\substack{j=n \\ \gcd(j,p)=1}}^{n+p^k-1} j\right),$$

so it is enough to check that

$$p^k | 1 + \prod_{\substack{j=n \\ \gcd(j,p)=1}}^{n+p^k-1} j.$$

But if $\pi : \mathbb{Z} \to \mathbb{Z}/p^k\mathbb{Z}$ is the natural reduction map, it is clear that

$$\pi\left(\prod_{\substack{j=n \\ \gcd(j,p)=1}}^{n+p^k-1} j\right) = \prod_{g \in G} g,$$

where $G = \left(\mathbb{Z}/p^k\mathbb{Z}\right)^*$. The elements g in the previous product come in pairs (g, g^{-1}), but one has to pay attention to the fact that one might have $g^2 = 1$.

However, as $p > 2$, this appears precisely when $g = 1$ or $g = -1$. Thus, the product of all g's equals -1 and we are done.

The previous lemma easily implies that $v_p(g(m) - g(n)) \geq v_p(m - n)$ for all distinct positive integers m and n. Choose any p-adic integer a and any sequence x_n of positive integers such that $\lim_{n\to\infty} x_n = a$ in $\mathbb{Z}_p$. Since $v_p(g(x_i) - g(x_j)) \geq v_p(x_i - x_j)$, it follows that the sequence $(g(x_n))_n$ is a Cauchy sequence and so it converges to some p-adic integer $g(a)$. If y_n is another sequence that converges to a, then applying the result we have just obtained to the sequence $x_1, y_1, x_2, y_2, \ldots$, we deduce that $g(y_n)$ converges to $g(a)$, i.e. $g(a)$ is independent of the choice of the sequence $(x_n)_n$. Thus, we obtain a map $\Gamma_p : \mathbb{Z}_p \to \mathbb{Z}_p$ which clearly extends g. Passing to the limit in the inequality $v_p(g(m) - g(n)) \geq v_p(m - n)$, we deduce that $v_p(\Gamma_p(x) - \Gamma_p(y)) \geq v_p(x - y)$ for all $x, y \in \mathbb{Z}_p$, showing that Γ_p is continuous. This proves the existence of Γ_p. The uniqueness part is a trivial consequence of the density of $\mathbb{N}$ in $\mathbb{Z}_p$. $\square$

The following proposition summarizes the basic properties of Γ_p. We will prove much deeper results in later sections, once we have developed enough tools.

Proposition 3.B.33. *1) For all positive integers n we have*

$$\Gamma_p(n+1) = (-1)^{n+1} \frac{n!}{\left[\frac{n}{p}\right]! \cdot p^{\left[\frac{n}{p}\right]}}.$$

2) $\Gamma_p(\mathbb{Z}_p) \subset \mathbb{Z}_p^*$.

3) If $\tau_p(x) = -x$ for $x \in \mathbb{Z}_p^*$ and $\tau_p(x) = -1$ for $x \in p\mathbb{Z}_p$, then

$$\Gamma_p(x+1) = \tau_p(x)\Gamma_p(x).$$

4) If $x \in \mathbb{Z}_p$ and $r(x) \in \{1, 2, \ldots, p\}$ is the unique integer such that

$$x - r(x) \in p\mathbb{Z}_p,$$

then

$$\Gamma_p(x) \cdot \Gamma_p(1-x) = (-1)^{r(x)}$$

Proof. 1) This follows immediately by definition of the p-adic Gamma function.

2) By construction, $v_p(\Gamma_p(n)) = 0$ for integers $n \geq 2$. As these integers form a dense subset of $\mathbb{Z}_p$ and as $v_p \circ \Gamma_p$ is continuous, 2) follows.

3) This follows immediately from the definition if x is a positive integer. The general case follows by density and continuity.

4) By density and continuity, it suffices to prove that

$$\Gamma_p(-n)\Gamma_p(n+1) = (-1)^{n+1-[n/p]}$$

for positive integers n. But multiplying the relations

$$\Gamma_p(1-j) = \tau_p(-j)\Gamma_p(-j)$$

from 3) yields

$$\begin{aligned} \frac{1}{\Gamma_p(-n)} &= \prod_{j=1}^{n} \tau_p(-j) \\ &= \prod_{p|j}(-1) \prod_{\gcd(p,j)=1} j \\ &= (-1)^{[n/p]}(-1)^{n+1}\Gamma_p(n+1) \end{aligned}$$

and the result follows. □

3.B.8 Mahler expansions and discrete antiderivatives

Exploiting Mahler's theorem, we develop basic p-adic calculus in this section. This will then be applied to the p-adic Gamma function and then to establish some fairly deep congruences. We start by proving that any continuous p-adic function has a continuous (discrete) antiderivative.

Theorem 3.B.34. *Let $f : \mathbb{Z}_p \to \mathbb{C}_p$ be a continuous function. There exists a unique continuous function $Sf : \mathbb{Z}_p \to \mathbb{C}_p$ such that $Sf(0) = 0$ and*

$$Sf(x+1) - Sf(x) = f(x)$$

for all $x \in \mathbb{Z}_p$. Moreover $a_n(Sf) = a_{n-1}(f)$ for $n \geq 1$, where $a_n(g)$ are Mahler's coefficients of g.

Proof. This is very easy using Mahler's theorem 3.B.27. Namely, look for

$$Sf(x) = \sum_{n \geq 0} b_n \binom{x}{n}$$

and observe that

$$Sf(x+1) - Sf(x) = \sum_{n \geq 1} b_n \binom{x}{n-1} = \sum_{n \geq 0} b_{n+1} \binom{x}{n}.$$

Taking into account the condition that $Sf(0) = 0$, we have $b_0 = 0$. Also, Sf satisfies the desired equation if and only if $b_{n+1} = a_n(f)$ for all $n \geq 0$. Thus any solution must satisfy $a_n(Sf) = a_{n-1}(f)$ for $n \geq 1$ and $a_0(Sf) = 0$, yielding uniqueness. But this also gives the existence, since if $a_n(f) \to 0$, then $a_{n-1}(f) \to 0$ and so $Sf(x) = \sum_{n \geq 0} b_n \binom{x}{n}$ is a continuous function if f is. □

Next, we define the notion of integrable and C^1 class functions. Unfortunately, there isn't really a very good notion of p-adic integration and each such construction has its limitations. For instance, the one we have chosen in this book has the drawback that not all continuous functions are integrable. But since we will deal only with sufficiently nice functions, this will be enough for our purposes.

Definition 3.B.35. 1) A function $f : \mathbb{Z}_p \to \mathbb{C}_p$ is called Volkenborn integrable (or simply integrable) if

$$\int_{\mathbb{Z}_p} f(x)dx = \lim_{n \to \infty} \frac{1}{p^n} \sum_{j=0}^{p^n - 1} f(j)$$

exists in $\mathbb{C}_p$.

2) Say a continuous function $f : \mathbb{Z}_p \to \mathbb{C}_p$ is of class C^1 if

$$\lim_{n \to \infty} n \cdot |a_n(f)|_p = 0.$$

These definitions might seem a bit strange at first. However, the average whose limit defines the integral of f should be seen as a Riemann sum. So Volkenborn integrable functions are precisely those for which Riemann sums converge. The definition of C^1 class functions is a bit more subtle, but one can prove (although this is nontrivial) that it agrees with the intuitive definition. We will prove part of this assertion when proving the next theorem. The following remark will play a very important role in the proof.

Remark 3.B.36. If $x \neq y \in \mathbb{Z}_p$, then

$$\frac{1}{x-y}\left(\binom{x}{n} - \binom{y}{n}\right) \in \frac{1}{\operatorname{lcm}(1,2,\ldots,n)}\mathbb{Z}_p.$$

Indeed, Vandermonde's identity (proposition 3.B.21) yields

$$\begin{aligned}\frac{1}{x-y}\left(\binom{x}{n} - \binom{y}{n}\right) &= \frac{1}{x-y}\cdot\sum_{j=0}^{n-1}\binom{y}{j}\cdot\binom{x-y}{n-j}\\ &= \sum_{j=0}^{n-1}\frac{1}{n-j}\binom{y}{j}\cdot\binom{x-y-1}{n-j-1},\end{aligned}$$

which immediately yields the result (using the fact that $\binom{x}{n} \in \mathbb{Z}_p$ if $x \in \mathbb{Z}_p$). Note that

$$|\operatorname{lcm}(1,2,\ldots,n)|_p = p^{-[\log_p n]} \geq \frac{1}{n},$$

so we obtain the very useful estimate

$$\left|\binom{x}{n} - \binom{y}{n}\right|_p \leq n|x-y|_p.$$

The following result is fairly technical, so the reader might want to skip the proof at a first reading.

Theorem 3.B.37. *A function f of class C^1 is integrable and*

$$\int_{\mathbb{Z}_p} f(x)dx = \sum_{n\geq 0}\frac{(-1)^n}{n+1}a_n(f).$$

Moreover, in this case we also have:

1) *f is continuously differentiable, i.e. $f'(x) = \lim_{u\to x} \frac{f(u)-f(x)}{u-x}$ exists in $\mathbb{C}_p$ for all x and $x \mapsto f'(x)$ is continuous.*

2) *We have*

$$S(f')(x) = \int_{\mathbb{Z}_p} (f(x+u) - f(u))du.$$

Proof. The observation that

$$\begin{aligned}\frac{1}{p^n}\sum_{j=0}^{p^n-1} f(j) &= \frac{1}{p^n} Sf(p^n)\\ &= \frac{1}{p^n}\sum_{k\geq 1} a_{k-1}(f)\binom{p^n}{k}\\ &= \sum_{k=1}^{\infty} \frac{a_{k-1}(f)}{k}\binom{p^n-1}{k-1},\end{aligned}$$

yields the estimate

$$\left|\frac{1}{p^n}\sum_{j=0}^{p^n-1} f(j) - \sum_{k\geq 0}\frac{(-1)^k}{k+1}a_k(f)\right|_p \leq \sup_{k\geq 1} x_{n,k},$$

where

$$x_{n,k} = \frac{|a_{k-1}(f)|_p}{|k|_p}\left|\binom{p^n-1}{k-1} - (-1)^{k-1}\right|_p.$$

Thus, to establish the first formula of the theorem, it is enough to check that $\sup_k x_{n,k}$ tends to 0 as $n \to \infty$. Clearly, $\frac{1}{|k|_p} \leq k$, so $x_{n,k} \leq k|a_{k-1}(f)|_p$. Moreover, since $(-1)^{k-1} = \binom{-1}{k-1}$, remark 3.B.36 yields $x_{n,k} \leq |a_{k-1}(f)|_p \frac{k^2}{p^n}$. Putting everything together, we deduce that

$$x_{n,k} \leq k|a_{k-1}(f)|_p \cdot \min\left(1, \frac{k}{p^n}\right).$$

Combined with $k|a_k(f)|_p \to 0$, this easily proves that $\sup_k x_{n,k} \to 0$, establishing the first formula.

1) By remark 3.B.36, there are polynomials $P_n \in \mathbb{Q}[X, Y]$ such that

$$\frac{\binom{u}{n} - \binom{x}{n}}{u - x} = P_n(u, x)$$

for all u, x and $|P_n(u, x)|_p \le n$. Since $n|a_n(f)|_p \to 0$, it follows that the series

$$\frac{f(u) - f(x)}{u - x} = \sum_{n \ge 0} a_n(f) P_n(u, x)$$

converges uniformly and its limit as $u \to x$ is $\sum_{n\ge 0} a_n(f) P_n(x, x)$, which is continuous (as the series converges uniformly).

2) First, we compute the Mahler coefficients of $g_x(u) = f(x + u) - f(u)$ by using Vandermonde's identity 3.B.21:

$$\begin{aligned} g_x(u) &= \sum_{n\ge 0} a_n(f) \binom{x+u}{n} - \sum_{n\ge 0} a_n(f) \binom{u}{n} \\ &= \sum_{n \ge 1} a_n(f) \sum_{i=1}^{n} \binom{x}{i} \binom{u}{n-i} \\ &= \sum_{k} a_k(g_x) \binom{u}{k}, \end{aligned}$$

where

$$a_k(g_x) = \sum_{n>k} a_n(f) \binom{x}{n-k}.$$

Thus $k|a_k(g_x)|_p \le \sup_{n>k} n|a_n(f)|_p$, so g_x is of class C^1 and we can express

$$G(x) = \int_{\mathbb{Z}_p} (f(x + u) - f(u)) du$$

in terms of Mahler coefficients of g_x, using the first formula of the theorem:

$$G(x) = \sum_{m=1}^{\infty} \left[\sum_{n=0}^{\infty} \frac{(-1)^n}{n+1} a_{m+n}(f) \right] \binom{x}{m}.$$

It is apparent from this formula that G is continuous. Finally, if $x \geq 1$ is an integer, we can write by definition

$$\begin{aligned} G(x) &= \lim_{n\to\infty} \frac{1}{p^n} \sum_{j=0}^{p^n-1} (f(x+j) - f(j)) \\ &= \lim_{n\to\infty} \frac{1}{p^n} \sum_{j=0}^{x-1} (f(p^n+j) - f(j)) \\ &= \sum_{j=0}^{x-1} f'(j) = Sf'(x). \end{aligned}$$

The conclusion follows now by continuity of G and Sf' together with the density of the set of positive integers in $\mathbb{Z}_p$. □

3.B.9 Application to the p-adic Γ-function

We are now ready to prove the following deep theorem, that will allow us to prove very difficult congruences:

Theorem 3.B.38. $\log_p \circ \Gamma_p$ *is an analytic function on* $p\mathbb{Z}_p$ *and has a power-series expansion*

$$\log_p \circ \Gamma_p(x) = \lambda_0 x - \sum_{n\geq 1} \frac{\lambda_n}{2n(2n+1)} x^{2n+1},$$

where

$$\lambda_0 = \int_{\mathbb{Z}_p^*} \log_p(u)du, \quad \lambda_n = \int_{\mathbb{Z}_p^*} x^{-2n} dx = \lim_{k\to\infty} \frac{1}{p^k} \sum_{\substack{i=1 \\ \gcd(i,p)=1}}^{p^k-1} i^{-2n}.$$

Proof. In this proof we will use the simpler notation $|x|$ for $|x|_p$. Using proposition 3.B.33, we deduce that if $f(x) = \log_p(\Gamma_p(x))$, then

$$f(x+1) - f(x) = 1_{|x|=1} \log_p(x).$$

This functional equation combined with the fact that $f(0) = 0$ (this follows from proposition 3.B.33) shows that $f = SA$, where $A(x) = 1_{|x|=1}\log_p(x)$. But it is easy to find an antiderivative of A: reasoning from the original power series, one chooses $B(x) = 1_{|x|=1}(x \cdot \log_p(x) - x)$ and checks that $B' = A$. Note that thanks to the hypothesis $x \in p\mathbb{Z}_p$, we have $|u|_p = 1$ if and only if $|u + x|_p = 1$, for all $u \in \mathbb{Z}_p$. Therefore, theorems 3.B.30 and 3.B.37 yield the integral representation of f:

$$\log_p(\Gamma_p(x)) = \int_{\mathbb{Z}_p} 1_{|u|=1}((x+u)\log_p(x+u) - (x+u) - u\log_p u + u)du.$$

Using the fact that $\log_p(x+u) = \log_p u + \log_p(1 + x/u)$ and expanding

$$\log_p(1 + x/u) = \sum_{n\geq 1} \frac{(-1)^{n-1}}{n} \cdot \frac{x^n}{u^n}$$

for $|u| = 1$ yields

$$\begin{aligned}
&(x+u)\log_p(x+u) - x - u\log_p u \\
&\quad = (x+u)\log_p u + (x+u)\cdot \sum_{n\geq 1} \frac{(-1)^{n-1}}{n} \cdot \frac{x^n}{u^n} - x - u\log_p u \\
&\quad = x\log_p u - x + \sum_{n\geq 1} \frac{(-1)^{n-1}}{n} \cdot \frac{x^{n+1}}{u^n} + x + \sum_{n\geq 2} \frac{(-1)^{n-1}}{n} \cdot \frac{x^n}{u^{n-1}}
\end{aligned}$$

and an easy calculation (based on the change of variable $n - 1 = j$ in the last sum) yields

$$1_{|u|=1}((x+u)\log_p(x+u) - (x+u) - u\log_p u + u) = x \cdot 1_{|u|=1}\log_p u + \sum_{n\geq 1} f_{n,x}(u),$$

where

$$f_{n,x}(u) = \frac{(-1)^{n-1}x^{n+1}}{n(n+1)} 1_{|u|=1}u^{-n}.$$

Of course, we integrate this, but the difficult point is to check that the integral commutes with the infinite sum. Again using theorems 3.B.30 and 3.B.37, we can write

$$\begin{aligned}\sum_{n\geq 1}\int_{\mathbb{Z}_p} f_{n,x}(u)du &= \sum_{n\geq 1}\sum_{k\geq 0}\frac{(-1)^k}{k+1}a_k(f_{n,x})\\ &= \sum_{k\geq 0}\frac{(-1)^k}{k+1}\sum_{n\geq 1}a_k(f_{n,x})\\ &= \sum_{k\geq 0}\frac{(-1)^k}{k+1}a_k\left(\sum_n f_{n,x}\right)\\ &= \int_{\mathbb{Z}_p}\left(\sum_n f_{n,x}\right),\end{aligned}$$

all manipulations being easily justified by the fact that $f_{n,x}$ takes small values (and so its Mahler coefficients are small). We finally deduce that

$$\log_p \Gamma_p(x) = x\cdot\lambda_0 + \sum_{n\geq 1}\int_{\mathbb{Z}_p} f_{n,x}(u)du.$$

It remains to simplify this a little bit thanks to the following

Lemma 3.B.39. *If g is of class C^1 and odd, then*

$$\int_{\mathbb{Z}_p} g(u)du = -\frac{g'(0)}{2}.$$

Proof. We leave this as an easy exercise to the reader. We give a hint, however: write

$$g(p^n - j) = -g(j-p^n) = -(g(j) - p^n g'(j) + o(p^n))$$

and observe that $\sum_{j=0}^{p^k-1} g'(j) = (Sg')(p^k)$. □

Applying this to the functions $f_{n,x}$ with n odd, we obtain that

$$\int_{\mathbb{Z}_p} f_{n,x}(u)du = 0$$

for n odd. Putting everything together, the result follows. □

We would like to have more information about the numbers λ_n that appear in the previous theorem. A first step in doing this is the following lemma:

Lemma 3.B.40. *We have*

$$\lambda_n - (1 - p^{2n-1}) \int_{\mathbb{Z}_p} x^{2n} dx \in \mathbb{Z}_p.$$

Proof. Using the fact that $g \to g^{-1}$ is a permutation of the finite group $(\mathbb{Z}/p^k\mathbb{Z})^*$, we deduce that

$$\sum_{\substack{i=1 \\ \gcd(i,p)=1}}^{p^k-1} i^{-2n} \equiv \sum_{\substack{i=1 \\ \gcd(i,p)=1}}^{p^k-1} i^{2n} = \sum_{i=0}^{p^k-1} i^{2n} - p^{2n} \sum_{i=0}^{p^{k-1}-1} i^{2n} \pmod{p^k\mathbb{Z}_p}.$$

Dividing by p^k and letting $k \to \infty$, the result follows. □

Next, let

$$B_n = \int_{\mathbb{Z}_p} x^n dx.$$

By linearity of the Volkenborn integral we can write for $n > 0$

$$\begin{aligned}
\sum_{k=0}^{n} \binom{n+1}{k} B_k &= \int_{\mathbb{Z}_p} \left(\sum_{k=0}^{n} \binom{n+1}{k} x^k \right) dx \\
&= \int_{\mathbb{Z}_p} ((x+1)^{n+1} - x^{n+1}) dx \\
&= \lim_{k\to\infty} \frac{1}{p^k} \sum_{j=0}^{p^k-1} ((j+1)^{n+1} - j^{n+1}) \\
&= \lim_{k\to\infty} p^{nk} \\
&= 0,
\end{aligned}$$

which immediately yields the exponential generating function of the sequence $(B_n)_n$:

$$\sum_{n\geq 0} \frac{B_n}{n!} X^n = \frac{X}{e^X - 1}.$$

This allows us to compute $B_0 = 1, B_1 = -\frac{1}{2}, B_2 = \frac{1}{6}$ and to deduce that all B_n are rational numbers. Actually, these numbers B_n are called Bernoulli numbers and appear all over the place in mathematics. Their rather subtle arithmetic properties will help us prove the following important

Theorem 3.B.41. *One has $p\lambda_n \in \mathbb{Z}_p$ for all $n \geq 1$. Moreover, if $p > 3$, then $\lambda_1 \in \mathbb{Z}_p$.*

Proof. Using the previous lemma and discussion, it suffices to prove that $pB_n \in \mathbb{Z}_p$ for all n (the second part follows from the observation that $B_2 = \frac{1}{6}$ is prime to p if $p > 3$, so we will not say more about it). We will prove this by induction on n. For $n = 1$, it is clear, so assume that it holds for $j < n$. Recall that

$$\frac{X}{e^X - 1} = \sum_{n \geq 0} \frac{B_n}{n!} X^n.$$

Let k be a positive integer. Multiplying this by e^{kX} yields the equality

$$\frac{e^{kX} X}{e^X - 1} = \sum_{n \geq 0} \frac{X^n}{n!} \sum_{i=0}^{n} \binom{n}{i} B_i k^{n-i}.$$

On the other hand,

$$\frac{e^{kX} X}{e^X - 1} = \frac{X}{e^X - 1} + X(1 + e^X + \cdots + e^{(k-1)X}).$$

Identifying the coefficients of X^{n+1} in this equality shows that for all $n \geq 1$

$$1 + 2^n + \cdots + (k-1)^n = \frac{1}{n+1} \sum_{i=0}^{n} \binom{n+1}{i} B_i k^{n+1-i}.$$

Take $k = p$ and note that $\frac{p^{n+1-i}}{n+1}\binom{n+1}{i} B_i = (pB_i)\binom{n}{i}\frac{p^{n-i}}{n+1-i} \in \mathbb{Z}_p$ for all $0 \leq i \leq n-1$. Combining this observation and the previous equality, we deduce that $pB_n \in \mathbb{Z}_p$, finishing the proof of the theorem. $\square$

3.B.10 Some deep congruences

p-adic numbers play a crucial role in establishing difficult congruences, which are almost impossible to get by other means. The following theorem is a very famous such example. We present here Robert's and Zuber's nice proof, following [68].

Theorem 3.B.42. *(Kazandzidis) If $p > 3$, then for all positive integers n, k we have*

$$\binom{pn}{pk} \equiv \binom{n}{k} \pmod{p^j},$$

where $j = 3 + v_p(nk(n-k)) + v_p(\binom{n}{k})$.

Proof. Proposition 3.B.33 shows that for all positive integers x we have

$$\Gamma_p(px) = (-1)^{px} \cdot \frac{(px)!}{p^x \cdot x!}.$$

Therefore, if we denote $x = kp$ and $y = (n-k)p$, we have

$$\frac{\binom{np}{kp}}{\binom{n}{k}} = \frac{\Gamma_p(x+y)}{\Gamma_p(x)\Gamma_p(y)}.$$

The definition of $\log_p(x)$ as a power series shows that $|x-1|_p = |\log_p(x)|_p$ for all $x \in 1 + p\mathbb{Z}_p$. So, if we let $f(x) = \log_p(\Gamma_p(x))$, the congruence we have to prove is equivalent to

$$v_p(f(x+y) - f(x) - f(y)) \geq v_p(xy(x+y)).$$

But theorem 3.B.38 yields

$$\begin{aligned} &v_p(f(x+y) - f(x) - f(y)) \\ &= v_p\left(\sum_{n\geq 1} \frac{\lambda_n}{2n(2n+1)}[(x+y)^{2n+1} - x^{2n+1} - y^{2n+1}]\right) \\ &\geq \inf_{n\geq 1} v_p(a_n), \end{aligned}$$

where

$$a_n = \frac{\lambda_n}{2n(2n+1)}[(x+y)^{2n+1} - x^{2n+1} - y^{2n+1}].$$

It is thus enough to prove that $v_p(a_n) \geq v_p(xy(x+y))$ for all $n \geq 1$. For $n = 1$, this follows from $\lambda_1 \in \mathbb{Z}_p$, which was proved in theorem 3.B.41. So, suppose that $n \geq 2$. Since $(X+1)^{2n+1} - X^{2n+1} - 1$ vanishes at 0 and -1, there is $f \in \mathbb{Z}[X]$ of degree $2n-2$ such that

$$(1+X)^{2n+1} - X^{2n+1} - 1 = X(1+X)f(X).$$

Since $y^{2n-2}f(x/y)$ is a homogeneous polynomial of degree $2n-2$ with integer coefficients in $x, y \in p\mathbb{Z}_p$, we deduce that

$$\begin{aligned} v_p((x+y)^{2n+1} - x^{2n+1} - y^{2n+1}) &= v_p(xy(x+y)) + v_p\left(y^{2n-2}f\left(\frac{x}{y}\right)\right) \\ &\geq v_p(xy(x+y)) + 2n - 2. \end{aligned}$$

It is thus enough to prove that

$$v_p(\lambda_n) + 2n - 2 \geq v_p(2n(2n+1)),$$

which follows easily from $v_p(\lambda_n) \geq -1$ (theorem 3.B.41) and the inequalities

$$v_p(2n(2n+1)) \leq \max(v_p(2n), v_p(2n+1)) \leq \log_5(2n+1) \leq 2n-3,$$

the last one being true for $n \geq 2$. □

Similar techniques apply to prove the following nice congruence, taken from [17].

Theorem 3.B.43. *If $p > 5$ and $n \geq 1$, then*

$$\frac{(np)!}{p^n \cdot n!} \equiv ((p-1)!)^n \pmod{p^{3+v_p(n^3-n)}}$$

and this does not hold modulo $p^{4+v_p(n^3-n)}$ unless p is a Wolstenlholme prime, i.e. $\binom{2p-1}{p-1} \equiv 1 \pmod{p^4}$.

There are similar sharp congruences for $p = 2, 3, 5$, for which we refer the reader to [17].

Proof. (sketch) As in the proof of the previous theorem, one starts by noting that

$$\Gamma_p(np) = (-1)^{np}\frac{(np)!}{p^n \cdot n!}, \quad \Gamma_p(p) = -(p-1)!,$$

so the congruence reduces to

$$v_p\left(\Gamma_p(np) - \Gamma_p(p)^n\right) \geq 3 + v_p(n^3 - n).$$

If $f(x) = \log_p(\Gamma_p(x))$, the equality $v_p(y-1) = v_p(\log_p y)$ (true for $y \in 1+p\mathbb{Z}_p$) reduces the problem to showing that

$$v_p(f(np) - nf(p)) \geq 3 + v_p(n^3 - n).$$

One uses again theorem 3.B.38, but one also has to analyze the term containing λ_2. As the details are routine, they are left to the reader. □

Here is yet another similar congruence, with the same idea.

Theorem 3.B.44. *(F.Rodrigo-Villegas) Let $(a_n)_n$ be a sequence of integers, finitely many of which are nonzero and such that $\sum_{n\geq 1} na_n = 0$. Define*

$$A(n) = \prod_{k\geq 1}(kn)!^{a_k}.$$

Then for any prime $p > 3$ and any $s \geq 1$ we have

$$\frac{A(p^s)}{A(p^{s-1})} \equiv 1 \pmod{p^{3s}}.$$

Proof. (sketch) The point is to study $\log_p\left(\frac{A(p^s)}{A(p^{s-1})}\right)$, by expressing it in terms of values of the function $x \mapsto \log_p \Gamma_p(x)$. For instance, if $s = 1$, then this equals $\sum_k a_k \log_p \Gamma_p(kp)$. One finishes as usual by using the analytic expansion of $f(x)$. □

Theorem 3.B.45. *Let n be a multiple of an odd prime p. Then we have an equality in $\mathbb{Q}_p$:*

$$\sum_{\substack{k=1 \\ \gcd(k,p)=1}}^{n-1} \frac{1}{k} = -\sum_{j\geq 1} \frac{\lambda_j}{2j} n^{2j},$$

where λ_j are as in theorem 3.B.38.

Proof. (sketch) Note that $f(x) = \log_p(\Gamma_p(x))$ is locally analytic on $\mathbb{Z}_p$, as follows from theorem 3.B.38 and from the functional equation

$$f(x+1) - f(x) = 1_{|x|_p=1} \log_p x.$$

It follows easily that f is differentiable and that $g = f'$ satisfies

$$g(x+1) - g(x) = 1_{|x|_p=1} \frac{1}{x}.$$

Thus

$$\sum_{k=1,\gcd(k,p)=1}^{n-1} \frac{1}{k} = g(n) - g(1).$$

Next, by differentiating the Taylor expansion of f in theorem 3.B.38, we obtain that for $x \in p\mathbb{Z}_p$

$$g(x) = \lambda_0 - \sum_{j\geq 1} \frac{\lambda_j}{2j} x^{2j}.$$

The result follows by combining these two identities (with $x = n$ in the second one) and by observing that $g(1) = g(0) = \lambda_0$. □

3.B.11 More on Bernoulli numbers

In this section we focus on some classical congruences satisfied by Bernoulli numbers. Recall from section 3.B.9 that they are defined by $B_n = \int_{\mathbb{Z}_p} x^n dx$. We saw that $(B_n)_n$ is a sequence of rational numbers, with exponential generating function

$$\sum_{n\geq 0} B_n \frac{X^n}{n!} = \frac{X}{e^X - 1}.$$

Lemma 3.B.39 shows that $B_n = 0$ for all odd numbers $n > 1$ (this can also be deduced from the generating function, by checking that the function $x \to \frac{x}{e^x-1} + \frac{x}{2}$ is even). We saw during the proof of theorem 3.B.41 that Bernoulli numbers satisfy the crucial identity

$$S_n(k) = 1^n + 2^n + \cdots + (k-1)^n = \frac{1}{n+1}\sum_{i=0}^{n}\binom{n+1}{i}B_i k^{n+1-i}.$$

This identity will play a crucial role in the proof of the following beautiful result, which considerably refines theorem 3.B.41.

Theorem 3.B.46. *(von Staudt-Clausen theorem) For all $n \geq 1$ we have*

$$B_{2n} + \sum_{p-1|2n} \frac{1}{p} \in \mathbb{Z}.$$

Proof. Let p be any prime. We need the following elementary result:

Lemma 3.B.47. *If n is even, then for all $k \geq 1$ we have $S_n(p^{k+1}) \equiv p \cdot S_n(p^k) \pmod{p^{k+1}}$.*

Proof. This is a simple application of the binomial formula:

$$\begin{aligned} S_n(p^{k+1}) &= \sum_{j=0}^{p^k-1}\sum_{l=0}^{p-1}(j + l\cdot p^k)^n \\ &\equiv \sum_{j=0}^{p^k-1}\sum_{l=0}^{p-1}(j^n + nlp^k \cdot j^{n-1}) \\ &\equiv pS_n(p^k) + \frac{n}{2}(p-1)p^{k+1}S_{n-1}(p^k) \\ &\equiv pS_n(p^k) \pmod{p^{k+1}}. \end{aligned}$$

Note that if p is odd, then the hypothesis that n is even is useless. □

Next, we claim that $\frac{S_{2n}(p)}{p} - B_{2n} \in \mathbb{Z}_p$ for all primes p and all n. Note that the previous lemma implies that $\frac{S_{2n}(p^{k+1})}{p^{k+1}} - \frac{S_{2n}(p^k)}{p^k} \in \mathbb{Z}_p$ for all $k \geq 1$, hence it is enough to show that we can find k such that $\frac{S_{2n}(p^k)}{p^k} - B_{2n} \in \mathbb{Z}_p$. But

$$\frac{S_{2n}(p^k)}{p^k} - B_{2n} = \frac{1}{2n+1} \sum_{j=0}^{2n-1} \binom{2n+1}{j} B_j p^{k(2n-j)}$$

is certainly in $\mathbb{Z}_p$ for k large enough, as each of the terms in the sum is in $\mathbb{Z}_p$. This proves the claim.

Finally, a standard argument[4] shows that $S_n(p) \equiv 0 \pmod{p}$ unless $p-1$ divides n, in which case $S_n(p) \equiv -1 \pmod{p}$. Combining this with the result of the previous paragraph, we deduce that $B_{2n} \in \mathbb{Z}_p$ unless $p-1|2n$, in which case $B_{2n} + \frac{1}{p} \in \mathbb{Z}_p$. All in all, this shows that the rational number $B_{2n} + \sum_{q-1|2n} \frac{1}{q}$ belongs to $\mathbb{Z}_p$ for all primes p and so it is an integer. The result follows. □

The following classical result is considerably more difficult to prove. The proof is actually rather mysterious, though elementary. It is highly related to the existence of p-adic analogues of the Riemann zeta function, though this may not be apparent at all...

Theorem 3.B.48. *(Kummer's congruences) Let m, n be positive integers and let p be a prime such that $p-1$ does not divide n, but $p-1$ divides $m-n$. Then $\frac{B_m}{m} - \frac{B_n}{n} \in p\mathbb{Z}_p$.*

Proof. It suffices to prove that $\frac{B_k}{k} \equiv \frac{B_{k+p-1}}{k+p-1} \pmod{p}$ whenever $p-1$ does not divide $k \geq 1$. Note that it is not even clear that $\frac{B_k}{k} \in \mathbb{Z}_p$, but this will be the case (of course, this is clear if k is odd, as then $B_k = 1$). Let us fix an integer $1 \leq a \leq p-1$ such that a is a primitive root modulo p. The following lemma is crucial.

---

[4]Let g be a primitive root mod p. If $p-1$ does not divide n, then $g^n \neq 1$ and

$$S_n(p) \equiv \sum_{i=0}^{p-2} g^{in} = \frac{g^{(p-1)n} - 1}{g^n - 1} = 0.$$

Lemma 3.B.49. *There exist $b_n \in \mathbb{Z}_p$ such that (as formal series)*

$$\frac{a}{e^{aX}-1} - \frac{1}{e^X-1} = \sum_{n\geq 0} b_n (e^X-1)^n.$$

Proof. By changing the variable $Y = e^X - 1$, this is equivalent to

$$\frac{a}{(1+Y)^a-1} - \frac{1}{Y} = \sum_{n\geq 0} b_n Y^n.$$

The binomial formula and the fact that $\gcd(a,p) = 1$ yield the existence of $g \in Y\mathbb{Z}_p[Y]$ such that $(1+Y)^a - 1 = aY(1+g(Y))$. We then have

$$\frac{a}{(1+Y)^a-1} - \frac{1}{Y} = \frac{1}{Y}\left(\frac{1}{1+g(Y)} - 1\right) = \sum_{n\geq 1} (-1)^n \frac{1}{Y} g(Y)^n.$$

The result follows, as $g \in Y\mathbb{Z}_p[Y]$. □

Let use denote $U_n = (a^n - 1)\cdot \frac{B_n}{n}$ and observe that by definition of $(B_n)_n$ we have

$$\frac{a}{e^{aX}-1} - \frac{1}{e^X-1} = \frac{1}{X}\sum_{n\geq 0} nU_n \frac{X^n}{n!} = \sum_{n\geq 0} U_{n+1}\frac{X^n}{n!}.$$

To fully exploit lemma 3.B.49, let us denote

$$s_{n,k} = k!\cdot [X^k](e^X-1)^n = \sum_{j=0}^{n} (-1)^{n-j}\binom{n}{j} j^k,$$

so that

$$(e^X-1)^n = \sum_{k\geq 0} s_{n,k}\frac{X^k}{k!}.$$

Replacing these relations in lemma 3.B.49 and identifying coefficients yields[5]

$$U_{n+1} = \sum_{j\geq 0} b_j \cdot s_{j,n}$$

[5] Note that there are no convergence issues, as $s_{n,k} = 0$ for $n > k$.

for all $n \geq 0$.

It is now easy to conclude. Recall that $p-1$ does not divide k, so p does not divide $a^k - 1$. The previous relation and the fact that $b_j, s_{j,n} \in \mathbb{Z}_p$ for all j, n show that $U_k \in \mathbb{Z}_p$ and so $\frac{B_k}{k} \in \mathbb{Z}_p$. The same argument shows that $\frac{B_{k+p-1}}{k+p-1} \in \mathbb{Z}_p$. Also, Fermat's little theorem yields $s_{j,n} \equiv s_{j,n+p-1} \pmod{p}$ for all $j \geq 0$ and all $n \geq 1$. Hence $U_n \equiv U_{n+p-1} \pmod{p}$ for all $n \geq 1$. As $a^{k+p-1} \equiv a^k \pmod{p}$, this congruence is equivalent to

$$(a^k - 1)\frac{B_k}{k} \equiv (a^k - 1)\frac{B_{k+p-1}}{k+p-1} \pmod{p}.$$

The desired result follows from this congruence and the fact that p does not divide $a^k - 1$. □

Remark 3.B.50. Here is the real meaning of this proof. As we have already said, the crucial fact is that $f_a(X) = \frac{a}{(1+X)^a - 1} - \frac{1}{X} \in \mathbb{Z}_p[[X]]$, say $f_a(X) = \sum_{n\geq 0} b_n X^n$. This allows us to define a measure μ_a on $\mathbb{Z}_p$ (i.e. continuous linear form on the space of continuous functions $g : \mathbb{Z}_p \to \mathbb{Q}_p$) by

$$\int_{\mathbb{Z}_p} g\mu_a = \sum_{n\geq 0} a_n(g) b_n$$

for all continuous functions $g : \mathbb{Z}_p \to \mathbb{Q}_p$. Here $a_n(g)$ is the nth Mahler coefficient of g and the series converges by Mahler's theorem (as $\lim_{n\to\infty} a_n(g) = 0$ and $b_n \in \mathbb{Z}_p$). Since $b_n \in \mathbb{Z}_p$ for all n, Mahler's theorem shows that

$$v_p\left(\int_{\mathbb{Z}_p} g\mu_a\right) \geq \min_{x\in\mathbb{Z}_p} v_p(g(x))$$

for all continuous maps g.

Note that $s_{n,k} = a_n(x^k)$, so the equality $U_{n+1} = \sum_{j\geq 0} b_j s_{j,n}$ established during the proof can be written as

$$U_{n+1} = \int_{\mathbb{Z}_p} x^n \mu_a.$$

Since $v_p(x^n - x^{n+p-1}) \geq 1$ for all $x \in \mathbb{Z}_p$ and $n \geq 1$, the previous paragraph shows that $U_n \equiv U_{n+p-1} \pmod{p}$ for all $n \geq 1$.

This new interpretation of the proof of Kummer's congruence yields other nontrivial congruences. Assume for instance that $m \equiv n \pmod{p^N(p-1)}$ and $m, n > N$. By Euler's theorem we have $v_p(x^m - x^n) \geq N+1$ for all $x \in \mathbb{Z}_p$ and we deduce that $U_m \equiv U_n \pmod{p^{N+1}}$. This observation is the beginning of the construction of the p-adic zeta function of Kubota and Leopoldt.

We end this addendum with an application of the previous results to harmonic numbers. It is standard that for any prime $p > 3$ we have

$$\sum_{k=1}^{p-1} \frac{1}{k} \equiv 0 \pmod{p^2} \quad \text{and} \quad \sum_{k=1}^{p-1} \frac{1}{k^2} \equiv 0 \pmod{p}.$$

The following result considerably refines these congruences.

Theorem 3.B.51. *For all primes $p > 3$ we have*

$$\sum_{k=1}^{p-1} \frac{1}{k^2} \equiv \frac{2p}{3} B_{p-3} \pmod{p^2} \quad \textit{and} \quad \sum_{k=1}^{p-1} \frac{1}{k} \equiv -\frac{p^2}{3} B_{p-3} \pmod{p^3}.$$

Proof. Euler's theorem yields

$$\sum_{k=1}^{p-1} \frac{1}{k^2} \equiv \sum_{k=1}^{p-1} k^{\varphi(p^2)-2} \pmod{p^2}.$$

We will use the following general result:

Lemma 3.B.52. *If $p > 3$ and $n \geq 1$, then*

$$1^n + 2^n + \ldots + (p-1)^n \equiv pB_n + \frac{np^2 B_{n-1}}{2} \pmod{p^2}.$$

Proof. The left-hand side is equal to

$$\frac{1}{n+1} \sum_{k=1}^{n+1} \binom{n+1}{k} B_{n+1-k} p^k = pB_n + \frac{np^2 B_{n-1}}{2} + \sum_{k=3}^{n+1} \binom{n}{k-1} B_{n+1-k} \frac{p^k}{k}.$$

It is thus enough to prove that $\binom{n}{k-1} B_{n+1-k} \frac{p^k}{k} \in p^2 \mathbb{Z}_p$ for all $3 \leq k \leq n+1$. As $pB_{n+1-k} \in \mathbb{Z}_p$ (by the von Staudt-Clausen theorem) and $\binom{n}{k-1} \in \mathbb{Z}_p$, it is enough to check that $\frac{p^{k-3}}{k} \in \mathbb{Z}_p$. This is easy and left to the reader (here one crucially uses the hypothesis $p > 3$). $\square$

Applying the previous lemma to $n = \varphi(p^2) - 2$ and noting that $B_{n-1} = 0$ (as n is even), we obtain

$$\sum_{k=1}^{p-1} k^{\varphi(p^2)-2} \equiv pB_{\varphi(p^2)-2} \pmod{p^2}.$$

It remains to use Kummer's congruence to obtain $B_{\varphi(p^2)-2} \equiv \frac{2}{3} B_{p-3} \pmod{p}$ and finally

$$\sum_{k=1}^{p-1} \frac{1}{k^2} \equiv \frac{2p}{3} B_{p-3} \pmod{p^2}.$$

One could apply a similar method for the second congruence, but we prefer to deduce it from the first one. Note that

$$2\sum_{k=1}^{p-1} \frac{1}{k} = \sum_{k=1}^{p-1} \left(\frac{1}{k} + \frac{1}{p-k} \right) = p \sum_{k=1}^{p-1} \frac{1}{k(p-k)}.$$

Using this identity and the congruence we have just proved, it is enough to prove that

$$\begin{aligned} &\sum_{k=1}^{p-1} \left(\frac{1}{k(p-k)} + \frac{1}{k^2} \right) \equiv 0 \pmod{p^2} \\ \Leftrightarrow &\sum_{k=1}^{p-1} \frac{1}{k^2(p-k)} \equiv 0 \pmod{p} \\ \Leftrightarrow &\sum_{k=1}^{p-1} \frac{1}{k^3} \equiv 0 \pmod{p}. \end{aligned}$$

The last congruence is also standard and it is proved using primitive roots mod p. $\square$

Chapter 4

Primes and Squares

This chapter is concerned with arithmetic properties of primes of the form $4k+1$. It is very elementary and most problems use the following two results.

Theorem 4.1. *(Fermat) Any prime number of the form $4k+1$ can be written as the sum of squares of two integers.*

There are many proofs of this classical result, see for instance the first example in chapter 4 of [3]. For a different proof, using properties of Gauss and Jacobi sums, see the addendum 9.A.

Proposition 4.2. *Let p be a prime of the form $4k+3$ and let a and b be integers such that p divides a^2+b^2. Then p divides a and b.*

Proof. If p does not divide a, there exists c such that $ca \equiv 1 \pmod{p}$. Then $(bc)^2 \equiv -1 \pmod{p}$, thus

$$(-1)^{\frac{p-1}{2}} \equiv (bc)^{p-1} \equiv 1 \pmod{p},$$

the last congruence being Fermat's little theorem. But this is clearly impossible, since by hypothesis $(-1)^{\frac{p-1}{2}} = -1$. The result follows. □

An easy consequence of proposition 4.2 is that $v_p(a^2+b^2)$ is even for any prime p of the form $4k+3$ and for any integers a, b. This is a very useful tool when studying some diophantine equations, as the following problems show.

1. Prove that the number $4mn - m - n$ cannot be a perfect square if m and n are positive integers.

Fermat

Proof. If $4mn - m - n = x^2$ for some integer x, then

$$(4m-1)(4n-1) = (2x)^2 + 1.$$

But there exists a prime $p \equiv 3 \pmod 4$ such that $p|4m-1$. Then p divides $(2x)^2+1$, contradicting proposition 4.2. □

2. Prove that the equation $y^2 = x^5 - 4$ has no integer solutions.

Balkan Olympiad 1998

Proof. Write the equation as $x^5 + 2^5 = y^2 + 6^2$. We claim that there is always a prime $p \equiv 3 \pmod 4$ such that $v_p(x^5+2^5)$ is odd. Since for any such prime p we have $v_p(y^2+6^2) \equiv 0 \pmod 2$, we will thus get a contradiction. Note that x is odd, otherwise $x = 2x_1$, $y = 2y_1$ and 8 divides y_1^2+1, which is impossible. If $x \equiv 1 \pmod 4$, there is a prime $p \equiv 3 \pmod 4$ such that $v_p(x+2) \equiv 1 \pmod 2$. One cannot have $p|x^4-2x^3+4x^2-8x+16$ (otherwise p would divide $5 \cdot 2^4$), so $v_p(x^5+2^5) = v_p(x+2) \equiv 1 \pmod 2$ and we are done. If $x \equiv -1 \pmod 4$, then $x^4-2x^3+\ldots+16 \equiv 3 \pmod 4$ and we can repeat the argument by taking a prime $p \equiv 3 \pmod 4$ such that $v_p(x^4-2x^3+\cdots+16) \equiv 1 \pmod 2$. The claim being proved, the result follows. □

Proof. We can work modulo 11: since $x^{11} \equiv x \pmod{11}$ for any x, we deduce that $x^5 \equiv 0, 1, -1 \pmod{11}$ for any x. On the other hand, the quadratic residues mod 11 are trivial to find: those are $0, 1, 4, 9, 5, 3$. One readily checks that the equation has no solution mod 11 using these observations. So, it does not have integral solutions either. □

3. Solve in integers the equation $x^2 = y^7 + 7$.

Titu Andreescu, USA TST 2008

Proof. We clearly have no solutions for $y < -1$, so let us suppose that $y+2 > 0$. Taking the equation modulo 4, we easily obtain that $y \equiv 1 \pmod 4$. The key point is to rewrite the equation as $x^2 + 11^2 = y^7 + 2^7$ or, by factoring the right-hand side, as

$$x^2 + 11^2 = (y+2)(y^6 - 2y^5 + 4y^4 - 8y^3 + 16y^2 - 32y + 64).$$

Since $y \equiv 1 \pmod 4$, we have $y + 2 \equiv 3 \pmod 4$, thus there exists a prime q such that $v_q(y+2)$ is odd. Note that q does not divide

$$y^6 - 2y^5 + 4y^4 - 8y^3 + 16y^2 - 32y + 64,$$

as otherwise q would divide $7 \cdot 64$ and $x^2 + 11^2$, a contradiction. Thus $v_q(y^7 + 2^7)$ is odd, which is impossible, as it equals $v_q(x^2 + 11^2)$ and $q \equiv 3 \pmod 4$. The result follows. $\square$

4. Find all pairs (m, n) of positive integers such that

$$m^2 - 1 | 3^m + (n! - 2)^m.$$

Gabriel Dospinescu

Proof. First, assume that $n > 2$. Then m cannot be odd, since otherwise 8 would divide $m^2 - 1$, but $3^m + (n! - 2)^m$ is odd. So m is even. But then $m^2 - 1 \equiv -1 \pmod 4$, so there exists a prime $p \equiv -1 \pmod 4$ dividing $m^2 - 1$. But then p divides $3^m + (n! - 2)^m$ and since m is even, this implies that p divides 3 and $n! - 2$. Thus 3 divides $n! - 2$, a contradiction.

Thus, we must have $n = 1$ or $n = 2$. If $n = 1$ then either $m^2 - 1 | 3^m + 1$ and m is even or $m^2 - 1 | 3^m - 1$ and m is odd. In the first case we can use the same argument as before to get a contradiction (choose a prime $p \equiv -1 \pmod 4$ dividing $m^2 - 1$), while the second case is impossible since $3^m - 1$ is not a multiple of 8 when m is odd. Thus $n = 2$ and $m^2 - 1 | 3^m$. Thus there is $k \leq m$ such that $(m-1)(m+1) = 3^k$. But then $m - 1, m + 1$ are powers of 3 which differ by 2. Thus clearly $m - 1 = 1$ and $m = 2$. We deduce that $m = n = 2$ is the only solution of the problem. $\square$

5. Find all pairs (x, y) of positive integers such that the number $\dfrac{x^2+y^2}{x-y}$ is a divisor of 1995.

Bulgaria 1995

Proof. Note that $1995 = 3 \cdot 5 \cdot 7 \cdot 19$. The key observation is that $3, 7, 19$ are all primes of the form $4k+3$. If p is any of these primes and if p divides $\frac{x^2+y^2}{x-y}$, then p divides x^2+y^2 and so it divides x, y. Writing $x = px_1, y = py_1$, we obtain

$$\frac{x^2+y^2}{x-y} = p\frac{x_1^2+y_1^2}{x_1-y_1}.$$

Doing this for every prime factor p of $\frac{x^2+y^2}{x-y}$ and noting that $x^2+y^2 = x-y$ has no solutions with $x, y \geq 1$, we just have to solve the equation $\frac{x^2+y^2}{x-y} = 5$. This is easy, since we can write it as $(2x-5)^2 + (2y+5)^2 = 50$ and since 50 can only be written as sum of two squares in two ways $1+49 = 25+25$. Thus $2y+5 \in \{-7, -5, -1, 1, 5, 7\}$ and since y is positive we must have $y = 1$. But then $x = 2$ or $x = 3$. Putting everything together, we deduce that the solutions are all pairs $(2k, k)$, $(3k, k)$ with $k \in \{1, 3, 7, 19, 21, 57, 133, 399\}$. □

6. Find all n-tuples $(a_1, a_2, \ldots, a_n)$ of positive integers such that

$$(a_1! - 1)(a_2! - 1) \ldots (a_n! - 1) - 16$$

is a perfect square.

Gabriel Dospinescu

Proof. Suppose that

$$(a_1! - 1)(a_2! - 1) \ldots (a_n! - 1) - 16 = k^2.$$

First, we claim that $a_i \in \{2, 3\}$ for all i. It is clear that $a_i \neq 1$ for all i, so assume that $a_i > 3$ for some i. Then $a_i! - 1 \equiv 3 \pmod 4$, thus there is a prime $p \equiv 3 \pmod 4$ such that p divides $a_i! - 1$. Then p divides $k^2 + 4^2$, a contradiction. Say m among the numbers a_i are equal to 3 and the remaining ones are equal to 2. The equation becomes $5^m - 16 = k^2$. As k is odd, a consideration mod 8 shows that m is even. But then $(5^{m/2}-k)(5^{m/2}+k) = 16$, which easily implies that $m = 2$. Thus the sequence $(a_1, a_2, \ldots, a_n)$ consists of two numbers equal to 3 and the remaining equal to 2. Clearly, all such sequences are solutions of the problem. $\square$

7. Prove that there are infinitely many pairs of consecutive numbers, no two of which have any prime factor of the form $4k + 3$.

Proof. Since no number of the form $n^2 + 1$ has a prime factor of the form $4k+3$, it is clear that the pairs $((n^2+1)^2, (n^2+1)^2+1)$ yield solutions of the problem. $\square$

The following problem is trickier and its solution uses the theory of Pell equations.

8. Let p be an odd prime. Prove that $p \equiv 1 \pmod 4$ if and only if there are integers x, y such that $x^2 - py^2 = -1$.

Proof. One direction follows directly from proposition 4.2, so let us assume that $p \equiv 1 \pmod 4$. The key point is to consider the positive Pell equation $x^2 - py^2 = 1$. By the general theory of Pell equations, this has a smallest nontrivial solution (x_0, y_0) (so $y_0 > 0$ is minimal among all solutions with $y \neq 0$). Note that x_0 is odd, since otherwise 4 would divide $y_0^2 + 1$. If $x_0 = 2a + 1$, we deduce that y_0 is even and $a^2 + a = pb^2$ for $b = \frac{y_0}{2}$. If p divides a, then $a = pc^2$ and $a + 1 = d^2$ for some integers c, d (because a and $a + 1$ are relatively prime), so that $d^2 - pc^2 = 1$. But obviously $c < y_0$, contradicting the minimality of (x_0, y_0). Thus p divides $a+1$ and we can write $a = c^2, a+1 = pd^2$ for some integers c, d. But then $c^2 - pd^2 = -1$ and we are done. $\square$

We continue with another beauty, which became classical in mathematical contests. It has the nice feature of having a completely elementary solution that uses quite a lot of different ideas.

9. Find all positive integers n such that the number $2^n - 1$ has a multiple of the form $m^2 + 9$.

IMO 1999 Shortlist

Proof. Assume that $2^n - 1$ divides $m^2 + 9$ for some integer m and that $n \geq 2$. Then the only prime factor of $2^n - 1$ which is 3 modulo 4 is 3. Indeed, if p is such a prime, then p divides $m^2 + 3^2$ and so p divides 3. Now, if n has a nontrivial odd divisor d, then $2^d - 1 \equiv -1 \pmod 4$ and 3 does not divide $2^d - 1$. Thus $2^d - 1$ has a prime factor $p \equiv 3 \pmod 4$ different from 3. Since $2^d - 1$ divides $2^n - 1$, Choosing $m = 3a$, it is enough to find a such that $(2^2+1)(2^4+1)...(2^{2^{k-1}} + 1)$ divides $a^2 + 1$. this is impossible, by the previous remark. Thus, n must be a power of 2. Conversely, assume that $n = 2^k$ and observe that

$$2^n - 1 = 3 \cdot (2^2 + 1)(2^4 + 1) \cdots (2^{2^{k-1}} + 1).$$

Choosing $m = 3a$, it is enough to find a such that $(2^2+1)(2^4+1) \cdots (2^{2^{k-1}} + 1)$ divides a^2+1. The crucial point is that the Fermat numbers $2^{2^i} + 1$ are pairwise relatively prime, so by the Chinese Remainder Theorem it is sufficient to prove that for any i there is a such that $a^2 + 1 \equiv 0 \pmod{2^{2^i} + 1}$. But this is clear, since $a = 2^{2^{i-1}}$ works. Thus, the answer to the problem is: all powers of 2. □

The next problems are concerned with properties of sums of two squares. A crucial fact is that the set of numbers which can be written as a sum of two squares of integers is closed under multiplication, by Lagrange's formula

$$(a^2 + b^2)(c^2 + d^2) = (ad + bc)^2 + (ac - bd)^2.$$

Actually, combining theorem 4.1 and proposition 4.2, it is easy to completely characterize the elements of this set: they are precisely the nonnegative integers n such that $v_p(n)$ is even for all primes $p \equiv 3 \pmod 4$.

10. Prove that a positive integer can be written as the sum of two perfect squares if and only if it can be written as the sum of the squares of two rational numbers.

Fermat

Proof. One direction being clear, let us prove that if n is the sum of the squares of two rational numbers, then it is also the sum of the squares of two integers. Thus, we know that $na^2 = b^2 + c^2$ for some integers a, b, c with $a \neq 0$. For any prime p, we deduce that

$$v_p(n) + 2v_p(a) = v_p(b^2 + c^2),$$

so that $v_p(n)$ has the same parity as $v_p(b^2 + c^2)$. Since for any prime $p \equiv 3 \pmod 4$ we have $v_p(b^2 + c^2) \equiv 0 \pmod 2$, it follows that for any such prime p we have $v_p(n) \equiv 0 \pmod 2$. The result now follows from the preliminary discussion. □

Remark 4.3. Actually, the following result holds: if $n \geq 2$ is an integer and if an integer x can be written as a sum of n squares of rational numbers, then it can also be written as a sum of n squares of integers. For $n = 3$, this follows from Davenport-Cassels' lemma ([3], chapter 13, example 12), while for $n \geq 4$, this follows from the famous theorem of Lagrange, according to which any nonnegative integer is the sum of four squares ([3], chapter 13, example 5).

11. Prove that each prime p of the form $4k+1$ can be represented in exactly one way as the sum of the squares of two integers, up to the order and signs of the terms.

Fermat

Proof. The fact that any such prime p is a sum of two squares is the content of theorem 4.1. Let us focus on the uniqueness part. Assume that we have $p = x^2 + y^2 = z^2 + t^2$. Then $(x - z)(x + z) = (t - y)(t + y)$. We will need the following very useful

Lemma 4.4. *If a, b, c, d are nonzero integers such that $ab = cd$, then there are integers m, n, p, q such that $a = mn, b = pq, c = mp, d = nq$.*

Proof. Let $\frac{a}{c} = \frac{d}{b} = \frac{n}{p}$ be the representation of the fraction $\frac{a}{c}$ in lowest terms. Since $ap = nc$, we have $p|c$, so we can write $c = mp$ for some m. Then also $a = mn$. Doing the same with $dp = nb$ yields the conclusion. □

Coming back to the problem, assume that $|x| \neq |z|$ and $|t| \neq |y|$. Then by the lemma we can find nonzero integers m_1, n_1, p_1, q_1 such that

$$x - z = m_1 n_1, \quad x + z = p_1 q_1, \quad t - y = m_1 p_1, \quad t + y = n_1 q_1.$$

But then

$$x = \frac{m_1 n_1 + p_1 q_1}{2}, \quad y = \frac{n_1 q_1 - m_1 p_1}{2},$$

so that using Lagrange's identity we obtain

$$4p = 4(x^2 + y^2) = (m_1^2 + q_1^2)(n_1^2 + p_1^2).$$

We may assume that p divides $m_1^2 + q_1^2$, so that $n_1^2 + p_1^2$ is equal to $1, 2$ or 4. As m_1, n_1, p_1, q_1 are nonzero, we obtain a contradiction unless $n_1 = p_1 = \pm 1$. But in this case we get $x - z = \pm(t - y)$ and $x + z = \pm'(t + y)$, thus

$$\{|x|, |z|\} = \{|t|, |y|\}$$

and we are done again. □

We continue with an easy exercise in Lagrange's formula.

12. Prove that the equation $3^k = m^2 + n^2 + 1$ has infinitely many solutions in positive integers.

Saint-Petersburg Olympiad

Proof. Guided by the formula

$$3^{2^a} - 1 = (2^2 + 2^2)(3^2 + 1) \cdots (3^{2^{a-1}} + 1),$$

we will choose $k = 2^a$. Since all factors in the product are sums of two squares and since the set of numbers which are sums of two squares is stable by multiplication, it follows that $3^{2^a} - 1$ is always a sum of two squares. Since it is trivially not a perfect square (it is of the form $3k + 2$), the conclusion follows. □

Proof. We will show that $3^{2k} = m^2 + n^2 + 1$ has infinitely many solutions. Indeed, start with the observation that $3^{2\cdot 1} = 2^2 + 2^2 + 1$. On the other hand, if (k, m, n) is a solution with $m \geq n$, then $(2k, 3^k m - n, 3^k n + m)$ is also a solution. Indeed, this follows from

$$(3^k m - n)^2 + (3^k n + m)^2 + 1 = (3^{2k} + 1)(m^2 + n^2) + 1 = 9^{2k} - 1 + 1 = 3^{4k}. \quad \square$$

In the following two problems we will use the fact that the density of the set of positive integers all of whose prime factors are of the form $4k + 1$ (or $4k + 3$) is zero. This is a nontrivial result, for a proof of which we refer to [3], chapter 4, example 10.

13. It is a long standing conjecture of Erdős that the equation

$$\frac{4}{n} = \frac{1}{x} + \frac{1}{y} + \frac{1}{z}$$

has solutions in positive integers for all positive integers n. Prove that the set of those n for which this statement is true has density 1.

Proof. We look for solutions with $y = z$ and $x = na$ for some positive integer a. The equation becomes $4xy = n(y + 2x)$ or, equivalently, $y(4a - 1) = 2na$. Thus, if we can find a prime factor p of n of the form $p = 4a - 1$, then we can take $y = \frac{2na}{p}$ and we have a solution in positive integers. Thus, it is enough to prove that the set of integers having at least one prime factor of the form $4k - 1$ has density 1, which has already been discussed. $\square$

14. Let T be the set of positive integers n for which the equation $n^2 = a^2 + b^2$ has solutions in positive integers. Prove that T has density 1.

Moshe Laub, AMM 6583

Proof. We will prove that $n \in T$ if and only if n has at least one prime factor of the form $4k + 1$. Suppose first that $n \in T$ and choose positive integers a, b such that $n^2 = a^2 + b^2$. If all prime factors p of n are of the form $4k - 1$, then for any such p we have $p | a^2 + b^2$, so that p divides a and p divides b. Dividing the previous relation by p^2 and repeating the argument, we deduce that $p^{v_p(n)}$

divides a and b. Since this happens for all $p|n$, it follows that n divides a and b, which is clearly impossible. Conversely, if n has a prime factor $p \equiv 1 \pmod 4$, by Fermat's theorem we can find integers c, d such that $p = c^2 + d^2$. We may assume that c, d are positive (they are nonzero since primes are not perfect squares). But then

$$n^2 = \left(\frac{n(c^2 - d^2)}{p}\right)^2 + \left(\frac{2ncd}{p}\right)^2$$

and so $n \in T$. Now, the density of those numbers which are not divisible by any prime of the form $4k + 1$ is 0 and we are done. $\square$

We continue with two very nice problems concerning primes of the form $4k+1$. The method used in the solution of the following problem is standard.

15. Let p be a prime number of the form $4k + 1$. Prove that

$$\sum_{j=1}^{\frac{p-1}{4}} \left\lfloor \sqrt{jp} \right\rfloor = \frac{p^2 - 1}{12}.$$

Proof. Write $p = 4k + 1$ and note that

$$\sum_{j=1}^{k} [\sqrt{jp}] = \sum_{j=1}^{k} \sum_{i^2 \le jp} 1 = \sum_{i=1}^{2k} \sum_{k \ge j \ge \frac{i^2}{p}} 1,$$

since the inequality $i \le \sqrt{jp}$ with $1 \le j \le k$ implies that $i \le 2k$. On the other hand, the condition $j \ge \frac{i^2}{p}$ is equivalent to $j \ge 1 + \left[\frac{i^2}{p}\right]$, since $\frac{i^2}{p}$ is not an integer. Thus we can also write

$$\sum_{j=1}^{k} [\sqrt{jp}] = \sum_{i=1}^{2k} \left(k - \left[\frac{i^2}{p}\right]\right) = 2k^2 - \sum_{i=1}^{2k} \left[\frac{i^2}{p}\right]$$

and it remains to prove that

$$\sum_{i=1}^{2k} \left[\frac{i^2}{p}\right] = \frac{2k^2 - 2k}{3}.$$

Note that

$$i^2 \pmod p = i^2 - p\left[\frac{i^2}{p}\right]$$

and since

$$\sum_{i=1}^{2k} i^2 = \frac{pk(2k+1)}{3},$$

it remains to prove that the sum of the quadratic residues mod p is pk. But if $x_1, x_2, \ldots, x_{\frac{p-1}{2}}$ are the nonzero quadratic residues mod p (any residue mod p is implicitly taken between 0 and $p-1$), then $p-x_1, p-x_2, \ldots, p-x_{2k}$ are a permutation of the x_i's (since -1 is a quadratic residue mod p, which follows from $p \equiv 1 \pmod 4$). Thus

$$\sum_{i=1}^{2k} x_i = \sum_{i=1}^{2k}(p - x_i)$$

and the result follows. □

The following functional equation is rather nonstandard.

16. Find all functions $f : \mathbb{Z}^+ \to \mathbb{Z}$ with the properties:

    1. $f(a) \geq f(b)$ whenever a divides b.
    2. for all positive integers a and b,

$$f(ab) + f(a^2 + b^2) = f(a) + f(b).$$

Gabriel Dospinescu, Mathlinks Contest

Proof. Since 1 divides any integer, it follows that $f(1) \geq f(x)$ for all x. Let $k = f(1)$.

The first step is to prove that $f(n)$ only depends on the prime factors of n and not their multiplicities, i.e.

$$f(n) = f\left(\prod_{p|n} p\right).$$

Indeed, consider two positive integers a, m and choose $b = am$. Thus

$$f(a^2m) + f(a^2(m^2+1)) = f(a) + f(am)$$

and by the first condition $f(a) \geq f(a^2(m^2+1))$ and $f(am) \geq f(a^2m)$. Thus both of these inequalities must be equalities and so $f(am) = f(a^2m)$ for any positive integers a, m. This immediately proves the claim.

In the second step, we prove that $f(n)$ does not depend on the prime factors of n that are congruent to 1 or 2 modulo 4. Indeed, the proof of the previous claim shows that for all n and x we have $f(n) = f(n(x^2+1))$. In particular, $f(n) = f(2n)$ and so we don't have to care about possible powers of 2 in the prime factorization of n. Also, if $p \equiv 1 \pmod 4$ and x is chosen such that $p | x^2 + 1$, then

$$f(n) \geq f(np) \geq f(n(x^2+1)) = f(n)$$

and so $f(np) = f(n)$. In conclusion, we have

$$f(n) = f\left(\prod_{p|n, p \in P_3} p\right),$$

where P_3 is the set of prime numbers of the form $4k+3$. Let $p_1, p_2, \ldots$ be the elements of P_3 and define $g(A) = f(\prod_{a \in A} p_a)$. We obtain a function g defined on the set of all subsets of $\mathbb{N}$ with integral values. Moreover, we claim that $g(\emptyset) = k$, $S_1 \subset S_2 \implies g(S_1) \geq g(S_2)$ and finally

$$g(S_1) + g(S_2) = g(S_1 \cup S_2) + g(S_1 \cap S_2)$$

The first two relations are obvious. For the third one, note that if the sets of prime factors congruent to 3 mod 4 of a and of b are A, B, then the set of prime factors of the form $4j+3$ of ab is exactly $A \cup B$ and the set of prime factors of the form $4j+3$ of a^2+b^2 is $A \cap B$. If $g(\{n\}) = k_n$ for some integers $k_n \leq k$, then an easy induction on $|A|$ shows that

$$g(A) = \sum_{a \in A} k_a - (|A| - 1)k.$$

Conversely, any choice of such k_n yields a corresponding g and the previous construction yields a solution f of the equation. This ends the solution. □

The following challenging problem requires some estimates about prime numbers that follow from Dirichlet's theorem, for which we refer the reader to addendum 7.A.

17. Prove that the equation $x^8 = n! + 1$ has only finitely many solutions in nonnegative integers.

Proof. Suppose that (x, n) is a solution of the equation $x^8 = n! + 1$. Then

$$n! = (x^2 - 1)(x^2 + 1)(x^4 + 1).$$

Let A_n be the set of prime numbers $p \leq n$ of the form $4k+3$. The key point is that no $p \in A_n$ can divide $x^2 + 1$ or $x^4 + 1$, so that $p^{v_p(n!)}$ must divide $x^2 - 1$. Thus we obtain

$$\sqrt[4]{n!} \geq x^2 - 1 \geq \prod_{p \in A_n} p^{v_p(n!)}.$$

Then, using that $v_p(n!) > n/p - 1$ (by Legendre's formula), we obtain

$$\frac{1}{4} n \ln n > \ln \sqrt[4]{n!} > \sum_{p \in A_n} \left(\frac{n}{p} - 1 \right) \ln p.$$

A classical inequality of Erdős (theorem 3.A.3, chapter 3) yields

$$\sum_{p \in A_n} \ln p < \ln \left(\prod_{p \leq n} p \right) < n \ln 4.$$

We deduce that

$$\sum_{p \in A_n} \frac{\ln p}{p} < \ln 4 + \frac{\ln n}{4}$$

for any solution (x, n).

Now, it remains to prove that there are only finitely many such integers n. This is a consequence of the proof of Dirichlet's theorem, which establishes, among many other things, that

$$\frac{1}{\ln n} \sum_{p \in A_n} \frac{\ln p}{p} \to \frac{1}{2}$$

for $n \to \infty$. See addendum 7.A for a proof. □

It is really amazing that the following result has a purely elementary proof. We present here the beautiful idea of John H.E.Cohn and we refer the reader to [18] for other similar results (including the fact that 1 and 144 are the only squares in the Fibonacci sequence).

18. Let $L_0 = 2$, $L_1 = 1$ and $L_{n+2} = L_{n+1} + L_n$ be Lucas's famous sequence. Then the only $n > 1$ for which L_n is a perfect square is $n = 3$.

Cohn's theorem

Proof. If x_1 and x_2 are the roots of the polynomial $X^2 - X - 1$, then $L_n = x_1^n + x_2^n$, which combined with $x_1x_2 = -1$ yields $L_{2n} = L_n^2 - 2(-1)^n$. This already shows that if L_n is a perfect square, then n is odd, for $y^2 \pm 2$ is never a perfect square.

The case when n is odd is much more subtle. There are two key properties of the Lucas numbers that make everything work. The first is that $L_k \equiv 3 \pmod 4$ whenever k is an even number not a multiple of 6 (use the previous formula for L_{2n} and the fact that L_n is odd whenever n is not a multiple of 3). Call such a number k good. The second key ingredient is the fact that L_k divides $L_{n+2k} + L_n$ for all good numbers k and all nonnegative integers n. This follows from $k \equiv 0 \pmod 2$, the equality $x_1x_2 = -1$ and the computation

$$L_{n+2k} + L_n = x_1^n(x_1^{2k} + 1) + x_2^n(x_2^{2k} + 1) =$$

$$x_1^{n+k}(x_1^k + x_2^k) + x_2^{n+k}(x_1^k + x_2^k) = L_kL_{n+k}.$$

Assume now that $n \equiv 1 \pmod 4$ and $n > 1$. Then we can write $n = 1 + 2 \cdot 3^r k$ for a nonnegative number r and a good number k. Applying the second key point 3^r times, we deduce that $L_n \equiv -L_1 \equiv -1 \pmod{L_k}$. As $L_k \equiv 3 \pmod 4$, it has a prime divisor of the form $4j + 3$ and so L_n cannot be a perfect square.

Assume finally that $n \equiv 3 \pmod 4$ and $n > 3$. Then we can write $n = 3 + 2 \cdot 3^r k$ for a nonnegative number r and a good number k. The same argument shows that $L_n \equiv -L_3 \equiv -4 \pmod{L_k}$ and we reach the same conclusion, as numbers of the form $x^2 + 4$ have no prime factors of the form $4j + 3$. □

4.1 Notes

Many of the solutions to the problems in this chapter were provided by the following people: Alexandru Chirvăsitu (problem 14), Daniel Harrer (problems 2, 3), Benjamin Gunby (problems 1, 6, 9), Fedja Nazarov (problems 12, 15), Gjergji Zaimi (problems 5, 7, 13, 16).

Chapter 5

T_2's Lemma

All of the following problems fall to a rather handy inequality, known as T_2's lemma even though it is a special case of the Cauchy-Schwarz inequality. This result says that for all real numbers $a_1, a_2, \ldots, a_n$ and all positive real numbers $x_1, x_2, \ldots, x_n$ the following inequality holds

$$\frac{a_1^2}{x_1} + \frac{a_2^2}{x_2} + \cdots + \frac{a_n^2}{x_n} \geq \frac{(a_1 + a_2 + \cdots + a_n)^2}{x_1 + x_2 + \cdots + x_n}.$$

To get the reader familiar with this trick, we start with a series of more direct applications.

1. Let $x_1, x_2, \ldots, x_n, y_1, y_2, \ldots, y_n$ be positive real numbers such that

$$x_1 + x_2 + \cdots + x_n \geq x_1 y_1 + x_2 y_2 + \cdots + x_n y_n.$$

Prove that

$$x_1 + x_2 + \cdots + x_n \leq \frac{x_1}{y_1} + \frac{x_2}{y_2} + \cdots + \frac{x_n}{y_n}.$$

Romeo Ilie, Romania 1999

Proof. Using T_2's lemma, we can write

$$\sum \frac{x_i}{y_i} = \sum \frac{x_i^2}{x_i y_i} \geq \frac{\left(\sum x_i\right)^2}{\sum x_i y_i} \geq \sum x_i,$$

the last inequality being exactly the hypothesis. □

2. Let a, b, c be nonzero real numbers such that $ab + bc + ca \geq 0$. Prove that

$$\frac{ab}{a^2+b^2} + \frac{bc}{b^2+c^2} + \frac{ca}{c^2+a^2} \geq -\frac{1}{2}.$$

Titu Andreescu

Proof. The trick is to add $\frac{1}{2}$ to each fraction, in order to exploit the identity

$$\frac{ab}{a^2+b^2} + \frac{1}{2} = \frac{(a+b)^2}{2(a^2+b^2)}.$$

With this observation, the inequality becomes

$$\sum \frac{(a+b)^2}{a^2+b^2} \geq 2$$

and it is then a trivial consequence of T_2's lemma, combined with the hypothesis $ab + bc + ca \geq 0$. □

3. Prove that for any positive real numbers a, b, c, d satisfying

$$ab + bc + cd + da = 1,$$

the following inequality holds

$$\frac{a^3}{b+c+d} + \frac{b^3}{c+d+a} + \frac{c^3}{d+a+b} + \frac{d^3}{a+b+c} \geq \frac{1}{3}.$$

IMO 1990 Shortlist

Proof. T_2's lemma gives a lower bound

$$\sum \frac{a^3}{b+c+d} \geq \frac{\left(\sum a^2\right)^2}{\sum a(b+c+d)}.$$

Since

$$\sum a(b+c+d) = \left(\sum a\right)^2 - \sum a^2 \leq 3\sum a^2,$$

it remains to prove that $\sum a^2 \geq 1$. But again by Cauchy-Schwarz we have

$$1 = (ab+bc+cd+da)^2 \leq (a^2+b^2+c^2+d^2)^2,$$

which proves that $\sum a^2 \geq 1$ and finishes the solution. □

4. Prove that if the positive real numbers a, b, c satisfy $abc = 1$, then

$$\frac{a}{b+c+1} + \frac{b}{c+a+1} + \frac{c}{a+b+1} \geq 1.$$

Vasile Cârtoaje, Gazeta Matematică

Proof. The solution using T_2's lemma is straightforward:

$$\sum \frac{a}{b+c+1} = \sum \frac{a^2}{a(b+c+1)} \geq \frac{(\sum a)^2}{2\sum ab + \sum a},$$

so that we only need to prove the inequality $\sum a^2 \geq \sum a$. This follows from $\sum a^2 \geq \frac{(\sum a)^2}{3}$ and $\sum a \geq 3$ (the first being a consequence of Cauchy-Schwarz, the second being the AM-GM inequality). □

5. Let a, b, c be real numbers such that

$$\frac{1}{a^2+1} + \frac{1}{b^2+1} + \frac{1}{c^2+1} \geq 2.$$

Prove that

$$ab+bc+ca \leq \frac{3}{2}.$$

Titu Andreescu

Proof. Observe that

$$\sum \frac{a^2}{a^2+1} = 3 - \sum \frac{1}{a^2+1} \le 1.$$

On the other hand, we have

$$\sum \frac{a^2}{a^2+1} \ge \frac{(a+b+c)^2}{a^2+b^2+c^2+3}.$$

Combining the two inequalities immediately yields the result. □

The following problems are a bit less straightforward. However, they do not require any heavy machinery.

6. Prove that for any positive real numbers a, b, c,

$$\frac{1}{a+b} + \frac{1}{b+c} + \frac{1}{c+a} + \frac{1}{2\sqrt[3]{abc}} \ge \frac{(a+b+c+\sqrt[3]{abc})^2}{(a+b)(b+c)(c+a)}.$$

Titu Andreescu, MOSP 1999

Proof. The solution using T_2's lemma is a bit tricky: the point is to look at the denominator of the right-hand side, because

$$(a+b)(b+c)(c+a) = (a+b)c^2 + (b+c)a^2 + (c+a)b^2 + 2abc.$$

This suggests writing the left-hand side in the following way

$$\frac{c^2}{c^2(a+b)} + \frac{a^2}{a^2(b+c)} + \frac{b^2}{b^2(a+c)} + \frac{(\sqrt[3]{abc})^2}{2abc}.$$

A direct application of T_2's lemma finishes the proof. □

7. Prove that for any positive real numbers a, b, c the following inequality holds

$$\left(\frac{a}{b+c}\right)^2 + \left(\frac{b}{c+a}\right)^2 + \left(\frac{c}{a+b}\right)^2 \ge \frac{3}{4} \cdot \frac{a^2+b^2+c^2}{ab+bc+ca}.$$

Gabriel Dospinescu

Proof. Using T_2's lemma, it is sufficient to prove the inequality

$$\frac{(a^2+b^2+c^2)^2}{a^2(b+c)^2+b^2(c+a)^2+c^2(a+b)^2} \geq \frac{3}{4} \cdot \frac{a^2+b^2+c^2}{ab+bc+ca}$$

Note however that the obvious application of T_2's lemma fails. The previous inequality is equivalent to

$$4(a^2+b^2+c^2)(ab+bc+ca) \geq 3(a^2(b+c)^2+b^2(c+a)^2+c^2(a+b)^2).$$

Unfortunately, the only reasonable way to prove this is to expand it in the form

$$\sum ab(a^2+b^2) \geq \frac{3}{2}\sum a^2b^2 + \frac{1}{2}abc\sum a.$$

Fortunately, this is trivial, since

$$\sum ab(a^2+b^2) \geq 2\sum a^2b^2 \text{ and } \sum a^2b^2 \geq abc\sum a. \qquad \square$$

It is possible to solve the following problem using T_2's lemma, but the proof is not really elegant. In the addendum we discuss some applications of Hölder's inequality, which makes this problem really easy.

8. Let a, b, c be positive reals such that $abc = 1$. Show that

$$\frac{1}{a^5(b+2c)^2} + \frac{1}{b^5(c+2a)^2} + \frac{1}{c^5(a+2b)^2} \geq \frac{1}{3}.$$

Titu Andreescu, USA TST 2010

Proof. Start by making the substitution

$$x = \frac{1}{a}, \quad y = \frac{1}{b}, \quad z = \frac{1}{c}.$$

The inequality becomes

$$\frac{x^3}{(z+2y)^2} + \frac{y^3}{(x+2z)^2} + \frac{z^3}{(y+2x)^2} \geq \frac{1}{3}.$$

Applying T_2's lemma, it is enough to prove that

$$3(x^{\frac{5}{2}}+y^{\frac{5}{2}}+z^{\frac{5}{2}})^2 \geq \sum x^2(z+2y)^2.$$

The right-hand side is equal to $5\sum x^2y^2+4\sum x$. Thus, we need to prove that

$$3\sum x^5+6\sum (xy)^{\frac{5}{2}} \geq 5\sum (xy)^2+4\sum x.$$

Note that

$$\sum (xy)^{\frac{5}{2}} = \sum (xy)^2\cdot\sqrt{xy} \geq \frac{1}{3}\sum\sqrt{xy}\cdot\sum (xy)^2 \geq \sum (xy)^2,$$

the first inequality being Chebyshev's, the second one by the AM-GM inequality and the fact that $xyz=1$. Thus, it suffices to prove that

$$3\sum x^5+\sum x^2y^2 \geq 4\sum x.$$

But $3x^5+y^2z^2 \geq 4x^{\frac{13}{4}}$ and so it is enough to prove that $\sum x^{\frac{13}{4}} \geq \sum x$. This follows from the power-mean inequality and the fact that $x+y+z\geq 3$. □

Proof. As in the previous solution, we reduce the problem to proving the following inequality

$$\frac{x^3}{(z+2y)^2}+\frac{y^3}{(x+2z)^2}+\frac{z^3}{(y+2x)^2} \geq \frac{1}{3}.$$

Using the AM-GM inequality, we can write

$$\frac{x^3}{(2y+z)^2}+\frac{2y+z}{27}+\frac{2y+z}{27} \geq \frac{x}{3}$$

and two similar inequalities. Adding them yields the following estimate

$$\frac{x^3}{(z+2y)^2}+\frac{y^3}{(x+2z)^2}+\frac{z^3}{(y+2x)^2} \geq \frac{x+y+z}{9}$$

and we end up using once more the AM-GM inequality. □

It is really not easy to prove the following inequality using T_2's lemma, but the trick is worth remembering.

9. Prove that for any positive real numbers a, b, c the following inequality holds

$$\frac{1}{3a+b}+\frac{1}{3b+c}+\frac{1}{3c+a}\geq\frac{1}{2a+b+c}+\frac{1}{2b+c+a}+\frac{1}{2c+a+b}.$$

M.O. Drâmbe

Proof. Choose three positive real numbers α, β, γ and use T_2's lemma in the form

$$\frac{\alpha}{3a+b}+\frac{\beta}{3b+c}+\frac{\gamma}{3c+a}\geq\frac{(\alpha+\beta+\gamma)^2}{a(3\alpha+\gamma)+b(3\beta+\alpha)+c(3\gamma+\beta)}.$$

Now, we impose the conditions

$$3\alpha+\gamma=2,\quad 3\beta+\alpha=1,\quad 3\gamma+\beta=1.$$

Solving this linear system yields the solution $\alpha=\frac{4}{7}, \beta=\frac{1}{7}, \gamma=\frac{2}{7}$. Therefore, we obtain the inequality

$$\frac{4}{7}\cdot\frac{1}{3a+b}+\frac{1}{7}\cdot\frac{1}{3b+c}+\frac{2}{7}\cdot\frac{1}{3c+a}\geq\frac{1}{2a+b+c}.$$

Proceeding in the same way with the two other terms of the left-hand side and adding up the resulting inequalities yields the desired result. □

Proof. We can also use the trick of integrating polynomial inequalities to deduce fractional inequalities. Namely, the inequality

$$x^3y+y^3z+z^3x\geq xyz(x+y+z)$$

can be easily proved using T_2's lemma, since it can be written in the form

$$\frac{x^2}{z}+\frac{y^2}{x}+\frac{z^2}{y}\geq x+y+z.$$

Using this inequality, we can write

$$t^{3a+b-1} + t^{3b+c-1} + t^{3c+a-1} \geq t^{2a+b+c-1} + t^{2b+c+a-1} + t^{2c+a+b-1}$$

for all $0 < t \leq 1$. Integrating this between 0 and 1 yields the desired inequality. □

Remark 5.1. The technique used in the second proof looks unusual. It is actually quite powerful and we refer the reader to [3], chapter 19 for many more applications.

The following two problems are closely related and use a rather useful inequality.

10. Prove that for all $n \geq 4$ and all $x_1, x_2, \ldots, x_n > 0$,

$$\frac{x_1}{x_n + x_2} + \frac{x_2}{x_1 + x_3} + \cdots + \frac{x_n}{x_{n-1} + x_1} \geq 2.$$

Tournament of the Towns 1982

Proof. We start in the usual way by using T_2's lemma:

$$\frac{x_1}{x_n + x_2} + \frac{x_2}{x_1 + x_3} + \cdots + \frac{x_n}{x_{n-1} + x_1} \geq$$

$$\frac{(x_1 + x_2 + \cdots + x_n)^2}{x_1(x_n + x_2) + x_2(x_1 + x_3) + \cdots + x_n(x_{n-1} + x_1)}.$$

Since

$$x_1(x_n + x_2) + x_2(x_1 + x_3) + \cdots + x_n(x_{n-1} + x_1) = 2(x_1x_2 + x_2x_3 + \cdots + x_nx_1),$$

it remains to prove the following very useful

Lemma 5.2. *If $n \geq 4$ and $x_1, x_2, \ldots, x_n$ are nonnegative real numbers, then*

$$(x_1 + x_2 + \cdots + x_n)^2 \geq 4(x_1x_2 + x_2x_3 + \cdots + x_nx_1).$$

The proof is a bit tricky. If n is even, the inequality follows trivially from the chain of inequalities

$$\begin{aligned} 4(x_1x_2 + \cdots + x_nx_1) &\le 4(x_1 + x_3 + \cdots + x_{n-1})(x_2 + x_4 + \cdots + x_n) \\ &\le (x_1 + x_2 + \cdots + x_n)^2, \end{aligned}$$

the first one being obvious and the second one being the AM-GM inequality.

For n odd, things are subtler, but we can reduce the problem to the case when n is even by the following mixing argument: we may assume that $x_1 \ge x_2$, so that we trivially have

$$x_1x_2 + x_2x_3 + x_3x_4 \le x_1x_2 + x_1x_3 + x_3x_4 \le x_1(x_2 + x_3) + (x_2 + x_3)x_4.$$

Thus, replacing $x_1, x_2, \ldots, x_n$ by $x_1, x_2 + x_3, x_4, \ldots, x_n$, we preserve the sum of the x_i's while not decreasing the quantity $x_1x_2 + \cdots + x_nx_1$. Since $n-1$ is even, everything follows from the previous step. □

11. Let $n \ge 4$ be an integer and let $a_1, a_2, \ldots, a_n$ be positive real numbers such that $a_1^2 + a_2^2 + \cdots + a_n^2 = 1$. Prove that

$$\frac{a_1}{a_2^2 + 1} + \frac{a_2}{a_3^2 + 1} + \cdots + \frac{a_n}{a_1^2 + 1} \ge \frac{4}{5}(a_1\sqrt{a_1} + a_2\sqrt{a_2} + \cdots + a_n\sqrt{a_n})^2.$$

Mircea Becheanu and Bogdan Enescu, Romanian TST 2002

Proof. Applying T_2's lemma we obtain

$$\sum \frac{a_i}{a_{i+1}^2 + 1} = \sum \frac{a_i^3}{a_i^2(a_{i+1}^2 + 1)} \ge \frac{\left(\sum a_i\sqrt{a_i}\right)^2}{\sum a_i^2(a_{i+1}^2 + 1)}.$$

Since $\sum a_i^2 = 1$, we have $\sum a_i^2 a_{i+1}^2 \le 1/4$ by lemma 5.2. The result follows. □

We give two proofs for the following beautiful problem. The first one is a standard application of T_2's lemma, the second one uses a very useful technique.

12. Prove that for any positive real numbers a, b, c, the following inequality holds
$$\frac{a}{\sqrt{a^2+8bc}}+\frac{b}{\sqrt{b^2+8ca}}+\frac{c}{\sqrt{c^2+8ab}}\geq 1.$$

Hojoo Lee, IMO 2001

Proof. The most natural application of T_2's lemma turns out to work very smoothly. Indeed,
$$\sum\frac{a}{\sqrt{a^2+8bc}}=\sum\frac{a^2}{a\sqrt{a^2+8bc}}\geq\frac{(a+b+c)^2}{\sum a\sqrt{a^2+8bc}}.$$

The only problem is to show that
$$\sum a\sqrt{a^2+8bc}\leq(a+b+c)^2.$$

This suggests using Cauchy-Schwarz and, indeed, we have
$$\left(\sum a\sqrt{a^2+8bc}\right)^2\leq\sqrt{\sum a}\cdot\sqrt{\sum a^3+24abc}.$$

Thus, the problem is solved if we can prove the inequality
$$\sum a^3+24abc\leq(a+b+c)^3.$$

Fortunately, after expansion, this becomes $\sum a(b-c)^2\geq 0$, which is obvious. $\square$

Proof. Make the substitution
$$x=\sqrt{\frac{a^2}{a^2+8bc}},\quad y=\sqrt{\frac{b^2}{b^2+8ac}},\quad z=\sqrt{\frac{c^2}{c^2+8ab}}.$$

Multiplying the relation $\frac{1}{x^2}-1=\frac{8bc}{a^2}$ and the two similar ones yields the relation
$$\left(\frac{1}{x^2}-1\right)\left(\frac{1}{y^2}-1\right)\left(\frac{1}{z^2}-1\right)=512.$$

The problem asks to prove that under this assumption we have $x+y+z \geq 1$. Assume that this is not the case, so $x+y+z<1$. Let

$$X = \frac{x}{x+y+z} > x, \quad Y = \frac{y}{x+y+z} > y, \quad Z = \frac{Z}{x+y+z} > z,$$

so that $X+Y+Z=1$ and

$$512 > \left(\frac{1}{X^2}-1\right)\left(\frac{1}{Y^2}-1\right)\left(\frac{1}{Z^2}-1\right).$$

We deduce that

$$\begin{aligned}&512X^2Y^2Z^2\\&\quad > (X+Y)(Y+Z)(Z+X)(2X+Y+Z)(2Y+X+Z)(2Z+X+Y),\end{aligned}$$

which is impossible since $X+Y \geq 2\sqrt{XY}$, $2X+Y+Z \geq 4\sqrt[4]{X^2YZ}$ and the similar inequalities obtained by cyclic permutations. □

There are two traps in the following problem, making the problem harder than it appears at first sight.

13. Let a, b, c be positive real numbers such that $ab+bc+ca=3$. Prove that

$$\frac{a}{2a+b^2} + \frac{b}{2b+c^2} + \frac{c}{2c+a^2} \leq 1.$$

T.Q. Anh

Proof. The inequality we have to establish seems to be opposite to the usual applications of T_2's lemma. This is actually not a very serious problem, since we can always change the terms of the sum a bit and change the sign of the inequality. Indeed, note that

$$\frac{a}{2a+b^2} = \frac{1}{2}\left(1-\frac{b^2}{2a+b^2}\right).$$

Thus, the inequality is equivalent to

$$\frac{b^2}{2a+b^2}+\frac{c^2}{2b+c^2}+\frac{a^2}{2c+a^2}\geq 1.$$

And now, another trap: one would be tempted to use T_2's lemma in the obvious form, but it is not difficult to check that this does not work. Instead, we take advantage of the hypothesis $ab+bc+ca=3$ to write

$$\sum\frac{b^2}{2a+b^2}=\sum\frac{(b\sqrt{b})^2}{2ab+b^3}\geq\frac{\left(\sum b\sqrt{b}\right)^2}{6+\sum b^3}.$$

Thus, it remains to prove that $\sum(ab)^{3/2}\geq 3$, which is immediate by the power-mean inequality and the hypothesis. □

The following problem is also quite tricky.

14. Determine the best constant k_n such that for all positive real numbers $a_1, a_2, \ldots, a_n$ satisfying $a_1a_2\cdots a_n=1$, the following inequality holds

$$\frac{a_1a_2}{(a_1^2+a_2)(a_2^2+a_1)}+\frac{a_2a_3}{(a_2^2+a_3)(a_3^2+a_2)}+\cdots+\frac{a_na_1}{(a_n^2+a_1)(a_1^2+a_n)}\leq k_n.$$

Gabriel Dospinescu, Mircea Lascu

Proof. The basic inequality that will be used is

$$(x^2+y)(y^2+x)\geq xy(1+x)(1+y).$$

This reduces immediately to $(x^2-y^2)(x-y)\geq 0$, which is clear. This already shows that for $n=2$ the maximum value is $k_n=\frac{1}{2}$, since $(1+x)(1+y)\geq 4$ if $xy=1$.

Let us assume now that $n>2$. We will prove that $k_n=n-2$. To prove that $k_n\geq n-2$, it is enough to exhibit sequences $a_1, a_2, \ldots, a_n$ for which the value of the expression

$$F(a_1, a_2, \ldots, a_n)=\sum_{i=1}^{n}\frac{a_ia_{i+1}}{(a_i^2+a_{i+1})(a_{i+1}^2+a_i)}$$

is close to $n-2$. This can be done easily, by taking $n-1$ of the variables equal to some x very close to 0, and the last variable equal to x^{1-n}. The difficult part is proving that $F(a_1, a_2, \ldots, a_n) \le n-2$ holds for any sequence $a_1, a_2, \ldots, a_n$ as in the statement. Using the basic inequality, this reduces to proving that

$$\sum_{i=1}^{n} \frac{1}{(1+a_i)(1+a_{i+1})} \le n-2$$

for any positive numbers a_i with $a_1 a_2 \cdots a_n = 1$. To prove this, write $a_i = \frac{x_i}{x_{i+1}}$, with $x_{n+1} = x_1$. Subtracting 1 from every fraction, we obtain the equivalent inequality

$$\sum_{i=1}^{n} \frac{x_i x_{i+1} + x_i x_{i+2} + x_{i+1}^2}{(x_i + x_{i+1})(x_{i+1} + x_{i+2})} \ge 2.$$

This is too complicated to try a Cauchy-Schwarz approach, but it simplifies a lot if we observe that

$$\frac{x_i x_{i+1} + x_i x_{i+2} + x_{i+1}^2}{(x_i + x_{i+1})(x_{i+1} + x_{i+2})} = \frac{x_i}{x_i + x_{i+1}} + \frac{x_{i+1}^2}{(x_i + x_{i+1})(x_{i+1} + x_{i+2})}.$$

Since

$$\sum \frac{x_i}{x_i + x_{i+1}} > \sum \frac{x_i}{x_1 + x_2 + \cdots + x_n} = 1,$$

it remains to prove the inequality

$$\sum \frac{x_{i+1}^2}{(x_i + x_{i+1})(x_{i+1} + x_{i+2})} \ge 1.$$

T_2's lemma reduces this to the easier inequality

$$\left(\sum x_i\right)^2 \ge \sum x_i^2 + 2\sum x_i x_{i+1} + \sum x_i x_{i+2}.$$

Expanding the left-hand side makes the previous inequality obvious and finishes the proof of the fact that $k_n = n-2$ for $n \ge 3$. □

The following difficult problem seems to be exactly the opposite of what is usually called a standard application of T_2's lemma. It turns out that we can actually apply T_2's lemma, but in a very nontrivial way.

15. Prove that for any positive real numbers a, b, c,

$$\frac{(2a+b+c)^2}{2a^2+(b+c)^2}+\frac{(2b+c+a)^2}{2b^2+(c+a)^2}+\frac{(2c+a+b)^2}{2c^2+(a+b)^2}\leq 8.$$

Titu Andreescu and Zuming Feng, USAMO 2003

Proof. Writing

$$x=\frac{b+c}{a},\quad y=\frac{a+c}{b},\quad z=\frac{a+b}{c},$$

we can write the inequality as

$$\frac{(2+x)^2}{2+x^2}+\frac{(2+y)^2}{2+y^2}+\frac{(2+z)^2}{2+z^2}\leq 8 \Longleftrightarrow \frac{1+2x}{x^2+2}+\frac{1+2y}{y^2+2}+\frac{1+2z}{z^2+2}\leq\frac{5}{2}$$

$$\Longleftrightarrow \frac{(x-1)^2}{x^2+2}+\frac{(y-1)^2}{y^2+2}+\frac{(z-1)^2}{z^2+2}\geq\frac{1}{2}.$$

In order to prove the last inequality, we use T_2's lemma in the obvious form and so it suffices thus to show that

$$\frac{(x+y+z-3)^2}{x^2+y^2+z^2+6}\geq\frac{1}{2}.$$

Expanding $(x+y+z-3)^2$, we reduce this to proving that

$$\sum x^2-12\sum x+4\sum xy+12\geq 0.$$

This is not obvious, but the observation that $x^2+4\geq 4x$ reduces it to proving that $\sum xy\geq 2\sum x$, which is problem 7 in chapter 1. □

Proof. This solution uses the linearization method. To simplify computations, we may assume (since the inequality is homogeneous) that $a+b+c=1$. Define the map

$$f(x)=\frac{(x+1)^2}{2x^2+(1-x)^2}.$$

The inequality can also be written as $f(a)+f(b)+f(c)\leq 8$. We will try to find u, v such that $f(x)\leq ux+v$ for any $x\in[0,1]$, with equality for $x=\frac{1}{3}$ (which is the obvious equality case in the original inequality). Thus, we should have $f\left(\frac{1}{3}\right) = \frac{u}{3} + v$ and also $f'\left(\frac{1}{3}\right) = u$, since $\frac{1}{3}$ would be a minimum of $ux + v - f(x)$ on $[0, 1]$. A straightforward, but tedious computation shows that $u = 4$ and $v = \frac{4}{3}$. We need to prove now that this pair (u, v) really works, i.e. that $f(x) \le ux + v$ holds for any $x \in [0, 1]$. Clearing denominators and expanding everything yields (after a tedious computation) the equivalent inequality

$$36x^3 - 15x^2 - 2x + 1 \ge 0.$$

But the conditions we imposed were made so that the left-hand side is divisible by $(3x - 1)^2$. Doing the euclidean division shows that the left-hand side is $(3x - 1)^2(4x + 1)$, which is obviously positive. Finally, we just have to add the inequalities $f(x) \le ux + v$ for $x \in \{a, b, c\}$ to end the proof. □

The following problems are harder than the previous ones. They are still based on T_2's lemma, but applied in more subtle ways and often combined with other tricks.

16. Let a, b, c, d be positive real numbers such that $abcd = 1$. Prove that

$$\frac{1}{(1+a)^2} + \frac{1}{(1+b)^2} + \frac{1}{(1+c)^2} + \frac{1}{(1+d)^2} \ge 1.$$

Vasile Cârtoaje

Proof. The proof using T_2's lemma is rather mysterious. By performing the substitution

$$a = \frac{y}{x}, \quad b = \frac{z}{y}, \quad c = \frac{t}{z}, \quad d = \frac{x}{t},$$

we obtain the equivalent inequality

$$\frac{x^2}{(x+y)^2} + \frac{y^2}{(y+z)^2} + \frac{z^2}{(z+t)^2} + \frac{t^2}{(t+x)^2} \ge 1.$$

Of course, an immediate application of T_2's lemma fails rather badly, so we need something more clever. Let us apply T_2's lemma for the first two and then for the last two terms of the inequality. If we try to prove the stronger inequality

$$\frac{(x+y)^2}{(x+y)^2+(y+z)^2}+\frac{(z+t)^2}{(z+t)^2+(t+x)^2}\geq 1,$$

we easily realize[1] that it is equivalent to $(y+z)(t+x) \leq (x+y)(z+t)$. Unfortunately there is no reason to have $(y+z)(t+x) \leq (x+y)(z+t)$. The miracle is that if this fails, then we can apply T_2's lemma for the first and fourth terms of the initial inequality and then for the middle terms. In this case, we obtain the stronger inequality

$$\frac{(x+t)^2}{(x+y)^2+(x+t)^2}+\frac{(y+z)^2}{(y+z)^2+(z+t)^2}\geq 1.$$

A similar argument shows that this holds precisely when $(x+y)(z+t) \leq (y+z)(x+t)$, in particular whenever $(y+z)(t+x) \leq (x+y)(z+t)$ fails. The result follows. □

Proof. This solution is not natural, either, but it is very elegant. We claim that for all positive numbers a, b we have

$$\frac{1}{(1+a)^2}+\frac{1}{(1+b)^2}\geq\frac{1}{1+ab}.$$

This is rather easy to prove, though it requires some nasty computations. Namely, by clearing denominators and performing the obvious simplifications, we reduce the problem to proving that

$$ab(a^2-ab+b^2)-2ab+1\geq 0,$$

which is equivalent to

$$ab(a-b)^2+(ab-1)^2\geq 0.$$

Once we have this inequality, we deduce that

$$\sum\frac{1}{(1+a)^2}\geq\frac{1}{1+ab}+\frac{1}{1+cd}=\frac{1}{1+ab}+\frac{ab}{1+ab}=1$$

and the result follows. □

[1] It is convenient to denote $a=\frac{y+z}{x+y}$ and $b=\frac{t+x}{t+z}$.

The following problem is a bit tricky, especially because of the strange hypotheses.

17. Let $n \geq 16$ be a positive integer and suppose that the positive numbers $a_1, a_2, \ldots, a_n$ satisfy $a_1 + a_2 + \cdots + a_n = 1$ and $a_1 + 2a_2 + \cdots + na_n = 2$. Prove that

$$(a_2 - a_1)\sqrt{2} + (a_3 - a_2)\sqrt{3} + \cdots + (a_n - a_{n-1})\sqrt{n} < 0.$$

Gabriel Dospinescu

Proof. The first step is to apply Abel's summation formula to rewrite the inequality as

$$a_1\sqrt{2} + a_2(\sqrt{3} - \sqrt{2}) + \cdots + a_{n-1}(\sqrt{n} - \sqrt{n-1}) > a_n\sqrt{n}.$$

Using the obvious bound $a_1\sqrt{2} > a_1(\sqrt{2} - 1)$ and the formula

$$\sqrt{k} - \sqrt{k-1} = \frac{1}{\sqrt{k} + \sqrt{k-1}},$$

an application of T_2's lemma shows that the left-hand side is greater than

$$\frac{(a_1 + a_2 + \cdots + a_{n-1})^2}{\sum_{i=1}^{n-1} a_i(\sqrt{i} + \sqrt{i+1})}.$$

On the other hand, Cauchy-Schwarz together with the hypothesis give the estimate

$$\sum_{i=1}^{n-1} a_i\sqrt{i} \leq \sqrt{\sum_{i=1}^{n-1} a_i} \cdot \sqrt{\sum_{i=1}^{n-1} ia_i} < \sqrt{2}$$

and also

$$\sum_{i=1}^{n-1} a_i\sqrt{i+1} < \sqrt{3}.$$

Therefore, taking into account that $a_1 + a_2 + \cdots + a_{n-1} = 1 - a_n$, we still need to prove the inequality

$$\frac{(1 - a_n)^2}{\sqrt{2} + \sqrt{3}} > a_n\sqrt{n}.$$

But since

$$1 = a_1 + 2a_2 + ... + na_n - (a_1 + ... + a_n) > (n-1)a_n,$$

we have $a_n < \frac{1}{n-1}$ and so it is enough to prove the previous inequality with a_n replaced by $\frac{1}{n-1}$. But this is immediate for $n \geq 16$, which ends the solution. □

We end this chapter with two challenging inequalities. The first one uses a combination of T_2's lemma and a rather subtle linearization technique.

18. Prove that for all positive numbers a, b, c the following inequality holds

$$\sqrt{\frac{a}{8b+c}} + \sqrt{\frac{b}{8c+a}} + \sqrt{\frac{c}{8a+b}} \geq 1.$$

Vo Quoc Ba Can

Proof. We use T_2's lemma in the form

$$\sqrt{\frac{a}{8b+c}} = \frac{(\sqrt{a})^2}{\sqrt{a(8b+c)}},$$

so it remains to prove that

$$\left(\sum \sqrt{a}\right)^2 \geq \sum \sqrt{a(8b+c)}.$$

Define $x = \sqrt{a}, y = \sqrt{b}, z = \sqrt{c}$, so that the inequality becomes

$$\sum x\sqrt{8y^2+z^2} \leq \left(\sum x\right)^2.$$

This is actually a very strong inequality and it is easy to see that it resists any attempt to prove it using Cauchy-Schwarz (even if its form invites us to use such a technique). We will use a linearization technique, by approximating first $\sqrt{8y^2+z^2}$ by a rational function in y, z. The crucial ingredient is the following estimate:

$$\sqrt{8y^2+z^2} \leq 3y + z - \frac{3yz}{2y+z}.$$

The hard point is to figure out that such an inequality holds, since proving it is a very easy matter. Note that the right-hand side is clearly positive, so by squaring and canceling similar terms we obtain the equivalent inequality

$$y^2 + 6yz + \frac{9y^2z^2}{(2y+z)^2} \geq \frac{6yz(3y+z)}{2y+z} \iff \frac{4y^2(y-z)^2}{(2y+z)^2} \geq 0.$$

Using this estimate we conclude:

$$\sum x\sqrt{8y^2+z^2} \leq 4\sum xy - 3xyz\sum \frac{1}{2y+z}$$

and since

$$\sum \frac{1}{2y+z} \geq \frac{3}{x+y+z},$$

it remains to prove that

$$4\sum xy - \frac{9xyz}{x+y+z} \leq \sum x^2 + 2\sum xy.$$

It is not difficult to check that the last inequality is actually equivalent to Schur's inequality, which finishes the proof of this hard problem. □

The next problem requires some preliminaries. We will prove the following beautiful discrete variant of a classical inequality of Wirtinger.

Theorem 5.3. *(Fan's inequality) If $x_1, x_2, \ldots, x_n$ are real numbers which add up to zero, then*

$$(x_1^2 + x_2^2 + \cdots + x_n^2)\cos\frac{2\pi}{n} \geq x_1x_2 + x_2x_3 + \cdots + x_nx_1.$$

Proof. We will actually mimic the proof of Wirtinger's inequality by using finite Fourier transforms (for more on this fascinating subject, the reader is invited to read the addendum 7.A). Namely, if $z_1, z_2, \ldots, z_n$ is a sequence of complex numbers, define

$$\hat{z}_j = \frac{1}{\sqrt{n}} \cdot \sum_{k=1}^{n} z_k e^{-\frac{2i\pi kj}{n}}.$$

Using the fact that

$$\sum_{k=1}^{n} e^{\frac{2i\pi kj}{n}} = 0$$

if and only if j is not a multiple of n (in which case it equals n), it is easy to check that we can recover our sequence from the sequence of its finite Fourier transforms using the inversion formula

$$z_j = \frac{1}{\sqrt{n}} \sum_{k=1}^{n} \hat{z}_k e^{\frac{2i\pi kj}{n}}.$$

We also have the following discrete version of Parseval's identity:

$$\sum_{k=1}^{n} |\hat{z}_k|^2 = \sum_{k=1}^{n} |z_k|^2.$$

Indeed,

$$\begin{aligned}
\sum_{k=1}^{n} |\hat{z}_k|^2 &= \frac{1}{n} \sum_j \sum_{k_1,k_2} z_{k_1}\overline{z_{k_2}} e^{-\frac{2i\pi(k_1-k_2)j}{n}} \\
&= \frac{1}{n} \sum_{k_1,k_2} z_{k_1}\overline{z_{k_2}} \sum_j e^{-\frac{2i\pi(k_1-k_2)j}{n}} \\
&= \sum_{k_1=k_2} z_{k_1}\overline{z_{k_2}} \\
&= \sum_k |z_k|^2.
\end{aligned}$$

Now, the proof of Fan's inequality is very easy: write the inequality in the form

$$\sum_k (x_k - x_{k+1})^2 \geq \left(2 - 2\cos\frac{2\pi}{n}\right) \sum_k x_k^2.$$

Note that the hypothesis $x_1+x_2+\cdots+x_n = 0$ can be written in the very simple form $\hat{x}_n = 0$. Now, let[2] $y_j = x_j - x_{j+1}$ and observe that $\hat{y}_j = (1 - e^{\frac{2i\pi j}{n}})\hat{x}_j$, as

[2] This is the analogue of the derivative of a function when establishing discrete inequalities.

follows from the definitions. Thus, using Parseval's identity we obtain

$$\begin{aligned}\sum_j y_j^2 &= \sum_j |\hat{y}_j|^2 \\ &= \sum_j |1 - e^{\frac{2i\pi j}{n}}|^2 |\hat{x}_j|^2 \\ &\geq \sum_{j=1}^{n-1} \left(2 - 2\cos\frac{2\pi}{n}\right) |\hat{x}_j|^2 \\ &= \left(2 - 2\cos\frac{2\pi}{n}\right) \sum_j |\hat{x}_j|^2\end{aligned}$$

and another application of Parseval's identity yields the desired inequality. □

We are now ready to prove the following version of Shapiro's inequality. The proof is based on a very tricky application of T_2's lemma combined with Fan's inequality.

19. Let $a_n = \frac{1}{\sqrt{2\cos\frac{2\pi}{n} - 1}}$. Prove that for all $x_1, x_2, \ldots, x_n \in \left[\frac{1}{a_n}, a_n\right]$, the Shapiro inequality holds:

$$\frac{x_1}{x_2 + x_3} + \frac{x_2}{x_3 + x_4} + \cdots + \frac{x_n}{x_1 + x_2} \geq \frac{n}{2}.$$

Vasile Cârtoaje, Gabriel Dospinescu

Proof. First, we write the inequality in the form

$$\sum_i \frac{x_i - \frac{x_{i+1} + x_{i+2}}{2a_n^2}}{x_{i+1} + x_{i+2}} \geq \frac{n}{2}\left(1 - \frac{1}{a_n^2}\right).$$

Note that

$$x_i \geq \frac{1}{a_n} \geq \frac{x_{i+1} + x_{i+2}}{2a_n^2}.$$

Using T_2's lemma, it suffices to prove that

$$\left(1-\frac{1}{a_n^2}\right)\left(\sum x_i\right)^2 \geq \frac{n}{2}\left(\sum x_i(x_{i+1}+x_{i+2}) - \frac{1}{2a_n^2}\sum (x_i+x_{i+1})^2\right).$$

A crucial step is to note that we have the identity

$$\sum x_i(x_{i+1}+x_{i+2}) = \sum (x_i+x_{i+1})(x_{i+1}+x_{i+2}) - \frac{1}{2}\sum (x_i+x_{i+1})^2,$$

which allows us to perform the substitution $x_i + x_{i+1} = 2b_i$ and reduce the problem to proving that

$$\left(1-\frac{1}{a_n^2}\right)\left(\sum b_i\right)^2 \geq n\left(2\sum b_i b_{i+1} - \left(1+\frac{1}{a_n^2}\right)\sum b_i^2\right)$$

for nonnegative real numbers b_i.

This simplifies drastically if we make another substitution

$$c_i = b_i - \frac{b_1+b_2+\cdots+b_n}{n}.$$

A rather tedious, but straightforward computation shows that the previous inequality is equivalent to

$$\left(1+\frac{1}{a_n^2}\right)\sum c_i^2 \geq 2\sum c_i c_{i+1}.$$

Of course, we have $c_1 + c_2 + \cdots + c_n = 0$. Finally, note that

$$1+\frac{1}{a_n^2} = 2\cos\frac{2\pi}{n}.$$

Thus, the result follows from Fan's inequality. □

5.1 Notes

We thank the following people for providing solutions: Alexandru Chirvăsitu (problem 16), Xiangyi Huang (problem 14), Michael Rozenberg (problem 15), Dusan Sobot (problems 1, 3), Gjergji Zaimi (problems 1, 10, 11).

Addendum 5.A Hölder's Inequality in Action

The purpose of this addendum is to present some applications of Hölder's inequality, which are quite similar to the problems considered in this chapter. Of course, Hölder's inequality is important because of its applications in measure theory, probability theory and analysis and not really for the amusing problems to be discussed here. Actually, we will not even deal with the classical version

$$(a_1^p + a_2^p + \cdots + a_n^p)^{\frac{1}{p}} (b_1^q + b_2^q + \cdots + b_n^q)^{\frac{1}{q}} \geq a_1 b_1 + a_2 b_2 + \cdots + a_n b_n,$$

which holds for any positive real numbers a_i, b_j, p, q such that $\frac{1}{p} + \frac{1}{q} = 1$, but rather with the following:

Theorem 5.A.1. *Let a_{ij} be positive real numbers. Then*

$$\prod_{j=1}^{k} (a_{j1} + a_{j2} + \cdots + a_{jn}) \geq \left(\sqrt[k]{a_{11} a_{21} \cdots a_{k1}} + \cdots + \sqrt[k]{a_{1n} a_{2n} \cdots a_{kn}}\right)^k.$$

Proof. This is a very easy application of the AM-GM inequality. Indeed, just add up the following inequalities

$$k \sqrt[k]{\frac{a_{11} a_{21} \cdots a_{k1}}{S_1 S_2 \cdots S_k}} \leq \frac{a_{11}}{S_1} + \cdots + \frac{a_{k1}}{S_k},$$

$$\vdots$$

$$k \sqrt[k]{\frac{a_{1n} a_{2n} \cdots a_{kn}}{S_1 S_2 \cdots S_k}} \leq \frac{a_{1n}}{S_1} + \cdots + \frac{a_{kn}}{S_k},$$

where $S_i = a_{i1} + a_{i2} + \cdots + a_{in}$. □

We are now able to obtain a generalization of T_2's lemma that turns out to be very handy in quite a lot of situations when T_2's lemma fails. There are of course much more general results than the following one, but our purpose is not to delve into the greatest generality.

Theorem 5.A.2. *Let $a_1, a_2, \dots, a_n$ and $x_1, x_2, \dots, x_n$ be positive real numbers and let $q > p$ be two positive integers. Then*

$$\frac{a_1^q}{x_1^p} + \frac{a_2^q}{x_2^p} + \cdots + \frac{a_n^q}{x_n^p} \geq n^{1+p-q} \frac{(a_1 + a_2 + \cdots + a_n)^q}{(x_1 + x_2 + \cdots + x_n)^p}.$$

Proof. Note that by the previous theorem we have

$$\begin{aligned}(x_1 + x_2 + \cdots + x_n)^p \cdot \left(\frac{a_1^q}{x_1^p} + \frac{a_2^q}{x_2^p} + \cdots + \frac{a_n^q}{x_n^p}\right) \\ = (x_1 + \cdots + x_n) \cdots (x_1 + \cdots + x_n) \cdot \left(\frac{a_1^q}{x_1^p} + \frac{a_2^q}{x_2^p} + \cdots + \frac{a_n^q}{x_n^p}\right) \\ \geq (a_1^{\frac{q}{p+1}} + a_2^{\frac{q}{p+1}} + \cdots + a_n^{\frac{q}{p+1}})^{p+1}\end{aligned}$$

and the result follows from the Power Mean inequality. □

The first theorem is particularly useful when working with sums of square (or cubic or ...) roots, which are a nightmare most of the times. Here are a few examples that will probably convince you how useful this result is. Most of the problems are quite hard to solve by other means.

A.2. Prove that for all positive real numbers a, b, c such that $abc = 1$ we have

$$\sqrt[3]{a^3 + 7} + \sqrt[3]{b^3 + 7} + \sqrt[3]{c^3 + 7} \leq 2(a + b + c).$$

Vasile Cârtoaje

Proof. This is quite a strong inequality, but the previous results are effective in such a context:

$$\begin{aligned}(\sqrt[3]{a^3 + 7} + \sqrt[3]{b^3 + 7} + \sqrt[3]{c^3 + 7})^3 \\ \leq (1 + 1 + 1)(a + b + c)((a^2 + 7bc) + (b^2 + 7ca) + (c^2 + 7ab))\end{aligned}$$

and the result follows immediately from the inequality

$$ab + bc + ca \leq a^2 + b^2 + c^2.$$ □

A.3. For any positive real numbers a, b, c the following inequality holds

$$\frac{b+c}{\sqrt{a^2+bc}}+\frac{c+a}{\sqrt{b^2+ca}}+\frac{a+b}{\sqrt{c^2+ab}}\geq 4.$$

Pham Kim Hung

Proof. Let S be the left-hand side. Using Hölder's inequality, we can write

$$S^2\cdot\left(\sum(b+c)(a^2+bc)\right)\geq 8(a+b+c)^3,$$

so it is enough to prove the stronger inequality

$$(a+b+c)^3\geq 2\sum(b+c)(a^2+bc).$$

An easy computation shows that this follows from Schur's inequality. □

A.4. Let a, b, c be positive reals such that $abc=1$. Show that

$$\frac{1}{a^5(b+2c)^2}+\frac{1}{b^5(c+2a)^2}+\frac{1}{c^5(a+2b)^2}\geq\frac{1}{3}.$$

Titu Andreescu, USA TST 2010

Proof. The substitution $a=\frac{1}{x}, b=\frac{1}{y}, c=\frac{1}{z}$ reduces the problem to

$$\frac{x^3}{(2y+z)^2}+\frac{y^3}{(2z+x)^2}+\frac{z^3}{(2x+y)^2}\geq 3.$$

By theorem 5.A.2,

$$\frac{x^3}{(2y+z)^2}+\frac{y^3}{(2z+x)^2}+\frac{z^3}{(2x+y)^2}\geq\frac{x+y+z}{9}$$

and the result follows from the AM-GM inequality. □

A.5. Prove that if $x_1, x_2, \ldots, x_n$ are positive real numbers with product 1, then

$$n^n(1+x_1^n)(1+x_2^n)\cdots(1+x_n^n) \geq \left(x_1+x_2+\cdots+x_n+\frac{1}{x_1}+\frac{1}{x_2}+\cdots+\frac{1}{x_n}\right)^n.$$

Gabriel Dospinescu

Proof. This follows easily by adding the inequalities

$$\sqrt[n]{(x_1^n+1)(x_2^n+1)...(x_n^n+1)} \geq x_i + x_1x_2...x_{i-1}x_{i+1}...x_n = x_i + \frac{1}{x_i}$$

obtained from Hölder's inequality. □

A.6. For any positive real numbers a, b, c the following inequality holds

$$\sqrt[3]{a \cdot \frac{a+b}{2} \cdot \frac{a+b+c}{3}} \geq \frac{a+\sqrt{ab}+\sqrt[3]{abc}}{3}.$$

Kiran Kedlaya

Proof. Using theorem 5.A.1, we obtain

$$(a+a+a)\left(a+\frac{a+b}{2}+b\right)(a+b+c) \geq \left(a+\sqrt[3]{\frac{ab(a+b)}{2}}+\sqrt[3]{abc}\right)^3.$$

One concludes by observing that $\sqrt[3]{\frac{ab(a+b)}{2}} \geq \sqrt{ab}$. □

A.7. Prove that for all real numbers a, b, c we have

$$2(1+a^2)(1+b^2)(1+c^2) \geq (1+ab+bc+ca)^2.$$

Michael Rozenberg

Proof. This is a pretty tricky application of Hölder's inequality:

$$\begin{aligned}4(1+a^2)^2(1+b^2)^2(1+c^2)^2 &= (a^2b^2+a^2+b^2+1)(b^2+c^2+b^2c^2+1)\\ &\quad\cdot(1+1+1+1)(a^2+a^2c^2+c^2+1)\\ &\geq (1+ab+bc+ca)^4.\end{aligned}$$

The result follows. □

A.8. Prove that for all positive real numbers a, b, c the following inequality holds

$$abc + \sqrt[3]{(a^3+1)(b^3+1)(c^3+1)} \geq ab+bc+ca.$$

Proof. Using Hölder's inequality, we can write

$$\sqrt[3]{(a^3+1)(b^3+1)(c^3+1)} \geq ab+c.$$

We would like to have

$$abc+ab+c \geq ab+bc+ca,$$

which is equivalent to $(a-1)(b-1) \geq 0$. There is no reason for this to hold, but at least one of the inequalities $(a-1)(b-1) \geq 0$, $(b-1)(c-1) \geq 0$ and $(a-1)(c-1) \geq 0$ holds, as two of the numbers $a-1, b-1, c-1$ must have the same sign. The result follows. □

A.9. Prove that for all positive real numbers a, b, c the following inequality holds

$$(a^5-a^2+3)(b^5-b^2+3)(c^5-c^2+3) \geq (a+b+c)^3.$$

Titu Andreescu, USAMO 2004

Proof. Of course, we cannot apply theorem 5.A.1 directly in this case, but the form of the inequality is a temptation to find a way to apply theorem 5.A.1. The key point is the inequality

$$a^5-a^2+3 \geq a^3+2,$$

which is equivalent to $(a^2-1)(a^3-1) \geq 0$ and thus obvious. Using this, the result follows immediately from theorem 5.A.1, by writing

$$a^3 + 2 = a^3 + 1^3 + 1^3. \qquad \square$$

A.10. Let a, b, c be the sides of a triangle. Prove that

$$\sqrt{\frac{abc}{b+c-a}} + \sqrt{\frac{abc}{c+a-b}} + \sqrt{\frac{abc}{a+b-c}} \geq a+b+c.$$

Titu Andreescu, Gabriel Dospinescu

Proof. This is a hard inequality. The point is to use Schur's inequality

$$a^2(a-b)(a-c) + b^2(b-a)(b-c) + c^2(c-a)(c-b) \geq 0,$$

which can also be written as

$$abc(a+b+c) \geq a^3(b+c-a) + b^3(c+a-b) + c^3(a+b-c)$$

with the theorem 5.A.2. Indeed, observe that we can write

$$\sum a^3(b+c-a) = \sum \frac{a^3}{\left(\sqrt{\frac{1}{b+c-a}}\right)^2} \geq \frac{(a+b+c)^3}{\left(\sum \sqrt{\frac{1}{b+c-a}}\right)^2}.$$

The result follows. $\square$

A.11. Prove that for all $a, b, c > 0$ we have

$$\left(\frac{a^2}{b} + \frac{b^2}{c} + \frac{c^2}{a}\right)^4 \geq 27 \cdot (a^4 + b^4 + c^4).$$

Proof. If X is the square root of the left-hand side, then Hölder's inequality yields

$$X \cdot (a^2b^2 + b^2c^2 + c^2a^2) \geq (a^2 + b^2 + c^2)^3,$$

so that after substituting $x = a^2, y = b^2$ and $z = c^2$, it is enough to prove the inequality

$$(x+y+z)^6 \geq 27(xy+yz+zx)^2(x^2+y^2+z^2).$$

But this is immediate from the AM-GM inequality and the identity

$$x^2+y^2+z^2+2(xy+yz+zx) = (x+y+z)^2.$$

□

A.12. Prove that for all positive real numbers a, b, c, d

$$\frac{a^3+1}{a^2+1} \cdot \frac{b^3+1}{b^2+1} \cdot \frac{c^3+1}{c^2+1} \cdot \frac{d^3+1}{d^2+1} \geq \frac{abcd+1}{2}.$$

Vasile Cârtoaje, Gazeta Matematica

Proof. This is also a very hard problem. Of course, one is again tempted to use the theorem, but one needs a trick. The key point is that for all positive numbers a we have

$$\frac{a^3+1}{a^2+1} \geq \sqrt[4]{\frac{a^4+1}{2}}.$$

Indeed, using the inequality

$$(a^2+1)^4 \leq (a+1)^2(a^3+1)^2,$$

which is just Cauchy-Schwarz, one reduces this to

$$2(a^3+1)^2 \geq (a+1)^2(a^4+1),$$

which, after division by $(a+1)^2$ and a small computation is equivalent to

$$(a-1)^4 \geq 0.$$

Once we have this key inequality, the rest is an immediate application of theorem 5.A.1.

□

Chapter 6

Some Classical Problems in Extremal Graph Theory

This elementary chapter is a variation on a classical topic in extremal graph theory: Turán's theorem on graphs without cliques. Recall that if G is a graph and k is a positive integer, then a k-clique is a set of k vertices, any two of which are connected. A graph is called k-free if it does not contain a k-clique. We then have the following standard result.

Theorem 6.1. *(Turán) The maximal number of edges in a k-free graph with n vertices is $\frac{k-2}{k-1} \cdot \frac{n^2 - r^2}{2} + \binom{r}{2}$, where r is the remainder of n when divided by $k-1$.*

In particular, the maximum number of edges in a k-free graph with n vertices is at most $\frac{k-2}{k-1} \cdot \frac{n^2}{2}$, which is really the estimate that we will constantly use. We start with a series of rather direct applications of these two theorems. The other problems are however different in nature and more difficult.

1. Let $x_1, x_2, \ldots, x_n$ be real numbers. Prove that there are at most $\frac{n^2}{4}$ pairs $(i, j) \in \{1, 2, \ldots, n\}^2$ such that $i < j$ and $1 < |x_i - x_j| < 2$.

   MOSP 2001

Proof. Consider the graph whose vertices are $1, 2, \dots, n$. Connect two vertices i, j by an edge if (i, j) satisfies $1 < |x_i - x_j| < 2$. We claim that this graph contains no triangle. If we manage to prove this, the result follows from Turán's theorem. Suppose that the graph contains a triangle, thus we can find distinct a, b, c such that

$$1 < |x_a - x_b| < 2, \quad 1 < |x_b - x_c| < 2, \quad 1 < |x_c - x_a| < 2.$$

By symmetry, we may assume that $x_a < x_b < x_c$. Then the previous inequalities become $x_b - x_a > 1, x_c - x_b > 1$, so $x_c - x_a > 2$, contradicting the inequality $|x_c - x_a| < 2$. The result follows. □

2. Prove that if n points lie on a unit circle, then at most $\dfrac{n^2}{3}$ segments connecting them have length greater than $\sqrt{2}$.

Poland 1997

Proof. Consider the graph whose vertices are the n points and connect two vertices if their distance is greater than $\sqrt{2}$. The point is that this graph contains no 4-clique. This is clear, since a chord of length $\sqrt{2}$ subtends a central angle of $\frac{\pi}{2}$. Thus, by Turán's theorem the number of vertices is at most $\left[\frac{n^2}{3}\right]$, finishing the proof. □

3. There are 1999 people participating in an exhibition. Out of any 50 people, at least two do not know each other. Prove that we can find at least 41 people who each know at most 1958 other people.

Taiwan 1999

Proof. This is a special case of the proof of Zarankiewicz's lemma. For the reader's convenience, let us recall the proof. Suppose that the conclusion does not hold, so we can find at least 1959 people, say $A_1, \dots, A_{1959}$, each having at least 1959 friends. Start with A_1. There is a person among $A_2, \dots, A_{1959}$ that knows A_1. Assuming that we found persons $A_{i_1}, \dots, A_{i_k}$ $(i_1 = 1)$ among $A_1, \dots, A_{1959}$, every two knowing each other, let us try to add a new person $A_{i_{k+1}}$ to the group. But there are at least $k \cdot 1959 - (k-1) \cdot 1999$ persons that know $A_{i_1}, \ldots, A_{i_k}$. Thus, if $k \cdot 1959 - (k-1) \cdot 1999 \geq 1$, we can always add one more person. If $k \leq 48$, then there are at least 79 people who know all of $A_{i_1}, \ldots, A_{i_k}$ and one of them is among $A_1, A_2, \ldots, A_{1995}$. If $k = 49$, then there are at least 39 people who know all of $A_{i_1}, \ldots, A_{i_k}$ and (although that person may not be among $A_1, A_2, \ldots, A_{1959}$) he completes a set of 50 all of whom know each other. □

4. We are given $5n$ points in a plane and we connect some of them so that $10n^2+1$ segments are drawn. We color these segments in 2 colors. Prove that we can find a monochromatic triangle.

Proof. Note that the corresponding graph (with vertices the $5n$ points and edges between points connected by a segment) contains a 6-clique. Indeed, otherwise by Turán's theorem it has at most $\frac{(5n)^2}{2} \cdot \frac{4}{5} = 10n^2$ edges, which contradicts the hypothesis. Now, consider a 6-clique G' of our graph G. The edges of G' are colored in two colors and we claim that there is always a monochromatic triangle in G'. This is standard, but we recall the proof. Pick a vertex v_1 of G'. By the pigeonhole principle, we may assume that v_1v_2, v_1v_3, v_1v_4 have the same color. If one of the edges v_2v_3, v_2v_4 or v_3v_4 has the same color as v_1v_2, we are done. Otherwise, the triangle $v_2v_3v_4$ is monochromatic and we win again. □

The following problem does not use Turán's theorem, but it is still a very classical topic, namely Ramsey's numbers.

5. A group of people is called n-balanced if the following two conditions are satisfied:

   a) among any three people, there are two who know each other;

   b) among any n people, there are at least two not knowing each other.

   Prove that there are always at most $\frac{(n-1)(n+2)}{2}$ people in an n-balanced group.

Dorel Miheţ, Romanian TST 2008

Proof. We will prove the result by induction. For $n = 2$, this is trivial, so assume that it holds for $n-1$. Consider an n-balanced group and pick an arbitrary person P. Let A be the set of friends of P and let B be the set of all the other persons in the group. By hypothesis, any two persons in B are friends and B has at most $n-1$ elements. By induction, A has at most $\frac{(n-2)(n+1)}{2}$ persons (by b) and the fact that P knows all elements of A, it follows that A is an $n-1$-balanced group). Thus there are at most

$$\frac{(n-2)(n+1)}{2} + 1 + (n-1) = \frac{(n-1)(n+2)}{2}$$

persons in the group, which is enough to prove the inductive step. □

The following problem and the method of proof are absolute classics.

6. Prove that a graph with n vertices and k edges has at least $\dfrac{k}{3n}(4k - n^2)$ triangles.

APMO 1989

Proof. Let $x_1, x_2, \ldots, x_n$ be the vertices of the graph G and let $d(x)$ be the degree of x. Observe that for a given edge $e = x_i x_j$ with endpoints x_i, x_j, there are at least $d(x_i) + d(x_j) - n$ triangles containing e. Indeed, there are $d(x_i) + d(x_j) - 2$ edges having as endpoints one of x_i, x_j and another vertex among the $n-2$ remaining vertices, so there are at least $d(x_i)+d(x_j)-2-(n-2)$ triangles containing x_i, x_j as vertices. Summing over all edges and taking into account that we count three times every triangle in this way, shows that the number of triangles is at least (we denote by $E(G)$ the set of edges of G)

$$\frac{1}{3} \sum_{e=x_i x_j \in E(G)} (d(x_i) + d(x_j) - n).$$

The previous sum is also equal to

$$\frac{1}{3}\left(\sum_{i=1}^{n} d(x_i)^2 - nk\right).$$

Applying Cauchy-Schwarz and taking into account that

$$\sum_{i=1}^{n} d(x_i) = 2k$$

yields the desired result. □

The following problem is very similar to the previous one.

7. A graph with n vertices and k edges has no triangles. Prove that we can choose a vertex such that the subgraph obtained by deleting this vertex and all its neighbors has at most $k\left(1 - \frac{4k}{n^2}\right)$ edges.

USAMO 1995

Proof. Each edge xy gets killed when we remove x and its neighbors, when we remove y and its neighbors, or when we remove any common neighbor of x and y and its neighbors. Thus, this edge is killed a total of $d(x) + d(y)$ times. Summing over all edges and taking into account that $d(x)$ appears as a summand exactly $d(x)$ times, we obtain that the average number of edges killed by removing a vertex and its neighbors is $\frac{1}{n}\sum_x d(x)^2$. Since

$$\sum_x d(x) = 2k,$$

we have

$$\frac{1}{n}\sum_x d(x)^2 \geq \left(\frac{2k}{n}\right)^2.$$

Thus, on average we kill at least $\frac{4k^2}{n^2}$ edges. The result follows. □

8. A graph G has n vertices and contains no complete subgraph with four vertices. Prove that G contains at most $\frac{n^3}{27}$ triangles.

Ivan Borsenco, Mathematical Reflections

Proof. The result is easy for $n = 2, 3, 4$, so let us assume that $n > 4$ and that the result holds for all $k < n$. We may assume that G contains a triangle ABC. Let G_1 be the subgraph formed by the vertices of G different from A, B, C. Since G_1 has no 4-clique, it has at most $\frac{(n-3)^2}{3}$ edges by Turán's theorem. By the inductive hypothesis, G_1 has at most $\frac{(n-3)^3}{27}$ triangles. Any other triangle in G consists either of a vertex of G_1 and an edge of ABC or of an edge of G_1 and a vertex of ABC. Moreover, any edge of G_1 forms a triangle with at most one vertex of ABC and any vertex of G_1 forms a triangle with at most one edge of ABC, because G contains no 4-clique. Thus G contains at most

$$\frac{(n-3)^3}{27} + \frac{(n-3)^2}{3} + n - 3 + 1 = \frac{n^3}{27},$$

establishing the inductive step. Note that the result is optimal if n is a multiple of 3, since we can consider the tripartite graph $K_{n/3,n/3,n/3}$. □

It is a standard result that a graph with $n \geq 4$ vertices and more than $\frac{n+n\sqrt{4n-3}}{4}$ edges has a four-cycle (see, for instance [3], example 3, chapter 22). The following nice problem shows that this result is almost optimal.

9. a) Let p be a prime. Consider the graph whose vertices are the ordered pairs (x, y) with $x, y \in \{0, 1, ..., p-1\}$ and whose edges join vertices (x, y) and (x', y') if and only if $xx' + yy' \equiv 1 \pmod{p}$. Prove that this graph does not contain a 4-cycle .

   b) Prove that for infinitely many n there is a graph G_n with n vertices and at least $\frac{n\sqrt{n}}{2} - n$ edges that does not contain a 4-cycle.

Hungary-Israel Competition 2001

Proof. One can give a rather down-to-earth proof of a), based on explicit computations, but we prefer the following approach. Consider the $\mathbb{F}_p = \mathbb{Z}/p\mathbb{Z}$-vector space $V = \mathbb{F}_p^2$ and the standard inner product $\langle (x, y), (x', y') \rangle = xx' + yy'$. We are asked to prove that we cannot find four distinct vectors $v_1, v_2, v_3, v_4 \in V$ such that $\langle v_i, v_{i+1} \rangle = 1$ for all $1 \leq i \leq 4$ (here $v_5 = v_1$). If such vectors existed, v_2 and v_4 would be both orthogonal to the nonzero vector $v_1 - v_3$ and so they would be contained in a line of V. Thus, we can find $\lambda \in \mathbb{F}_p$ such that $v_4 = \lambda v_2$. Then $\lambda = 1$, as $\langle v_1, v_4 \rangle = \langle v_1, v_2 \rangle = 1$. We conclude that $v_2 = v_4$, a contradiction.

For the second part, take any prime p and set $n = p^2$. The graph G_n in the first part of the problem has n vertices. Let us evaluate the number of edges. If $v = (x, y)$ is a nonzero vector in V, then the equation $\langle w, v \rangle = 1$ has p solutions. Thus the number of edges is $\frac{p(p^2-1)}{2}$ and it is immediate to check that this is greater than $\frac{n\sqrt{n}}{2} - n = \frac{p^3}{2} - p^2$. The result follows. □

We continue with two very nice problems concerning graphs with $n^2 + 1$ edges and $2n$ vertices. Note that this is the first case when Turán's theorem ensures the existence of a triangle. The following results show that we can do better.

10. Prove that a graph with $2n$ vertices and $n^2 + 1$ edges contains two triangles sharing a common edge.

Chinese TST 1987

Proof. We will prove the result by induction. For $n = 2$, this is trivial. Assume now that the result holds for n and consider a graph G with $2n + 2$ vertices and $n^2 + 2n + 2$ edges. By Turán's theorem and the pigeonhole principle, we can find a triangle xyz such that $d(x) \equiv d(y) \pmod 2$. Consider the $2n$ points different from x, y. If there are at least $n^2 + 1$ edges among them, we are done by the inductive hypothesis. If not, then there are at least $2n + 2$ edges whose endpoints are x or y. But since $d(x) \equiv d(y) \pmod 2$, it follows that there are actually at least $2n + 2$ edges whose endpoints are x or y and which are different from the edge xy. Let A_1 be the set of vertices $v \neq x, y$ that are connected to x and let A_2 be the set of vertices $v \neq x, y$ that are connected to y. Then the previous result implies that $|A_1| + |A_2| \geq 2n + 2$. Since $|A_1 \cup A_2| \leq 2n$, it follows that $|A_1 \cap A_2| \geq 2$ and so we can find two distinct vertices $z \neq t$ that are connected to both x, y. Then the triangles xyz and xyt share a common edge and we are done. □

11. A graph has $2n$ vertices and $n^2 + 1$ edges. Prove that it contains at least n triangles.

Proof. Again, the proof is by induction. For $n = 2$ this is trivial, so assume that the result holds for n and consider a graph with $2n + 2$ vertices and $n^2 + 2n + 2$ edges. By Turán's theorem we can find a triangle $x_1x_2x_3$. Let A_i be the set of vertices $v \neq x_1, x_2, x_3$ that are connected to x_i. There are obviously at least

$$S = |A_1 \cap A_2| + |A_2 \cap A_3| + |A_3 \cap A_1|$$

triangles different from $x_1x_2x_3$, since any element of $A_i \cap A_j$ forms a triangle with the vertices x_ix_j. Thus, if we have $S \geq n$, we are done. Assume that $S \leq n - 1$. Then, using the inclusion-exclusion principle, it follows that

$$2n - 1 \geq |A_1 \cup A_2 \cup A_3| \geq |A_1| + |A_2| + |A_3| - n + 1,$$

thus one of the numbers $|A_1| + |A_2|, |A_2| + |A_3|, |A_3| + |A_1|$ is smaller than $2n - 1$, say $|A_1| + |A_2|$. But this means that there are at least

$$n^2 + 2n + 2 - (2n + 1) = n^2 + 1$$

edges among the vertices $A_3, A_4, \ldots, A_{2n+2}$. Thus, by induction we find at least n triangles among these vertices and adding the triangle $A_1A_2A_3$ yields at least $n + 1$ triangles. The inductive step is thus proved and the problem solved. □

The following problems are more difficult. The next one concerns coverings of the edges of a graph by cliques.

12. There are n aborigines on an island. Any two of them are either friends or enemies. One day they receive an order saying that all citizens should make and wear a necklace with zero or more stones so that

    i) for any pair of friends there exists a color such that each of the two persons has a stone of that color;

    ii) for any pair of enemies there does not exist such a color.

    What is the least number of colors of stones required (considering all possible relationships between the inhabitants of the island)?

    Belarus 2001

Proof. Let G be the graph whose vertices are the aborigines, two of them being connected if they are friends. Let B_i be the set of aborigines who receive a bead of the i-th color. Then by condition ii) each B_i forms a clique in G and by i) each edge is in one of these cliques. Conversely, if we have a collection of k cliques that cover every edge of G, then we can give a bead of color i to every aborigine in the i-th clique and satisfy the conditions of the order.

The previous paragraph shows that we need to find the smallest number $f(n)$ of cliques that cover the edges of any graph on n vertices G. Clearly $f(2) = 1$ and $f(3) = 2$. To find an optimal graph, consider a complete bipartite graph on n vertices. The key point is that such a graph has no triangles, so if we want to cover its edges by cliques, we need at least $E(G)$ cliques, where $E(G)$ is the number of edges. Therefore, we have a clear bound on $f(n)$, namely $f(n) \geq \left[\frac{n^2}{4}\right]$.

The hard part is to prove the opposite inequality. We will prove this by induction, going from n to $n+2$ (which, combined with the first two values of $f(n)$, will be enough to conclude). Consider a graph G with $n+2$ vertices. We may assume that at least two vertices are connected, say x, y. Consider the graph G_1 obtained by deleting x, y and all edges adjacent to these two vertices. By induction, we know that we can cover its edges by $f(n)$ cliques. If v is a vertex of G_1, consider the subgraph spanned by x, y, v. By adding either a triangle or an edge, we can cover all edges of this subgraph incident to v. Finally, by adding one more edge to cover the edge xy, we covered all edges of G by using at most $f(n) + n + 1$ cliques (by the way, all the cliques used were only edges or triangles). Therefore

$$f(n+2) \leq f(n) + n + 1.$$

Since

$$\left[\frac{(n+2)^2}{4}\right] = \left[\frac{n^2}{4}\right] + n + 1,$$

it follows that $f(n) \leq \left[\frac{n^2}{4}\right]$. Therefore the answer is $\left[\frac{n^2}{4}\right]$. □

It is not hard to guess the answer of the following problem, but proving that this is the correct answer is quite a pain in the neck.

13. What is the least number of edges in a connected n-vertex graph such that any edge belongs to a triangle?

Paul Erdős, AMM E 3255

Proof. Let $f(n)$ be the desired number and let $g(n) = [\frac{3n-2}{2}]$. We will prove that $f(n) = g(n)$ for all n. To save words, call good a connected graph in which every edge belongs to a triangle.

We can easily establish that $f(n) \leq g(n)$ by explicit constructions: if $n = 2k+1$, consider k triangles sharing a common vertex, this being the only common point of any two triangles. This is a good graph with $3k$ edges, so $f(2k+1) \leq 3k = g(2k+1)$. If $n = 2k$, start from the above good graph with order $2k-1$, choose an edge AB and add a vertex X and edges AX, BX. We obtain a good graph with n vertices and $g(n)$ edges.

The hard point is proving that any good graph on n vertices has at least $g(n)$ edges. We will prove this by strong induction. For $n = 3, 4$, this is easily checked. The crucial observation that makes the induction work is that a good graph with n vertices and less than $g(n)$ edges must have a vertex of degree 2. This is trivial, since by connectedness all vertices have degree at least 1 and since every edge is in a triangle, every vertex must have degree at least 2. However, the sum of the degrees of all vertices is twice the number of edges, thus smaller than $3n$.

Suppose now that $f(j) = g(j)$ for all $j < n$ and let us prove that $f(n) \geq g(n)$. Suppose G is a connected graph with $f(n)$ edges such that each edge belongs to a triangle and suppose $f(n) < g(n)$. There is a vertex x of G of degree 2, which by assumption must be in some triangle xyz. Suppose the edge yz is in a second triangle. Then the graph G' obtained from G by removing x and the edges xy and xz is connected, has $n-1$ vertices, every edge belongs to a triangle, and $f(n) - 2 < g(n) - 2 \leq g(n-1)$ edges. This is a contradiction. Thus yz is not contained in another triangle. Form a new graph G'' by deleting the vertex x and collapsing the vertices y and z into a single new vertex w, where a vertex v of G'' is joined to w if in G it was joined to either of y or z. Clearly G'' is a connected graph with $n-2$ vertices and every edge of G'' is contained in a triangle. In forming G'', we lost three edges xy, yz, and xz. We would also lose an edge if some vertex v were adjacent to both y and z, but by assumption this does not occur. Thus G'' has $f(n)-3 < g(n)-3 = g(n-2)$ edges. Again this contradicts the induction hypothesis. Thus we must have $f(n) \geq g(n)$. □

The following result is classical and very nice. We give two proofs, the first one being an explicit inductive construction, the second one using the powerful probabilistic method of Erdős. We refer the reader to the addendum 6.B for more details on this versatile tool.

14. Prove that for every n there is a graph with no triangles and whose chromatic number is at least n.

Mycielski's theorem

Before starting the proof, let us recall some definitions. A k-proper coloring of a graph G is a coloring of its vertices with at most k colors such that each vertex receives one color and no two adjacent vertices have the same color. The chromatic number of a graph is the smallest number k for which G has a proper k-coloring.

Proof. We will prove this by induction on n. For $n = 1$ everything is clear, so assume that we have a graph G with no triangles and chromatic number at least n and let us construct another graph with no triangles and chromatic number at least $n+1$. Let $\chi(G)$ be the chromatic number of G. We may assume that $\chi(G) = n$ (otherwise keep G for the inductive step). Let $v_1, \ldots, v_k$ be the vertices of G and consider the following new graph G': the set of vertices consists of $v_1, \ldots, v_k$, together with vertices $x_1, \ldots, x_k, y$ (such that the $2k+1$ vertices thus obtained are distinct). Connect x_i with all neighbors of v_i and connect y with $x_1, \ldots, x_k$. Also, keep the edges in G. We claim that G' is triangle free and $\chi(G') \geq n+1$. It is clear by construction that G' has no triangles. Also, we easily have $\chi(G') \leq \chi(G) + 1$, since any proper n-coloring of G extends to a proper $n+1$-coloring of G' (assign the same color to x_i as to v_i and assign the color $n+1$ to y).

The difficult part is proving that this is actually an equality. Suppose thus that G' has chromatic number at most n and take any proper n-coloring of G'. We will construct a proper $n-1$-coloring of G, which will contradict the fact that G has chromatic number n. Assume, without loss of generality, that y has color n and let A be the set of vertices of G whose color is n and for each $v_i \in A$ change its color with that of x_i. We claim that this gives a proper $n-1$-coloring of G. Note that no two vertices of A are adjacent, as all these vertices have the same color. Now, if $v_i \in A$ is adjacent to $v_j \notin A$, then v_j is adjacent to x_i, so that they have different colors. Consequently, we found a $n-1$-proper coloring of G, which is impossible. We deduce that the chromatic number of G' is at least $n+1$. □

Proof. We will actually prove a stronger result, which is a classical theorem of Erdős. We will use the probabilistic method, for which the reader is referred to the addendum 6.B.

Theorem 6.2. *For any positive integers k, n there exists a graph with chromatic number greater than k and such that each cycle has length greater than n.*

Take N sufficiently large and consider random graphs with N vertices $G_{N,p}$, each edge appearing independently, with probability $p = N^{c-1}$ for some $0 < c < \frac{1}{n}$. If X is the number of cycles of length at most n in $G_{N,p}$ (which cycles we will call short in the sequel), then clearly

$$E[X] \le \sum_{i=3}^{n} N^i \cdot p^i < \frac{N^{cn}}{1 - N^{-c}}$$

since there are at most N^i cycles of length i, each appearing with probability p^i. Since the last quantity is smaller than $\frac{N}{4}$ for sufficiently large N, it follows by Markov's inequality that

$$P\left(X \ge \frac{N}{2}\right) \le \frac{1}{2}.$$

Let $\alpha(G)$ be the size of the largest independent set in G (the independence number of the graph) and let $\chi(G)$ be the chromatic number of G. Then it is easy to see that $\chi(G) \ge \frac{N}{\alpha(G)}$ if G has N vertices. Taking $a = [\frac{3}{p} \ln N]$, the probability that $\alpha(G) \ge a$ is at most $\binom{N}{a} \cdot (1-p)^{\binom{a}{2}}$ (one can choose a

independent vertices in $\binom{N}{a}$ ways and the probability that a given set of a vertices is independent is certainly $(1-p)^{\binom{a}{2}})$. An easy computation shows that the last quantity tends to 0 as $N \to \infty$. Thus for N sufficiently large we have $P(\alpha(G) \geq a) < \frac{1}{2}$ and so there is a graph $G_{p,N} = G$ with $X < \frac{N}{2}$ and $\alpha(G) < a$. Now, delete one vertex from each short cycle arbitrarily. The remaining graph G_1 has at least $N/2$ vertices, no cycles of length at most n and independence number smaller than a. Thus

$$\chi(G_1) \geq \frac{N/2}{3N^{1-c}\ln N} > k$$

for sufficiently large N. The result follows. □

We end this chapter with three gems in extremal graph theory, all taken from mathematical contests.

15. For a pair $A = (x_1, y_1)$ and $B = (x_2, y_2)$ of points on the coordinate plane, let

$$d(A, B) = |x_1 - x_2| + |y_1 - y_2|.$$

We call a pair (A, B) of (unordered) points harmonic if $1 < d(A, B) \leq 2$. Determine the maximum number of harmonic pairs among 100 points in the plane.

USA TST 2006

Proof. Consider the graph with vertices on the 100 points and whose edges connect points of harmonic pairs. The key point is that this graph contains no 5-clique. Indeed, if $P_i(x_i, y_i)$ are five vertices, any two of which are connected, then for all $i \neq j$ we have $1 < d(P_i, P_j) \leq 2$. Order these points such that $x_1 \leq x_2 \leq \cdots \leq x_5$. It is easy to see that among the numbers $y_1, y_2, \ldots, y_5$ one can find three forming a monotonic sequence. Call these three numbers $y_{i_1}, y_{i_2}, y_{i_3}$. Then

$$d(P_{i_1}, P_{i_2}) + d(P_{i_2}, P_{i_3}) = d(P_{i_1}, P_{i_3}).$$

This is however impossible, since $d(P_{i_1}, P_{i_2}), d(P_{i_2}, P_{i_3}), d(P_{i_3}, P_{i_1}) \in (1, 2]$. This finishes the proof of the fact that our graph contains no K_5. By Turán's theorem, it has at most $\frac{3}{4} \cdot \frac{100^2}{2} = 3750$ edges and so there are always at most 3750 harmonic pairs.

To finish the proof, it remains to exhibit a configuration having 3750 harmonic pairs. It is enough to distribute the 100 points near the points $(0, \pm\frac{3}{4})$ and $(\pm\frac{3}{4}, 0)$, having 25 points near each of these four points. □

16. For a finite graph G let $f(G)$ be the number of triangles formed by the edges of G and let $g(G)$ be the number of tetrahedra formed by the edges of G. Find the least constant c such that $g(G)^3 \leq c \cdot f(G)^4$ for any finite graph G.

IMO Shortlist 2004

Proof. Let us begin by considering the case of a complete graph K_n. Then obviously

$$f(G) = \binom{n}{3} \text{ and } g(G) = \binom{n}{4}.$$

Thus

$$c \geq \frac{g(K_n)^3}{f(K_n)^4} = \frac{\binom{n}{4}^3}{\binom{n}{3}^4}$$

and this happens for all n. Taking the limit as $n \to \infty$, we deduce that $c \geq \frac{3}{32}$.

The hard part is to prove that the value $\frac{3}{32}$ actually holds for arbitrary graphs G. To do this, we will first consider the two dimensional version of the problem, which is comparing the number of triangles and edges in a graph. The reason is simple: computing tetrahedra in a graph G comes down to computing triangles in all subgraphs induced by the vertices of G and taking the sum of these numbers of triangles (and then dividing by 4, since each tetrahedra is counted four times).

Therefore, consider first any graph G with n vertices $x_1, x_2, \ldots, x_n$ and e edges. Let us bound the number of triangles in terms of e. If $d(x_i)$ is the degree of the vertex x_i, then there are at most $\binom{d(x_i)}{2} \leq \frac{d(x_i)^2}{2}$ triangles containing the vertex x_i. But the number of triangles containing the vertex x_i is also obviously bounded by the number of edges of G, that is e. Thus for each vertex x_i, there are at most $d(x_i) \cdot \sqrt{\frac{e}{2}}$ triangles containing the vertex

x_i. Summing over i and taking into account that we count each triangle three times in this way, we obtain that

$$f(G) \le \frac{1}{3} \cdot \sqrt{\frac{e}{2}} \cdot \sum d(x_i) = \frac{2}{3} e \sqrt{\frac{e}{2}}.$$

Note that the constant appearing in this estimate in optimal, by taking for $G = K_n$ and then letting $n \to \infty$. This already shows that we are on the right path, since we solved the two dimensional version of the problem.

To solve the original problem, all we have to do it to repeat some of the arguments in the previous paragraph. Namely, take a graph G with vertices $x_1, \ldots, x_n$ and fix a vertex x_i. The number of tetrahedra containing x_i as a vertex is the number of triangles in the subgraph induced by the vertices connected to x_i. Let e_i be the number of edges in this subgraph, then the previous paragraph shows that there are at most $\frac{2}{3} e_i \sqrt{\frac{e_i}{2}}$ tetrahedra containing x_i as vertex. Also, note that we have $3f(G) = \sum e_i$, since for each vertex x_i there are e_i triangles containing x_i as vertex (and we count each triangle three times this way). However, one still needs an observation to end the proof: we gave an estimate for the number of tetrahedra having x_i as vertex in terms of e_i, but there is also an obvious estimate: this number is at most $f(G)$. Therefore, there are at most $\sqrt[3]{\frac{2}{9} f(G)} e_i$ tetrahedra containing x_i. Summing over i and taking into account that we count each tetrahedron four times, we finally obtain

$$4g(G) \le \sum \sqrt[3]{\frac{2}{9} f(G)} e_i = \sqrt[3]{\frac{2}{9} f(G)} \cdot 3f(G).$$

Taking the cube of the last inequality yields the desired estimate

$$g^3(G) \le \frac{3}{32} \cdot f(G)^4$$

and ends the proof. □

17. Let k be a positive integer. A graph whose vertex set is the set of positive integers does not contain any complete $k \times k$ bipartite subgraph. Prove

that there are arbitrarily long arithmetic progression of positive integers such that no two elements of the progression are joined by an edge in this graph.

KöMaL

Proof. We will first prove the following result, of independent interest:

Lemma 6.3. *Let $a \geq 2$ and let $K_{a,a}$ be the complete bipartite graph with a vertices in each half of the partition. There is a constant $c > 0$ such that a graph with n vertices and without $K_{a,a}$ has at most $cn^{2-\frac{1}{a}}$ edges.*

Proof. Let G be such a graph and call $1, 2, \ldots, n$ its vertices. Let d_i be the degree of vertex i. Let us count pairs $(v, \{v_1, \ldots, v_a\})$, where $\{v_1, \ldots, v_a\}$ is a set of a (distinct) vertices, all connected to v. For each $v = i$, there are precisely $\binom{d_i}{a}$ sets $\{v_1, \ldots, v_a\}$ sharing a pair with v. Thus we find $\sum_{i=1}^{n} \binom{d_i}{a}$ pairs in total. On the other hand, for each set of a vertices $\{v_1, \ldots, v_a\}$ there are at most $a-1$ vertices v sharing a pair with $\{v_1, \ldots, v_a\}$. Thus

$$\sum_{i=1}^{n} \binom{d_i}{a} \leq (a-1)\binom{n}{a}.$$

Let q be the number of edges in the graph and let $f(x) = \binom{x}{a}$ if $x > a-1$ and $f(x) = 0$ for $0 \leq x \leq a-1$. Then f is convex and since $\sum d_i = 2q$, we deduce that

$$(a-1)\binom{n}{a} \geq nf\left(\frac{2q}{n}\right).$$

This yields the rough estimate

$$a \cdot n^a \geq n\left(\frac{2q}{n} - a + 1\right)^a,$$

from which the result follows immediately. □

Coming back to the proof, fix an integer $m > 1$ and suppose that among any m vertices forming an arithmetic progression, at least two are connected.

Fix a large integer N and consider all arithmetic progressions with m terms, all among $1, 2, \ldots, N$. There are at least

$$\sum_{a=1}^{N-1}\left\lceil\frac{N-a}{m-1}\right\rceil > \sum_{a=1}^{N-1}\left(\frac{N-a}{m-1}-1\right) = \frac{N(N-1)}{2(m-1)} - N + 1$$

such arithmetic progressions, since each is determined by a pair (a, d) of positive integers such that $a + (m-1)d \le N$. This is greater than $c(m)N^2$ for N sufficiently large, where $c(m) > 0$ is a constant depending only on m. Since each of these progressions contributes an edge to the graph G_N (the subgraph of G induced by the vertices $1, 2, \ldots, N$), and since each edge is counted at most m^2 times, we deduce that G_N has at least $\frac{c(m)}{m^2}N^2$ edges for all sufficiently large N. Since $\frac{c(m)}{m^2}N^2 > cN^{2-\frac{1}{k}}$ for all sufficiently large N (and this for any constant $c > 0$), it follows by the previous lemma that G_N contains a $K_{k,k}$ for all sufficiently large N. Since this contradicts the hypothesis on G, the conclusion follows. □

6.1 Notes

We would like to thank the following people for providing solutions to some of the problems presented in this chapter: Alon Amit (problem 12), Alexandru Chirvăsitu (problem 10), Xiangyi Huang (problem 3), Szymon Kubicius (problem 4), Joel Brewster Lewis (problems 2, 7, 15), Fedja Nazarov (problem 14), Richard Stong (problem 13), Gjergji Zaimi (problem 5).

Addendum 6.A Some Pearls of Extremal Graph Theory

The purpose of this addendum is to present some beautiful theorems from extremal graph theory, that nicely complement the more elementary results discussed in chapter 6. The main result is the famous Szemerédi-Trotter theorem, a deep result in incidence geometry which has a lot of wonderful geometric consequences. For instance, it is the basic tool in dealing with natural (but very hard) questions such as: if we are given n points in the plane, how many distinct distances do they always determine, what is the greatest number of segments of length 1 (or triangles of area 1) they determine, etc. Thanks to a brilliant observation of Elekes, it also plays an important role in additive combinatorics, giving nontrivial bounds on the so-called sum-product problem, a major research problem in modern combinatorics.

The results discussed in this addendum found wide ranges of applications and their extensions are a very hot topic of research. See the papers [1], [11], [15], [22], [23], [36], [74], [72], [73], [75], [80], [81], and the excellent book [82] for more details, as we will only scratch the surface of the subject (but hopefully that will be enough to convince the reader of the beauty of these results). Also, some proofs in this addendum use the probabilistic method, so the reader is invited to take a look at addendum 6.B for details on this method and for probabilistic vocabulary.

6.A.1 The Szemerédi-Trotter theorem

The famous Szemerédi-Trotter theorem deals with the number of intersections between geometric objects, giving a fairly nontrivial (and even sharp up to absolute constants) upper bound for the number of incidences between a family of points and a family of curves. Since we do not want to delve into subtle topological considerations, let us agree that curves will always mean arcs of circles or polygonal lines, as this appears in all applications we have in mind.

Theorem 6.A.1. *(Szemerédi-Trotter) Consider n points $P_1, P_2, \ldots, P_n$ and m curves $C_1, C_2, \ldots, C_m$ in the plane. Suppose that C_i and C_j have at most* *one common point for all $i \neq j$. Then the number of pairs (i,j) such that $P_i \in C_j$ is at most $m + 4\max(n, (mn)^{\frac{2}{3}})$.*

In applications, the following immediate consequence is very handy:

Corollary 6.A.2. *Let S be a set of n points in the plane, L a collection of curves such that any two curves have at most one common point. Suppose that each curve in L contains at least $k \geq 2$ points of S. Then the number of pairs of a point in S and a curve in L containing that point is at most $2^9 \cdot \max\left(n, \frac{n^2}{k^2}\right)$.*

Proof. Let $l = |L|$ and let p be the number of such pairs. The theorem gives $p \leq l + 4n^{\frac{2}{3}} \max(n^{\frac{1}{3}}, l^{\frac{2}{3}})$. By hypothesis, we have $p \geq lk$, therefore $p - l \geq p\left(1 - \frac{1}{k}\right) \geq \frac{p}{2}$. Hence our inequality becomes $p \leq 8n^{\frac{2}{3}} \max(n^{\frac{1}{3}}, (p/k)^{\frac{2}{3}})$. From this the result follows by considering two cases. □

It is important to note that corollary 6.A.2 (and also Szemerédi-Trotter's theorem) is sharp up to a constant. Suppose that $2k^2 \leq n$ and consider the set of points (x, y), where $1 \leq x \leq k, 1 \leq y \leq \frac{n}{k}$ and x, y are integers. Also, consider the set of lines $y = mx + b$, where m, b are positive integers such that $1 \leq b \leq \frac{n}{2k}$ and $1 \leq m \leq \frac{n}{2k^2}$. It is easy to check that any such line contains exactly k such points (namely all points $(i, mi + b)$ with $1 \leq i \leq k$). Moreover, there are about n points and about $\frac{n^2}{4k^3}$ lines. Also, if $k \geq \sqrt{n}$, one can simply pick about n/k lines and put k points on each of them.

The original proof [81] of theorem 6.A.1 was very intricate. In a beautiful paper [80], Székely gave an amazing proof using a graph-theoretic result on crossing numbers. We will follow this path, but in order to do that we first need some terminology. Let G be a graph and consider an injective map that sends vertices of G to points of the plane. Draw an arc (in the plane) between any two points that come from the endpoints of an edge of G, such that except for the endpoints this arc does not meet any vertex. Call such a map a drawing of G. A crossing (or crossing point) is simply the intersection of two arcs not at the image of a vertex. Such a point belongs therefore to at least two arcs, but it is not an image of a vertex of the graph (with a piece of paper and a pencil in front of you, all these definitions should be obvious!). The number of crossings is the total number of crossing points, counted with multiplicities (i.e. if a crossing point belongs to k arcs, its multiplicity is $\binom{k}{2}$). It is also the number of pairs of edges with no common endpoints and whose associated arcs intersect each other. After this dry list of obvious definitions, let us glorify one that will be very helpful in what follows:

Definition 6.A.3. The crossing number of a graph G, denoted $c(G)$, is the minimal number of crossings over all possible drawings of G.

Note that, by definition, a graph is planar if and only if its crossing number is 0. The key ingredient in Székely's proof of theorem 6.A.1 is the following deep theorem [1], that will be proved in the next section.

Theorem 6.A.4. *(Ajtai, Chvátal, Newborn, Szemerédi, Leighton) Let G be a simple graph with e edges and n vertices. If $e \geq 4n$, then $c(G) \geq \frac{e^3}{64n^2}$.*

Let us see why theorem 6.A.4 implies theorem 6.A.1. We may assume that every curve C_i contains at least one of the points $P_1, P_2, \ldots, P_n$. Consider the graph G whose vertices are $P_1, P_2, \ldots, P_n$ and whose edges connect adjacent points on some C_i (i.e. two points P_j, P_k belonging to some C_i, such that there is no other point P_l between them on C_i). Let e be the number of edges of G and let I be the number of incidences between points and curves (i.e. the number of pairs (i, j) such that $P_i \in C_j$). As every curve contains at least one point among $P_1, P_2, \ldots, P_n$, we have $e = I - m$ (since if a curve contains s points, it yields $s - 1$ edges of G and edges coming from two different curves are distinct). Now, since two curves intersect in at most one point, it follows that $c(G) \leq m^2$. If $e \leq 4n$, we deduce that $I \leq m + 4n$ and we are done. Otherwise, the previous theorem yields $m^2 \geq \frac{e^3}{64n^2}$, thus $e \leq 4(nm)^{\frac{2}{3}}$ and so $I \leq m + 4(nm)^{\frac{2}{3}}$. The result follows.

6.A.2 Proof of theorem 6.A.4 and a generalization

The proof of theorem 6.A.4 is simply beautiful: we will start with a rather weak inequality obtained from Euler's formula and, using the probabilistic method, we will improve it drastically by averaging. Let us start with the weak inequality:

Lemma 6.A.5. *If G is a graph with e edges and n vertices, then $c(G) \geq e-3n$.*

Proof. First, we reduce the proof to the case when G is planar. To do this, remove edges of G which cross other edges until this is no longer possible, so you end up with a graph G' which is planar. Suppose that you removed k edges of G. Each of them removes at least one crossing, so that if we accept the truth of the lemma for G', we can write

$$c(G) \geq c(G') + k \geq e(G') + k - 3n = e - 3n.$$

Now, assume that G is planar, so $c(G) = 0$. We need to prove that $e \leq 3n$. If $n \leq 2$, this is clear, so suppose that $n \geq 3$. Let f be the number of (possibly infinite) faces of G. By Euler's formula we have $n + f = e + 2$. As every face has at least three edges, we have $3f \leq 2e$. We easily deduce that $e \leq 3n - 6$, finishing the proof of the lemma.

Finally, let us recall the proof of Euler's formula: if you start with a single vertex and try to reconstruct the graph, then every time you add an edge, you either add a vertex or a face, so the quantity $n + f - e - 2$ does not change. Since initially it is obviously zero, it is zero all the time. □

Now, we use the probabilistic method to improve the previous inequality. Take an arbitrary number $p \in (0, 1]$ and consider a random induced subgraph H of G, by picking each vertex of G independently and with probability p. As the probability of a given vertex to be in H is p, by linearity of expectation we have $E[v(H)] = np$, where $v(H)$ is the number of vertices of H. Also, since an edge appears in H with probability p^2, we have $E[e(H)] = p^2 e$, where e (respectively $e(H)$) is the number of edges of G (respectively H).

The previous lemma and linearity of expectation yield

$$E[c(H)] \geq E[e(H)] - 3E[v(H)].$$

We cannot easily express $E[c(H)]$ in terms of $c(G)$, but we can at least say that $E[c(H)] \leq p^4 c(G)$. Indeed, take a drawing of G with exactly $c(G)$ crossings and observe that the probability that a crossing survives in H is p^4. The conclusion is that

$$p^4 c(G) \geq E[c(H)] \geq p^2 e - 3np.$$

As p was arbitrary, it is tempting to choose a value that will optimize the previous inequality. This value is $p = \frac{9n}{2e}$, but unfortunately it is not necessarily smaller than 1. However, this is far from being a subtle issue: simply choose something a bit smaller, namely $p = \frac{4n}{e}$. This finishes the proof of the theorem.

In geometric applications, it is useful to have a more flexible version of theorem 6.A.4, which allows graphs with multiple edges. This is the objective of the following theorem of Székely [80].

Theorem 6.A.6. *Let $G = (V, E)$ be a multigraph with n vertices and e edges. If the maximal multiplicity of the edges of G is m, then $e < 32nm$ or $c(G) \geq \frac{e^3}{2^{16}n^2m}$.*

Proof. The idea is to reduce everything to the case of simple graphs, where theorem 6.A.4 applies. The details are however a bit involved. For each $1 \leq i \leq \log_2 m + 1$, let $G_i = (V, E_i)$ be the subgraph of G using only those edges with multiplicity between 2^{i-1} and 2^i (more precisely, two vertices a, b are joined by k edges in G_i if they are joined by k edges in G and $k \in [2^{i-1}, 2^i)$). Let e_i be the number of edges in G_i, without multiplicity. In order to apply theorem 6.A.4, we will restrict to those i for which $e_i \geq 4n$. Let S be the set of these i. As G_i has maximal multiplicity at most 2^i, we have (under the assumption that $e \geq 32mn$, which will be made from now on)

$$\sum_{i \in S} |E(G_i)| = e - \sum_{i \notin S} |E(G_i)| \geq e - \sum_{i=1}^{1+[\log_2(m)]} 4n \cdot 2^i \geq e - 16nm \geq \frac{e}{2}.$$

We claim that for $i \in S$ we have

$$c(G_i) \geq \frac{4^{i-1}e_i^3}{64n^2}.$$

Indeed, pick an arbitrary drawing of G_i and for every pair of vertices of G_i with a multi-edge between them, select 2^{i-1} of these edges and then arbitrarily and independently only one of them. We get a family of $(2^{i-1})^{e_i}$ simple graphs, each having $e_i \geq 4n$ edges (as $i \in S$) and so with crossing number at least $\frac{e_i^3}{64n^2}$. Thus, over this whole family of graphs we obtain at least $2^{(i-1)e_i}\frac{e_i^3}{64n^2}$

crossings. However, one has to pay attention to the fact that each crossing is counted in $2^{(i-1)(e_i-2)}$ members of the family, so in total we obtain at least

$$\frac{2^{(i-1)e_i}}{2^{(i-1)(e_i-2)}}\frac{e_i^3}{64n^2}=\frac{4^{i-1}e_i^3}{64n^2}$$

crossings in G_i, proving the claim.

Finally, by definition of the crossing number, it is clear that

$$c(G)\geq\sum_i c(G_i)\geq\sum_{i\in S}c(G_i)\geq\sum_{i\in S}\frac{4^{i-1}e_i^3}{64n^2}=\frac{1}{2^8n^2}\sum_{i\in S}4^ie_i^3.$$

Now, since

$$\sum_{i\in S}2^ie_i\geq\sum_{i\in S}|E(G_i)|\geq\frac{e}{2},$$

it is natural to use Hölder's inequality in the form

$$\left(\sum_{i\in S}4^ie_i^3\right)\left(\sum_{i\in S}2^{i/2}\right)^2\geq\left(\sum_{i\in S}2^ie_i\right)^3\geq\frac{e^3}{8}.$$

Combining this with the easy estimate

$$\sum_{i\in S}2^{i/2}\leq\sum_{i=1}^{1+[\log_2(m)]}2^{i/2}<4\sqrt{2m}$$

yields

$$c(G)>\frac{1}{2^8n^2}\frac{e^3}{8\cdot 32m}=\frac{e^3}{2^{16}n^2m},$$

finishing the proof. □

6.A.3 An application to additive combinatorics

In [23] Elekes made a nice connection between theorem 6.A.1 and a famous problem of Erdős and Szemerédi, the sum-product problem. This asks for a sharp lower bound of the expression $\max(|A+A|, |A \cdot A|)$ over all sets of real numbers A with n elements. Here

$$A + A = \{a + b | a, b \in A\} \text{ and } A \cdot A = \{a \cdot b | a, b \in A\}.$$

In particular, they made the very deep conjecture that for any $\varepsilon > 0$ there exists $c_\varepsilon > 0$ such that for all sets A with sufficiently many elements we have

$$\max(|A+A|, |A \cdot A|) \geq c_\varepsilon \cdot |A|^{2-\varepsilon}.$$

The following result treats the case when $2 - \varepsilon = \frac{5}{4}$. In [72], Solymosi proves the case $2 - \varepsilon = \frac{14}{11}$.

Theorem 6.A.7. *(Elekes) For any finite set of real numbers A we have*

$$|A + A| \cdot |A \cdot A| \geq \frac{1}{32}|A|^{5/2}.$$

Proof. Consider $P = (A + A) \times (A \cdot A)$ as a set of points in $\mathbb{R}^2$. Let L be the set of lines $l_{a,b}$ with equations $y = a(x - b)$ ranging over all choices of elements a, b of A. Note that all lines $l_{a,b}$ with $a \in A - \{0\}$ and $b \in A$ are distinct, so $|L| \geq |A|(|A| - 1)$. Also, $l_{a,b}$ is incident with $(b + c, a \cdot c) \in P$, for any $c \in A$. Thus we have at least $|A| \cdot |L|$ incidences between L and P. Thus, by the Szemerédi-Trotter theorem, we have

$$|A| \cdot |L| \leq |L| + 4\max(|P|, (|P| \cdot |L|)^{2/3}).$$

If $|P| \geq |L|^2$, then clearly

$$|P| \geq |A|^2 \cdot (|A| - 1)^2 \geq \frac{|A|^4}{4}$$

and we are done. Otherwise, the previous inequality yields

$$|P| \geq \left(\frac{|A| - 1}{4}\right)^{3/2} \sqrt{|L|} \geq \sqrt{|A|} \cdot \frac{(|A| - 1)^2}{8} \geq \frac{1}{32}|A|^{5/2}$$

and the result follows. □

The variant of the sum-product problem in which instead of sets of real numbers one considers subsets of a finite field has generated a huge amount of work. The proofs are in general very technical, as the analogue of the Szemerédi-Trotter theorem over finite fields is wrong. One has nevertheless the following deep theorem [11].

Theorem 6.A.8. *(Bourgain, Katz, Tao, Glibichuk, Konyagin)*
Let $\varepsilon > 0$. There exist positive constants C, δ such that for any prime p and any $A \subset \mathbb{F}_p$ with $|A| < p^{1-\varepsilon}$ we have

$$\max(|A+A|, |A \cdot A|) \geq C|A|^{1+\delta}.$$

The analogue of the Erdős-Szemerédi conjecture for finite fields would be

$$\max(|A+A|, |A \cdot A|) \geq c(\varepsilon)\min(|A|^{2-\varepsilon}, q^{1-\varepsilon})$$

for subsets A of $\mathbb{F}_q$. This is far from being settled. The following theorem can be deduced from the sum-product theorem.

Theorem 6.A.9. *(Bourgain, Katz, Tao) Let $0 < \alpha < 2$. There exists $\varepsilon > 0$ and $C > 0$ such that there are at most $Cn^{\frac{3}{2}-\varepsilon}$ incidences between $n \leq p^{\alpha}$ points and $n \leq p^{\alpha}$ lines in $\mathbb{F}_p$.*

Note that the previous theorem no longer holds if one considers subsets A of $\mathbb{F}_q$, where q is not a prime number. It suffices to consider a nontrivial subfield of $\mathbb{F}_q$, for which $|A+A| = |A|$ and $|A \cdot A| = |A|$. Also, it is necessary to have some upper bound on $|A|$, since otherwise $A = \mathbb{F}_p^*$ would be again a counterexample for p large enough.

6.A.4 Some geometric applications

In this section, we present some nice geometric applications of Szemerédi-Trotter's theorem. The first result deals with the number of triangles of unit area spanned by n points in the plane. It is not difficult to construct examples in which there are at least $cn^2 \log n$ unit-area triangles, for an absolute constant $c > 0$, but it is rather hard to give nontrivial upper bounds for the number of such triangles. We will use an easy geometric argument combined with corollary 6.A.2 to give such an upper bound.

Theorem 6.A.10. *There exists an absolute constant $c > 0$ with the following property: the number of triangles of area 1 with vertices among n points in the plane is smaller than $cn^{\frac{7}{3}}$.*

Proof. Let P be a set of n points in the plane. The essential observation is the following: if a, b are points of P, then all points $p \in P$ such that the triangle abp has area 1 are on the union of two fixed lines $L_{a,b}$, $L'_{a,b}$, parallel to ab. This is immediate. Now, consider those pairs (a, b) for which the lines $L_{a,b}$ and $L'_{a,b}$ contain at most $n^{\frac{1}{3}}$ points of P apiece. Each such pair determines at most $2n^{1/3}$ unit-area triangles and so these pairs contribute at most $2n^2 \cdot n^{1/3} = 2n^{7/3}$ unit-area triangles. On the other hand, by corollary 6.A.2, there are $O(n^{4/3})$ incidences between lines l containing at least $n^{1/3}$ points of P and points of P. However, given such a line l, there are $O(n)$ pairs (a, b) such that $l = L_{a,b}$ (as for any a and l there are at most two points b such that $l = L_{a,b}$). Thus pairs (a, b) for which at least one of the lines $L_{a,b}$ and $L'_{a,b}$ contains more than $n^{1/3}$ points of P also determine $O(n \cdot n^{4/3}) = O(n^{7/3})$ unit-area triangles. The result follows. □

Using a lot more work, one can improve the previous result to $O(n^{9/4+\varepsilon})$ for any $\varepsilon > 0$ as shown in [22]. The exact order of growth of the number of unit-area triangles is unknown. If instead one considers the number of triangles with perimeter 1, the same kind of reasoning (but working with incidences between ellipses and points) yields a bound $O(n^{16/7})$.

The second application is a famous result of Beck. It is again an easy consequence of corollary 6.A.2.

Theorem 6.A.11. *(Beck) For any n points in the plane, either more than $n \cdot 2^{-14}$ of them are on a line or these points determine at least $2^{-29}n^2$ different lines.*

Proof. Let S be a set of n points in the plane. Call a line average if it contains between 2^{14} and $n/2^{14}$ points of S. We claim that there are at most $n^2/2$ pairs of points of S that determine average lines. Indeed, for each $i \in [14, \log_2 \frac{n}{2^{14}}]$ there are at most $2^9 \cdot \max\left(\frac{n}{2^i}, \frac{n^2}{8^i}\right)$ lines that contain between 2^i and 2^{i+1} points of S, by corollary 6.A.2. Any such line contains at most 4^{i+1} pairs of points of S. Thus the number of pairs of points that lie on average lines is bounded by

$$2^9 \sum_{14 \le i \le \log_2 \frac{n}{2^{14}}} 4^{i+1} \max\left(\frac{n}{2^i}, \frac{n^2}{8^i}\right) \le 2^{11} \left(\sum_{14 \le i \le \log_4 n} \frac{n^2}{2^i} + \sum_{\log_4 n < i \le \log_2 \frac{n}{2^{14}}} 2^i \cdot n \right)$$

$$< 2^{11} \left(\sum_{i \ge 14} \frac{n^2}{2^i} + \sum_{i \le \log_2 \frac{n}{2^{14}}} 2^i n \right) < \frac{n^2}{4} + \frac{n^2}{4} = \frac{n^2}{2},$$

proving the claim.

So we have at least $n^2/2$ pairs of points of S that do not determine an average line. Now, we have two possibilities: either some line contains more than $n/2^{14}$ points of S, in which case the result is proved, or actually the $n^2/2$ pairs of points determine only lines that contain less than 2^{14} points of S. But then each such line contains at most 2^{28} pairs of points of S, so there must be at least $n^2/2^{29}$ such lines. So again the result is proved. □

We continue with another rather natural question: what is the maximal number of unit-distances spanned by n points in the plane? Erdős proved the lower bound $n^{1+\frac{c}{\log \log n}}$ and the upper bound $cn^{3/2}$ and conjectured that the true order of growth should be about $n^{1+\frac{c}{\log \log n}}$. This is far from being proved, but the following theorem gives a better upper bound. In dimension 3, Clarkson proved that the number of unit distances is $O(n^{\frac{3}{2}+\varepsilon})$ for all $\varepsilon > 0$, while Erdős, Pach and Hickerson gave examples with at least $cn^{4/3}$ unit distances. Also, note that already in dimension 4 one might have cn^2 unit distances, as shown by placing n points on $x_1^2 + x_2^2 = 1/2$ and n points on $x_3^2 + x_4^2 = 1/2$.

Theorem 6.A.12. *(Spencer, Szemerédi, Trotter) Let S be a set of n points in the plane. Then there are at most $16n^{4/3}$ ordered pairs of points (P, Q) in S such that $PQ = 1$.*

Proof. Draw a unit circle around each point of S and consider the multigraph G with vertex set S and in which $P, Q \in S$ are joined if they are adjacent on one of these circles. Note that there might be more than one edge between two vertices and that G might have loops (if there are circles containing exactly one point of S). If q is the number of unit distances between points in S, then G has q edges, for if a circle centered at $P \in S$ contains x_i points of S, these points contribute x_i edges to G. Now, remove all circles containing at most two points of S, so we get a new multigraph with at least $q - 2n$ edges. We claim that the maximal multiplicity in this multigraph is at most 2. Indeed, if there are at least 3 edges between $P \neq Q \in S$, then there are at least three circles containing P, Q, so the circles of radius 1 centered at P, Q have at least three common points, a contradiction. Next, for every pair of vertices with exactly two edges between them, remove one edge (arbitrarily), so that we end up with a simple graph having at least $\frac{q}{2} - n$ edges. Now, if $q > 10n$, then this simple graph has more than $\max(\frac{q}{4}, 4n)$ edges and so at least $(q/4)^3/(64n^2)$ crossings by theorem 6.A.4. But since any two circles have at most two common points, there cannot be more than $2\binom{n}{2} < n^2$ crossings. The conclusion is that if $q > 10n$, then $q^3 \leq 4096n^4$ and the result follows (if $q \leq 10n$, everything is clear). $\square$

We end this addendum with a rather technical result concerning a famous question of Erdős: what is the least number of distinct distances determined by n points in the plane? Though rather innocent-looking, this is an extremely difficult problem. Erdős gave examples of configurations which determine at most $\frac{cn}{\sqrt{\log n}}$ distances, for some absolute constant c. For more than thirty years, the best result was Moser's bound $cn^{2/3}$, which is a rather tricky, but very elementary argument. Note that this bound also follows from the previous theorem: there are at least $\frac{n^2}{2}$ ordered pairs of points and the previous theorem (combined with a rescaling argument) shows that each distance appears at most $16n^{4/3}$ times, thus the number of distinct distances is at least $\frac{n^{2/3}}{32}$. It was only in 1984 that Chung [15] improved Moser's bound to $cn^{5/7}$. The next big step was done in the paper [16], where bounds of the form $\frac{n^{4/5}}{(\log n)^{c_1}}$ are proved by rather difficult arguments. After a huge effort, a major breakthrough was made in [36], where Guth and Katz prove that n points always determine at least $\frac{cn}{\log n}$ distinct distances. Here, we prove a much weaker estimate, but which is already highly nontrivial (following [80]).

Theorem 6.A.13. *(Székely) Any set of n points in the plane determines at least $cn^{4/5}$ distinct distances, for an absolute constant $c > 0$.*

Proof. From now on c is an absolute constant that will change throughout the proof without changing its name.

Fix n distinct points $P_1, P_2, \dots, P_n$ in the plane and let t be the number of distinct distances among them. Let $d_1, d_2, \dots, d_t$ be these distances. We may assume that $t < n^{\frac{9}{10}}$, as otherwise we are done. Around each point P_i draw t circles, having radii $d_1, d_2, \dots, d_t$. Consider now the multigraph whose vertices are $P_1, \dots, P_n$ and whose edges are arcs connecting adjacent vertices on these circles. Note that there are at least $n(n-1)$ edges in this multigraph (for each P_i, there are $n-1$ points on the union of the circles centered at P_i and if a circle contains s points, it contributes s edges). As usual, we remove edges that come from circles containing at most two points. It is easy to see that this removes at most $2nt$ edges, so we still have $n^2 + o(n^2)$ edges.

Unfortunately, the maximal multiplicity might be too big for the previous theorems to yield the desired estimate. The crucial point is to show that there cannot be too many edges with very high multiplicity. This is the aim of the following lemma. Before stating it, call a pair of points k-rich if there are at least k edges between them.

Lemma 6.A.14. *There are at most $\frac{ctn^2}{k^2} + ctn \log n$ edges joining k-rich pairs of points.*

Proof. By definition, this number of edges is at most the number of pairs (d, e), where d is the symmetry axis of an edge e of G such that d contains at least k vertices. If $2^i \leq \sqrt{n}$, there are at most $\frac{cn^2}{8^i}$ lines containing at least 2^i vertices, by corollary 6.A.2. If such a line contains u points then, by definition of t, this line bisects at most $2tu$ edges. So, as long as we are counting pairs (d, e) such that d contains between k and $4\sqrt{n}$ vertices, the total number of such pairs is at most

$$\sum_{k/2 < 2^i \leq 4\sqrt{n}} t2^{i+1} \cdot \frac{cn^2}{8^i}$$

and an easy computation shows that this is bounded by $\frac{ctn^2}{k^2}$. On the other hand, for each $a \geq 4\sqrt{n}$, it is easy to see that there are at most $\frac{cn}{a}$ lines containing between a and $2a$ vertices. These lines yield at most

$$\sum_{2\sqrt{n}<2^i\leq n} \frac{cn}{2^i} t 2^{i+1} < ctn \log n$$

pairs. Combining these two observations finishes the proof. □

Finally, the previous lemma shows that there is an absolute constant c_1 such that if we delete all $c_1\sqrt{t}$-rich edges, the resulting multigraph G_1 has at least cn^2 edges. As clearly G_1 has crossing number at most $2n^2t^2$, it follows from theorem 6.A.6 that

$$2n^2t^2 \geq \frac{cn^6}{n^2 c_1 \sqrt{t}}$$

and the result follows. □

Addendum 6.B Probabilities in Combinatorics

This addendum presents some applications of probabilities in combinatorics, or what is commonly called the probabilistic method. This is one of the most powerful tools in modern combinatorics, even though, just as with the pigeonhole principle, the underlying idea is extremely easy. The probabilistic method was used at first by Erdős, who proved a great deal of fairly nontrivial results using rather simple probabilistic arguments. For a much deeper study of the method, a canonical reference is the excellent book [2] by Alon and Spencer. Many of the examples we are going to discuss are taken from this book (however, some of them are left as exercises in the book and are not really easy...).

We begin by recalling some useful notions from probability theory. In discrete combinatorics, one works with finite probability spaces, which makes the discussion much more elementary, avoiding subtle issues from measure theory. A finite probability space is the data of a finite set Ω (called the sample space) and of a map (called the probability distribution) $P : \Omega \to [0, 1]$ such that

$$\sum_{\omega \in \Omega} P(\omega) = 1.$$

One obvious such map is the constant map $\frac{1}{|\Omega|}$, which is called the uniform distribution. One defines for any subset A of Ω, which we call an event, its probability as

$$P(A) = \sum_{x \in A} P(x).$$

It is very easy to check that $P(A \cup B) + P(A \cap B) = P(A) + P(B)$ and that the probability of the complement of A is $1 - P(A)$. Given a probability space, a random variable X is simply a map $X : \Omega \to \mathbb{R}$. The expectation (or mean value) of X is

$$E[X] = \sum_{\omega \in \Omega} X(\omega) P(\omega) = \sum_{x \in \mathbb{R}} x \cdot P(X = x).$$

Note that the second sum is finite, as the image of X is finite. Another extremely important notion is that of independence. Some events $A_1, A_2, \dots, A_k$ are called independent if for all $I \subset \{1, 2, \dots, k\}$ we have

$$P(\cap_{i\in I} A_i) = \prod_{i\in I} P(A_i).$$

Two random variables X, Y are called independent if the events $X = a$ and $Y = b$ are independent for all a, b. It is an easy exercise to check that in this case $E[XY] = E[X]E[Y]$.

The following theorem consists of two elementary, but incredibly powerful principles. The first one is basically the principle of the probabilistic method: if you want to show the existence of an object with properties $P_1, \dots, P_k$, it is enough to prove that there is a probability space (Ω, P) such that the sum of the probabilities that an object does not have property P_i is smaller than 1. Of course, the difficult point is constructing the good probability space (though in practice the construction is naturally imposed by the statement of the problem we are trying to solve) and estimating these probabilities. The second part of the theorem is called linearity of expectation and it is also a very powerful result, as we will soon see.

Theorem 6.B.1. *Let (Ω, P) be a finite probability space.*

1) For any subsets $A_1, A_2, \dots, A_k$ of Ω we have

$$P(\cup_{i=1}^k A_i) \le \sum_{i=1}^k P(A_i),$$

with equality if the events are pairwise disjoint. So, if $\sum_{i=1}^k P(A_i) < 1$, then $\cup A_i \neq \Omega$.

2) If $X_1, X_2, \dots, X_k$ are random variables, then

$$E[X_1 + X_2 + \cdots + X_k] = E[X_1] + E[X_2] + \cdots + E[X_k].$$

Proof. 1) Let $B_1 = A_1$ and let B_i be the complement of $\cup_{j=1}^{i-1} A_j$ in A_i for all $i \ge 2$. Then clearly $\cup_j B_j = \cup_j A_j$ and the B_j's are pairwise disjoint. But then it follows that $P(B_j) \leq P(A_j)$ (as $B_j \subset A_j$ and P takes nonnegative values) and

$$P(\cup_j A_j) = P(\cup_j B_j) = \sum P(B_j) \leq \sum P(A_j).$$

The rest is immediate.

2) is an easy consequence of the definition of expectation. □

When using the probabilistic method, one is naturally confronted with estimating probabilities, which sometimes can be quite painful. A simple example is that if (Ω, P) is a probability space and if X is a random variable on Ω, then the definition gives that there exists $\omega \in \Omega$ such that $X(\omega) \geq E[X]$. The following two inequalities are also very basic, but useful tools in estimating probabilities:

Theorem 6.B.2. *1) (Markov's inequality) If X is a random variable taking nonnegative values and if $a > 0$, then*

$$P(X \geq a) \leq \frac{1}{a} E[X].$$

2) (Chebyshev's inequality) Let X be a random variable and let $a > 0$. Then

$$P(|X - E[X]| \geq a) \leq \frac{1}{a^2}(E[X^2] - E[X]^2).$$

Proof. For the first part, simply note that

$$E[X] = \sum_x P(X = x)x \geq \sum_{x \geq a} P(X = x)a = aP(X \geq a).$$

For the second part, using Markov's inequality, we can write

$$P(|X - E[X]| \geq a) = P((X - E[X])^2 \geq a^2) \leq \frac{1}{a^2} E[(X - E[X])^2]$$

and an easy computation using linearity of expectation yields the result. □

As usual, knowledge comes with practice, so we will spend the remainder of this chapter by giving a lot of applications of the previous theorems. We start with some applications of the sub-additivity of probabilities. One can prove the following inequality using standard techniques, but there is also a very elegant probabilistic proof:

B.2. Let $x_{ij} \in [0,1]$ for $1 \le i, j \le n$. Prove that

$$\prod_{j=1}^{n}\left(1-\prod_{i=1}^{m} x_{ij}\right)+\prod_{i=1}^{m}\left(1-\prod_{j=1}^{n}(1-x_{ij})\right) \geq 1.$$

Proof. Consider a random binary matrix (a_{ij}) such that $P(a_{ij} = 1) = x_{ij}$ and all these events are independent. Then $\prod_{j=1}^{n}(1-\prod_{i=1}^{m} x_{ij})$ is simply the probability that each column contains at least a zero, while the expression $\prod_{i=1}^{m}\left(1-\prod_{j=1}^{n}(1-x_{ij})\right)$ is the probability that each row contains at least a 1. Since all binary matrices have either at least one zero in every column or at least one one in every row, the result follows. □

Another application of theorem 6.B.1 is the following nice problem:

B.3. Let S be a finite set of points in a plane, no three of which are collinear. For each convex polygon P with vertices in S, let $a(P)$ be the number of vertices of P[1] and let $b(P)$ be the number of points of S lying outside P (i.e. outside its interior or border). Prove that for all real numbers x,

$$\sum_{P} x^{a(P)}(1-x)^{b(P)} = 1,$$

where here the sum is taken over possibly degenerate convex polygons(polygons with 2, 1, or 0 vertices), too

IMO Shortlist 2006

[1]With the convention that $a(P) = 0, 1, 2$ if P is $\emptyset$, a point of S, respectively a segment connecting two points of S.

Proof. It is enough to check the equality for $x \in (0,1)$, since a polynomial vanishing on $(0,1)$ vanishes everywhere. Consider a random two-coloring of S such that the probability that a point is black is x and all these events are independent. For a given P, the quantity $x^{a(P)}(1-x)^{b(P)}$ is simply the probability that the vertices of P are black and the points outside P are white. As these events are disjoint, the sum over all P is the probability that there is a polygon with black vertices and such that all points outside it are white. But this probability is 1, since the convex hull of the black points is such a polygon (note that there might be $0, 1$ or 2 black points and this is why the sum is also taken over degenerate polygons). □

We continue with two problems which use the principle of the probabilistic method, namely the fact that if the sum of probabilities of some events is (strictly) less than 1, then there is a point in the probability space not belonging to any of these events.

B.4. Prove that there is a four-coloring of the set $M = \{1, 2, \ldots, 1987\}$ such that M contains no monochromatic arithmetic progression with 10 terms.

IMO Shortlist 1987

Proof. Pick a random coloring and observe that the probability that M contains a monochromatic progression of length at least 10 is bounded by $N/4^9$, where N is the number of progressions of length 10 contained in M. So the problem is solved if we prove that $N < 4^9$. But there are $\lfloor \frac{1987-i}{9} \rfloor$ progressions of length 10 contained in M and whose first term is i. So

$$N \leq \sum_{i=1}^{1987} \frac{1987-i}{9} < \frac{1}{18} \cdot 2000^2.$$

So, it remains to check that $2^8 \cdot 5^6 < 9 \cdot 2^{19}$, which follows from

$$5^6 = (5^3)^2 < (2^7)^2 = 2^{11} \cdot 8 < 9 \cdot 2^{11}.$$ □

Recall that the m-th Ramsey number $R(m)$ is the smallest positive integer n such that in any coloring of the edges of K_n (complete graph with n vertices) with two colors, one can find a monochromatic K_m subgraph. Note that it is not obvious that this is well defined, but one can prove without too much difficulty that any coloring of the edges of $K_{\binom{2m-2}{m-1}}$ with two colors contains a monochromatic K_m, so that $R(m)$ is well-defined and at most $\binom{2m-2}{m-1}$ (which grows roughly as 4^m). The following result of Erdős from 1947 is an absolute classic. Amazingly, even after more than 60 years, it remains close to the best known lower bound (and 4^m remains close to the best known upper bound).

B.5. If $2^{\binom{m}{2}-1} > \binom{n}{m}$, then $R(m) > n$. In particular, $R(m) > 2^{\frac{m}{2}}$ for all $m \geq 4$.

Erdős

Proof. By definition, saying that $R(m) > n$ is the same as saying that we can find a coloring of the edges of K_n with no monochromatic K_m. Consider a random coloring of the edges of K_n with two colors, each having probability $1/2$. If S is an m-element subset of the vertices of K_n, let A_S be the event that the corresponding subgraph of K_n is monochromatic. We want to prove that there is an element in the probability space (i.e. a 2-coloring) which belongs to none of the events A_S. It is enough to check that

$$\sum_S P(A_S) < 1.$$

However, it is clear that we have $\binom{n}{m}$ choices of sets S and for each of them

$$P(A_S) = 2^{1-\binom{m}{2}}.$$

Thus, as long as $2^{\binom{m}{2}-1} > \binom{n}{m}$, we can find the desired coloring. The second part of the theorem is now easy: taking $n = [2^{m/2}]$, we have (for $m \geq 4$)

$$\binom{n}{m} = \frac{n(n-1)\cdots(n-m+1)}{m!} < \frac{n^m}{m!} < (n/2)^m \leq 2^{\frac{m^2}{2}-m} < 2^{\binom{m}{2}-1}. \quad \square$$

Here is a more delicate problem using similar ideas, but for which it is more difficult to find a good probabilistic interpretation.

B.6. Let $A_1, A_2, \ldots, A_n$ and $B_1, B_2, \ldots, B_n$ be distinct subsets of $\mathbb{N}$ such that $A_i \cap B_i = \emptyset$ for all i and $(A_i \cap B_j) \cup (A_j \cap B_i) \neq \emptyset$ for all $i \neq j$. Prove that for all $p \in [0,1]$

$$\sum_{i=1}^{n} p^{|A_i|}(1-p)^{|B_i|} \leq 1.$$

Tusza

Proof. Let X be the union of all A_i and B_i and consider a random subset S of X such that the events $x \in S$ for $x \in X$ are independent and of probability p (more formally, let Ω be the set of all subsets of X and define

$$P(S) = p^{|S|}(1-p)^{|X|-|S|},$$

which is a probability measure on Ω, because $\sum_{S \subset X} P(S) = 1$ by the binomial theorem). Consider the event E_i: $A_i \subset S \subset X - B_i$. By hypothesis, no two events occur at the same time, thus

$$\sum_i P(E_i) = P(\cup_i E_i) \leq 1.$$

On the other hand, we have

$$P(E_i) = \sum_{A_i \subset S \subset X - B_i} p^{|S|}(1-p)^{|X|-|S|} = p^{|A_i|}(1-p)^{|B_i|},$$

the last equality being an easy computation left to the reader. Inserting this in the previous inequality yields the desired result. □

Remark 6.B.3. If all A_i have a elements and all B_i have b elements, we deduce from Tusza's inequality (by taking $p = \frac{a}{a+b}$) that $n \leq \frac{(a+b)^{a+b}}{a^a \cdot b^b}$.

We continue with a series of applications of the linearity of expectation and Chebyshev's inequality. First, a very simple problem:

B.7. Let $p_n(k)$ be the number of permutations of $\{1, 2, \ldots, n\}$ having exactly k fixed points. Prove that

$$\sum_{k=0}^{n} k p_n(k) = n!.$$

IMO 1987

Proof. Let σ have the uniform distribution on the set of permutations of $\{1, 2, \ldots, n\}$. If $X(\sigma)$ is the number of fixed points of σ, then

$$\frac{1}{n!} \sum_{k=0}^{n} k p_n(k) = E[X].$$

On the other hand,

$$X = \sum_{i=1}^{n} X_i,$$

where $X_i(\sigma) = 1_{\sigma(i)=i}$. Then

$$E[X_i] = P(\sigma(i) = i) = \frac{1}{n}$$

and the conclusion follows by linearity of expectation. □

The next two applications are absolute classics.

B.8. Any graph with q edges contains a bipartite subgraph with at least $q/2$ edges.

Erdős

Proof. Pick a random subset S of the set V of vertices, by including a vertex in S independently with probability $1/2$. Let $X(S)$ be the number of edges xy for which exactly one of x, y is in S. For a given edge xy, the probability that exactly one of x, y is in S is $1/2$, so by linearity of expectation $E[X] = q/2$. Thus one can find S such that $X(S) \geq q/2$. By construction, the subgraph comprising all edges with exactly one vertex in S is a solution. □

For what follows, in a tournament there is exactly one match between each pair of players and there is no draw. A Hamiltonian path in a tournament is a permutation σ of the players such that player $\sigma(i)$ beats player $\sigma(i+1)$ for all i.

B.9. There exists a tournament with n players which has at least $\frac{n!}{2^{n-1}}$ Hamiltonian paths.

Szele

Proof. Pick a random tournament (in which the results of each match occur with equal probability), and let X be the number of Hamiltonian paths. If σ is a permutation of the players, let $X_\sigma(T)$ be 1 if σ induces a Hamiltonian path in the tournament T and 0 otherwise. It is clear that, as random variables, we have $X = \sum_\sigma X_\sigma$. It is equally clear that $E[X_\sigma] = 2^{1-n}$ (since the result of the matches between $\sigma(i)$ and $\sigma(i+1)$ is imposed for all i). Thus $E[X] = \frac{n!}{2^{n-1}}$ and the result follows. □

A good exercise for the reader is to prove that the following result implies (a weak form of) Turán's famous theorem on graphs without n-complete subgraphs.

B.10. Let $d_1, d_2, \ldots, d_n$ be the degrees of the vertices of a graph. Prove that one can find a subset S of vertices such that

1) S has at least $\sum_{i=1}^{n} \frac{1}{d_i+1}$ elements.

2) There are no edges between vertices in S.

Proof. Consider a random permutation σ of the vertices of our graph G, all permutations having equal probability $\frac{1}{n!}$. Let A_i be the event: $\sigma(i) < \sigma(j)$ for any neighbor j of i. We claim that $P(A_i) = \frac{1}{d_i+1}$. Indeed, we need to find the number of permutations σ of the vertices such that $\sigma(i) < \sigma(j)$ for any neighbor j of i. If $y_1, \ldots, y_{d_i}$ are the neighbors of i, there are $\binom{n}{d_i+1}$ possibilities for the set $\{\sigma(i), \sigma(y_1), \ldots, \sigma(y_{d_i})\}$, $d_i!$ ways to permute the elements of this set (and not $(d_i+1)!$, since $\sigma(i)$ is the smallest element of the set) and finally $(n-d_i-1)!$ ways to permute the remaining vertices. So

$$P(A_i) = \binom{n}{d_i+1}\frac{(n-d_i-1)!d_i!}{n!} = \frac{1}{d_i+1},$$

as claimed. If X is the random variable $X(\sigma) = \sum_{i=1}^n 1_{\sigma\in A_i}$, then

$$E[X] = \sum_{i=1}^n P(A_i) = \sum_{i=1}^n \frac{1}{d_i+1}.$$

Hence one can find σ such that

$$X(\sigma) \geq \sum_{i=1}^n \frac{1}{d_i+1}.$$

It is clear that the set of vertices i such that $\sigma \in A_i$ satisfies both properties. □

We continue with a beautiful proof, due to Lubell, of a famous theorem of Sperner on maximal anti-chains.

B.11. Let $\mathcal{F}$ be a family of subsets of $\{1,2,\ldots,n\}$ which does not contain two elements A, B such that $A \subset B$. Then $|\mathcal{F}| \leq \binom{n}{[\frac{n}{2}]}$.

Sperner's theorem

Proof. Consider a random permutation σ of $\{1,2,\ldots,n\}$ (with uniform distribution) and let A_i be the event that $\{\sigma(1),\sigma(2),\ldots,\sigma(i)\} \in \mathcal{F}$. The hypothesis on $\mathcal{F}$ implies that the random variable $X(\sigma) = \sum_{i=1}^n 1_{\sigma\in A_i}$ satisfies $X(\sigma) \leq 1$ for all σ, so that its expectation $E[X] \leq 1$. On the other hand, $E[X] = \sum_{i=1}^n P(A_i)$ and the probability that $\{\sigma(1),\ldots,\sigma(i)\}$ is a given set with i elements is $\frac{i!(n-i)!}{n!}$, thus $P(A_i) = \frac{n_i}{\binom{n}{i}}$, where n_i is the number of subsets with i elements in $\mathcal{F}$. Putting this together yields the beautiful inequality

$$\sum_{i=1}^n \frac{n_i}{\binom{n}{i}} \leq 1,$$

from which the result follows, as $\binom{n}{[\frac{n}{2}]}$ is the largest among the binomial coefficients. □

Actually, the proof shows the following more general result, known as the Lubell-Yamamoto-Meshalkin-inequality: let $A_1, A_2, \ldots, A_k$ be subsets of $\{1, 2, \ldots, n\}$ such that no A_i is a subset of any A_j. Then

$$\frac{1}{\binom{n}{|A_1|}} + \frac{1}{\binom{n}{|A_2|}} + \cdots + \frac{1}{\binom{n}{|A_k|}} \leq 1.$$

The following famous theorem of Bollobás generalizes this result further:

B.12. Let $A_1, A_2, \ldots, A_n$ and $B_1, B_2, \ldots, B_n$ be distinct sets of positive integers such that $A_i \cap B_i = \emptyset$ for all i, but $A_i \cap B_j \neq \emptyset$ for all $i \neq j$. Then

$$\sum_{i=1}^{n} \frac{1}{\binom{|A_i|+|B_i|}{|A_i|}} \leq 1.$$

Bollobás

Proof. We may assume that all A_i's and B_j's are subsets of $\{1, 2, \ldots, N\}$. Consider the uniform distribution on the set S_N of all permutations of $\{1, 2, \ldots, N\}$. Let E_i be the event consisting of all permutations σ with the following property: all elements of A_i come before all elements of B_i in the list $\sigma(1), \sigma(2), \ldots, \sigma(n)$. It is easy to see that the hypothesis implies that the events E_i are pairwise disjoint. Thus $\sum_i P(E_i) \leq 1$. It remains to notice that $P(E_i) = \frac{1}{\binom{|A_i|+|B_i|}{|A_i|}}$, as any subset with $|A_i|$ elements of $A_i \sqcup B_i$ is equally likely to form the first $|A_i|$ elements in the list $\sigma(1), \sigma(2), \ldots, \sigma(n)$. The result follows. □

We take a break here to discuss a very nice consequence of Bollobás' inequality. Recall that an r-uniform hypergraph is a pair (V, E), where V is a set and E is a collection of subsets of V, each subset having r elements. Let K_t^r be the complete r-uniform hypergraph on t vertices.

B.13. Let $G = (V, E)$ be a r-uniform hypergraph on n vertices. Suppose that G contains no K_t^r, but that if we add any r-element set to E, at least one K_t^r appears. Then

$$|E| \geq \binom{n}{r} - \binom{n-t+r}{r}.$$

Bollobás

Proof. Let $[n] = \{1, 2, \ldots, n\}$ and let $[n]^{(r)}$ be the collection of r-element subsets of $[n]$. If $A \in [n]^{(r)} - E$, consider a copy $C(A)$ of K_t^r that appears by adding A to E. Then $A \in C(A)$, yet A' is not in $C(A)$ if $A' \neq A$ is in $[n]^{(r)} - E$. So, if $\{A_1, \ldots, A_m\} = [n]^{(r)} - E$ and B_i is the complement of $C(A_i)$, then $(A_i)_i$ and $(B_j)_j$ satisfy the hypotheses of Bollobás' inequality, yielding $m \leq \binom{n-t+r}{r}$. The result follows. □

We continue with a very nice result from additive number theory.

B.14. Let $A \subset \mathbb{Z}/n^2\mathbb{Z}$ be a subset with n elements. Prove that there exists a subset $B \subset \mathbb{Z}/n^2\mathbb{Z}$ with n elements such that $|A + B| \geq \frac{n^2}{2}$.

IMO Shortlist 1999

Proof. Pick a random collection of n elements of $\mathbb{Z}/n^2\mathbb{Z}$, each of the n elements being taken with probability $1/n^2$ and all choices being independent. Consolidate the distinct elements among the n chosen ones in a set B, which may have less than n elements. Consider then the random variable $X = |A + B|$. As

$$X = \sum_{i \in \mathbb{Z}/n^2\mathbb{Z}} 1_{i \in A+B},$$

we have by linearity of expectation

$$E[X] = \sum_{i \in \mathbb{Z}/n^2\mathbb{Z}} P(i \in A + B).$$

On the other hand, the probability that $i \notin A+B$ is clearly the n-th power of the probability that a given integer is not in A, that is

$$P(i \in A+B) = 1 - \left(1 - \frac{|A|}{n^2}\right)^n = 1 - \left(1 - \frac{1}{n}\right)^n.$$

We deduce that

$$E[X] = n^2\left(1 - \left(1 - \frac{1}{n}\right)^n\right)$$

and the result follows from the inequality $\left(1 - \frac{1}{n}\right)^n < \frac{1}{2}$. □

Often, the choices of the probability space and distribution are imposed by common sense. However, in order to get better quantitative results, one uses sometimes less natural probability distributions. The following problems illustrate this:

B.15. Let V be a set with n elements and let F be a family of m subsets of V, each having three elements. Prove that if $3m \geq n$, then there exists $S \subset V$ having at least $\frac{2n}{3}\sqrt{\frac{n}{3m}}$ elements and such that no element of F is contained in S.

Proof. Choose $p \in [0,1]$ and pick a random subset S of V such that $P(v \in S) = p$ for all $v \in V$, all the events being independent. If $n(S)$ is the number of elements of F contained in S, then linearity of expectation yields

$$E[|S| - n(S)] = E[|S|] - E[n(S)] = np - mp^3.$$

This is maximal for $p = \sqrt{\frac{n}{3m}} \in [0,1]$ and is equal to $\frac{2n}{3}\sqrt{\frac{n}{3m}}$. Thus, we can find S such that $|S| - n(S) \geq \frac{2n}{3}\sqrt{\frac{n}{3m}}$. Choose such S. For each $e \in F$ with all of its three elements in S, delete arbitrarily one of these three elements. We thus end up with at least $|S| - n(S)$ elements of V which obviously have the desired property. □

B.16. The minimal degree of the vertices of a graph G with n vertices is $d > 1$. Prove that there exists a subset S of the vertices such that

1) $|S| \leq n \cdot \frac{1+\ln(d+1)}{d+1}$.

2) Any vertex of G is either in S or neighbor of a vertex in S.

Noga Alon

Proof. Consider a random subset S of the set of vertices V such that each vertex is in S independently with probability $p = \frac{\ln(1+d)}{1+d}$. If S' is the set of vertices which do not belong to S and which have no neighbor in S, then $S \cup S'$ satisfies the second condition. It is thus enough to show that

$$E[|S \cup S'|] \leq n \cdot \frac{1 + \ln(d+1)}{d+1}.$$

However,

$$\begin{aligned} E[|S \cup S'|] &\leq E[|S|] + E[|S'|] \\ &= \sum_v P(v \in S) + \sum_v P(v \in S') \\ &\leq pn + n(1-p)^{d+1} \\ &\leq n \cdot \frac{1 + \ln(d+1)}{d+1}, \end{aligned}$$

the last inequality being equivalent (after dividing by n, canceling p and taking logarithms) to $\log(1-p) \leq -p$, which is well-known. □

A very beautiful but rather subtle application of the probabilistic method is the following gem due to Erdős:

B.17. Let A be a finite set of nonzero integers. Then A contains a subset B of cardinality greater than $\frac{|A|}{3}$ such that the equation $x + y = z$ has no solutions in $B \times B \times B$.

Erdős

Proof. The proof is quite tricky. Choose a prime $p = 3k + 2$ large enough, say such that $p > 2|a|$ for all $a \in A$. The subset $S = \{k+1, k+2, \ldots, 2k+1\}$ of $\mathbb{F}_p$ is a sum-free subset, i.e. the equation $x + y = z$ has no solutions with

$x, y, z \in S$. Pick $x \in \mathbb{F}_p^*$ randomly with uniform probability and consider the set

$$A_x = \{a \in A | ax \pmod p \in S\}.$$

This is easily seen to be a sum-free subset of A. To prove that one can choose x such that $|A_x| > \frac{|A|}{3}$, it is enough to compute the expected value of $|A_x|$. Linearity of expectation shows that

$$E[|A_x|] = \sum_{a \in A} P(ax \pmod p \in S)$$

and it is clear that for any $a \in A$ we have $P(ax \pmod p \in S) = \frac{|S|}{p-1}$, as $(ax)_{x \in \mathbb{F}_p^*}$ is a permutation of $\mathbb{F}_p^*$. Thus $E[|A_x|] \geq \frac{k+1}{3k+1}|A|$ and we are done. $\square$

We end this chapter with a very beautiful application of Chebyshev's inequality. Before attacking this problem, let us recall a few basic properties of the variance. Note that for any constant c and any random variable X we have $\mathrm{Var}(cX) = c^2\mathrm{Var}(X)$. Also, it is not difficult to check that if X, Y are independent variables, then $\mathrm{Var}(X+Y) = \mathrm{Var}(X) + \mathrm{Var}(Y)$. Finally, if X is 0 with probability p and 1 with probability $1-p$, then $\mathrm{Var}(X) = p(1-p)$.

B.18. Let $v_1, v_2, \dots, v_n$ be n vectors in the plane, whose coordinates are integers of absolute value less that $\frac{1}{100}\sqrt{\frac{2^n}{n}}$. Prove that there are disjoint subsets I, J of $\{1, 2, \dots, n\}$ such that

$$\sum_{i \in I} v_i = \sum_{j \in J} v_j.$$

Proof. Note first of all that it is enough to find two distinct such subsets I, J, for then it is enough to take out of each the common elements of I, J. Now, assume that we cannot find two such distinct subsets and write $v_i = (x_i, y_i)$. Consider a random binary sequence $(a_1, a_2, \dots, a_n)$, each a_i being 0 or 1 with probability 1/2 (all these events being independent). Consider the random variables $X = \sum_{i=1}^n a_i x_i$ and $Y = \sum_{i=1}^n a_i y_i$. The point is that the values taken by (X, Y) are all different and that a lot of values are concentrated near the expectation of (X, Y). To be more precise, let $m_x = E[X]$, $m_y = E[Y]$ and observe that the discussion preceding the problem yields

$$\text{Var}(X) = \text{Var}\left(\sum a_i x_i\right) = \sum \frac{x_i^2}{4} < \frac{2^n}{40000},$$

by hypothesis. Let $\sigma^2 = \frac{2^n}{40000}$. Using Chebyshev's inequality, we obtain

$$P(|X - m_x| \geq 2\sigma) \leq \frac{1}{4}, \quad P(|Y - m_y| \geq 2\sigma) \leq \frac{1}{4},$$

hence

$$P(|X - m_x| \leq 2\sigma, |Y - m_y| \leq 2\sigma) \geq \frac{1}{2}.$$

This means that at least half of the (pairwise distinct) points (X, Y) are located in the square $|X - m_x| \leq 2\sigma, \quad |Y - m_y| \leq 2\sigma$. But this square contains at most $(1 + 4\sigma)^2$ lattice points, so

$$(1 + 4\sigma)^2 \geq 2^{n-1},$$

which is easily seen to be impossible (as say $25\sigma^2 \geq (1 + 4\sigma)^2$ for $n \geq 16$) unless $n = 1$, in which case we are easily done. $\square$

Chapter 7

Complex Combinatorics

There is a class of combinatorial problems, most of the times with number-theoretic flavor, which have very elegant solutions using finite Fourier transforms or versions of it. The main observation is the identity

$$1_{n\equiv a \pmod{k}} = \frac{1}{k}\sum_{j=0}^{k-1} z^{j(n-a)},$$

which holds for any primitive root of unity z of order k. This gives a rather powerful approach to problems concerning the distribution mod k of the sums of subsets of a given set or to tiling problems. Actually, this relation is a very special case of a much broader theory, that of representations of finite groups. The case of finite abelian groups is particularly easy and useful in combinatorial problems and we refer the reader to addendum 7.A for a more detailed discussion. For the next problems the previous relation is actually sufficient.

7.1 Tiling and coloring problems

We start with two tiling problems which have rather elegant solutions using complex numbers. Of course, the idea is similar to the usual coloring method, but we think it is neater.

1. Let k be an integer greater than 2. For which odd positive integers n can we tile a $n \times n$ table by $1 \times k$ or $k \times 1$ rectangles such that only the central unit square is uncovered?

Gabriel Dospinescu

Proof. Assign coordinates to the squares of the table, with $(0,0)$ assigned to the upper left corner. Put the number z^{x+y} in square (x,y), where z is a primitive k-th root of unity. Then all $1 \times k$ and $k \times 1$ tiles will cover a total value of 0. Thus the total value of the whole board must be 0. On the other hand, this total value is precisely $\left(\sum_{i=0}^{n-1} z^i\right)^2 - z^{n-1}$. Thus we must have $\left(\sum_{i=0}^{n-1} z^i\right)^2 - z^{n-1} = 0$ so for some $\varepsilon \in \{-1,1\}$ we have $\frac{z^n-1}{z-1} = \varepsilon \cdot z^{\frac{n-1}{2}}$. This can be also written as $(z^{\frac{n-1}{2}} - \varepsilon)(z^{\frac{n+1}{2}} + \varepsilon) = 0$, which immediately implies that n is congruent to 1 or $-1 \pmod k$. The converse is easily seen to hold, for if $n \equiv 1 \pmod k$ then we can tile both the center column and the center row (excluding the center square) using the rectangles, leaving four rectangles of dimensions multiples of k. And if $n \equiv -1 \pmod k$, then we can split the table into four rectangles each with a dimension a multiple of k and so again the tiling is possible. Therefore the answer to the problem is: all n for which $n \equiv \pm 1 \pmod k$. □

2. Let $n \geq 2$ be an integer. At each point (i,j) having integer coordinates we write the number $i + j \pmod n$. Find all pairs (a,b) of positive integers such that any residue modulo n appears the same number of times on the sides (except for the vertices) of the rectangle with vertices $(0,0)$, $(a,0)$, (a,b), $(0,b)$ and also any residue modulo n appears the same number of times in the interior of this rectangle.

Bulgaria 2001

Proof. First, let us get rid of the easy case $n = 2$. As the interior of the rectangle must have an even number of lattice points, clearly one of a, b must be odd. But if one of a, b is odd, then clearly all conditions are satisfied, so that in this case the solutions are all pairs (a,b) with a, b not simultaneously even.

Now, let us treat the more difficult case $n > 2$. Let z be a primitive root of unity of order n and put the number z^{i+j} in the square (i, j). By hypothesis, the sum of the numbers associated to the squares of the rectangle is 0, so that $\sum_{i=1}^{a-1} \sum_{j=1}^{b-1} z^{i+j} = 0$. But this factors as

$$\sum_{i=1}^{a-1} \sum_{j=1}^{b-1} z^{i+j} = z^2 \cdot \frac{z^{a-1} - 1}{z - 1} \cdot \frac{z^{b-1} - 1}{z - 1},$$

so that necessarily one of a, b is congruent to 1 modulo n. By symmetry, we may assume that $a \equiv 1 \pmod{n}$. Now, since every residue modulo n appears the same number of times on the sides of the rectangle, we also have

$$(z^a + 1) \left(\sum_{j=1}^{b-1} z^j \right) + (z^b + 1) \left(\sum_{i=1}^{a-1} z^i \right) = 0.$$

As $a \equiv 1 \pmod{n}$, the previous equality becomes $(z+1) \left(\sum_{j=1}^{b-1} z^j \right) = 0$. But $n > 2$, so that $z + 1 \neq 0$ and so we must have $\sum_{j=1}^{b-1} z^j = 0$ and $b \equiv 1 \pmod{n}$. Since it is clear that for $a \equiv b \equiv 1 \pmod{n}$ all conditions are satisfied, it follows that this is the solution if $n > 2$. □

The next two problems have very beautiful solutions using the methods of this chapter and some classical algebraic identities.

3. Each element of $M = \{1, 2, \ldots, n\}$ is colored in either red, blue or yellow. Let A be the set of triples $(x, y, z) \in M \times M \times M$ such that n divides $x + y + z$ and x, y, z have the same color. Let B be the set of triples $(x, y, z) \in M \times M \times M$ such that n divides $x + y + z$ and x, y, z have pairwise different colors. Prove that $2|A| \geq |B|$.

Chinese TST 2010

Proof. For simplicity write 1, 2, 3 for the colors and let

$$f_j(X) = \sum_{i \in M, c(i) = j} X^i,$$

where $c(i) = j$ if i has color j. The identity

$$1_{n \equiv a \pmod{k}} = \frac{1}{k} \sum_{j=0}^{k-1} z^{j(n-a)}$$

yields

$$|A| = \sum_{j=1}^{3} \sum_{c(x)=c(y)=c(z)=j} \frac{1}{n} \sum_{k=0}^{n-1} e^{\frac{2i\pi k}{n}(x+y+z)} = \frac{1}{n} \sum_{k=0}^{n-1} \sum_{j=1}^{3} f_j(e^{\frac{2i\pi k}{n}})^3$$

and

$$|B| = \frac{6}{n} \sum_{k=0}^{n-1} \prod_{j=1}^{3} f_j(e^{\frac{2i\pi k}{n}}).$$

Of course, this method has some limitations, because it is rather difficult to compare complex numbers. The miracle is in the formula

$$x^3 + y^3 + z^3 - 3xyz = \frac{1}{2}(x+y+z)((x-y)^2 + (y-z)^2 + (z-x)^2)$$

and especially in the fact that

$$\sum_{j=1}^{3} f_j(e^{\frac{2i\pi k}{n}}) = \sum_{j=1}^{n} e^{\frac{2i\pi kj}{n}} = 0$$

unless $k = 0$, when it equals n.

Putting these observations together, we deduce that

$$2|A| - |B| = (f_1(1) - f_2(1))^2 + (f_2(1) - f_3(1))^2 + (f_1(1) - f_3(1))^2 \geq 0. \quad \square$$

The following problem is very similar, but more technically involved.

4. Color the numbers $1, 2, \ldots, N$ using 3 colors such that there are at most $\frac{N}{2}$ numbers of each color. Let A be the set of 4-tuples $(a, b, c, d) \in \{1, 2, \ldots, N\}^4$ such that $a+b+c+d \equiv 0 \pmod{N}$ and a, b, c, d have the same color. Let B be the set of 4-tuples $(a, b, c, d) \in \{1, 2, \ldots, N\}^4$ such that $a + b + c + d \equiv 0 \pmod{N}$, a, b and c, d have the same color, but these colors are distinct. Prove that $|A| \leq |B|$.

KöMaL

Proof. Let us write $z_k = e^{\frac{2i\pi k}{N}}$ and $c(j)$ for the color of number j. Define $f_j(X) = \sum_{c(a)=j} X^a$. As in the previous solution, one ends up with the formula

$$|B| - |A| = \frac{1}{N}\sum_{k=0}^{N-1}(2f_1(z_k)^2 f_2(z_k)^2 + 2f_2(z_k)^2 f_3(z_k)^2 + 2f_3(z_k)^2 f_1(z_k)^2 - f_1(z_k)^4 - f_2(z_k)^4 - f_3(z_k)^4).$$

Readers familiar with Heron's formula have already noticed that

$$2(a^2b^2 + b^2c^2 + c^2a^2) - (a^4 + b^4 + c^4) = (a+b+c)(b+c-a)(c+a-b)(a+b-c).$$

Also, as in the previous solution we have

$$f_1(z_k) + f_2(z_k) + f_3(z_k) = \sum_{a=1}^{N} e^{\frac{2i\pi ak}{N}}$$

and this is nonzero if and only if $k = 0$, in which case it equals N. The result follows from these remarks and the hypothesis, which ensures that

$$f_1(1) + f_2(1) \geq f_3(1)$$

and similar inequalities. □

7.2 Counting problems

In this section we combine the technique of generating functions with the fundamental relation

$$1_{n \equiv a \pmod{k}} = \frac{1}{k}\sum_{j=0}^{k-1} z^{j(n-a)}.$$

This mixture is very powerful and applies very well for counting problems with arithmetical flavor, as roots of unity can detect congruences very efficiently.

5. Three persons A, B, C play the following game: a subset with k elements of the set $\{1, 2, \ldots, 1986\}$ is selected randomly, all selections having the same probability. The winner is A, B, or C, according to whether the sum of the elements of the selected subset is congruent to 0, 1, or 2 modulo 3. Find all values of k for which A, B, C have equal chances of winning.

IMO 1987 Shortlist

Proof. Let z be a primitive third root of unity and consider the polynomial

$$P(X) = (1 + Xz)(1 + Xz^2) \cdots (1 + Xz^{1986}) = \sum_{j=0}^{1986} a_j X^j.$$

Note that

$$P(X) = \sum_{I \subset \{1,2,\ldots,1986\}} z^{m(I)} X^{|I|},$$

where $m(I)$ is the sum of elements of I. Thus

$$a_j = \sum_{|I|=j} z^{m(I)} = \sum_{|I|=j,3|m(I)} 1 + \left(\sum_{|I|=k,3|m(I)-1} 1 \right) z + \left(\sum_{|I|=k,3|m(I)-2} 1 \right) z^2$$

and this is zero if and only if

$$\sum_{|I|=j,3|m(I)} 1 = \sum_{|I|=j,3|m(I)-1} 1 = \sum_{|I|=j,3|m(I)-2} 1.$$

Therefore, the problem asks precisely for those k such that $a_k = 0$. Now, fortunately the polynomial P has a very simple form, since

$$P(X) = ((1 + Xz)(1 + Xz)(1 + Xz^2))^{662} = (1 + X^3)^{662},$$

so the nonzero coefficients are precisely those whose index is a multiple of 3. Thus the answer to the problem is: those $k \equiv 1, 2 \pmod 3$. □

We present two solutions for the following problem: the first one is the standard proof using complex numbers, while the second one is a very elegant probabilistic proof.

6. We roll a regular die n times. What is the probability that the sum of the numbers shown is a multiple of 5?

IMC 1999

Proof. We need to count the number of sequences $(x_1, x_2, \ldots, x_n)$ with $1 \leq x_i \leq 6$ and such that 5 divides $x_1 + x_2 + \cdots + x_n$. Let a_j be the number of sequences $(x_1, x_2, \ldots, x_n)$ such that $1 \leq x_i \leq 6$ and $x_1 + x_2 + \cdots + x_n \equiv j \pmod 5$. Then for each fifth root of unity z we have

$$a_0 + a_1 z + \cdots + a_4 z^4 = \sum_{1 \leq x_1, \ldots, x_n \leq 6} z^{x_1 + x_2 + \cdots + x_n} = (z + z^2 + \cdots + z^6)^n.$$

This equals z^n unless $z = 1$, in which case it equals 6^n. Adding these relations for all fifth roots of unity z gives $5a_0 = 6^n + z^n + z^{2n} + z^{3n} + z^{4n}$, for some primitive root of unity of order 5. If n is a multiple of 5, $z^n + \cdots + z^{4n} = 4$ and the probability is $\frac{a_0}{6^n} = \frac{6^n + 4}{5 \cdot 6^n}$. If not, then

$$z^n + \cdots + z^{4n} = \frac{z^n - z^{5n}}{1 - z^n} = -1,$$

so that the probability is $\frac{6^n - 1}{5 \cdot 6^n}$. □

Proof. We use a simple fact from probability. Suppose X is an integer-valued random variable which is uniformly distributed mod m (that is, takes on every value mod m with probability $1/m$) and suppose Y is any other integer-valued random variable independent of X. Then $X + Y$ is also uniformly distributed mod m.

To use this we break a roll of a die into two steps. We first flip a coin with a $5/6$ probability of giving a heads. If we get a heads, then we roll a fair 5-sided die with the numbers $1, 2, 3, 4, 5$ on it. If we get a tails, then we just get the number 6. If any one of our n coin flips gives a heads (which occurs with probability $1 - \left(\frac{1}{6}\right)^n$), then one of our summands is uniformly distributed mod

5 and hence the sum is uniformly distributed mod 5. Otherwise, all the coin flips were tails and hence all the dice were 6's, so that the total is $n \pmod 5$. Thus the probability of getting a sum which is a multiple of 5 is $\frac{1}{5}\left(1-\left(\frac{1}{6}\right)^n\right)$ for n not a multiple of 5 and $\frac{1}{5}\left(1-\left(\frac{1}{6}\right)^n\right)+\left(\frac{1}{6}\right)^n$ for n a multiple of 5. $\square$

The following problem is a good opportunity to recall a very useful result: if p is a prime and if $a_0, a_1, \dots, a_{p-1}$ are rational numbers such that $a_0 + a_1 z + \dots + a_{p-1}z^{p-1} = 0$ for some primitive root z of order p of unity, then $a_0 = a_1 = \dots = a_{p-1}$. This is an immediate consequence of the irreducibility of $1 + X + \dots + X^{p-1}$ over the rational numbers.

7. Let $p > 2$ be a prime. How many subsets of $\{1, 2, \dots, p-1\}$ have the sum of their elements divisible by p?

Ivan Landjev, Bulgaria TST 2006

Proof. Let z be a primitive root of order p of unity and consider the sum

$$S = \sum_{A \subset \{1,2,\dots,p-1\}} z^{m(A)},$$

where $m(A) = \sum_{a \in A} a$. If x_j is the number of subsets $A \subset \{1, 2, \dots, p-1\}$ such that $m(A) \equiv j \pmod p$, then clearly

$$S = x_0 + x_1 z + x_2 z^2 + \dots + x_{p-1} z^{p-1}.$$

On the other hand, we can explicitly compute S, since

$$S = \prod_{i=1}^{p-1}(1 + z^i) = \prod_{\zeta}(1 + \zeta),$$

the product being taken over all roots ζ of the polynomial

$$\frac{X^p - 1}{X - 1} = 1 + X + \dots + X^{p-1}.$$

We deduce that

$$\prod_{\zeta}(1 + \zeta) = \frac{(-1)^p - 1}{-1 - 1} = 1.$$

So $x_0 - 1 + x_1 z + \cdots + x_{p-1} z^{p-1} = 0$, which implies that $x_0 - 1 = x_1 = \cdots = x_{p-1} = k$ for some k. Since $x_0 + x_1 + \cdots + x_{p-1}$ is simply the number of subsets of $\{1, 2, \ldots, p-1\}$, that is 2^{p-1}, we deduce that $kp + 1 = 2^{p-1}$ and so $k = \frac{2^{p-1}-1}{p}$. Since $x_0 = 1 + k$, the problem is solved. Note that we included the empty set when counting x_0 (by convention the sum of the elements of the empty set is zero). □

The following problem is technically more involved, but uses the same ideas as before. The special case when n is a prime was problem 6 of IMO 1996.

8. Prove that the number of subsets with n elements of $\{1, 2, \ldots, 2n\}$ whose sum is a multiple of n is

$$\frac{(-1)^n}{n} \cdot \sum_{d|n} (-1)^d \varphi\left(\frac{n}{d}\right) \binom{2d}{d}.$$

Proof. Consider the polynomial

$$f(X, Y) = (1 + XY)(1 + X^2 Y) \cdots (1 + X^{2n} Y).$$

If $m(A)$ is the sum of the elements of A, we also have

$$f(X, Y) = \sum_{a=0}^{2n} \left(\sum_{|A|=a} X^{m(A)} \right) Y^a,$$

where all sets A in this solution are subsets of $\{1, 2, \ldots, 2n\}$. Taking for X the roots of unity of order n, say $z_j = e^{\frac{2i\pi j}{n}}$, the fundamental relation implies that

$$\frac{1}{n}(f(z_1, Y) + \cdots + f(z_n, Y)) = \sum_{a=0}^{2n} x_a \cdot Y^a,$$

where x_a is the number of subsets A with a elements and such that $m(A) \equiv 0 \pmod{n}$. We deduce that x_n, which is the number of sets A we are trying to find, is also the coefficient of Y^n in the polynomial $\frac{1}{n}(f(z_1,Y)+\cdots+f(z_n,Y))$. Now, fix j and let us compute

$$f(z_j,Y)=(1+z_jY)(1+z_j^2Y)\cdots(1+z_j^{2n}Y).$$

We can write $j=du, n=dv$, where $d=\gcd(n,j)$ and then we clearly have

$$f(z_j,Y)=((1+Y)(1+e^{\frac{2i\pi u}{v}}Y)\cdots(1+e^{\frac{2i\pi(v-1)u}{v}}Y))^{2d}=(1-(-Y)^v)^{2d}.$$

So the coefficient of Y^n is $(-1)^{d+n}\binom{2d}{d}$. Since for any $d|n$ there are exactly $\varphi(n/d)$ integers $1\le j\le 2n$ such that $\gcd(j,n)=d$, we deduce that the coefficient of Y^n in $\frac{1}{n}(f(z_1,Y)+\cdots+f(z_n,Y))$ is exactly

$$\frac{(-1)^n}{n}\cdot\sum_{d|n}(-1)^d\varphi\left(\frac{n}{d}\right)\binom{2d}{d},$$

which finishes the solution. □

We continue with a very nice problem concerning the number of solutions mod p of a quadratic equation in several variables. Actually, the same methods also apply in some other situations, but the general problem of estimating the number of solutions mod p of systems of polynomial equations mod p is very deep. See the addendum of this chapter for more details.

9. Let p be an odd prime. Find the number of 6-tuples (a,b,c,d,e,f) of integers between 0 and $p-1$ such that

$$a^2+b^2+c^2\equiv d^2+e^2+f^2 \pmod p.$$

MOSP 1997

Proof. Let z be a primitive root of order p of unity. Since $\sum_{k=0}^{p-1}z^{kx}=0$ if x is not a multiple of p and equals p otherwise, the desired number of 6-tuples is

$$S=\frac{1}{p}\sum_{a,b,c,d,e,f\in\mathbb{Z}/p\mathbb{Z}}\sum_{k=0}^{p-1}z^{k(a^2+b^2+c^2-d^2-e^2-f^2)}.$$

Note that

$$S = \frac{1}{p}\sum_{k=0}^{p-1} \sum_{a,b,c,d,e,f\in\mathbb{Z}/p\mathbb{Z}} z^{k(a^2+b^2+c^2-d^2-e^2-f^2)}$$

$$= \frac{1}{p}\sum_{k=0}^{p-1}\left(\sum_{a\in\mathbb{Z}/p\mathbb{Z}} z^{ka^2}\right)^3 \cdot \left(\sum_{d\in\mathbb{Z}/p\mathbb{Z}} z^{-kd^2}\right)^3.$$

In the previous sum, there is one obvious term: the one for $k = 0$, which gives us p^6. Also, for each $1 \le k \le p-1$ we have

$$\left(\sum_{a\in F_p} z^{ka^2}\right) \cdot \left(\sum_{d\in\mathbb{Z}/p\mathbb{Z}} z^{-kd^2}\right) = \sum_{a,d\in\mathbb{Z}/p\mathbb{Z}} z^{k(a^2-d^2)} = \sum_{x,y\in\mathbb{Z}/p\mathbb{Z}} z^{kxy},$$

by the change of variable $a - d = x, a + d = y$ (which recovers a, d uniquely because 2 is invertible modulo p). On the other hand, for a fixed $x \neq 0$, we have

$$\sum_{y\in\mathbb{Z}/p\mathbb{Z}} z^{kxy} = \sum_{y\in\mathbb{Z}/p\mathbb{Z}} z^{ky} = 0,$$

as the map $y \to kxy \pmod p$ is a bijection of $\mathbb{Z}/p\mathbb{Z}$. Therefore, $\sum_{x,y\in\mathbb{Z}/p\mathbb{Z}} z^{kxy} = p$, which shows that

$$\left(\sum_{a\in\mathbb{Z}/p\mathbb{Z}} z^{ka^2}\right) \cdot \left(\sum_{d\in\mathbb{Z}/p\mathbb{Z}} z^{-kd^2}\right) = p.$$

Combining the previous paragraphs yields the answer to the problem, namely $p^5 + (p-1)p^2$. □

Proof. First look at $a^2 - d^2 = (a-d)(a+d) \bmod p$. Since p is an odd prime, we can choose $x = a - d$ and $y = a + d$ arbitrarily and then solve uniquely for a and d via $a = (x+y)/2$, $d = (y-x)/2$. If x is nonzero, then since p is prime, xy takes on every value exactly once whereas if $x = 0$ then xy is always zero.

Thus $a^2 - d^2 = xy$ takes on the value zero $2p - 1$ times and takes on every nonzero value $p - 1$ times. Therefore it is described by the generating function

$$2p - 1 + (p-1)X + (p-1)X^2 + \cdots + (p-1)X^{p-1}$$
$$= p + (p-1)(1 + X + X^2 + \cdots + X^{p-1})$$

in $\mathbb{Z}[X]$ modulo $X^p - 1$. Note that modulo $X^p - 1$, the polynomial

$$1 + X + X^2 + \cdots + X^{p-1}$$

has the feature that

$$f(X)(1 + X + X^2 + \cdots + X^{p-1}) = f(1)(1 + X + X^2 + \cdots + X^{p-1})$$

for any polynomial $f(X)$ (as $X - 1$ divides $f(X) - f(1)$). Thus the generating function for the values taken on by $(a^2 - d^2) + (b^2 - e^2) + (c^2 - f^2)$ is

$$\left(p + (p-1)(1 + X + X^2 + \cdots + X^{p-1})\right)^3$$
$$= p^3 + (p^5 - p^2)(1 + X + X^2 + \cdots + X^{p-1}).$$

From this we read off the number of 6-tuples for which $a^2 + b^2 + c^2 \equiv d^2 + e^2 + f^2 \pmod{p}$ as $p^5 + p^3 - p^2$. □

We continue with a very nice problem, in which one uses a mixture of the previous techniques, algebraic manipulations and a rather tricky number-theoretic argument.

10. Let p be an odd prime. Prove that the $2^{\frac{p-1}{2}}$ numbers $\pm 1 \pm 2 \pm \cdots \pm \frac{p-1}{2}$ represent each nonzero residue class mod p the same number of times. Compute this number.

R. L. McFarland, AMM 6457

Proof. Let $z = e^{\frac{2i\pi}{p}}$ and write

$$S = \sum_{\varepsilon_i \in \{-1,1\}} z^{\varepsilon_1 + 2\varepsilon_2 + \cdots + \frac{p-1}{2}\varepsilon_{\frac{p-1}{2}}} = a_0 + a_1 z + \cdots + a_{p-1} z^{p-1}$$

for some integers a_i. Since z^x only depends on $x \pmod p$, it is clear that a_i is exactly the number of ways residue i is represented by the numbers $\pm 1 \pm 2 \pm \cdots \pm \frac{p-1}{2}$. Thus, the problem asks us to prove that $a_1 = a_2 = \cdots = a_{p-1}$ and to find this common value.

The point is that S has a nice closed expression, since it obviously factors as

$$S = \prod_{j=1}^{\frac{p-1}{2}} (z^j + z^{-j}).$$

On the other hand,

$$\prod_{j=1}^{p-1} (z^j + z^{-j}) = S \cdot \prod_{j=\frac{p+1}{2}}^{p-1} (z^j + z^{-j}) = S \cdot \prod_{j=1}^{\frac{p-1}{2}} (z^{p-j} + z^{j-p}) = S^2$$

and

$$\prod_{j=1}^{p-1} (z^j + z^{-j}) = \frac{1}{z^{\frac{p(p-1)}{2}}} \cdot \prod_{j=1}^{p-1} (1 + z^{2j}) = \prod_{j=1}^{p-1} (1 + z^j) = 1.$$

The last relation uses the fact that $x \to 2x$ is a bijection of the nonzero remainders modulo p (as p is odd) and that

$$\prod_{j=1}^{p-1} (1 + z^j) = 1,$$

which follows from

$$\prod_{j=1}^{p-1} (X - z^j) = \frac{X^p - 1}{X - 1}$$

by taking $X = -1$.

The previous computation shows that $S^2 = 1$, so that $S = \pm 1$ is definitely an integer. But then then relation $a_0 - S + a_1 z + \cdots + a_{p-1} z^{p-1} = 0$ implies that $a_0 - S = a_1 = \cdots = a_{p-1}$. In particular, $a_1 = \cdots = a_{p-1}$ and the first part of problem is solved.

On the other hand, we clearly have

$$a_0 + a_1 + \cdots + a_{p-1} = 2^{\frac{p-1}{2}},$$

which combined with $a_0 - S = a_1 = \cdots = a_{p-1}$ and with $S = \pm 1$ shows that

$$S \equiv 2^{\frac{p-1}{2}} \pmod{p} \equiv (-1)^{\frac{p^2-1}{8}} \pmod{p},$$

so that $S = (-1)^{\frac{p^2-1}{8}}$ and the value we are looking for is

$$a_1 = a_2 = \cdots = a_{p-1} = \frac{2^{\frac{p-1}{2}} - (-1)^{\frac{p^2-1}{8}}}{p}.$$

We used here a standard result in quadratic residues, saying that Legendre's symbol $\left(\frac{2}{p}\right) = (-1)^{\frac{p^2-1}{8}}$. □

It is rather difficult to approach the following problem directly with the methods of this chapter. However, a small observation reduces the problem to a more familiar one.

11. a) Let n be an odd integer. Find the number of sequences $(a_0, a_1, \ldots, a_n)$ such that $a_i \in \{1, 2, \ldots, n\}$ for all i, $a_n = a_0$ and $a_i - a_{i-1} \not\equiv i \pmod{n}$ for all $i = 1, 2, \ldots, n$.

    b) Let n be an odd prime. Find the number of sequences $(a_0, a_1, \ldots, a_n)$ such that $a_i \in \{1, 2, \ldots, n\}$ for all i, $a_n = a_0$ and $a_i - a_{i-1} \not\equiv i, 2i \pmod{n}$ for all $i = 1, 2, \ldots, n$.

Reid Barton, USA TST 2004

Proof. a) Call a sequence as in the problem admissible. Considering

$$b_i = a_i - a_{i-1} - i \pmod{n}$$

for $1 \le i \le n$, the condition becomes $b_i \neq 0 \pmod{n}$. Note that

$$b_1 + b_2 + \cdots + b_n \equiv -(1 + 2 + \cdots + n) \equiv 0 \pmod{n},$$

since n is odd. But conversely, if we have a sequence $b_1, b_2, \dots, b_n$ of elements of $\mathbb{Z}/n\mathbb{Z}$ satisfying $b_i \neq 0 \pmod{n}$ and $b_1+b_2+\cdots+b_n \equiv 0 \pmod{n}$, then we can find precisely n admissible sequences giving rise to $b_1, b_2, \dots, b_n$. Indeed, choose any $a_0 \in \{1, 2, \dots, n\}$ and take for a_k the remainder modulo n in $\{1, 2, \dots, n\}$ of the number $a_0 + 1 + 2 + \cdots + k + b_1 + b_2 + \cdots + b_k$. Thus, if $f(n)$ is the number of sequences $(b_1, b_2, \dots, b_n) \in (\mathbb{Z}/n\mathbb{Z} - \{0\})^n$ such that $b_1 + b_2 + \cdots + b_n \equiv 0 \pmod{n}$, then the desired answer is $nf(n)$.

Now, we evaluate $f(n)$ in the standard way: let $z = e^{\frac{2i\pi}{n}}$, so that

$$\begin{aligned} f(n) &= \frac{1}{n} \sum_{b_1,\dots,b_n \in \mathbb{Z}/n\mathbb{Z}-\{0\}} \sum_{k=0}^{n-1} z^{k(b_1+b_2+\cdots+b_n)} \\ &= \frac{1}{n} \sum_{k=0}^{n-1} \sum_{1 \le b_i \le n-1} z^{kb_1} \cdot \dots \cdot z^{kb_n} \\ &= \frac{1}{n} \sum_{k=0}^{n-1} \left(\sum_{b=1}^{n-1} z^{kb} \right)^n . \end{aligned}$$

Since

$$\sum_{b=1}^{n-1} z^{kb} = n - 1$$

for $k = 0$ and it equals $\frac{1-z^{nk}}{1-z^k} - 1 = -1$ for $1 \le k \le n-1$, we deduce that

$$f(n) = \frac{(n-1)^n - (n-1)}{n},$$

so the answer to part (a) is $(n-1)^n - (n-1)$.

b) The same argument as in the first paragraph of (a) shows that it is enough to count the number $g(n)$ of sequences $(b_1, b_2, \dots, b_n)$ with $b_i \in \mathbb{Z}/n\mathbb{Z}$, $b_i \not\equiv 0, i \pmod{n}$ and $b_1 + b_2 + \cdots + b_n \equiv 0 \pmod{n}$. The desired answer will be $ng(n)$.

We compute in the same way

$$\begin{aligned} g(n) &= \frac{1}{n} \sum_{b_i \neq 0, i} \sum_{k=0}^{n-1} z^{k(b_1+b_2+\cdots+b_n)} \\ &= \frac{1}{n} \sum_{k=0}^{n-1} \sum_{b_i \neq 0, i} z^{kb_1} \cdots z^{kb_n} \\ &= \frac{1}{n} \sum_{k=0}^{n-1} \prod_{i=1}^{n} \left(\sum_{b \neq 0, i \pmod{n}} z^{kb} \right). \end{aligned}$$

For $k = 0$ the corresponding product equals trivially $(n-2)^n \cdot (n-1)$ (the factor $n-1$ comes from the sum associated to $i = n$). Now fix $1 \leq k \leq n-1$. For all $1 \leq i < n$ we have

$$\sum_{b \neq 0, i \pmod{n}} z^{kb} = 1 + z^k + \cdots + z^{(n-1)k} - (1 + z^{ki}) = -(1 + z^{ki}).$$

For $i = n$ the corresponding sum is -1. Thus

$$\prod_{i=1}^{n} \left(\sum_{b \neq 0, i \pmod{n}} z^{kb} \right) = -\prod_{i=1}^{n-1} (1 + z^{ki}).$$

Now, since n is a prime and $1 \leq k < n$, we have $\gcd(k, n) = 1$, so the numbers z^{ki} with $1 \leq i \leq n-1$ are precisely the n-th roots of unity different from 1. Thus

$$\prod_{i=1}^{n-1} (1 + z^{ki}) = \prod_{u^n=1, u \neq 1} (1 + u) = 1,$$

the last equality being a consequence of the equality

$$\prod_{u^n=1, u \neq 1} (X - u) = \frac{X^n - 1}{X - 1}$$

and of the fact that n is odd. We deduce that

$$\prod_{i=1}^{n}\left(\sum_{b\neq 0,i \pmod n} z^{kb}\right) = -1$$

for all $1 \leq k \leq n-1$ and so the answer to the problem is

$$g(n) = \frac{(n-1)(n-2)^{n-1}-(n-1)}{n}. \qquad \square$$

The following problem is hard, even though the first steps are rather clear. The difficulty lies in the algebraic technicalities.

12. Let $p > 3$ be a prime number.

    If X is a nonempty subset of $\{0, 1, \ldots, p-1\}$, let $f(X)$ be the number of sequences $(a_1, a_2, \ldots, a_{p-1})$ such that $a_j \in X$ for all j and p divides $\sum_{j=1}^{p-1} ja_j$. Prove that $f(\{0,1,3\}) \geq f(\{0,1,2\})$, with equality if and only if $p = 5$.

    IMO 1999 Shortlist

Proof. Let us first find a formula for $f(X)$. Letting $z = e^{\frac{2i\pi}{p}}$, the usual argument yields

$$\begin{aligned} f(X) &= \frac{1}{p}\sum_{k=0}^{p-1}\sum_{a_1,\ldots,a_{p-1}\in X} z^{k(a_1+2a_2+\cdots+(p-1)a_{p-1})} \\ &= \frac{1}{p}\sum_{k=0}^{p-1}\left(\sum_{x\in X} z^{kx}\right)\cdot\dots\cdot\left(\sum_{x\in X} z^{k(p-1)x}\right). \end{aligned}$$

In particular,

$$f(\{0,1,2\}) = \frac{1}{p}\sum_{i=0}^{p-1}\left(\prod_{j=1}^{p-1}(1+z^{ij}+z^{2ij})\right).$$

Since $\{z^{ij}|1 \le j \le p-1\} = \{z^j|1 \le j \le p-1\}$ for all $i \ne 0 \pmod p$ and since $p > 3$, it follows that

$$f(\{0,1,2\}) = \frac{1}{p}\left(3^{p-1} + (p-1)\prod_{j=1}^{p-1}\frac{\zeta^{3j}-1}{\zeta^j-1}\right) = 1 + \frac{3^{p-1}-1}{p}.$$

We also have

$$f(\{0,1,3\}) = \frac{1}{p}\sum_{i=0}^{p-1}\left(\prod_{j=1}^{p-1}(1+\zeta^{ij}+\zeta^{3ij})\right).$$

So, if α, β, γ are the roots of x^3+x+1, we have (using again that for all $i \ne 0 \pmod p$ the numbers $(z^{ij})_j$ are a permutation of the numbers $(z^j)_j$)

$$\begin{aligned} f(\{0,1,3\}) &= \frac{1}{p}\left(3^{p-1} + (p-1)\prod_{j=1}^{p-1}(\zeta^j-\alpha)(\zeta^j-\beta)(\zeta^j-\gamma)\right) \\ &= \frac{3^{p-1}+(p-1)a_p}{p}, \end{aligned}$$

where

$$a_n = \frac{(\alpha^n-1)(\beta^n-1)(\gamma^n-1)}{(\alpha-1)(\beta-1)(\gamma-1)}.$$

It is thus enough to prove that $a_p \ge 1$, with equality precisely for $p = 5$. However, this sequence satisfies a linear recursive relation with characteristic polynomial

$$\begin{aligned} &(X-\alpha)(X-\beta)(X-\gamma)(X-\alpha\cdot\beta)(X-\beta\cdot\gamma)(X-\alpha\cdot\gamma)(X-\alpha\cdot\beta\cdot\gamma) \\ &\quad = (X+1)(X^3+X+1)(X-\alpha\cdot\beta)(X-\beta\cdot\gamma)(X-\alpha\cdot\gamma). \end{aligned}$$

We can easily compute that

$$\begin{aligned} (X-\alpha\cdot\beta)(X-\beta\cdot\gamma)(X-\alpha\cdot\gamma) &= \left(X+\frac{1}{\gamma}\right)\left(X+\frac{1}{\alpha}\right)\left(X+\frac{1}{\beta}\right) \\ &= X^3 - X^2 - 1 \end{aligned}$$

and so the characteristic polynomial of the sequence $(a_n)_n$ is

$$(x^3+x+1)(x^3-x^2-1)(x+1) = x^7 - 2x^3 - 2x^2 - 2x - 1.$$

Thus, there exists a constant C such that for all n we have

$$a_{n+7} = 2(a_{n+3} + a_{n+2} + a_{n+1}) + a_n + C$$

It is not difficult to compute that the first terms of the sequence $(a_n)_n$ are $0, 1, 1, 3, 1, 1, 3, 8, \ldots$ and that the sequence is increasing after the sixth term (and $C = -2$). The result follows easily. $\square$

Remark 7.1. Except for the technical combinatorial part, which, as we have seen, is quite standard, the difficult point of the problem is establishing the inequality $a_p \geq 1$, with equality precisely for $p = 5$. Proving that $a_p \geq 1$ is however rather easy, without the computation of the characteristic polynomial. Indeed, note that a_p is a symmetric polynomial with integer coefficients in the roots of $X^3 + X + 1$, so it is an integer. On the other hand, the polynomial function $x^3 + x + 1$ is increasing on the whole real line, so among α, β, γ precisely one is real, say α. Moreover, it is easy to check that $-1 < \alpha < 0$. Thus $\frac{1-\alpha^p}{1-\alpha} > 0$. Also,

$$\frac{(1-\beta^p)(1-\gamma^p)}{(1-\beta)(1-\gamma)} = \left|\frac{1-\beta^p}{1-\beta}\right|^2 > 0,$$

so that $a_p > 0$. It follows that $a_p \geq 1$. However, we know no easy proof of the fact that $a_p > 1$ for $p > 5$.

7.3 Miscellaneous problems

13. Let a_k, b_k, c_k be integers, $k = 1, 2, \ldots, n$ and let $f(x)$ be the number of ordered triples (A, B, C) of disjoint subsets (not necessarily nonempty) of the set $S = \{1, 2, \ldots, n\}$ whose union is S and for which

$$\sum_{i \in S \setminus A} a_i + \sum_{i \in S \setminus B} b_i + \sum_{i \in S \setminus C} c_i \equiv x \pmod 3.$$

Suppose that $f(0) = f(1) = f(2)$. Prove that there exists $i \in S$ such that $3 \mid a_i + b_i + c_i$.

Gabriel Dospinescu

Proof. Observe that

$$F(X) = \prod_{i=1}^{n} \left(X^{a_i+b_i} + X^{b_i+c_i} + X^{c_i+a_i} \right)$$
$$= \sum_{\{A,B,C\}} X^{\sum_{i\in S/A} a_i + \sum_{i\in S/B} b_i + \sum_{i\in S/C} c_i}$$

Thus, the equality $f(0) = f(1) = f(2) = 3^{n-1}$ becomes

$$F(X) \equiv 3^{n-1}(1 + X + X^2) \pmod{X^3 - 1}$$

Since $1 + X + X^2 | X^3 - 1$ this means in particular that $1 + X + X^2 | F(X)$. So there exists i such that $z^{a_i+b_i} + z^{b_i+c_i} + z^{c_i+a_i} = 0$, where z is a primitive third root of unity. This means that $\{a_i + b_i, b_i + c_i, c_i + a_i\}$ is equivalent to a permutation of $\{0, 1, 2\}$ (mod 3) and so $3 | a_i + b_i + c_i$. The conclusion follows. □

Readers who are not familiar with basic linear algebra will have a rather hard time trying to solve the following problem.

14. Let p be an odd prime and $n \geq 2$. For a permutation σ of the set $\{1, 2, \ldots, n\}$ define

$$S(\sigma) = \sigma(1) + 2\sigma(2) + \cdots + n\sigma(n).$$

Let A_j be the set of even permutations σ such that $S(\sigma) \equiv j \pmod{p}$ and let B_j be the set of odd permutations σ for which $S(\sigma) \equiv j \pmod{p}$. Prove that $n > p$ if and only if A_j and B_j have the same number of elements for all j.

Gabriel Dospinescu

Proof. Consider the matrix $A = \left(X^{ij}\right)$, whose determinant is given by

$$\det A = \sum_{\sigma \text{ even}} X^{S(\sigma)} - \sum_{\sigma \text{ odd}} X^{S(\sigma)}.$$

On the other hand, we also have a simple closed form for $\det A$, given by Vandermonde's formula:

$$\det A = \prod_{1 \le i < j \le n} \left(X^j - X^i\right).$$

The first formula and the definition of A_j, B_j imply the following equality in $\mathbb{Z}[X]$

$$\sum_{j=0}^{p-1} (A_j - B_j) X^j \equiv \prod_{1 \le i < j \le n} \left(X^j - X^i\right) \pmod{X^p - 1}.$$

Therefore, the problem comes down to showing that

$$n > p \iff \prod_{1 \le i < j \le n} \left(X^j - X^i\right) \equiv 0 \pmod{X^p - 1}.$$

This is actually very easy. Indeed, if $n > p$ then clearly

$$X^p - 1 | X^{p+1} - X | \prod_{1 \le i < j \le n} \left(X^j - X^i\right).$$

For the other direction, take z any primitive root of order p of unity. Then by hypothesis $\prod_{1 \le i < j \le n} (z^j - z^i) = 0$, showing that for some $i < j$ we have $z^{j-i} = 1$. Since this implies that $j - i$ is a multiple of p, thus at least p, we are done. □

The following problem is very challenging. Here, it is very hard to find the correct approach, especially because the problem suggests number theory.

15. Is there a positive integer k such that $p = 6k + 1$ is a prime and $\binom{3k}{k} \equiv 1 \pmod{p}$?

USA TST 2010

Proof. The answer is negative, but the proof is far from being obvious. Suppose that $p = 6k+1$ satisfies $\binom{3k}{k} \equiv 1 \pmod{p}$. Let g be a primitive root mod p and let $z = g^6$, so that z has order k mod p. Consider the sum

$$S = \sum_{i=0}^{k-1}(1+z^i)^{3k} = \sum_{i=0}^{k-1}\left(\sum_{j=0}^{3k}\binom{3k}{j}z^{ij}\right) = \sum_{j=0}^{3k}\binom{3k}{j}\sum_{i=0}^{k-1}z^{ij}.$$

Since z has order k mod p, the sum $\sum_{i=0}^{k-1} z^{ij}$ is 0 unless j is a multiple of k, which happens if and only if $j = 0, k, 2k, 3k$. Hence

$$\begin{aligned} S &\equiv \left(\binom{3k}{0} + \binom{3k}{k} + \binom{3k}{2k} + \binom{3k}{3k}\right)k \\ &= \left(2 + 2\binom{3k}{k}\right)k \pmod{p}. \end{aligned}$$

Since $\binom{3k}{k} \equiv 1 \pmod{p}$, we deduce that $S \equiv 4k \pmod{p}$. On the other hand,

$$(1+z^i)^{3k} \equiv (1+z^i)^{\frac{p-1}{2}} \equiv -1, 0, 1 \pmod{p}.$$

It is now immediate to check that we cannot have k remainders mod p, each of them $-1, 0$ or 1, adding up to $4k$ modulo p. This contradiction shows that no such k can be found and solves the problem. □

We present two difficult solutions for the following beautiful, but very challenging problem.

16. Let p be an odd prime and let a, b, c, d be integers not divisible by p such that

$$\left\{\frac{ra}{p}\right\} + \left\{\frac{rb}{p}\right\} + \left\{\frac{rc}{p}\right\} + \left\{\frac{rd}{p}\right\} = 2$$

for all integers r not divisible by p (here $\{\cdot\}$ is the fractional part). Prove that at least two of the numbers $a+b, a+c, a+d, b+c, b+d, c+d$ are divisible by p.

Kiran Kedlaya, USAMO 1999

Proof. Let $r(x) \in \{0, 1, \dots, p-1\}$ be the remainder of x mod p and observe that $r(x) = p \cdot \{\frac{x}{p}\}$. Thus the hypothesis can also be written

$$r(an) + r(bn) + r(cn) + r(dn) = 2p$$

for any $\gcd(n, p) = 1$. Let z be a primitive root of unity of order p and let a', b', c', d' be the inverses of a, b, c, d modulo p. For any m not a multiple of p we have $z^{mnaa'} = z^{mn}$, so[1] we have $\sum r(an) z^{mnaa'} = 2pz^{mn}$. Fix a number m relatively prime to p and let n run over a system of representatives of the nonzero classes mod p. Then $r(mn)$ runs over $1, 2, \dots, p-1$ and so does $r(an)$, as a and m are not multiples of p. Hence if we take the sum over n of the previous relation, we obtain

$$\sum \left(\sum_{j=1}^{p-1} j z^{mja'} \right) = 2p \sum_{j=1}^{p-1} z^j = -2p.$$

It is easy to check the identity

$$1 + 2X + \cdots + nX^{n-1} = \frac{nX^{n+1} - (n+1)X^n + 1}{(X-1)^2},$$

which easily implies that

$$\sum_{j=1}^{p-1} j z^{ma'j} = \frac{p}{z^{a'm} - 1}$$

and similarly for b, c, d. Thus, we can write

$$\sum \frac{1}{1 - z^{a'm}} = 2$$

and this for all m relatively prime to p. Clearing denominators and canceling equal terms yields (after a tedious computation) an equivalent equality

$$2 + \sum z^{(a'+b'+c')m} = \sum z^{a'm} + 2z^{(a'+b'+c'+d')m},$$

[1] All unspecified sums are over a, b, c, d.

which continues to hold (trivially) when m is a multiple of p.

Finally, we add these relations for $0 \le m \le p-1$. Note that

$$\sum_{m=0}^{p-1} z^{ma'} = 0$$

and similarly for b', c', d' and that

$$\sum_{m=0}^{p-1} z^{Nm} \ge 0$$

for any N (simply because the corresponding sum equals p when $z^N = 1$ and 0 otherwise). We deduce that

$$2\sum_{m=0}^{p-1} z^{m(a'+b'+c'+d')} \ge 2p,$$

which implies that $a'+b'+c'+d'$ is a multiple of p (otherwise the left-hand side would be 0). Therefore, by the previous key equality, we also have

$$\sum z^{-a'm} = \sum z^{a'm}$$

for any m relatively prime to p. By multiplying the previous relation by $z^{-ma'}$, we get

$$\begin{aligned} 1+&z^{m(b'-a')} + z^{(c'-a')m} + z^{m(d'-a')} \\ &= z^{-2a'm} + z^{-(a'+b')m} + z^{-(a'+c')m} + z^{-(a'+d')m}, \end{aligned}$$

and a similar argument (add the equations and observe that the left-hand side is at least p) shows that at least one of $a'+b', a'+c', a'+d'$ is a multiple of p. Say $p|a'+b'$, then also $p|c'+d'$ (as $p|a'+b'+c'+d'$) and so $p|a+b$ and $p|c+d$, finishing the proof of this hard problem. □

Proof. This second proof uses Lagrange's interpolation theorem, for which we refer the reader to chapter 11. As in the previous solution,

$$r(x) \in \{0, 1, \ldots, p-1\}$$

is the remainder of x modulo p. Define

$$Q(x) = \frac{2r(x) - r(2x)}{p},$$

so that $Q(x) = 0$ if $0 \le r(x) \le (p-1)/2$ and $Q(x) = 1$ if $(p-1)/2 < r(x) < p$. Call (a, b, c, d) good if it satisfies the relation in the problem, which can also be written as $r(ka) + r(kb) + r(kc) + r(kd) = 2p$ for all $1 \le k < p$. Note that if (a, b, c, d) is good, then so is (ka, kb, kc, kd) for any k which is not a multiple of p. Also, note that if (a, b, c, d) is good, then

$$Q(a) + Q(b) + Q(c) + Q(d) = 2.$$

Combining these two observations, we deduce that

$$Q(ka) + Q(kb) + Q(kc) + Q(kd) = 2$$

for all $1 \le k < p$.

By Lagrange's interpolation theorem, there exists a polynomial $P(X)$ of degree at most $p-2$ such that $P(x) \equiv Q(x) \pmod{p}$ for all $x \not\equiv 0 \pmod{p}$. Note that if $R(X) = P(X+1) - P(X)$, then $R(x) \equiv 0 \pmod{p}$ when

$$x = 1, \ldots, \frac{p-3}{2}, \frac{p+1}{2}, \ldots, p-2$$

and $R((p-1)/2) \not\equiv 0 \pmod{p}$, so $\deg R = p-3$ and $\deg P = p-2$. On the other hand, the polynomial $S(X) = P(Xa) + P(Xb) + P(Xc) + P(Xd)$ is congruent to 2 mod p for $p-1$ values of the variable. Since $\deg S \le p-2$, $S-2$ must be the zero polynomial mod p. Imposing that the coefficient of X^{p-2} vanishes in S, we obtain the key relation

$$a^{p-2} + b^{p-2} + c^{p-2} + d^{p-2} \equiv 0 \pmod{p},$$

which can be also written (by Fermat's little theorem)

$$\frac{1}{a}+\frac{1}{b}+\frac{1}{c}+\frac{1}{d}=0\in\mathbb{F}_p.$$

Finally, since $r(a)+r(b)+r(c)+r(d)=2p$, we have $a+b+c+d=0$ in $\mathbb{F}_p$. Combining the two relations yields

$$\frac{1}{a}+\frac{1}{b}+\frac{1}{c}=\frac{1}{a+b+c},$$

which readily becomes (after clearing denominators) $(a+b)(b+c)(c+a)=0$. By symmetry, we may assume that $a+b=0$ in $\mathbb{F}_p$. But then $c+d=0$ also and we are done. □

We end this chapter with a very deep result, whose proof is however elementary (and follows [52]). It improves previous results of Brown and Buhler [12], Frankl, Graham and Rödl [33] and finally Ruzsa [52]. For more details on some arguments concerning orthogonality relations and properties of characters, we refer the reader to addendum 7.A.

17. Let p be a prime and let $S\subset(\mathbb{Z}/3\mathbb{Z})^d$ be a subset containing no line of the affine space $(\mathbb{Z}/3\mathbb{Z})^d$. Prove that S has at most $\frac{2\cdot 3^d}{d}$ elements. However, prove that we can find such a set with at least $3^{\frac{2d}{3}-1}$ elements.

Meshulam-Roth theorem

Proof. Let $f(d)$ be the largest cardinality of a set $S\subset\mathbb{F}_3^d$ (we write $\mathbb{F}_3$ for $\mathbb{Z}/3\mathbb{Z}$) containing no line. Note that three distinct points of $\mathbb{F}_3^d$ adding up to the zero vector form a line and conversely, the sum of the points of a line is the zero vector. So, $f(d)$ is also the largest cardinality of a set S containing no three elements that add up to 0. Since $\mathbb{F}_3^d$ and $F=\mathbb{F}_{3^d}$ (the field with 3^d elements) are isomorphic as additive groups, we may work from now on with subsets S of F.

In particular, if S is such a set, the orthogonality relations for characters of abelian groups (theorem 7.A.5) yield

$$\frac{1}{3^d}\sum_{\chi}\left(\sum_{s\in S}\chi(s)\right)^3=\frac{1}{3^d}\sum_{s_1,s_2,s_3\in S}\sum_{\chi}\chi(s_1+s_2+s_3)=|S|,$$

the sum being taken over all characters χ of $F = \mathbb{F}_{3^d}$. The whole point is to be able to estimate the previous quantity for each character χ. Let $g(d) = \frac{f(d)}{3^d}$ and take a subset S as in the theorem, but with $|S| = f(d) = 3^d g(d)$. Note that

$$\sum_{x \in F} (1_{x \in S} - g(d-1))\chi(x) = \sum_{s \in S} \chi(s)$$

if χ is not trivial and it equals $\sum_{s \in S} \chi(s) - 3^d g(d-1)$ otherwise (again by orthogonality of characters). We deduce that

$$\begin{aligned} |S| &= g(d-1)|S|^2 + \frac{1}{3^d} \sum_{\chi} \left(\sum_{s \in S} \chi(s) \right)^2 \left(\sum_{x \in F} (1_{x \in S} - g(d-1))\chi(x) \right) \\ &\geq g(d-1)|S|^2 - \frac{1}{3^d} \sum_{\chi} \left| \sum_{s \in S} \chi(s) \right|^2 \cdot \left| \sum_{x \in F} (1_{x \in S} - g(d-1))\chi(x) \right|. \end{aligned}$$

Here comes the crucial estimate:

Lemma 7.2. *For all characters χ we have*

$$\left| \sum_{x \in F} (1_{x \in S} - g(d-1))\chi(x) \right| \leq 3^d g(d-1) - |S|.$$

Proof. If χ is nontrivial, let H_0 be its kernel, while if χ is trivial, let H_0 be any hyperplane of F (where F is seen as $\mathbb{F}_3$-vector space of dimension d). Take H_1, H_2 two affine hyperplanes parallel to H_0 so that the H_i's form a partition of F. Note that by definition χ is constant on each H_i, say $\chi(x) = z_i$ for $x \in H_i$ (where of course $|z_i| = 1$). Then

$$\begin{aligned} \left| \sum_{x \in F} (1_{x \in S} - g(d-1))\chi(x) \right| &= \left| \sum_{i=0}^{2} z_i (|S \cap H_i| - f(d-1)) \right| \\ &\leq \sum_{i=0}^{2} \left| |S \cap H_i| - f(d-1) \right|. \end{aligned}$$

But clearly $S \cap H_i$ does not contain three elements adding up to 0 and since H_i is $(d-1)$-dimensional, we deduce that $|S \cap H_i| \le f(d-1)$ (by definition of $f(d-1)$). Therefore, the previous estimate yields

$$\left| \sum_{x \in F} (1_{x \in S} - g(d-1))\chi(x) \right| \le 3^d g(d-1) - \sum_{i=0}^{2} |S \cap H_i|$$
$$= 3^d g(d-1) - |S|,$$

which is precisely what we wanted. □

Coming back to the proof and using the lemma and Plancherel's identity (theorem 7.A.5)

$$\sum_{\chi} \left| \sum_{s \in S} \chi(s) \right|^2 = 3^d |S|,$$

we deduce the estimate (do not forget that S was chosen such that $|S| = 3^d g(d) = f(d)$)

$$|S| \ge g(d-1)|S|^2 - |S|(3^d g(d-1) - |S|),$$

which implies that

$$g(d) \le \frac{g(d-1) + 3^{-d}}{g(d-1) + 1}.$$

Since $g(1) = \frac{2}{3}$, it is immediate to check that the last estimate implies $g(d) \le \frac{2}{d}$, which is precisely what we wanted.

Finally, let us show that $f(3d) \ge 9^d$, which will trivially imply the remaining part of the theorem. Let $x = 3^d + 1$ and consider

$$S = \{(a, a^x) | a \in \mathbb{F}_{9^d}\} \subset \mathbb{F}_{9^d} \times \mathbb{F}_{3^d}.$$

This has 9^d elements and we claim that it does not contain three elements adding up to 0. Note that $a^x \in \mathbb{F}_{3^d}$ if $a \in \mathbb{F}_{9^d}$, because $(a^x)^{3^d - 1} = 1$ if $a \ne 0$. If (a_j, a_j^x) add up to 0, we obtain $a_1 + a_2 + a_3 = 0$ and $a_1^x + a_2^x + a_3^x = 0$. But then

$$a_1^x + a_2^x + (a_1 + a_2)^x = 0.$$

On the other hand, since we work in characteristic 3, we have

$$\begin{aligned}(a_1+a_2)^x &= (a_1+a_2)(a_1+a_2)^{3^d}\\ &= (a_1+a_2)(a_1^{3^d}+a_2^{3^d})\\ &= a_1^x+a_2^x+a_1a_2^{x-1}+a_2a_1^{x-1}.\end{aligned}$$

We deduce that

$$(a_1-a_2)^x = (a_1-a_2)(a_1^{x-1}-a_2^{x-1}) = 0,$$

a contradiction. □

7.4 Notes

We thank the following people for providing solutions to the problems discussed in this chapter: Mitchell Lee (problems 1, 2, 10, 15), Richard Stong (problems 6, 9), Qiaochu Yuan (problems 5, 6, 8), Victor Wang (problem 16), Alex Zhu (problems 15, 16), Gjergji Zaimi (problems 12, 13, 14).

Addendum 7.A Finite Fourier Analysis

The identity

$$\frac{1}{n}\sum_{k=0}^{n-1} e^{\frac{2i\pi k(a-b)}{n}} = 1_{a\equiv b \pmod n}$$

(for all integers a, b and all positive integers n) is at the origin of a lot of beautiful mathematical results, as we could see in chapter 7, but it is just a special case of a broader theory. In the first part of this addendum we will see a vast generalization of this relation, through Fourier analysis on finite abelian groups. Though rather elementary, these results are extremely useful in number theory and combinatorics. On the other hand, they are the very first component of a much broader picture, that of harmonic analysis. Things are much easier for finite abelian groups, since one avoids quite a lot of technicalities that appear when dealing with harmonic analysis on other groups (such as $\mathbb{R}$, the unit circle in $\mathbb{C}$ or, much harder, non-abelian and non-compact topological groups). For instance, all convergence issues are automatic (as all sums we deal with are finite), while the integration theory has no subtlety (on the other hand, if one wants to do harmonic analysis on $\mathbb{R}$, one has to be very careful with these issues and one has to develop a rather powerful integration theory first). Also, the fact that the group is abelian simplifies the problem considerably, basically because all irreducible complex representations of such a group are one-dimensional. We will also recall the main features of Fourier analysis on finite general groups, without proving anything, since this is already the content of a course in representation theory. Next, we deal with a classical, but amazing application of these ideas, namely Dirichlet's theorem. This allows us to say a few words about L-functions of Dirichlet characters. This is again just the tip of a huge iceberg, which is far from being understood, even though progress is constantly being made.

7.A.1 Dual group

From now on, let G be a finite abelian group with n elements. We want to define the Fourier transform of a function $f : G \to \mathbb{C}$. Recall that if f is an integrable function on $\mathbb{R}$, with complex values, its Fourier transform is defined by

$$\widehat{f}(x) = \int_{-\infty}^{\infty} f(y) e^{2i\pi xy} dy.$$

The presence of the characters $y \to e^{2i\pi xy}$ of $\mathbb{R}$ will be our guide to Fourier analysis on abelian groups.

Remark 7.A.1. When dealing with abelian groups (which will always be the case for us), one also denotes by $+$ the internal operation of G. This is quite intuitive when dealing with groups such as $\mathbb{Z}/n\mathbb{Z}$, but definitely not suitable for groups such as $(\mathbb{Z}/n\mathbb{Z})^*$. Our convention will be the following: when dealing with an abstract abelian group G, we will use multiplicative notation for the internal operation of the group, while in concrete examples we will choose the most intuitive notation, depending on the situation. Hopefully, this will not create too much confusion...

A character of G is a morphism of groups $\chi : G \to \mathbb{C}^*$. So, according to whether we use additive or multiplicative notation for the internal operation of G, a character satisfies $\chi(x+y) = \chi(x) \cdot \chi(y)$ for all $x, y \in G$ (respectively $\chi(xy) = \chi(x) \cdot \chi(y)$). The character is called trivial if $\chi(g) = 1$ for all $g \in G$. Let $\widehat{G}$ be the set of all characters of G. It becomes a group with respect to the obvious multiplication $(\chi_1 \cdot \chi_2)(g) = \chi_1(g)\chi_2(g)$ and it is called the dual group of G. Note that for all $\chi \in \widehat{G}$ and $g \in G$ we have $\chi(g)^n = \chi(g^n) = 1$, because $g^n = 1$ by Lagrange's theorem. In particular, $|\chi(g)| = 1$ and $\chi^{-1}(g) = \overline{\chi(g)}$ for all $g \in G$ and $\chi \in \widehat{G}$. The idea of harmonic analysis is that all information about the space of $\mathbb{C}$-valued functions on G is encoded in $\widehat{G}$.

Example 7.A.2. Take $n \geq 2$ and $G = \mathbb{Z}/n\mathbb{Z}$. What is the dual group of G? We saw that $\chi(1)$ is an nth root of unity. And clearly χ is uniquely determined by $\chi(1)$, as G is generated by 1. Conversely, if z is an n-th root of unity, $x \to z^x$ defines a character of G (by z^x we mean z^a for any lifting a of x; this does not depend on the choice of a, as $z^n = 1$). We deduce that $\widehat{G}$ is isomorphic to $\mathbb{Z}/n\mathbb{Z}$, even though this isomorphism depends on the choice of a primitive nth root of 1, so it is not really canonical.

It is an easy exercise to check that passing to duals is compatible with direct products (i.e. the dual of $G \times H$ is $\widehat{G} \times \widehat{H}$). Since any finite abelian group is a direct product of cyclic groups (this is a nontrivial, but absolutely classical result), it follows from this remark and the previous example that for any finite abelian group G, its dual is isomorphic to G. In particular, G and its dual have the same number of elements, which is really not obvious at first sight. We also deduce from this observation and example 7.A.2 that if $x \in G - \{1\}$, then there exists $\chi \in \widehat{G}$ such that $\chi(x) \neq 1$ (if you think about this for a moment, you will realize that it is nontrivial!). Thus, the map $g \to (\chi \to \chi(g))$ is an injective homomorphism $G \to \widehat{\widehat{G}}$, realizing G as subgroup of $\widehat{\widehat{G}}$. Since $|\widehat{\widehat{G}}| = |G|$, the previous injection has to be an isomorphism. So G is canonically isomorphic to its double dual. Let us now glorify what we have just proved:

Theorem 7.A.3. *If G is a finite abelian group, then $\widehat{G}$ is an abelian group isomorphic to G and $\widehat{\widehat{G}}$ is canonically isomorphic to G.*

Let $F(G, \mathbb{C})$ be the $\mathbb{C}$-vector space of all maps $f : G \to \mathbb{C}$. It is a $\mathbb{C}$-vector space of dimension $|G|$, since the map $F(G, \mathbb{C}) \to \mathbb{C}^{|G|}$ sending f to $(f(g))_{g \in G}$ is obviously a $\mathbb{C}$-linear isomorphism. If $f, g \in F(G, \mathbb{C})$, let

$$\langle f, g \rangle = \frac{1}{|G|} \sum_{x \in G} f(x)\overline{g(x)}.$$

It is easy to check that this is an inner product on the $\mathbb{C}$-vector space $F(G, \mathbb{C})$. We are now ready to prove the main theorems of Fourier analysis on G.

Theorem 7.A.4. *The elements of $\widehat{G}$ form an orthonormal basis of $F(G, \mathbb{C})$.*

Proof. We split the proof in several steps.

Step 1. We prove that $\langle \chi_1, \chi_2 \rangle = 1_{\chi_1 = \chi_2}$, i.e. that $(\chi)_{\chi \in \widehat{G}}$ is an orthonormal set. If $\chi_1 = \chi_2$, everything follows from the fact that $\chi(g)$ has magnitude 1 for any g and any character χ. Assume that $\chi = \frac{\chi_1}{\chi_2}$ is not trivial. Then

$$\langle \chi_1, \chi_2 \rangle = \frac{1}{|G|} \sum_{x \in G} \chi(x).$$

Let $S = \sum_{x \in G} \chi(x)$. Then for all $g \in G$ we have

$$\chi(g)S = \sum_{x \in G} \chi(gx) = \sum_{x \in G} \chi(x) = S,$$

because $x \to gx$ is a permutation of G. Since χ is not trivial, there is g such that $\chi(g) \neq 1$ and the previous identity yields $S = 0$ and so $\langle \chi_1, \chi_2 \rangle = 0$.

Step 2. We claim that for all $x \in G - \{1\}$ we have $\sum_{\chi \in \widehat{G}} \chi(x) = 0$. The same argument as in the first step shows that it is enough to prove that there exists $\chi \in \widehat{G}$ such that $\chi(x) \neq 1$. This is nontrivial, but has already been explained.

Step 3. Since $(\chi)_{\chi \in \widehat{G}}$ is an orthonormal set, it is linearly independent. But since it has the same cardinality as the dimension of the vector space $F(G, \mathbb{C})$ (namely $|G|$, by theorem 7.A.3), basic theory of vector spaces shows that it has to be a basis of $F(G, \mathbb{C})$. The result follows. □

In practice, one really uses the following consequence of the previous theorem:

Theorem 7.A.5. *For any finite abelian group G, the following relations hold:*

1) (orthogonality relations) For all $\chi, \chi_1, \chi_2 \in \widehat{G}$ and $g, h \in G$

$$\frac{1}{|G|} \sum_{x \in G} \chi_1(x)\overline{\chi_2(x)} = 1_{\chi_1 = \chi_2}, \quad \frac{1}{|G|} \sum_{\chi \in \widehat{G}} \chi(g)\overline{\chi(h)} = 1_{g=h}.$$

2) (Fourier inversion) For all $f \in F(G, \mathbb{C})$ we have $f = \sum_{\chi \in \widehat{G}} \langle f, \chi \rangle \chi$.

3) (Plancherel's identity) For all $f \in F(G, \mathbb{C})$

$$\frac{1}{|G|} \sum_{x \in G} |f(x)|^2 = \sum_{\chi \in \widehat{G}} |\langle f, \chi \rangle|^2.$$

Proof. 1) has already been proved while proving the previous theorem. For 2), note that for any $x \in G$ we have, by the orthogonality relations

$$\left(\sum_{\chi} \langle f, \chi \rangle \cdot \chi \right)(x) = \frac{1}{|G|} \sum_{\chi} \chi(x) \sum_{y} f(y)\overline{\chi(y)}$$

$$= \frac{1}{|G|} \sum_{y} f(y) \sum_{\chi} \chi(x/y) = f(x),$$

from which the result follows. Finally, using 2) and the previous theorem, we can write

$$\frac{1}{|G|}\sum_{x\in G}|f(x)|^2 = \langle f,f\rangle = \sum_{\chi_1}\sum_{\chi_2}\langle f,\chi_1\rangle\cdot\overline{\langle f,\chi_2\rangle}\cdot\langle\chi_1,\chi_2\rangle = \sum_{\chi\in\widehat{G}}|\langle f,\chi\rangle|^2. \quad \square$$

Remark 7.A.6. More generally, we have

$$\langle f,g\rangle = \sum_{\chi\in\widehat{G}}\langle f,\chi\rangle\overline{\langle g,\chi\rangle}$$

for all $f, g \in F(G,\mathbb{C})$, as can be deduced by an obvious adaptation of the proof of Plancherel's identity.

By analogy with the usual formula in classical Fourier analysis

$$\widehat{f}(n) = \frac{1}{2\pi}\int_0^{2\pi} f(y)e^{-iny}dy,$$

we write $\widehat{f}(\chi) = \langle f,\chi\rangle$ and call it the χ-Fourier coefficient of f.

7.A.2 A glimpse of the non-abelian case

One may ask whether the above theory can be adapted to the case when the group G is still finite, but no longer abelian. The answer is positive, but things are much more complicated than in the abelian case. In order to develop Fourier analysis on any finite group G, one has to consider all complex finite dimensional representations of G. By definition, such a representation consists of a finite dimensional $\mathbb{C}$-vector space V on which the group G acts linearly, i.e. for each element $g \in G$ one has an automorphism still denoted g of V such that $gh = g\circ h$ (in the left-hand side gh is the automorphism associated to $gh \in G$, while in the right-hand side we have a composition of automorphisms). In more down to earth terms, a representation of G is a morphism[2] $\rho : G \to \mathrm{GL}_n(\mathbb{C})$ for some $n \geq 1$. Two representations $\rho_1, \rho_2 : G \to \mathrm{GL}_n(\mathbb{C})$ are called isomorphic if there exists $P \in \mathrm{GL}_n(\mathbb{C})$ such that $\rho_2(g) = P\rho_1(g)P^{-1}$ for all $g \in G$.

[2]Recall that $\mathrm{GL}_n(\mathbb{C})$ is the group of matrices $A \in M_n(\mathbb{C})$ with nonzero determinant.

If $n = 1$, we call ρ a character (this is compatible with the definition of a character given in the previous section). Such a representation V is called irreducible if no proper subspace of V is stable under all automorphisms g, with $g \in G$. For instance, a character is an irreducible representation, as there is no proper subspace at all! Moreover, the only irreducible representations of a finite abelian group are the characters, so the new theory will be compatible with the theory developed in the previous section. To prove the previous assertion, one has to use a basic result from linear algebra, stating that any commutative family of endomorphisms of a finite dimensional vector space over an algebraically closed field has a common eigenvector. So, if G is an abelian group acting irreducibly on V, there is a common eigenvector v for all $g \in G$. But then $\mathbb{C}v$ is a nonzero subspace of V stable under all $g \in G$ and so we must have $\mathbb{C}v = V$; hence V is a character.

We have the following very basic theorem due to Maschke:

Theorem 7.A.7. *Any finite dimensional $\mathbb{C}$-representation V of a finite group is a direct sum of irreducible representations, that is there exist sub-vector-spaces $V_1, \ldots, V_k$ of V stable under G, which are irreducible representations and such that $V = \oplus_i V_i$.*

Now, the role of the dual $\widehat{G}$ of G in the abelian case is played by the set $\widehat{G}$ of (isomorphism classes of) irreducible representations of G. It turns out that this is a finite set and by the previous discussion its definition agrees with the usual definition if G is abelian. The set of maps $F(G, \mathbb{C})$ is naturally a $\mathbb{C}$-algebra and as such it is isomorphic to the group algebra $\mathbb{C}[G]$. By definition, the elements of the last object can be uniquely written $\sum_{g \in G} a_g \cdot g$ with $a_g \in \mathbb{C}$ (so the elements of G form a $\mathbb{C}$-basis of $\mathbb{C}[G]$) and multiplication is defined by

$$\left(\sum_{g \in G} a_g \cdot g\right) \cdot \left(\sum_{g \in G} b_g \cdot g\right) = \sum_{g \in G} \left(\sum_{hk=g} a_h b_k\right) \cdot g.$$

Note that $\mathbb{C}[G]$ itself is a representation of G (each element of G acting as multiplication by g). The following theorem summarizes the properties of $\mathbb{C}[G]$.

Theorem 7.A.8. *1) $\widehat{G}$ is a finite set and $\mathbb{C}[G]$ is isomorphic as a representation of G to $\oplus_{V\in\widehat{G}}(\dim V)V$, where $nV = V\oplus V\oplus\cdots\oplus V$ (n times). In particular,*

$$|G| = \sum_{V\in\widehat{G}} (\dim V)^2.$$

2) The center of $\mathbb{C}[G]$ consists of all maps $f : G \to \mathbb{C}$ such that $f(g) = f(hgh^{-1})$ for all $g, h \in G$. Its dimension over $\mathbb{C}$ is equal to $|\widehat{G}|$ and also to the number of conjugacy classes[3] of G.

In the abelian case, we defined an inner product on $\mathbb{C}[G]$ and we showed that the characters of G formed an orthonormal basis of $\mathbb{C}[G]$. All this can be done in the non-abelian case, though things are usually more difficult to prove. Namely, define for $f_1, f_2 \in \mathbb{C}[G]$

$$\langle f_1, f_2\rangle = \frac{1}{|G|}\sum_{g\in G} f_1(g)\overline{f_2(g)}.$$

Now, to any representation $\rho : G \to GL_n(\mathbb{C})$ one can associate its character, which is the element of $\mathbb{C}[G]$ defined by $\chi_\rho(g) = \mathrm{Tr}(\rho(g))$. The main theorem of Fourier theory on finite groups is then the following:

Theorem 7.A.9. *1) If V_1, V_2 are two representations of G such that $\chi_{V_1} = \chi_{V_2}$ as elements of $\mathbb{C}[G]$, then V_1 and V_2 are isomorphic (and conversely).*

2) $(\chi_V)_{V\in\widehat{G}}$ is an orthonormal basis of the center of $\mathbb{C}[G]$ for the previously defined inner product. Moreover, a representation V is irreducible if and only if $\langle\chi_V, \chi_V\rangle = 1$.

3) If $g, h \in G$, then $\sum_{V\in\widehat{G}} \chi_V(g)\overline{\chi_V(h)}$ is equal to the cardinality of the centralizer of g in G if g, h are conjugate in G and it is equal to 0 otherwise.

[3]The conjugacy class of an element $g \in G$ is $\{hgh^{-1} | h \in G\}$.

7.A.3 Some concrete examples

Let us specialize the previously quite abstract (at first...) theory to a very concrete situation that appears frequently in number theory. Let N be an integer greater than 1 and let

$$G = (\mathbb{Z}/N\mathbb{Z})^*$$

be the abelian group of invertible residue classes mod N. A character of G is called a Dirichlet character of modulus N or simply a Dirichlet character mod N. If χ is such a character, we set

$$\chi(n) = 1_{\gcd(n,N)=1} \cdot \chi(n)$$

for all integers n, obtaining in this way an N-periodic function. We will focus on a more restricted class of characters modulo N, namely the primitive ones. Let us recall what that means. If d divides N and if χ_d is a character mod d, then χ_d yields a character mod N simply by composing it with the natural map $(\mathbb{Z}/N\mathbb{Z})^* \to (\mathbb{Z}/d\mathbb{Z})^*$. We say that a character mod N is primitive if it is not obtained in this way, for any proper divisor d of N and any χ_d. A more practical and useful criterion is the following: a character χ mod N is primitive if and only if for any proper divisor d of N there exists $n \equiv 1 \pmod{d}$ such that $\gcd(n, N) = 1$ and $\chi(n) \neq 1$. We leave to the reader the easy task of checking this.

Let us see what happens when we apply the abstract theory to this situation. Let a be an integer prime to N and let f be the map $f(n) = 1_{n\equiv a \pmod{N}}$. It is naturally a map on G and it is clear from the definitions that for all characters χ of G we have $\widehat{f}(\chi) = \overline{\chi(a)}$. So, using Fourier's inversion formula, we obtain

$$1_{n\equiv a \pmod{N}} = \frac{1}{\varphi(N)} \sum_{\chi} \overline{\chi(a)}\chi(n),$$

the sum being taken over all Dirichlet characters mod N. This relation holds for all n prime to N and plays a crucial role in the proof of the famous Dirichlet's theorem, to be discussed in the next section.

Gauss sums and Fourier coefficients of Dirichlet characters

To any Dirichlet character χ mod N we attached an N-periodic function on $\mathbb{Z}$, defined by $\chi(n) = 1_{\gcd(n,N)=1} \cdot \chi(n)$. Any N-periodic function on $\mathbb{Z}$ induces a map on $\mathbb{Z}/N\mathbb{Z}$ and we can look at its Fourier coefficients with respect to this finite abelian group. As we have already seen, the characters of $\mathbb{Z}/N\mathbb{Z}$ are identified with $\mathbb{Z}/N\mathbb{Z}$ (we identify a and the character $x \to e^{\frac{2i\pi ax}{N}}$). Via this identification, the Fourier coefficients of a map f on $\mathbb{Z}/N\mathbb{Z}$ will be denoted

$$\widehat{f}(r) = \frac{1}{N} \sum_{x \in G} f(x) e^{\frac{-2i\pi rx}{N}}.$$

The following result gives further information about these coefficients when f comes from a Dirichlet character:

Proposition 7.A.10. *Let χ be a Dirichlet character mod N.*

1) For any a relatively prime to N we have

$$\widehat{\chi}(a) = \overline{\chi(a)} \widehat{\chi}(1).$$

2) If χ is primitive, then $\widehat{\chi}(a) = 0$ whenever $\gcd(a, N) > 1$ and we have

$$|\widehat{\chi}(a)| = \frac{1}{\sqrt{N}}$$

for $\gcd(a, N) = 1$.

Proof. 1) Since $\chi(x) = 0$ whenever $\gcd(x, N) > 1$, we have

$$\widehat{\chi}(a) = \frac{1}{N} \sum_{x \in G} \chi(x) \zeta^{ax},$$

where $\zeta = e^{-\frac{2i\pi}{N}}$. As $x \to ax$ is a permutation of G, we have

$$\chi(a)\widehat{\chi}(a) = \frac{1}{N} \sum_{x \in G} \chi(ax) \zeta^{ax} = \frac{1}{N} \sum_{x \in G} \chi(x) \zeta^{x} = \widehat{\chi}(1)$$

and the result follows from $|\chi(a)| = 1$.

2) Write $N = dv$ and $a = du$ with $d > 1$ and $\gcd(u, v) = 1$. Let

$$\zeta = e^{-\frac{2i\pi u}{v}},$$

a primitive root of unity of order v. Note that

$$\begin{aligned}
\widehat{\chi}(a) &= \frac{1}{N} \sum_{x \,(\mathrm{mod}\ N)} \chi(x)\zeta^x \\
&= \frac{1}{N} \sum_{j=0}^{N-1} \chi(j)\zeta^j \\
&= \frac{1}{N} \sum_{k=0}^{d-1}\sum_{j=0}^{v-1} \chi(j+kv)\zeta^j \\
&= \frac{1}{N} \sum_{j \,(\mathrm{mod}\ v)} \left(\sum_{k \,(\mathrm{mod}\ d)} \chi(j+kv) \right) \zeta^j.
\end{aligned}$$

It is thus enough to prove that

$$S_j = \sum_{k \pmod d} \chi(j+kv)$$

vanishes for all j. Now, since χ is primitive, there exists $n \equiv 1 \pmod v$ such that $\gcd(n, N) = 1$ and $\chi(n) \neq 1$. It is easy to see that $(n(j+kv) \pmod N)_{k \pmod d}$ is simply a permutation of $(j+kv \pmod N)_{k \pmod d}$. Thus

$$\chi(n)S_j = \sum_{k \,(\mathrm{mod}\ d)} \chi(n(j+kv)) = \sum_{k \,(\mathrm{mod}\ d)} \chi(j+kv) = S_j$$

and since $\chi(n) \neq 1$, we have $S_j = 0$. This proves the first part of 2).

Finally, using Plancherel's identity (theorem 7.A.5) for $\mathbb{Z}/N\mathbb{Z}$ and the information we have already gathered, we deduce that

$$\begin{aligned}\frac{\varphi(N)}{N} &= \sum_{x \pmod N} |\chi(x)|^2 \\ &= \sum_{a \pmod N} |\widehat{\chi}(a)|^2 \\ &= \sum_{\gcd(a,N)=1} |\overline{\chi(a)}\widehat{\chi}(1)|^2 \\ &= \varphi(N)|\widehat{\chi}(1)|^2,\end{aligned}$$

from where we obtain $|\widehat{\chi}(1)| = \frac{1}{\sqrt{N}}$. Using again part 1), the result follows. □

7.A.4 Some applications

We have developed enough theory in the previous sections to be able to prove quite a few nontrivial results. The first one is very elementary, but tricky and taken from [71]. The interested reader will find in loc.cit many beautiful applications of finite Fourier analysis.

Proposition 7.A.11. *Let A be a finite set of integers and let $f : A \to \mathbb{Z}/p\mathbb{Z}$ be a map. Then for any positive integer k there exist at least $\frac{|A|^{2k}}{p}$ $(2k)$-tuples $(a_1, \ldots, a_{2k}) \in A^{2k}$ such that*

$$f(a_1) + f(a_2) + \cdots + f(a_k) \equiv f(a_{k+1}) + f(a_{k+2}) + \cdots + f(a_{2k}) \pmod p.$$

Proof. Let $N(j)$ be the number of $2k$-tuples $(a_1, \ldots, a_{2k}) \in A^{2k}$ such that

$$f(a_1) + f(a_2) + \cdots + f(a_k) \equiv f(a_{k+1}) + f(a_{k+2}) + \cdots + f(a_{2k}) + j \pmod p.$$

Clearly $\sum_{j=0}^{p-1} N(j) = |A|^{2k}$. We will prove that $N(0) \geq N(j)$ for all j, which will be enough to deduce the desired result. Next, note that by the orthogonality relations we can write

$$\begin{aligned} N(j) &= \sum_{a_1,\ldots,a_{2k}\in A} \frac{1}{p}\sum_{b=0}^{p-1} e^{\frac{2i\pi}{p}b(f(a_1)+\cdots-f(a_{k+1})-\cdots-f(a_{2k})-j)} \\ &= \frac{1}{p}\sum_{b=0}^{p-1} e^{-\frac{2i\pi bj}{p}} \left(\sum_{a\in A} e^{\frac{2i\pi bf(a)}{p}}\right)^k \cdot \left(\sum_{a\in A} e^{-\frac{2i\pi bf(a)}{p}}\right)^k \\ &= \frac{1}{p}\sum_{b=0}^{p-1} e^{-\frac{2i\pi bj}{p}} \left|\sum_{a\in A} e^{\frac{2i\pi bf(a)}{p}}\right|^{2k}. \end{aligned}$$

But it is apparent in this last expression that $N(j) \leq N(0)$, finishing the proof. □

The method used to prove the following result is extremely useful in additive combinatorics.

Theorem 7.A.12. *(D.Hart, A.Iosevich)*

Let q be a prime power, d a positive integer and let $A \subset \mathbb{F}_q^d$ be a set with more than $q^{\frac{d+1}{2}}$ elements. Then for any $x \in \mathbb{F}_q^*$ one can find $a, b \in A$ such that $a \cdot b = x$ (here $\cdot$ is the standard inner product in $\mathbb{F}_q^d$).

Proof. The idea is to express the number of solutions of the equation $a \cdot b = x$ using the orthogonality relations, then analyze the error term. Let χ be a nontrivial additive character of $\mathbb{F}_q$ and let $n(x)$ be the number of solutions of the equation $a \cdot b = x$ with $a, b \in A$. The orthogonality relations yield

$$n(x) = \sum_{a,b\in A} \frac{1}{q}\sum_{c\in\mathbb{F}_q} \chi(c(a\cdot b - x)) = \frac{|A|^2}{q} + R,$$

where

$$R = \frac{1}{q}\sum_{a\in A}\sum_{b\in A}\sum_{c\in\mathbb{F}_q^*} \chi(c(a\cdot b - x)).$$

Now, using the Cauchy-Schwarz inequality and once again the orthogonality relations, we obtain

$$
\begin{aligned}
R^2 &\le \frac{|A|}{q^2}\sum_{a\in A}\left|\sum_{b\in A}\sum_{c\neq 0}\chi(c(a\cdot b - x))\right|^2\\
&\le \frac{|A|}{q^2}\sum_{a\in\mathbb{F}_q^d}\sum_{b_1,b_2\in A}\sum_{c_1,c_2\in\mathbb{F}_q^*}\chi(a\cdot(c_1b_1-c_2b_2))\chi(x(c_2-c_1))\\
&= \frac{|A|}{q^2}\sum_{b_1,b_2\in A}\sum_{c_1,c_2\in\mathbb{F}_q^*}\chi(x(c_2-c_1))\sum_{a\in\mathbb{F}_q^d}\chi(a\cdot(c_1b_1-c_2b_2))\\
&= q^{d-2}|A|\sum_{b_1,b_2\in A}\sum_{c_1,c_2\in\mathbb{F}_q^*}\chi(x(c_2-c_1))1_{c_1b_1=c_2b_2}.
\end{aligned}
$$

Using the orthogonality relations and the substitution $s_1 = \frac{c_1}{c_2}$, $s_2 = c_2$, we can write

$$
\begin{aligned}
&\sum_{b_1,b_2\in A}\sum_{c_1,c_2\in\mathbb{F}_q^*}\chi(x(c_2-c_1))1_{c_1b_1=c_2b_2}\\
&\qquad= \sum_{b_1,b_2\in A}\sum_{s_1\in\mathbb{F}_q^*}1_{s_1b_1=b_2}\sum_{s_2\in\mathbb{F}_q^*}\chi(xs_2(1-s_1))\\
&\qquad= \sum_{b_1,b_2\in A}(q-1)1_{b_1=b_2} - \sum_{b_1,b_2\in A}\sum_{s_1\neq 1}1_{s_1b_1=b_2}\\
&\qquad\le (q-1)\sum_{b_1,b_2\in A}1_{b_1=b_2}\\
&\qquad< q|A|.
\end{aligned}
$$

Combining this with the first formula for $n(x)$, we deduce that $n(x) > 0$ if $|A| > q^{\frac{d+1}{2}}$, from where the result follows. □

Corollary 7.A.13. *Let $A \subset \mathbb{F}_q$ be a subset with more than $q^{\frac{1}{2}+\frac{1}{2d}}$ elements. Then for any $x \in \mathbb{F}_q^*$ there exist $a_1, \ldots, a_d \in A$ and $b_1, \ldots, b_d \in A$ such that $x = a_1b_1 + a_2b_2 + \cdots + a_db_d$.*

Proof. Apply the previous theorem to the subset $A \times A \times \cdots \times A$ of $\mathbb{F}_q^d$. □

The following bound on character sums is a basic tool in analytic number theory.

Theorem 7.A.14. *(Polya-Vinogradov) Let χ be a primitive character modulo N. Then for all positive integers m, n we have*

$$\left|\sum_{m \le j < n} \chi(j)\right| \le \sqrt{N} \log N.$$

Before giving the proof, let us mention that Schur proved that if χ is a primitive character mod N, then

$$\max_M \left|\sum_{n \le M} \chi(n)\right| > \frac{1}{2\pi}\sqrt{N},$$

so the Polya-Vinogradov inequality is not far from being optimal.

Proof. Using Fourier's inversion formula and proposition 7.A.10, we can write

$$\chi(j) = \sum_{(a,N)=1} \widehat{\chi}(a) e^{\frac{2i\pi aj}{N}}.$$

Adding this over j, using the triangle inequality and proposition 7.A.10, we deduce that

$$|\sum_{m \le j < n} \chi(j)| = \left|\sum_{(a,N)=1} \widehat{\chi}(a) \cdot \frac{e^{\frac{2i\pi an}{N}} - e^{\frac{2i\pi am}{N}}}{e^{\frac{2i\pi a}{N}} - 1}\right| \le \frac{1}{\sqrt{N}} \sum_{(a,N)=1} \frac{1}{|\sin \frac{\pi a}{N}|}.$$

An easy convexity argument shows that $\sin x \ge \frac{2}{\pi}x$ for $0 \le x \le \frac{\pi}{2}$. Applying this to $x = \frac{\pi a}{N}$ (respectively $x = \frac{\pi(N-a)}{N}$) if $a \le \frac{N}{2}$ (respectively $a > \frac{N}{2}$), we deduce that

$$|\sum_{m \le j < n} \chi(j)| \le \sqrt{N} \cdot \sum_{a \le N/2} \frac{1}{a} \le \sqrt{N} \log N$$

and the result follows. □

Here is a nice application of the previous theorem, taken from [59]. We refer the reader to that paper for many refinements and further discussion:

Theorem 7.A.15. *(Murty) Let p be a prime and let q be a prime divisor of $p-1$. Let a be an integer such that $a^{\frac{p-1}{q}} \equiv 1 \pmod{p}$. Then there exists an integer x such that $|x| < \frac{p^{3/2}\log p}{q}$ and $x^q \equiv a \pmod{p}$.*

In [59], Murty proves that the hypothesis that q is prime is not necessary and that we can strengthen the conclusion by asking that $|x| < cp^{3/2}/q$ for a suitable absolute constant c. This requires some stronger estimates on character sums.

Proof. Consider a parameter $p > T > 1$ and look at the number of solutions of the congruence $x^q \equiv a \pmod{p}$ with $x \in [1, T]$. Using the orthogonality relations, we write this as

$$\begin{aligned} S &= \sum_{n \le T} 1_{n^q \equiv a \pmod{p}} \\ &= \sum_{n \le T} \frac{1}{p-1} \sum_{\chi \pmod{p}} \overline{\chi(a)}\chi(n^q) \\ &= \frac{1}{p-1} \sum_{\chi} \overline{\chi(a)} \sum_{n \le T} \chi^q(n). \end{aligned}$$

In the previous sum we will distinguish those characters χ such that $\chi^q = 1$ from the others. Indeed, if $\chi^q = 1$, then $\sum_{n \le T} \chi^q(n) = T$ and $\chi(a) = 1$ (since the hypothesis on a implies that a is a qth power in $\mathbb{F}_p^*$), while if $\chi^q \ne 1$, Polya-Vinogradov's theorem gives

$$\left| \sum_{n \le T} \chi^q(n) \right| \le \sqrt{p} \log p.$$

Thus, since there are precisely q characters χ such that $\chi^q = 1$, the previous remarks yield

$$\left| S - \frac{qT}{p-1} \right| \le \frac{p-1-q}{p-1} \sqrt{p} \log p < \sqrt{p} \log p.$$

Now, everything follows from the simple observation that if $S \neq 0$, then there exists $1 \leq n \leq T$ such that $n^q \equiv a \pmod{p}$, while if $S = 0$, then

$$T < \frac{p^{\frac{3}{2}} \log p}{q} - 1. \qquad \square$$

If p is a prime, let $n_2(p)$ be the smallest positive integer a such that a is a quadratic non-residue mod p. It is an easy exercise to check that $n_2(p) < 1 + \sqrt{p}$, but it is much more challenging to find nontrivial bounds for $n_2(p)$. The next result gives such an upper bound, by combining the Polya-Vinogradov inequality with Mertens' theorems.

Theorem 7.A.16. *(Vinogradov) If p is a sufficiently large prime number, then there exists $1 \leq a < p^{\frac{1}{2\sqrt{e}}} \log^2 p$ such that*

$$\left(\frac{a}{p}\right) = -1.$$

Proof. Let $m = [\sqrt{p} \log^2 p]$ and suppose that

$$\left(\frac{a}{p}\right) = 1$$

for all $1 \leq a \leq X = \left[p^{\frac{1}{2\sqrt{e}}} \log^2 p\right]$. Let N be the number of quadratic non-residues among $\{1, 2, \ldots, m\}$. By Polya-Vinogradov's inequality

$$|m - 2N| = \left|\sum_{x=1}^{m} \left(\frac{x}{p}\right)\right| \leq \sqrt{p} \log p,$$

so $N > \frac{m}{2} - \frac{1}{2}\sqrt{p}\log p$. On the other hand, any quadratic non-residue mod p must have a prime factor greater than X. Thus

$$\frac{m}{2} - \frac{1}{2}\sqrt{p}\log p < N \leq \sum_{X < q \leq m} \frac{m}{q}.$$

Mertens' theorem (theorem 3.A.5 and the remark following it) yields

$$\sum_{X<q\le m}\frac{m}{q}=m\log\frac{\log m}{\log X}+O\left(\frac{m}{\log X}\right).$$

Note that $\frac{m}{\log X}=O(\sqrt{p}\cdot\log p)$ and

$$\begin{aligned}\log\frac{\log m}{\log X}&=\log\frac{\frac{1}{2}\log p+2\log\log p}{\frac{1}{2\sqrt{e}}\log p+2\log\log p}+O\left(\frac{1}{X\log p}\right)\\&=\frac{1}{2}+\log\frac{1+4\frac{\log\log p}{\log p}}{1+4\sqrt{e}\frac{\log\log p}{\log p}}+O\left(\frac{1}{X\log p}\right)\\&=\frac{1}{2}-\frac{4(\sqrt{e}-1)\log\log p}{\log p}+O\left(\frac{1}{\log p}\right).\end{aligned}$$

Combining these estimates yields

$$\frac{m}{2}-\frac{1}{2}\sqrt{p}\log p<\frac{m}{2}-4(\sqrt{e}-1)m\cdot\frac{\log\log p}{\log p}+O(\sqrt{p}\log p),$$

which is not possible for p large enough. □

7.A.5 Dirichlet's theorem

We will use the previous tools and a bit of complex analysis to give a proof of the famous

Theorem 7.A.17. *Let a and N be relatively prime positive integers. For $s\to 1^+$ we have*

$$\sum_{p\equiv a \pmod N}\frac{1}{p^s}=\frac{1}{\varphi(N)}\cdot\log\left(\frac{1}{s-1}\right)+O(1).$$

In particular, there are infinitely many primes $p\equiv a \pmod N$.

The proof of theorem 7.A.17 combines Fourier analysis on $G = (\mathbb{Z}/N\mathbb{Z})^*$ and very subtle estimates of the L function of a Dirichlet character near 1. To start with, define the complex L-function of a Dirichlet character χ mod N by (recall that $\chi(n) = 0$ for $\gcd(n, N) > 1$)

$$L(s, \chi) = \sum_{n \geq 1} \frac{\chi(n)}{n^s}.$$

As $|\chi(n)| \leq 1$ for all n, this series converges uniformly on compact subsets of $\mathrm{Re}(s) > 1$, so $L(s, \chi)$ defines a holomorphic function in this region. Moreover, a simple argument going back to Euler and using the unique factorization theorem and the fact that $\frac{1}{1-x} = \sum_{n \geq 0} x^n$ for $|x| < 1$ shows that for $\mathrm{Re}(s) > 1$ we have

$$L(s, \chi) = \prod_p \frac{1}{1 - \chi(p)p^{-s}}.$$

We easily deduce from this that $L(s, \chi)$ does not vanish if $\mathrm{Re}(s) > 1$. Moreover, if we choose a branch of the complex logarithm, we can write

$$\log L(s, \chi) = -\sum_p \log(1 - \chi(p)p^{-s}) = \sum_p \sum_{n \geq 1} \frac{\chi(p^n)p^{-ns}}{n}.$$

Dirichlet's key insight was to use the Fourier transform to express the condition that $n \equiv a \pmod N$ in an analytic way. More precisely, multiplying the previous relation by $\overline{\chi(a)}$, summing over χ and using the orthogonality relations, we obtain for $s > 1$

$$\begin{aligned} \frac{1}{\varphi(N)} \sum_{\chi \,(\mathrm{mod}\ N)} \overline{\chi(a)} \log L(s, \chi) &= \sum_p \sum_{\substack{n \geq 1 \\ p^n \equiv a \,(\mathrm{mod}\ N)}} \frac{1}{np^{ns}} \\ &= \sum_{p \equiv a \,(\mathrm{mod}\ N)} \frac{1}{p^s} + O(1), \end{aligned}$$

the last estimate being a consequence of the inequalities

$$0 < \sum_p \sum_{\substack{n \geq 2 \\ p^n \equiv a \,(\mathrm{mod}\ N)}} \frac{1}{np^{ns}} \leq \sum_p \sum_{n \geq 2} \frac{1}{p^n} = \sum_p \frac{1}{p(p-1)} < \infty.$$

The crucial point is that $\log L(s,\chi)$ remains bounded as $s \to 1^+$ precisely when χ is nontrivial and that $\log L(s,1)$ (where 1 is the trivial character) is relatively easy to handle and yields the factor $\log \frac{1}{s-1}$. If the second part is rather easy to prove, the first part is a very deep result, equivalent to the non-vanishing of $L(s,\chi)$ at $s=1$ whenever χ is nontrivial. Dirichlet's proof was very roundabout, but we have quite a few different ways to prove this nowadays. First, let us deal with the "easy" part, which will also be important in the proof of the hard part.

Theorem 7.A.18. *$L(s,\chi)$ extends to a function on $\mathrm{Re}(s) > 0$, which is holomorphic except possibly at $s=1$. If χ is nontrivial, this function is holomorphic at $s=1$, but if χ is trivial, we have*

$$\lim_{s\to 1^+}(s-1)L(s,1) = \prod_{p|N}\left(1-\frac{1}{p}\right).$$

Proof. The key ingredient is Abel summation. Suppose that χ is nontrivial and note that for all n we have $|\chi(1)+\chi(2)+\cdots+\chi(n)| \le N$, which follows easily from the orthogonality relations (theorem 7.A.5). An easy computation shows that

$$n^{-s} - (n+1)^{-s} = sn^{-s-1} + O(n^{-s-2}),$$

which is uniform for s in compact sets. Thus the series

$$\sum_{n\ge 1}(\chi(1)+\chi(2)+\cdots+\chi(n))\left(n^{-s}-(n+1)^{-s}\right)$$

converges uniformly on compact subsets of $\mathrm{Re}(s) > 0$. Moreover, by Abel summation, the sum of the series is $L(s,\chi)$ if $\mathrm{Re}(s) > 1$, which yields the holomorphic extension of $L(s,\chi)$ to $\mathrm{Re}(s) > 0$.

On the other hand, if χ is trivial, the inclusion-exclusion principle yields

$$L(s,\chi) = \prod_{p|N}\left(1-\frac{1}{p^s}\right)\cdot\zeta(s), \quad \zeta(s) = \sum_{n\ge 1}\frac{1}{n^s}.$$

Since

$$\int_n^{n+1}\left(n^{-s}-t^{-s}\right)dt = \frac{1}{n^s} + \frac{(n+1)^{1-s}-n^{1-s}}{s-1},$$

we get

$$\zeta(s) = \frac{1}{s-1} + \sum_{n\geq 1} \int_n^{n+1} \left(n^{-s} - t^{-s}\right) dt.$$

But for $t \in [n, n+1]$ we have

$$|t^{-s} - n^{-s}| = \left| \int_n^t -sx^{-s-1} dx \right| \leq \int_n^t \frac{|s|}{x^{\mathrm{Re}(s)+1}} dx \leq \frac{|s|}{n^{1+\mathrm{Re}(s)}},$$

yielding the uniform convergence of the series

$$g(s) = \sum_{n\geq 1} \int_n^{n+1} \left(n^{-s} - t^{-s}\right) dt$$

on all compact subsets of $\mathrm{Re}(s) > 0$. Thus g is holomorphic on $\mathrm{Re}(s) > 0$ and since $\zeta(s) = \frac{1}{s-1} + g(s)$, the result follows. □

Taking into account the previous theorem and discussion, theorem 7.A.17 is a consequence of the difficult

Theorem 7.A.19. *If χ is nontrivial, then $L(1, \chi) \neq 0$, so $\log L(s, \chi)$ remains bounded for $s \to 1^+$.*

Proof. We saw that for all $s > 1$ we have

$$\frac{1}{\varphi(N)} \sum_{\chi \,(\mathrm{mod}\ N)} \overline{\chi(a)} \log L(s, \chi) = \sum_p \sum_{\substack{n\geq 1 \\ p^n \equiv a \,(\mathrm{mod}\ N)}} \frac{1}{np^{ns}} \geq 0,$$

so by taking $a = 1$ we obtain

$$\prod_{\chi \,(\mathrm{mod}\ N)} L(s, \chi) \geq 1.$$

But by the previous theorem all $L(s, \chi)$ with χ nontrivial are holomorphic at $s = 1$ and $L(s, 1)$ has a simple pole at $s = 1$. Combining this observation with the previous inequality, we deduce that there is at most one nontrivial character χ such that $L(1,\chi) = 0$ (otherwise the product of $L(s,\chi)$ would vanish for $s \to 1^+$). Assume that χ is such a character, then clearly

$$L(1,\overline{\chi}) = 0,$$

so that we must have $\chi = \overline{\chi}$, that is χ takes only values ± 1 and 0. Thus, it remains to prove that $L(1,\chi) \neq 0$ whenever χ is a nontrivial real character. We will present Paul Monsky's elementary proof from [54].

Consider the function

$$f(x) = \sum_{n\geq 1} \chi(n) \frac{x^n}{1-x^n},$$

defined for $x \in [0,1)$ (the series is obviously absolutely convergent for such x). We claim that f is unbounded as $x \to 1^-$. Indeed, we can write

$$f(x) = \sum_{n\geq 1} \chi(n) \sum_{j\geq 1} x^{nj} = \sum_{n\geq 1} \left(\sum_{d|n} \chi(d) \right) x^n,$$

all manipulations of the series being justified by the absolute convergence. Let $c_n = \sum_{d|n} \chi(d)$. If p divides N, then $c_{p^k} = 1$ for all $k \geq 1$, as $\chi(p) = 0$. Also, it is easy to see that $c_{mn} = c_m c_n$ for $\gcd(m,n) = 1$. An immediate computation shows that $c_{l^k} \geq 0$ for all primes l and all k and using the previous observation we deduce that $c_n \geq 0$ for all n. As infinitely many c_n's are equal to 1 and all c_n are nonnegative, $\sum_n c_n x^n$ is not bounded as $x \to 1^-$ and the claim is proved.

Now, assuming that $L(1,\chi) = 0$, we will prove that f is bounded as $x \to 1^-$, which will finish the proof. If

$$b_n(x) = \frac{1}{n(1-x)} - \frac{x^n}{1-x^n},$$

then the hypothesis $L(1,\chi) = 0$ reduces the boundedness of $f(x)$ as $x \to 1$ to the boundedness of $\sum_n b_n(x)\chi(n)$ as $x \to 1$. The miracle is that the sequence

$(b_n(x))_n$ is nondecreasing, as one easily checks using the AM-GM inequality that

$$b_n(x) - b_{n+1}(x) = \frac{1}{1-x}\left(\frac{1}{n(n+1)} - \frac{x^n}{(1+x+\cdots+x^{n-1})(1+x+\cdots+x^n)}\right)$$

is nonnegative. Next, using Abel's summation formula, the monotonicity of the $b_n(x)$ and the fact that $|\sum_{i=1}^n \chi(i)|$ are bounded, it is easy to see that f is bounded. But this contradicts the result of the previous paragraph, ending the proof of Dirichlet's theorem. □

Once we know that $L(1,\chi) \neq 0$ for all nontrivial characters χ, it is a natural question to ask how far it is from being zero. If we look carefully at the proof of the previous theorem, we see that it gives $L(1,\chi) > 0$ for any nontrivial real Dirichlet character. The following deep theorem gives much more. The proof is much more involved:

Theorem 7.A.20. *(Siegel) For any $\varepsilon > 0$ there exists $c(\varepsilon) > 0$ such that $L(1,\chi) > \frac{c(\varepsilon)}{N^\varepsilon}$ for any real primitive Dirichlet character modulo N.*

7.A.6 A useful consequence

The purpose of this section is to give an analogue of Mertens' theorems for primes in arithmetic progressions. This uses the proof of Dirichlet's theorem and the standard technique of Abel summation. Recall that Λ is Von Mangoldt's function, defined by $\Lambda(n) = \log p$ if there is a prime p and $k \geq 1$ such that $n = p^k$, and $\Lambda(n) = 0$ otherwise. Its usefulness in number theory comes from the relation $\sum_{d|n} \Lambda(d) = \log n$, which will be heavily used in the proof of the following result.

Theorem 7.A.21. *Let a, N be relatively prime positive integers. Then*

$$\sum_{\substack{p\leq x \\ p\equiv a \pmod N}} \frac{\log p}{p} = \frac{\log x}{\varphi(N)} + O(1).$$

Proof. Since

$$\sum_{k\geq 2}\sum_{p}\frac{\log p}{p^k}<\infty,$$

we have

$$\sum_{\substack{p\leq x\\ p\equiv a\,(\mathrm{mod}\ N)}}\frac{\log p}{p}=\sum_{\substack{n\leq x\\ n\equiv a\,(\mathrm{mod}\ N)}}\frac{\Lambda(n)}{n}+O(1).$$

Using the orthogonality relations, we can write

$$\sum_{\substack{n\leq x\\ n\equiv a\,(\mathrm{mod}\ N)}}\frac{\Lambda(n)}{n}=\frac{1}{\varphi(N)}\sum_{\chi}\overline{\chi(a)}\sum_{n\leq x}\frac{\chi(n)}{n}\Lambda(n),$$

so everything comes down to the study of

$$F(\chi,x)=\sum_{n\leq x}\frac{\chi(n)}{n}\Lambda(n).$$

If χ is trivial, Mertens' theorem 3.A.4 yields

$$F(1,x)=\sum_{\substack{p^k\leq x\\ \gcd(p,N)=1}}\frac{\log p}{p^k}=\sum_{p\leq x}\frac{\log p}{p}+O(1)=\log x+O(1).$$

Suppose now that χ is nontrivial. We will prove that $F(\chi,x)$ is bounded as $x\to\infty$, which will finish the proof of the theorem. Exploiting the relation $\sum_{d|n}\Lambda(d)=\log n$, let us consider the expression

$$E(x)=\sum_{n\leq x}\frac{\chi(n)}{n}\log n.$$

Since $x\to\frac{\log x}{x}$ is decreasing for $x\geq 3$ and since the partial sums of the sequence $(\chi(n))_n$ are bounded, Abel summation shows that $E(x)=O(1)$. On the other hand

$$E(x)=\sum_{d_1d_2\leq x}\frac{\chi(d_1d_2)}{d_1d_2}\Lambda(d_1)=\sum_{d_1\leq x}\frac{\chi(d_1)\Lambda(d_1)}{d_1}\sum_{d_2\leq x/d_1}\frac{\chi(d_2)}{d_2}.$$

Another Abel summation (using the fact that the partial sums $\sum_{k=1}^{n}\chi(k)$ are bounded) shows that

$$\sum_{j\le x}\frac{\chi(j)}{j}=L(1,\chi)-\sum_{j>x}\frac{\chi(j)}{j}=L(1,\chi)+O\left(\frac{1}{x}\right).$$

We deduce that

$$E(x)=F(\chi,x)L(1,\chi)+O\left(\frac{1}{x}\sum_{d\le x}\Lambda(d)\right).$$

Since by theorem 7.A.19 we have $L(1,\chi)\neq 0$, in order to prove that

$$F(\chi,x)=O(1)$$

it is enough to prove that

$$\frac{1}{x}\sum_{d\le x}\Lambda(d)=O(1).$$

But

$$\sum_{d\le x}\Lambda(d)=\sum_{p^k\le x}\log p\le\sum_{p\le x}\log p\cdot\frac{\log x}{\log p}=\log x\cdot\pi(x)=O(x),$$

using theorem 3.A.2. The result follows. □

7.A.7 An "elementary proof" of Dirichlet's theorem

The proof of Dirichlet's theorem that we presented used rather heavily basic properties of holomorphic functions, but it turns out that the ideas used in the proof of theorem 7.A.21 may be combined with careful estimates to obtain an entirely "elementary" proof of Dirichlet's theorem, i.e. completely avoiding the use of complex analysis. We will sketch such a proof in this section, but we must emphasize that the complex analytic approach is much more powerful and conceptual.

We will actually give an elementary proof of theorem 7.A.21, which obviously implies Dirichlet's theorem. We will use the notations of arguments of the proof of theorem 7.A.21. The only "non-elementary" result used in the proof of this theorem is the fact that $L(1,\chi) \neq 0$ for all nontrivial characters χ. Monsky's proof (see the proof of theorem 7.A.19) deals with the case when χ is real in an elementary (but very tricky) way, so it remains to deal with the case when χ is not real. Let k be the number of nontrivial characters χ for which $L(1,\chi) = 0$ and assume that $k \geq 1$. Then Monsky's result implies that $k \geq 2$ (since if $L(1,\chi) = 0$ then $L(1,\overline{\chi}) = 0$). Either a direct computation or a simple application of Möbius' inversion formula yields the identity

$$\sum_{d|n} \mu(d) \cdot \log \frac{x}{d} = 1_{n=1} \cdot \log x + \Lambda(n),$$

from which a standard computation yields

$$\begin{aligned} F(\chi, x) &= \sum_{n \leq x} \frac{\chi(n)}{n} \Lambda(n) \\ &= -\log x + \sum_{d_1 d_2 \leq x} \mu(d_1) \cdot \frac{\chi(d_1 d_2)}{d_1 d_2} \cdot \log \frac{x}{d_1} \\ &= -\log x + \sum_{d_1 \leq x} \mu(d_1) \log \frac{x}{d_1} \cdot \frac{\chi(d_1)}{d_1} \cdot \sum_{d_2 \leq \frac{x}{d_1}} \frac{\chi(d_2)}{d_2}. \end{aligned}$$

The argument used in the proof of theorem 7.A.21 shows that $F(\chi, x)$ is bounded if $L(1,\chi) \neq 0$. Assume that $L(1,\chi) = 0$. Then Abel's summation formula shows that $\sum_{d>x} \frac{\chi(d)}{d} = O(1/x)$, so the previous formula yields

$$\begin{aligned} F(\chi, x) &= -\log x + O\left(\frac{1}{x} \sum_{d_1 \leq x} \log \frac{x}{d_1}\right) \\ &= -\log x + O\left(\frac{[x] \log x - \log[x]!}{x}\right) \\ &= -\log x + O(1). \end{aligned}$$

The conclusion is that

$$\sum_{\chi} F(\chi, x) = (1-k)\log x + O(1).$$

But this contradicts the fact that $k \geq 2$ and (again by the orthogonality relations)

$$\sum_{\chi} F(\chi, x) = \varphi(N) \cdot \sum_{\substack{n \leq x \\ n \equiv 1 \,(\mathrm{mod}\, N)}} \frac{\Lambda(n)}{n} \geq 0.$$

The reader has probably noticed that the crucial part of this argument was an "elementary" version of the argument used in the beginning of the proof of theorem 7.A.19.

Chapter 8

Formal Series Revisited

In this chapter, we discuss applications of generating functions in combinatorics or combinatorial number theory. This is a rather vast subject and we will only be able to scratch its surface through some nice examples. The idea is very simple (even though in practice things are less clear): given a sequence $(a_n)_n$ of complex numbers (but we may allow this sequence to take values in any ring, in theory), one is sometimes able to extract information about the sequence from its generating function $\sum_{n\geq 0} a_n X^n$ in an easier way than by dealing with the sequence itself. For instance, if the sequence satisfies a rather complicated recursive relation, it is very hard to study the sequence directly, but quite often it turns out that its generating function satisfies some differential equations or some regularity properties that allow us to study the sequence. Also, in counting problems it is very often much easier to find the generating function of the desired number of objects to count than to find directly that number. Of course, one sometimes needs quite a lot of imagination to extract the information from the generating function, but it should be a principle that if one knows the generating function, then one knows a lot of things about the sequence. It is important to note that when dealing with generating functions one neglects all convergence and analytic issues. However, once one knows the generating function, one usually uses analytic arguments to study the sequence.

Before passing to concrete problems, let us recall a few things about operations with formal series. Addition and multiplication are defined just as for polynomials. It is an easy exercise to check that a formal series has an inverse (for multiplication) if and only if the constant term is nonzero. A more subtle problem is the composition of formal series. Here, one has to be a bit careful, as simple things such as $\sum_n \frac{(X+1)^n}{n!}$ do not make sense formally. To do things properly, the best way is to introduce the X-adic valuation on formal series: if $f = a_0 + a_1X + \cdots$ is a formal series with complex coefficients (more generally, any field), let $v_X(f)$ be the least nonnegative integer n such that $a_n \neq 0$. Thus $f \cdot X^{-v_X(f)}$ is an invertible power series and one easily checks using this that $v_X(fg) = v_X(f) + v_X(g)$ and $v_X(f+g) \geq \min(v_X(f), v_X(g))$. Thus v_X behaves as the p-adic valuation.

Definition 8.1. Say a sequence of formal series f_n converges towards a formal series f if $v_X(f - f_n)$ tends to ∞. If $f = a_0 + a_1X + \cdots$, this means that for all i, the coefficient of X^i in f_n is a_i for all sufficiently large n. One says that $f = \sum_{n\geq 0} f_n$, respectively $f = \prod_{n\geq 0} f_n$ if the sequence $(f_0 + f_1 + \cdots + f_n)_n$ (respectively $(f_0 f_1 \cdots f_n)_n$) converges to f.

It is an easy exercise for the reader to check that $\sum_{n\geq 0} f_n$ converges to a formal series if and only if $v_X(f_n)$ tends to infinity (the analogous result holds for p-adic numbers). Also, if f_n are formal series such that $f_n(0) = 0$, then $\prod_{n\geq 0}(1 + f_n)$ converges if and only if $v_X(f_n)$ tends to ∞. We can now explain the composition of two formal series. Assume that f, g are formal series such that $g(0) = 0$ and $f = a_0 + a_1X + a_2X^2 + \cdots$. The composition $f \circ g$ is then by definition the formal series

$$f \circ g = a_0 + a_1 g + a_2 g^2 + \cdots .$$

The series converges, because $v_X(g^n) \geq n$. All other operations on formal series (such as differentiation, integration) do not involve any difficulty and have the properties that we imagine.

Before passing to problems, let us recall the following useful extension of the binomial formula: for any rational number α we have

$$(1+X)^\alpha = \sum_{n\geq 0} \binom{\alpha}{n} X^n,$$

where $\binom{\alpha}{n} = \frac{\alpha(\alpha-1)\cdots(\alpha-n+1)}{n!}$. A very useful special case is $\alpha = -m$, where m is a positive integer. In this case the binomial formula becomes

$$\frac{1}{(1-X)^m} = \sum_{n\geq 0} \binom{m+n-1}{m-1} X^n.$$

Finally, we will sometimes use the classical notation $[X^n]f$ for the coefficient of X^n in f.

8.1 Counting problems

The first exercise is quite technical, but there is no serious difficulty in it and the result is quite beautiful. Recall that if x_n is a sequence of complex numbers, an equivalent for it is a sequence y_n in closed form such that $\frac{x_n}{y_n}$ converges to 1.

1. Let $a_1, a_2, \ldots, a_n$ be relatively prime positive integers. Find an equivalent as $k \to \infty$ for the number of positive integral solutions of the equation $a_1x_1 + a_2x_2 + \cdots + a_nx_n = k$.

Proof. The generating function of the sequence y_k, the number of positive integral solutions of the equation $a_1x_1 + a_2x_2 + ... + a_nx_n = k$, is

$$\begin{aligned} f(X) &= \sum_{k\geq 1} y_k X^k \\ &= \sum_{x_1,x_2,\ldots,x_n\geq 1} X^{a_1x_1+\cdots+a_nx_n} \\ &= \left(\sum_{x_1\geq 1} X^{a_1x_1}\right)\cdots\left(\sum_{x_n\geq 1} X^{a_nx_n}\right) \\ &= \prod_{i=1}^{n} \frac{X^{a_i}}{1-X^{a_i}}. \end{aligned}$$

This is a rational function and in order to analyze sequence y_k we will decompose it into simpler pieces. Namely, if $1 = z_1, z_2, \ldots, z_r$ are the distinct roots of the polynomial $\prod_{i=1}^{n}(1-X^{a_i})$, with multiplicities $m_1, m_2, \ldots, m_r$, then by general theory there exist constants C_{ij} such that

$$f(X) = C_{00} + \sum_{i=1}^{r}\sum_{j=1}^{m_i} \frac{C_{ij}}{(X-z_i)^j}.$$

Using the binomial formula, we obtain

$$\frac{1}{(X-z_i)^j} = (-z_i^{-1})^j \cdot \left(1-\frac{X}{z_i}\right)^{-j} = (-z_i^{-1})^j \cdot \sum_{k\geq 0} z_i^{-k}\binom{j+k-1}{k}X^k.$$

Inserting this in the formula for f and re-arranging terms yields

$$\begin{aligned} y_k &= \sum_{i=1}^{r}\sum_{j=1}^{m_i}(-1)^j z_i^{-(k+j)} C_{ij}\binom{j+k-1}{k} \\ &= \sum_{i=1}^{r}\sum_{j=1}^{m_i}(-1)^j z_i^{-(k+j)} \frac{C_{ij}}{(j-1)!}(k+1)\cdots(k+j-1). \end{aligned}$$

We claim that the dominant term in the previous sum corresponds to $i=1$ and $j=m_1=n$. First, note that $m_i<n$ for all $i\neq 1$. Indeed, each of $X^{a_j}-1$ has simple roots, so the only possibility for $m_i\geq n$ would be that z_i is a root of all $X^{a_k}-1$, thus also of $X-1$ and so $i=1$. Thus the contribution to y_k of all terms with $i\neq 1$ is dominated by ck^{n-2} for some constant c. A similar argument shows that the contribution of the terms with $i=1$ and $j<n$ is also dominated by a constant times k^{n-2}. Thus the main contribution is that of the term with $i=1$ and $j=n$, which is $\frac{(-1)^n}{(n-1)!}(k+1)\cdots(k+n-1)C_{1n}$. To find C_{1n}, note that

$$C_{1n} = \lim_{x\to 1} f(x)(x-1)^n = (-1)^n \lim_{x\to 1}\prod_{i=1}^{n}\frac{1-x}{1-x^{a_i}} = \frac{(-1)^n}{a_1a_2\cdots a_n}.$$

Combining the previous observations yields the equivalent $\frac{k^{n-1}}{(n-1)!a_1\cdots a_n}$ for y_k. □

The following problem is a bit trickier, but it shows how formal series in several variables appear quite naturally.

2. For positive integers m and n, let $f(m,n)$ denote the number of n-tuples $(x_1, x_2, \ldots, x_n)$ of integers such that $|x_1| + |x_2| + \cdots + |x_n| \le m$. Show that $f(m,n) = f(n,m)$.

Putnam 2005

Proof. We will use a two-variable generating function in this problem: with the convention $f(m,0) = f(0,n) = 1$, we have the chain of equalities

$$
\begin{aligned}
F(X,Y) &= \sum_{m,n\ge 0} f(m,n)X^mY^n = \sum_{m,n\ge 0} X^mY^n \sum_{|k_1|+\cdots+|k_n|\le m} 1 \\
&= \sum_{n\ge 0} Y^n \sum_{k_1,\ldots,k_n\in\mathbb{Z}} \frac{X^{|k_1|+\cdots+|k_n|}}{1-X} = \sum_{n\ge 0} \frac{Y^n}{1-X}\cdot\left(\sum_{k\in\mathbb{Z}} X^{|k|}\right)^n \\
&= \frac{1}{1-X}\sum_{n\ge 0} Y^n\left(1+\frac{2X}{1-X}\right)^n = \frac{1}{1-X}\sum_{n\ge 0}\left(Y\cdot\frac{1+X}{1-X}\right)^n \\
&= \frac{1}{1-X-Y-XY}.
\end{aligned}
$$

Since $F(X,Y) = F(Y,X)$, it is clear that $f(m,n) = f(n,m)$ for all m,n. □

3. For $n \ge 3$ and $A \subset \{1, 2, \ldots, n\}$, say A is even if the sum of the elements of A is an even number. Otherwise, we say that A is odd. By convention, the empty set is even.

   a) Find the number of even, respectively odd subsets of $\{1, 2, \ldots, n\}$.

   b) Find the sum of the elements of the even, respectively odd subsets of $\{1, 2, \ldots, n\}$.

Romanian TST 1994

Proof. The generating function for the sums of the elements of the subsets of $\{1, 2, \ldots, n\}$ is $\prod_{i=1}^{n}(1 + X^i)$, because of the obvious identity

$$\prod_{i=1}^{n}(1 + X^i) = \sum_{A} X^{m(A)},$$

where $m(A) = \sum_{a\in A} a$. Let $E(n)$, respectively $O(n)$ be the sets of even, respectively odd subsets of $\{1, 2, \ldots, n\}$. Taking $X = -1$ in the previous identity yields $0 = |E(n)| - |O(n)|$. Since we clearly have $|E(n)| + |O(n)| = 2^n$, we deduce that the number of odd, respectively even subsets is 2^{n-1}. To answer the second question, we differentiate the identity, to get

$$\sum_{i=1}^{n} iX^{i-1}\prod_{j\neq i}(1 + X^j) = \sum_{A} m(A)X^{m(A)-1}.$$

Taking $X = -1$ yields

$$\sum_{A\in O(n)} m(A) = \sum_{A\in E(n)} m(A).$$

On the other hand, taking $X = 1$ gives

$$\sum_{A\in O(n)} m(A) + \sum_{A\in E(n)} m(A) = n(n + 1)2^{n-2}.$$

We deduce that the answer to the second question is $n(n + 1)2^{n-3}$ for both quantities involved. □

Remark 8.2. Here is an alternative proof for the second part.

If $a_n = \sum_{A\in O(n)} m(A)$ and $b_n = \sum_{A\in E(n)} m(A)$, then

$$a_n = a_{n-1} + b_{n-1} + n \cdot 2^{n-2}, \quad b_n = b_{n-1} + a_{n-1} + n \cdot 2^{n-2}$$

for n odd, as the odd subsets of $\{1, 2, \ldots, n\}$ are either odd subsets of $\{1, 2, \ldots, n - 1\}$ or the union of $\{n\}$ and of an even subset of $\{1, 2, \ldots, n - 1\}$. A similar analysis shows that

$$a_n = 2a_{n-1} + n \cdot 2^{n-2}, \quad b_n = 2b_{n-1} + n \cdot 2^{n-2}$$

for even n. By induction, we deduce that $a_n = b_n$ for all n and then solving the previous recursive relation (which becomes $a_n = 2a_{n-1} + n \cdot 2^{n-2}$) yields $a_n = 2^{n-3}n(n+1)$.

We continue with a very nice problem, though quite simple.

4. How many polynomials P with coefficients 0, 1, 2, or 3 satisfy $P(2) = n$, where n is a given positive integer?

Romanian TST 1994

Proof. Let a_n be the number of such polynomials. Then a_n is also the number of solutions of the equation $x_0+2x_1+\cdots+x_k2^k = n$ in integers $x_i \in \{0,1,2,3\}$, for varying k. Thus, the generating function is

$$\sum_{n\geq 0} a_n X^n = \prod_{k\geq 0}\left(1 + X^{2^k} + X^{2\cdot 2^k} + X^{3\cdot 2^k}\right).$$

This is also equal to

$$\prod_{n\geq 0} \frac{1-X^{2^{n+2}}}{1-X^{2^n}} = \frac{1-X^4}{1-X}\cdot\frac{1-X^8}{1-X^2}\cdot\frac{1-X^{16}}{1-X^4}\cdots,$$

which simplifies drastically to

$$\sum_{n\geq 0} a_n X^n = \frac{1}{(1-X)(1-X^2)}.$$

Now, we can decompose the last rational fraction in simple elements, following the standard procedure. A simple computation yields

$$\frac{1}{(1-X)(1-X^2)} = \frac{1}{4}\left(\frac{1}{1-X} + \frac{1}{1+X}\right) + \frac{1}{2(1-X)^2}.$$

Expanding this once more, we deduce that the answer is $\left\lfloor \frac{n}{2} \right\rfloor + 1$.

Note that we could have avoided the simple elements decomposition, because $\frac{1}{(1-X)(1-X^2)}$ is simply the generating function for the number of ways to express n as $a+2b$, where a, b are nonnegative integers. There are $\left\lfloor \frac{n}{2} \right\rfloor + 1$ possible choices for b, each of which leaves one choice for a. So there are $\left\lfloor \frac{n}{2} \right\rfloor + 1$ ways to express n as $a + 2b$ and the conclusion follows. □

Remark 8.3. Here is another nice solution: if $P(X) = a_0 + a_1X + \cdots + a_kX^k$ satisfies $P(2) = n$, define

$$f(P) = \left\lfloor \frac{a_0}{2} \right\rfloor + \left\lfloor \frac{a_1}{2} \right\rfloor \cdot 2 + \cdots + \left\lfloor \frac{a_k}{2} \right\rfloor \cdot 2^k.$$

We obtain a map f from the set of polynomials P as in the statement of the problem, to values in $\{0, 1, \dots, \lfloor \frac{n}{2} \rfloor\}$. It is easy to check that this map is a bijection.

In the next two problems, one combines an easy combinatorial argument with a generating functions argument. Such mixtures appear very often in combinatorial problems and in this case generating functions do the dirty work of solving quite complicated recursive relations. Let us start with an absolute classic.

5. In how many different ways can we parenthesize a non-associative product $x_1x_2\cdots x_n$?

Catalan

Proof. Let a_n be the desired answer. Note that in a product of k, respectively $n-k$ factors, we can put parentheses in a_k, respectively a_{n-k} ways. Looking at the position at which the first parenthesis ends, we deduce the recursive relation $a_1 = 1$ and $a_n = \sum_{k=1}^{n-1} a_k a_{n-k}$.

Next, define $f(X) = \sum_{n\geq 1} a_n X^n$ and observe that the recursive relation implies the chain of equalities

$$f(X)^2 = X^2 + \sum_{n\geq 3} \left(\sum_{k=1}^{n-1} a_k a_{n-k} \right) X^n = f(X) - X,$$

which implies (taking into account that $f(0) = 0$) that

$$\begin{aligned} f(X) &= \frac{1 - \sqrt{1-4X}}{2} \\ &= \frac{1}{2} - \frac{1}{2}(1-4X)^{\frac{1}{2}} \\ &= \frac{1}{2} - \frac{1}{2} \cdot \sum_{n\geq 0} (-1)^n \binom{1/2}{n} 4^n X^n. \end{aligned}$$

Now,

$$\binom{1/2}{n}(-1)^n \cdot 4^n = (-1)^{n-1} \cdot \frac{1 \cdot 3 \cdot \dots \cdot (2n-3)}{n!}(-1)^n \cdot 2^n = -\frac{2}{n}\binom{2n-2}{n-1}$$

and so, using the previous relation yields $a_n = \frac{1}{n}\binom{2n-2}{n-1}$. □

6. Let $F(n)$ be the number of functions $f : \{1, 2, \dots, n\} \to \{1, 2, \dots, n\}$ with the property that if i is in the range of f, then so is j, for all $j \le i$. Prove that

$$F(n) = \sum_{k \ge 0} \frac{k^n}{2^{k+1}}.$$

L. Lovasz, Miklos Schweitzer Competition

Proof. The key point to obtain a recursive relation for the sequence $F(n)$ is to look at $|f^{-1}(1)|$. If $|f^{-1}(1)| = j$, then the j elements of $f^{-1}(1)$ can be chosen in $\binom{n}{j}$ ways and f can be defined in $F(n-j)$ ways on the remaining elements. Thus, there are $\binom{n}{j}F(n-j)$ maps f satisfying the property in the statement of the problem and for which $|f^{-1}(1)| = j$. Summing over all possible values of j, we cover all functions f satisfying the property in the statement, so

$$F(n) = \sum_{j=1}^{n} \binom{n}{j} F(n-j) = \sum_{j=0}^{n-1} \binom{n}{j} F(j).$$

Here we took by convention $F(0) = 1$. Considering the exponential generating function $f(X) = \sum_{n=0}^{\infty} \frac{F(n)}{n!} X^n$, the previous relation yields

$$f(X) = 1 + (e^X - 1)f(X) \Longrightarrow f(X) = \frac{1}{2 - e^X}.$$

Next, we can write

$$f(X) = \frac{1}{2} \frac{1}{1 - \frac{e^X}{2}} = \frac{1}{2} \cdot \sum_{n \ge 0} \frac{e^{nX}}{2^n}.$$

Expanding now

$$e^{nX} = 1 + \frac{nX}{1!} + \frac{(nX)^2}{2!} + \cdots$$

and identifying coefficients in the previous equality yields the desired formula for $F(n)$. Note[1] that we cheated a bit, since $\sum_{n\geq 0} \frac{e^{nX}}{2^n}$ does not converge X-adically, but this can be easily fixed: consider the function $f(z) = \sum_{n=0}^{\infty} \frac{F(n)}{n!} z^n$, defined in a neighborhood of 0 in $\mathbb{C}$. This series converges absolutely and the previous computations show that $f(z) = \frac{1}{2-e^z}$, if z is close enough to 0. But then we can expand

$$\frac{1}{1-\frac{e^z}{2}} = \sum_{n\geq 0} \frac{e^{nz}}{2^n},$$

where this time the series converges in $\mathbb{C}$. Expanding e^{nz}, collecting terms and identifying coefficients yields the result. □

We end this section with two more challenging problems. The first one is a very classical result in the theory of finite fields.

7. Let N_n be the number of irreducible monic polynomials of degree n with coefficients in $\mathbb{Z}/p\mathbb{Z}$. Then for all n we have $\sum_{d|n} dN_d = p^n$.

Proof. Write $\mathbb{F}_p = \mathbb{Z}/p\mathbb{Z}$ and consider the generating function

$$F = \sum_{f\in\mathbb{F}_p[X]} X^{\deg f},$$

the sum being taken over monic polynomials $f \in \mathbb{F}_p[X]$. As there are p^n monic polynomials of degree n with coefficients in $\mathbb{F}_p$, we have

$$F = \sum_{n\geq 0} p^n X^n = \frac{1}{1-pX}.$$

[1]We thank Richard Stong for pointing this out.

On the other hand, the unique factorization of monic polynomials into products of irreducible monic polynomials yields

$$F = \prod_h (1 + X^{\deg h} + X^{2\deg h} + \cdots) = \prod_h \frac{1}{1 - X^{\deg h}},$$

the product being taken over the irreducible monic polynomials h. Taking the formal logarithm yields

$$\log \frac{1}{1 - pX} = \sum_h \log \frac{1}{1 - X^{\deg h}} = \sum_{n \geq 0} N_n \log \frac{1}{1 - X^n}.$$

The desired formula is easily obtained from this equality and the classical expansion

$$\log \frac{1}{1 - X} = \sum_{k \geq 1} \frac{X^k}{k}. \qquad \square$$

Remark 8.4. Using Möbius' inversion formula and the result of the previous problem, we obtain

$$N_n = \frac{1}{n} \sum_{d|n} \mu(d) p^{\frac{n}{d}}.$$

It is fairly easy to see from here that $N_n > 0$, so there exists at least one irreducible polynomial of degree n over $\mathbb{F}_p$. This result is absolutely not trivial. There is also an arithmetic solution of the previous problem, the idea being to prove the stronger result

$$\prod_{\substack{f \in \mathbb{F}_p[X], \\ \deg f | n}} f = X^{p^n} - X,$$

where the product is taken only over those irreducible monic polynomials $f \in \mathbb{F}_p[X]$ whose degree divides n.

8. Let x and y be noncommutative variables. Express in terms of n the constant term of the expression $(x + y + x^{-1} + y^{-1})^n$.

M. Haiman, D. Richman, AMM 6458

Proof. Note that variables cancel in variable/inverse pairs so the constant term is zero for n odd. Let $a_{m,n}$ be the number of products $u_1u_2\cdots u_n$ where each u_i is one of $\{x,y,x^{-1},y^{-1}\}$ and after cancellation we are left with a word of length m. Note that $a_{0,n}$ is the desired constant term. By convention we set $a_{-1,n}=0$. Starting from a word of length $m>0$ there are three ways we can add one more term on the right and produce a word of length $m+1$ and one way we can add a cancelling term and produce a word of length $m-1$. Therefore we have

$$a_{m,n+1}=\begin{cases}3a_{m-1,n}+a_{m+1,n} & m\neq 1\\ 4a_{0,n}+a_{2,n} & m=1\end{cases}$$

Letting $A_n=\sum_{m=0}^{n}a_{m,n}X^m$ we can rewrite this as

$$A_{n+1}=(3X+X^{-1})A_n+a_{0,n}(X-X^{-1}).$$

This is a first order linear difference equation in A_n and it can be solved by standard techniques to give

$$A_n=(3X+X^{-1})^n+\sum_{m=0}^{n-1}a_{0,n-m-1}(X-X^{-1})(3X+X^{-1})^m.$$

By definition it is clear that A_n is a polynomial in X, therefore the coefficient of any negative power of X in this series must be zero. In particular, setting $n=2s+1$ and looking at the coefficient of X^{-1} gives

$$3^s\binom{2s+1}{s}=\sum_{m=0}^{s}a_{0,2(s-m)}\left[3^m\binom{2m}{m}-3^{m-1}\binom{2m}{m-1}\right].$$

Forming generating functions of this identity gives

$$\sum_{s=0}^{\infty}3^s\binom{2s+1}{s}T^s=\sum_{m=0}^{\infty}\left[3^m\binom{2m}{m}-3^{m-1}\binom{2m}{m-1}\right]T^m\cdot\sum_{r=0}^{\infty}a_{0,2r}T^r.$$

Let $a(T)=\sum_{r=0}^{\infty}a_{0,2r}T^r$. Recognizing the other sums via

$$\sum_{m=0}^{\infty}3^m\binom{2m}{m}T^m=\sum_{m=0}^{\infty}(-12)^m\binom{-1/2}{m}T^m=\frac{1}{\sqrt{1-12T}},$$

$$\binom{2s+1}{s} = \frac{1}{2}\binom{2s+2}{s+1} \text{ and } \binom{2m}{m-1} = \binom{2m+1}{m} - \binom{2m}{m},$$

we can write this as

$$\frac{1}{6T\sqrt{1-12T}} - \frac{1}{6T} = \left(\frac{1}{\sqrt{1-12T}} - \frac{1-6T}{18T\sqrt{1-12T}} + \frac{1}{18T}\right) a(T).$$

Solving for $a(T)$ and simplifying gives

$$a(T) = \frac{3}{1+2\sqrt{1-12T}} = \frac{2\sqrt{1-12T}-1}{1-16T}.$$

Extracting the coefficient of T^r using the binomial formula gives

$$\begin{aligned} a_{0,2r} &= 2\sum_{k=0}^{r} 2^{4(r-k)}(-12)^k \binom{1/2}{k} - 2^{4r} \\ &= 2^{4r} - \sum_{k=1}^{r} \frac{2^{4(r-k)+2}3^k}{k}\binom{2k-2}{k-1}. \end{aligned}$$

□

8.2 Proving identities using generating functions

Generating functions are a very powerful method for proving combinatorial identities. Namely, we compute the generating functions of both sides of the identity we are given and we prove that they are the same. Here are a few examples.

9. Prove that for all positive integers n,

$$\sum_{k=1}^{n} \binom{n+k-1}{2k-1} = F_{2n},$$

where F_n is the Fibonacci sequence (with $F_1 = F_2 = 1$).

Iranian Olympiad 2008

Proof. The generating function of the left-hand side is

$$\sum_{n\geq 1} X^n \cdot \sum_{k=1}^{n} \binom{n+k-1}{2k-1} = \sum_{k\geq 1}\sum_{n\geq k} X^n \binom{n+k-1}{2k-1}.$$

On the other hand, we have

$$\sum_{n\geq k} X^n \binom{n+k-1}{2k-1} = X^k \sum_{n\geq 0} \binom{n+2k-1}{2k-1} X^n = X^k(1-X)^{-2k},$$

by the binomial formula.

So the generating function of the left-hand side is

$$\sum_{k\geq 1} X^k(1-X)^{-2k} = \frac{X}{X^2-3X+1}.$$

It remains to compute the generating function of the sequence F_{2n}. This is very easy, since we know that $F_n = \frac{x^n-y^n}{x-y}$, where x, y are the roots of the equation $t^2-t-1=0$. Thus

$$\sum_{n\geq 1} X^n F_{2n} = \frac{1}{x-y}\left(\frac{x^2X}{1-x^2X} - \frac{y^2X}{1-y^2X}\right)$$

$$= \frac{X(x^2-y^2)}{(x-y)(1-x^2X)(1-y^2X)}.$$

Since $x+y=1, xy=-1$, we easily deduce that

$$(1-x^2X)(1-y^2X) = X^2-3X+1$$

and the result follows. □

10. Let n and k be positive integers. For any sequence of nonnegative integers $(a_1, a_2, \ldots, a_k)$ adding up to n, compute the product $a_1a_2\cdots a_k$. Prove that the sum of all these products is

$$\frac{n(n^2-1^2)(n^2-2^2)\cdots(n^2-(k-1)^2)}{(2k-1)!}.$$

Proof. Let $f(n,k)$ be the desired sum of products and consider the generating function

$$
\begin{aligned}
g(X) &= \sum_{n\geq 0} f(n,k)X^n = \sum_{n\geq 0}\sum_{a_1+\cdots+a_k=n} a_1a_2\cdots a_kX^n \\
&= \sum_{a_1,a_2,\ldots,a_k\geq 0} a_1X^{a_1}\cdot\cdots\cdot a_kX^{a_k} = \left(\sum_{i\geq 0} iX^i\right)^k \\
&= \left(\frac{X}{(1-X)^2}\right)^k = X^k(1-X)^{-2k}.
\end{aligned}
$$

Expanding $(1-X)^{-2k}$ thanks to the binomial formula and using the previous relation, we deduce that

$$f(n,k) = (-1)^{n-k}\binom{-2k}{n-k} = \binom{n+k-1}{2k-1}.$$

On the other hand, it is easy to check that

$$\frac{n(n^2-1^2)(n^2-2^2)\cdots(n^2-(k-1)^2)}{(2k-1)!} = \binom{n+k-1}{2k-1},$$

from where the result follows. □

Remark 8.5. Here are a few other such identities that can be easily proved using generating functions:

a) $\sum_{a_1+a_2+\cdots+a_k=n} a_1a_2\cdots a_k = F_{2n}$, where the sum is taken over all ordered partitions of n and F_n is the nth Fibonacci number.

b) $\sum_{a_1+a_2+\cdots+a_k=n}(2^{a_1-1}-1)(2^{a_2-1}-1)\cdot\cdots\cdot(2^{a_k-1}-1) = F_{2n-2}$.

We continue with a very nice combinatorial identity, for which we give two proofs: a natural one using generating functions and a more subtle, purely combinatorial one.

11. Let m, n be positive integers with $m \geq n$, and let S be the set of all sequences of positive integers $(a_1, a_2, \ldots, a_n)$ such that $a_1+a_2+\cdots+a_n = m$. Show that

$$\sum_{(a_1,\ldots,a_n)\in S} 1^{a_1}2^{a_2}\cdots n^{a_n} = \sum_{i=1}^{n}(-1)^{n-i}\binom{n}{i}i^m.$$

Palmer Mebane, USA TST 2010

Proof. The solution using generating functions is rather straightforward. Indeed, note that

$$\begin{aligned}\sum_{m\geq n} X^m \cdot \left(\sum_{a_1+\cdots+a_n=m} 1^{a_1}2^{a_2}\cdots n^{a_n}\right) &= \sum_{a_1,\ldots,a_n\geq 1} X^{a_1}(2X)^{a_2}\cdots(nX)^{a_n}\\ &= \left(\sum_{a_1\geq 1} X^{a_1}\right)\cdot\cdots\cdot\left(\sum_{a_n\geq 1}(nX)^{a_n}\right)\\ &= \frac{X}{1-X}\cdot\frac{2X}{1-2X}\cdot\cdots\cdot\frac{nX}{1-nX}.\end{aligned}$$

On the other hand, we have

$$\begin{aligned}\sum_{m\geq n} X^m\left(\sum_{i=1}^{n}(-1)^{n-i}\binom{n}{i}i^m\right) &= \sum_{i=1}^{n}(-1)^{n-i}\binom{n}{i}\sum_{m\geq n}(iX)^m\\ &= \sum_{i=1}^{n}(-1)^{n-i}\binom{n}{i}\frac{(iX)^n}{1-iX}.\end{aligned}$$

Thus, it is enough to prove that

$$\frac{n!}{(1-X)(1-2X)\cdots(1-nX)} = \sum_{i=1}^{n}(-1)^{n-i}\binom{n}{i}\cdot\frac{i^n}{1-iX}.$$

But the theory of simple fractions decomposition shows that there exist $A_1, A_2, \ldots, A_n$ such that

$$\frac{n!}{(1-X)(1-2X)\cdots(1-nX)} = \sum_{i=1}^{n}\frac{A_i}{1-iX}.$$

Multiplying this by $1 - iX$ and taking $X \to \frac{1}{i}$ yields the expression of A_i and a trivial computation shows that $A_n = (-1)^{n-i}\binom{n}{i}i^n$. □

Proof. Define $k = m - n$ and let T be the set of sequences of nonnegative integers $(b_1, b_2, \ldots, b_n)$ such that $b_1 + b_2 + \cdots + b_n = k$. The substitution $b_i = a_i - 1$ transforms the desired identity into

$$n! \sum_T 1^{b_1} 2^{b_2} \cdots n^{b_n} = \sum_{j=1}^{n} (-1)^{n-j} j^{k+n} \binom{n}{j}.$$

To prove this, we will count in two ways the colorings of $k + n$ objects with n colors such that each color is used at least once.

The first method of counting is to line up the objects in some order and to consider the first appearance of each color. Suppose object c_i is the first appearance of the ith color (in order of occurrence). Let $b_i = c_{i+1} - c_i - 1$ (with $c_{n+1} = n + k + 1$) be the number of objects between c_i and c_{i+1}. Any choice of $(b_1, b_2, \ldots, b_n)$ with $b_1 + b_2 + \cdots + b_n = k$ also gives a valid choice of the c_i's via $c_i = i + \sum_{j=1}^{i} b_j$. Now for the b_i objects in between the first appearance of color i and color $i + 1$, there are i ways to color each one since only i colors have appeared so far. Therefore, for each choice of $(b_1, b_2, \ldots b_n)$ the number of ways is $1^{b_1} 2^{b_2} \cdots n^{b_n}$. Summing over all possible cases and taking into account that the order the colors appear in can be rearranged in $n!$ ways, we get $n! \sum_T 1^{b_1} 2^{b_2} \cdots n^{b_n}$ ways to color our objects.

On the other hand, there are n^{k+n} ways to color $k + n$ objects with n colors and there are $\binom{n}{n-1}(n-1)^{k+n}$ ways to do the coloring with $n - 1$ of the n colors, $\binom{n}{n-2}(n-2)^{k+n}$ ways with at most $n - 2$ colors and so on. A standard inclusion exclusion argument shows that the number of admissible colorings is

$$\binom{n}{n} n^{k+n} - \binom{n}{n-1}(n-1)^{k+n} + \cdots + (-1)^{n-2}\binom{n}{2} 2^{k+n} + (-1)^{n-1}\binom{n}{1}$$

and we are done. □

8.3 Recurrence relations

Generating functions are an extremely powerful tool to deal with the complicated recurrence relations that appear quite often in enumerative combinatorics. The reader is warned that the problems in this section are rather technical and difficult.

12. Let $A_1 = \emptyset$, $B_1 = \{0\}$ and

$$A_{n+1} = \{1 + x |\ x \in B_n\}, \quad B_{n+1} = (A_n \setminus B_n) \cup (B_n \setminus A_n).$$

Find all positive integers n such that $B_n = \{0\}$.

Chinese Olympiad

Proof. Using the relations given in the statement of the problem, it is immediate to check that

$$1_{B_{n+1}}(k) \equiv 1_{B_n}(k) + 1_{B_{n-1}}(k-1) \pmod 2,$$

where $1_A(x) = 1$ if $x \in A$ and 0 otherwise. This relation suggests considering the sequence of polynomials defined by

$$P_0 = 0, \quad P_1 = 1, \quad P_n(X) = P_{n-1}(X) + XP_{n-2}(X).$$

Looking at the two recursive relations, it is clear that, modulo 2, the coefficient of X^k in $P_n(X)$ is simply $1_{B_n}(k)$. To solve this recursion, it is convenient to introduce a new variable T and set $X = T + T^2$. To avoid confusion let $Q_n(T) = P_n(T + T^2)$. Then this relation becomes

$$Q_n(T) = Q_{n-1}(T) + T(1+T)Q_{n-2}(T).$$

This relation is easy to solve mod 2 giving

$$Q_n(T) = (T+1)^n + T^n \in \mathbb{F}_2[T].$$

It is obvious either from the original recursion or because $Q_n(T) \equiv Q_n(T+1) \pmod 2$, that $Q_n(T)$ can actually be written mod 2 as a polynomial $P_n(T+T^2)$.

Explicitly finding P_n might be tedious, however we do not need to do so since we only care about when $B_n = \{0\}$ or equivalently $P_n(X) = 1$ or $Q_n(T) = 1$. It is easy to see from the formula above that this occurs if and only if n is a power of 2. More explicitly, from Legendre's formula it follows that the smallest $k > 0$ for which $\binom{n}{k}$ is odd is $k = 2^{v_2(n)}$. Therefore $Q_n(T)$ has degree $n - 2^{v_2(n)}$ as a polynomial in T and $P_n(X)$ has degree $\frac{1}{2}\left(n - 2^{v_2(n)}\right)$ as a polynomial in X. Therefore $Q_n(T) = 1$ if and only if n is a power of 2. $\square$

Proof. As above, we consider the sequence of polynomials defined by $P_0 = 0$, $P_1 = 1$ and

$$P_n(X) = P_{n-1}(X) + XP_{n-2}(X)$$

and we observe that, modulo 2, the coefficient of X^k in $P_n(X)$ is simply $1_{B_n}(k)$. We consider the generating function $P(X,Y) = \sum_{n=0}^{\infty} P_n(X)Y^n$ and observe that

$$\begin{aligned}(1 - Y - XY^2)P(X,Y) &= P_0(X) + (P_1(X) - P_0(X))Y \\ &\quad + \sum_{n=2}^{\infty}(P_n(X) - P_{n-1}(X) - XP_{n-2}(X))Y^n = Y,\end{aligned}$$

so we get

$$\begin{aligned}P(X,Y) &= \frac{Y}{1 - Y - XY^2} \\ &= \sum_{m=0}^{\infty} Y^{m+1}(1 + XY)^m \\ &= \sum_{m=0}^{\infty}\sum_{j=0}^{m}\binom{m}{j}X^jY^{m+j+1} \\ &= \sum_{n=0}^{\infty}\sum_{j}\binom{n-j-1}{j}X^jY^n.\end{aligned}$$

Thus the problem is reduced to finding all n such that $\binom{n-j-1}{j}$ is even for all $n > j > 0$. We will prove that this happens if and only if n is a power of 2.

Consider the representation of n in the form $o2^r$ where o is an odd number. If n is not a power of 2, then $o \geq 3$ and we have $\binom{n-2^r-1}{2^r}$ is odd. If n is a power of 2, then we need to prove that $\binom{n-j-1}{j}$ is even no matter which j we choose. Using the Legendre formula we are done if we can find an m for which $\left\lfloor \frac{n-j-1}{2^m} \right\rfloor > \left\lfloor \frac{n-2j-1}{2^m} \right\rfloor + \left\lfloor \frac{j}{2^m} \right\rfloor$. We can choose one more than the binary logarithm of the highest power of 2 that divides j. Thus the conditions are that n is a power of 2. □

The following problem is very technical and its solution is too. But these kind of problems appear quite often in real life and we think it is rather important to be able to deal with them.

13. Suppose that $a_0 = a_1 = 1$ and $(n+3)a_{n+1} = (2n+3)a_n + 3na_{n-1}$ for $n \geq 1$. Prove that all terms of this sequence are integers.

KöMaL

Proof. The first step is to consider the generating function

$$f(X) = \sum_{n\geq 0} a_n X^n.$$

Multiplying by X^n the recursive relation and adding these relations yields

$$\sum_{n\geq 1}(n+3)a_{n+1}X^n = \sum_{n\geq 1}(2n+3)a_nX^n + \sum_{n\geq 1} 3na_{n-1}X^n.$$

Now, since $f'(X) = \sum_{n\geq 1} na_nX^{n-1}$, it is very easy to express each of the previous sums in terms of f and f'. More precisely, a straightforward computation yields

$$\sum_{n\geq 1}(n+3)a_{n+1}X^n = f'(X) + \frac{2}{X}(f(X)-1) - 3,$$

$$\sum_{n\geq 1}(2n+3)a_nX^n = 2Xf'(X) + 3f(X) - 3 \text{ and}$$

$$\sum_{n\geq 1} 3na_{n-1}X^n = 3Xf(X) + 3X^2f'(X).$$

Replacing these expressions in the previous equality and collecting similar terms, we end up with the differential equation

$$f'(X)(X - 2X^2 - 3X^3) + f(X)(2 - 3X - 3X^2) - 2 = 0.$$

Now comes the technical part, which we will skip: solving this differential equation. There are standard methods to solve this, but unfortunately when one uses them in this case, one obtains rather horrendous expressions. Thus, we leave to the reader to convince himself that the resolution of this equation yields

$$f(X) = \frac{1 - X - \sqrt{1 - 2X - 3X^2}}{2X^2}.$$

And now? It is absolutely not clear that the coefficients in the Taylor expansion of f are integers! The tricky point is to write

$$(1 - X - 2X^2 f(X))^2 = 1 - 2X - 3X^2$$
$$\iff 1 + (X - 1)f(X) + X^2 f(X)^2 = 0.$$

And now we are saved, because if we identify the coefficients in the previous relation, we obtain the following recursive relation

$$a_{n+2} = a_{n+1} + \sum_{k=0}^{n} a_k a_{n-k}.$$

And since the first terms are integers, the previous relation shows inductively that all terms of the sequence are integers. Another elegant way, found by Richard Stong, is to note that

$$\begin{aligned} f(X) &= \frac{1-X}{2X^2}\left(1 - \sqrt{1 - \frac{4X^2}{(1-X)^2}}\right) \\ &= \sum_{k=0}^{\infty} \frac{1}{k+1}\binom{2k}{k}\frac{X^{2k}}{(1-X)^{2k+1}} \\ &= \sum_{n=0}^{\infty}\left[\sum_{k=0}^{\lfloor n/2 \rfloor} \frac{1}{k+1}\binom{2k}{k}\binom{n}{2k}\right] X^n. \end{aligned}$$

and that $\frac{1}{k+1}\binom{2k}{k} = \binom{2k}{k} - \binom{2k}{k+1}$ is an integer (this is the famous Catalan number). We leave as a challenge to the reader to prove directly from the recursive relation that

$$a_{n+2} = a_{n+1} + \sum_{k=0}^{n} a_k a_{n-k}$$

or that

$$a_n = \sum_{k=0}^{\lfloor n/2 \rfloor} \frac{1}{k+1}\binom{2k}{k}\binom{n}{2k}.$$

He will probably appreciate the power of generating functions (as far as we know, the proofs by induction are hideous, to say the least...). □

Remark 8.6. The numbers appearing in the previous problem are called Motzkin numbers and they have nice combinatorial interpretations. Here is one of them: a Motzkin path is a lattice path from $(0,0)$ to $(n,0)$ with steps $(1,0)$, $(1,1)$ and $(1,-1)$, never going below the x-axis. Then a_n is the number of Motzkin paths of length n. This is a beautiful and not obvious exercise that we leave to the reader. Another good exercise is to prove that a_n is also the number of paths on $\mathbb{N}$ with n steps, each step being $-1, 0$ or 1, starting and ending at 0. To make everything precise, let us recall that if A is a subset of $\mathbb{Z}^n$, then a lattice path of length l from $x \in \mathbb{Z}^n$ to $y \in \mathbb{Z}^n$, with steps in A is a sequence $v_0 = x, v_1, \ldots, v_l = y$ of elements of $\mathbb{Z}^n$ such that $v_i - v_{i-1} \in A$ for all i.

We give three different solutions for the following beautiful problem. The first two proofs are a bit technical, but quite natural. The third one is very short and elegant, but not very natural.

14. Consider $(b_n)_{n\geq 1}$ a sequence of integers such that $b_1 = 0$ and define $a_1 = 0$ and $a_n = nb_n + a_1 b_{n-1} + \cdots + a_{n-1}b_1$ for all $n \geq 2$. Prove that $p|a_p$ for any prime number p.

KöMaL

Proof. Considering the generating functions

$$A(X) = \sum_{n\geq 2} a_n X^n, \quad B(X) = \sum_{n\geq 2} b_n X^n,$$

the recursive relation can also be written in the form

$$A(X) = XB'(X) + A(X)B(X),$$

from where we obtain the fundamental identity

$$A(X) = \frac{XB'(X)}{1 - B(X)} = -X\frac{d}{dX}\log(1 - B(X)).$$

Now, we have the following very useful result:

Lemma 8.7. *Any $f \in \mathbb{Z}[[X]]$ with constant term 1 can be written in the form*

$$f = \prod_{i=1}^{\infty} \left(1 - a_i X^i\right)^{-1}$$

with $a_i \in \mathbb{Z}$.

Proof. Write

$$f(X) = 1 + f_1 X + f_2 X^2 + \cdots$$

for some integers f_i. Looking at the coefficient of X we obtain that $a_1 = f_1$. In general, we find a_n in terms of $a_1, \ldots, a_{n-1}$ by imposing the condition that

$$(1 - a_1 X) \cdots (1 - a_n X^n) f(X) \equiv 1 \pmod{X^n}.$$

This expresses a_n as a polynomial with integer coefficients in $a_1, \ldots, a_{n-1}$ and $f_1, \ldots, f_n$, from where the conclusion follows immediately by induction. □

Using this result, let us write

$$1 - B(X) = \prod_{i=1}^{\infty} \left(1 - c_i X^i\right)^{-1}$$

with $c_i \in \mathbb{Z}$. Then

$$\begin{aligned} A(X) &= \sum_{i \geq 1} X \frac{d}{dX} \log(1 - c_i X^i) \\ &= -\sum_{i \geq 1} \frac{i c_i X^i}{1 - c_i X^i} \\ &= -\sum_{i \geq 1} (i c_i X^i + i c_i^2 X^{2i} + \cdots). \end{aligned}$$

Since a_p is the coefficient of X^p in $A(X)$, the only contribution comes from the terms with $i = 1$ and $i = p$ and since the contribution for $i = p$ is clearly a multiple of p, it is enough to check that $c_1 = 0$. But this is clear, since $1 - B(X) \equiv 1 \pmod{X^2}$. □

Proof. Let us modify the definitions of the generating functions a bit, by imposing different initial terms. Namely, define $a_0 = a_1 = 0, b_0 = -1, b_1 = 0$ and

$$A(X) = \sum_{i=0}^{\infty} a_i X^i, \quad B(X) = \sum_{i=0}^{\infty} -b_i X^i.$$

Then the recursive relation satisfied by the sequences implies that

$$A(X)B(X) = -XB'(X),$$

so that

$$\frac{-A(X)}{X} = \frac{B'(X)}{B(X)}.$$

Integrating both sides we get

$$\sum_{i=2}^{\infty} \frac{-a_i X^i}{i} = \log\left(1 - (b_2 X^2 + b_3 X^3 + \cdots)\right).$$

Since

$$\log(1 - X) = -X - \frac{X^2}{2} - \frac{X^3}{3} - \cdots,$$

we deduce that

$$\sum_{i=2}^{\infty} \frac{a_i X^i}{i} = \sum_{i=1}^{\infty} \frac{(b_2 X^2 + b_3 X^3 + \cdots)^i}{i}.$$

Looking at the coefficients of X^p in both sides shows that $\frac{a_p}{p}$ is the coefficient of X^p in $\sum_{i=1}^{\lfloor \frac{p}{2} \rfloor} \frac{(b_2 X^2 + b_3 X^3 + \cdots)^i}{i}$. Since the denominator of this coefficient is clearly relatively prime to p, it follows that $p|a_p$. □

Proof. Let $x_1, x_2, \ldots, x_p$ be the roots of the polynomial

$$X^p + b_1 X^{p-1} + b_2 X^{p-2} + b_3 X^{p-3} + \cdots + b_p.$$

Comparing the identities $a_1 = b_1$ and $a_n = nb_n + a_1 b_{n-1} + \cdots + a_{n-1} b_1$ for all $n \geq 2$ with Newton's identities, we easily deduce that

$$-a_i = x_1^i + x_2^i + \cdots + x_p^i$$

for every $i \in \{1, 2, \ldots, p\}$. On the other hand, $b_1 = 0$ implies that

$$x_1 + x_2 + \cdots + x_p = 0.$$

The desired result follows then from corollary 9.15. □

8.4 Additive properties

When studying additive properties of sets $A \subset \mathbb{Z}$, it is very useful to consider the generating function $f = \sum_{a \in A} X^a$. For instance, the square of f encodes the number of solutions to the equation $a + b = n$, with $a, b \in A$. This rather innocent-looking observation yields quite a lot of nontrivial results and the purpose of this section is to present a few of them.

Quite often, it is more convenient to reduce the generating function modulo suitable primes, as this simplifies a lot the computations: it is a very fortunate feature of $\mathbb{F}_p$ that $(1+X)^p = 1 + X^p$ in $\mathbb{F}_p[X]$.

15. Let A be a finite set of nonnegative integers. Define a sequence of sets by: $A_0 = A$ and for all $n \geq 0$, an integer a is in A_{n+1} if and only if exactly one of the integers $a-1$ and a is in A_n. Prove that for infinitely many positive integers k, A_k is the union of A with the set of numbers of the form $k+a$ with $a \in A$.

Putnam Competition 2000

Proof. The key point is to note that the definition of A_{n+1} in terms of A_n can be expressed algebraically by

$$\sum_{a \in A_{n+1}} X^a = (1+X) \sum_{a \in A_n} X^a$$

for all n, the equality being in $\mathbb{F}_2[X]$. We deduce that

$$\sum_{a \in A_n} X^a = (1+X)^n \sum_{a \in A} X^a$$

for all n. On the other hand, the condition that A_n is the union of A and $A+n$ can be also written (when $n > \max A$) in the form

$$\sum_{a \in A_n} X^a = (1+X^n) \sum_{a \in A} X^a.$$

Thus, it suffices to find infinitely many n such that $(1+X)^n = 1+X^n$ in $\mathbb{F}_2[X]$. Simply take for n a power of 2 greater than $\max A$ and we will have $A_n = A \cup (A+n)$. $\square$

The following problem is an absolute classic.

16. Prove that if we partition the set of nonnegative integers into a finite number of infinite arithmetic progressions, then there will be two of them having the same common difference.

Proof. Suppose that $\mathbb{N}$ is partitioned into the arithmetic progressions $a_i\mathbb{N}+b_i$, with $a_i, b_i \geq 0$ and $a_1 \leq a_2 \leq \cdots \leq a_n$. We will actually prove that $a_{n-1} = a_n$.

Suppose that this is not the case and note that the partition hypothesis yields an identity of formal series

$$\frac{1}{1-X} = \sum_{i=1}^{n}\sum_{k\geq 0} X^{a_ik+b_i} = \sum_{i=1}^{n} \frac{X^{b_i}}{1-X^{a_i}}.$$

However, the right-hand side has a pole at a primitive a_n^{th} root of unity, while the left-hand side definitely does not have such a pole. □

The following problem is much trickier.

17. Let p be a prime and let $n \geq p$ and $a_1, a_2, \ldots, a_n$ be integers. Define $f_0 = 1$ and f_k the number of subsets $B \subset \{1, 2, \ldots, n\}$ having k elements and such that p divides $\sum_{i\in B} a_i$. Show that $f_0 - f_1 + f_2 - \cdots + (-1)^n f_n$ is a multiple of p.

Saint Petersburg 2003

Proof. Adding to all of the a_i's a suitable large multiple of p (which does not affect the hypothesis or the conclusion), we may assume that the a_i's are positive. The point is to consider the remainder of

$$\begin{aligned} f(X) &= (1-X^{a_1})(1-X^{a_2})\cdots(1-X^{a_n}) \\ &= \sum_{k=0}^{n}(-1)^k \sum_{\substack{B\subset A \\ |B|=k}} X^{m(B)} \in \mathbb{F}_p[X] \end{aligned}$$

modulo $X^p - 1$, where $m(B)$ is the sum of the elements of B. On the one hand, the hypothesis $n \geq p$ and the fact that $1 - X$ divides $1 - X^{a_i}$ imply that $1 - X^p = (1 - X)^p$ (remember that we are working in $\mathbb{F}_p[X]$) divides $f(X)$. On the other hand, we have $X^N \equiv X^M \pmod{X^p - 1}$ if $N \equiv M \pmod{p}$, thus

$$f(X) \equiv \sum_{j=0}^{p-1}\left(\sum_{k=0}^{n}(-1)^k \sum_{|B|=k, m(B)\equiv j \pmod{p}} 1\right) X^j \pmod{X^p-1}.$$

Since this polynomial is 0, it follows that its constant term is 0 in $\mathbb{F}_p$. But this is precisely saying that p divides $f_0 - f_1 + f_2 - \cdots + (-1)^n f_n$. □

The following is also a rather tricky problem. We found it in the wonderful little book [62]. It was also proposed in a Chinese Team Selection Test in 2002.

18. For which positive integers n can we find real numbers $a_1, a_2, \ldots, a_n$ such that

$$\{|a_i - a_j| \mid 1 \le i < j \le n\} = \left\{1, 2, \ldots, \binom{n}{2}\right\}?$$

Proof. It is clear that if a_i are such numbers, then all their differences are integers, thus we may actually assume that they are integers. Let

$$f(X) = X^{a_1} + X^{a_2} + \cdots + X^{a_n},$$

so that the hypothesis can be written

$$f(X)f\left(\frac{1}{X}\right) = n + \sum_{i=1}^{\binom{n}{2}} X^i + \sum_{i=1}^{\binom{n}{2}} X^{-i} = n - 1 + \frac{X^{\binom{n}{2}+1} - X^{-\binom{n}{2}}}{X - 1}.$$

The point is to look at values of f at points on the unit circle, since for $|z| = 1$ we have $f\left(\frac{1}{z}\right) = f(\bar{z}) = \overline{f(z)}$, thus $f(z)f\left(\frac{1}{z}\right) = |f(z)|^2 \ge 0$. We deduce that for any $z = e^{ix}$ we have

$$n - 1 + \frac{z^{\binom{n}{2}+1} - z^{-\binom{n}{2}}}{z - 1} \ge 0.$$

However, an easy computation shows that this is equivalent to

$$n - 1 + \frac{\sin(n^2 - n + 1)\frac{x}{2}}{\sin\frac{x}{2}} \ge 0.$$

We will take x such that $(n^2 - n + 1)\frac{x}{2} = \frac{3\pi}{2}$ to deduce that

$$n - 1 > \frac{1}{\sin\frac{3\pi}{2(n^2-n+1)}} > \frac{2(n^2 - n + 1)}{3\pi}.$$

However, it is immediate to check that the last inequality cannot hold unless $n \le 4$. And indeed, for $n \le 4$ such sequences exist. For $n = 1, 2$ things are clear, while for $n = 3$ and $n = 4$ one can take the sequences $1, 2, 4$, respectively, $1, 2, 5, 7$. $\square$

Proof. It is easy to build examples with $n = 1, 2, 3$, and 4. Suppose $n \geq 5$ and without loss of generality that $a_1 < a_2 < \cdots < a_n$. Then the largest difference must be $a_n - a_1 = \binom{n}{2}$ and since

$$a_n - a_1 = \sum_{k=1}^{n-1} (a_{k+1} - a_k) \geq \sum_{k=1}^{n-1} k = \binom{n}{2}$$

we see that the differences $a_{k+1} - a_k$ for $1 \leq k \leq n-1$ must be a permutation of $1, 2, \ldots, n-1$. In particular, all differences of a_i's with indices two apart are at least n. Suppose $a_{k+1} - a_k = 1$, then if they are present, both $a_{k+2} - a_k = 1 + (a_{k+2} - a_{k+1})$ and $a_{k+1} - a_{k-1} = 1 + (a_k - a_{k-1})$ would be at most n, a contradiction since they are at least n and distinct. Thus the difference of 1 can only occur at one of the ends $a_2 - a_1$ or $a_n - a_{n-1}$ and it must be adjacent to the difference of $n-1$. Since this accounts for the difference of n all remaining differences of a_i's two apart must be at least $n+1$. Therefore repeating the above argument shows that the difference of 2 must also occur at one of the ends and can only be adjacent to the difference of $n-1$. But this is impossible since the ends are at least $n - 1 \geq 4$ apart. □

The following gem is one of our favorite problems.

19. Find all positive integers n with the following property: for any real numbers $a_1, a_2, \ldots, a_n$, knowing the numbers $a_i + a_j$, $i < j$ (but not knowing which number corresponds to which sum) determines the values $a_1, a_2, \ldots, a_n$ uniquely.

Selfridge and Straus

Proof. We will eventually prove that the answer is: all positive integers but the powers of 2.

First, let us give a nice counter-example when $n = 2^k$. Let $A_1 = \{0, 3\}$ and $B_1 = \{1, 2\}$ and define inductively

$$A_{j+1} = A_j \cup (2^{j+1} + B_j), \quad B_{j+1} = B_j \cup (2^{j+1} + A_j).$$

It is an amusing exercise left to the reader to check that A_j, B_j have 2^j elements and that

$$\{a_1 + a_2 | a_1 \neq a_2 \in A_j\} = \{b_1 + b_2 | b_1 \neq b_2 \in B_j\}$$

for all j. Actually, A_j and B_j are precisely the sets of numbers having at most $j+1$ digits when written in base 2 and whose sum of digits is even (respectively odd). All this can be easily proved by induction and shows that A_k, B_k are a counter-example for $n = 2^k$.

Let us prove now that if n is not a power of 2 and if the collections $(a_i + a_j)_{i<j}$ and $(b_i + b_j)_{i<j}$ are identical, then the collections $(a_i)_i$ and $(b_j)_j$ are identical. This is the hard part. We may always assume that a_i, b_i are positive real numbers, by adding to each of them the same large integer.

Assume first that the a_i's and b_i's are integers and consider the polynomials

$$f(X) = \sum_{i=1}^{n} X^{a_i}, \quad g(X) = \sum_{i=1}^{n} X^{b_i}.$$

Then the hypothesis implies the equality

$$f(X)^2 - f(X^2) = g(X)^2 - g(X^2).$$

If $f = g$, then we are done, so assume that $h = f - g$ is not the zero polynomial. Write $h(X) = (X-1)^k p(x)$ for some polynomial p with $p(1) \neq 0$ and some $k \geq 1$ (note that $f(1) = g(1) = n$). Then

$$(X-1)^k p(X)(f(X) + g(X)) = (X^2 - 1)^k p(X^2),$$

which, after division by $(X-1)^k$, can be written as

$$p(X)(f(X) + g(X)) = (X+1)^k p(X^2).$$

Taking $X = 1$ in this identity and dividing by $p(1) \neq 0$, we obtain that $n = 2^{k-1}$. Since this is a contradiction, it follows that $f = g$ and the result is proved for integers. It follows trivially that the result also holds for rational numbers (by multiplying them by a suitable positive integer we can make all of them integers).

In order to prove the general case, we will use Dirichlet's approximation theorem. Suppose that n is not a power of 2 and that the two collections of numbers $(a_i + a_j)_{i<j}$ and $(b_i + b_j)_{i<j}$ are the same. Pick any $\varepsilon < \frac{1}{4}$.

By Dirichlet's approximation theorem, we can find integers p, q_i, r_i such that $p \neq 0$, $|pa_i - q_i| < \varepsilon$ and $|pb_i - r_i| < \varepsilon$ for all $1 \leq i \leq n$. By the triangle inequality, any equality $a_i + a_j = b_i + b_j$ forces $q_i + q_j = r_i + r_j$ (indeed, we have

$$q_i + q_j - (r_i + r_j) = q_i - pa_i + q_j - pa_j - (r_i - pb_i) - (r_j - pb_j)$$

and each term has magnitude smaller than 1/4). Hence the two collections $(q_i + q_j)_{i<j}$ and $(r_i + r_j)_{i<j}$ are the same and so (by what has already been done) we have an equality of polynomials $\prod_{i=1}^n (X - q_i) = \prod_{i=1}^n (X - r_i)$. This can also be written as

$$\prod_{i=1}^n \left(X - a_i + \frac{pa_i - q_i}{p} \right) = \prod_{i=1}^n \left(X - b_i + \frac{pb_i - r_i}{p} \right).$$

By taking ε smaller and smaller, we obtain sequences u_n, v_n which converge to 0 and such that $\prod_{i=1}^n (X - a_i + u_n) = \prod_{i=1}^n (X - b_i + v_n)$ for all n. It is clear that this forces $\prod_{i=1}^n (X - a_i) = \prod_{i=1}^n (X - b_i)$, which is the desired result. □

The following alternative proof is a very elegant argument due to Selfridge and Straus ([70]):

Proof. Assume that n is not a power of 2. Let us prove that the knowledge of the multiset $(a_i + a_j)_{i<j}$ uniquely determines the symmetric elementary sums of the a_i's (this in turns uniquely determines the multiset $(a_i)_i$, yielding the desired result). Using Newton's formulae, it is enough to prove that we can recover the sums $S_k = a_1^k + a_2^k + \dots + a_n^k$ from the multiset $(a_i + a_j)_{i<j}$, and this for every k (it would be enough to take $k \le n$). Note that

$$\begin{aligned}
\sum_{i \neq j} (a_i + a_j)^k &= \sum_{1 \le i,j \le n} (a_i + a_j)^k - 2^k S_k \\
&= \sum_{i,j=1}^n \sum_{l=0}^k \binom{k}{l} a_i^{k-l} a_j^l - 2^k S_k \\
&= \sum_{l=0}^k \binom{k}{l} S_{k-l} S_l - 2^k S_k \\
&= (2n - 2^k) S_k + \sum_{l=1}^{k-1} \binom{k}{l} S_{k-l} S_l.
\end{aligned}$$

As $2n-2^k \neq 0$ for all k, the previous relations show (by induction on k) that S_k can be uniquely determined as a linear combination of the sums $\sum_{i>j}(a_i+a_j)^k$ and $S_0, S_1, \ldots, S_{k-1}$. This is precisely what we needed. □

The last two problems of this chapter are very hard. The first one requires a few facts about polynomials over finite fields, for which we refer the reader to the addendum 9.A.

20. Prove that there exists a subset S of $\{1, 2, \ldots, n\}$ such that $0, 1, 2, ..., n-1$ all have an odd number of representations as $x - y$ with $x, y \in S$, if and only if $2n - 1$ has a multiple of the form $2 \cdot 4^k - 1$.

Miklos Schweitzer Competition

Proof. Define $f(X) = \sum_{s\in S} X^{s-1}$. The fact that S satisfies the property in the statement is equivalent to the following equality in $\mathbb{F}_2[X]$

$$X^{n-1} f(X) f\left(\frac{1}{X}\right) = 1 + X + \cdots + X^{2n-2}.$$

Indeed, the coefficient of X^a in $f(X)f\left(\frac{1}{X}\right)$ is the number of representations of a as a difference of two elements of S. Conversely, if we can find a polynomial $f \in \mathbb{F}_2[X]$ satisfying the previous equality, we can define a set S with the desired property simply by saying that $s \in S$ if and only if X^{s-1} has coefficient 1 in f.

Thus, we must find n for which one can find $f \in \mathbb{F}_2[X]$ satisfying the previous relation. A first crucial remark is that $1 + X + \cdots + X^{2n-2}$ has no multiple root in the algebraic closure of $\mathbb{F}_2$. Indeed, this is already the case with $X^{2n-1} - 1$, since this polynomial is relatively prime to its derivative. Next, the roots of f are precisely the inverses of the roots of $X^{n-1} f\left(\frac{1}{X}\right)$. We deduce that no irreducible factor of $1 + X + \cdots + X^{2n-2}$ can vanish at z and $1/z$ for any root z of $1 + X + \cdots + X^{2n-2}$. But conversely, if this property holds, we can find such f. Indeed, we can then pair the irreducible factors of $1 + X + \cdots + X^{2n-2}$ such that in each pair the roots of the first polynomial are the inverses of the roots of the second polynomial. One can then take for f the product of the first components of these pairs.

Now, consider the permutation $z \to z^2$ of the roots of $X^{2n-1} - 1$.

If $C_1, \dots, C_l$ are the cycles of this permutation, the irreducible factors of $X^{2n-1} - 1$ are the polynomials $\prod_{z \in C_i}(X - z)$. We deduce that the existence of S is equivalent to: for any root $z \neq 1$ of $X^{2n-1} - 1$, there is no k such that $z^{2^k} = 1/z$. This is equivalent to $(X^{2n-1} - 1, X^{2^k+1} - 1) = X - 1$ for all $k \geq 0$ and so therefore $\gcd(2n-1, 2^k+1) = 1$ for all $k \geq 1$. Hence the least s such that $2n - 1|2^s - 1$ is odd and writing $s = 2k + 1$ we are done. □

Rather strong analytic skills are required to prove the following result, which we found in chapter III of the excellent book [62]. It is there considered an "appetizer."

21. Let A be an infinite set of positive integers. Let x_n be the number of pairs $(a, b) \in A \times A$ such that $a < b$ and $a + b = n$. Prove that the sequence $(x_n)_n$ is not eventually constant.

Donald J. Newman

Proof. Consider the generating function $f(X) = \sum_{a \in A} X^a$. Note that $f(z)$ converges for all $|z| < 1$ and $z \to f(z)$ is a continuous function on the open unit disk. Moreover, the hypothesis that x_n is eventually constant is equivalent to the existence of a constant c and of a polynomial P such that

$$f(X)^2 - f(X^2) = \frac{c}{1 - X} + P(X).$$

We will prove that this cannot happen. The idea is to look at the behavior of f on the real axis, close to 1 and then at its average behavior on circles of radius tending to 1. To avoid introducing too many functions, we will write $f >> g$ (when f, g are positive functions on $(0, 1)$) if there exists a constant $c > 0$ such that $f(r) > cg(r)$ for all r in a neighborhood of 1. In particular, we have $|P| << 1$ and since $f(x^2) > 0$ for $0 < x < 1$, we deduce that $f(x)^2 \geq \frac{c}{1-x} + P(x)$ and so $f(x) >> \frac{1}{\sqrt{1-x}}$. So f grows quite fast on the real axis, near 1.

Now we integrate the relation satisfied by f on circles of radius close to 1. Using the triangle inequality, we deduce that for all $r \in (0, 1)$

$$\frac{1}{2\pi} \int_0^{2\pi} |f(re^{ix})|^2 dx \leq \frac{1}{2\pi} \int_0^{2\pi} |f(r^2 e^{ix})| dx + \sup_{|z| \leq 1} |P(z)| + cF(r),$$

where $F(r) = \frac{1}{2\pi}\int_0^{2\pi} \frac{dx}{|1-re^{ix}|}$. Using Parseval's identity we obtain

$$\frac{1}{2\pi}\int_0^{2\pi} |f(re^{ix})|^2 dx = \sum_{a\in A} r^{2a} = f(r^2).$$

On the other hand, Cauchy-Schwarz inequality and Parseval's identity yield

$$\frac{1}{2\pi}\int_0^{2\pi} |f(r^2e^{ix})| dx \leq \sqrt{\frac{1}{2\pi}\int_0^{2\pi} |f(r^2e^{ix})|^2 dx} = \sqrt{f(r^4)} \leq \sqrt{f(r^2)}.$$

Putting these remarks together, we obtain the estimate:

$$f(r^2) \leq \sqrt{f(r^2)} + C_1 + C_2F(r)$$

for some constants $C_1, C_2 > 0$ independent of r. This immediately yields $f(r^2) << 1 + F(r)$. If we combine this with the result established in the first paragraph, we deduce that $\frac{1}{\sqrt{1-r^2}} << 1 + F(r)$. It remains to prove that this is not the case. Note that for $0 \leq x \leq \pi$ we have

$$|1 - re^{ix}|^2 = (1-r)^2 + 4r\sin^2(x/2) \geq (1-r)^2 + 4rx^2/\pi^2,$$

hence

$$\begin{aligned} F(r) &= \frac{1}{\pi}\int_0^{\pi} \frac{dx}{|1-re^{ix}|} \\ &\leq \frac{1}{\pi}\int_0^{\pi} \frac{dx}{\sqrt{(1-r)^2 + 4rx^2/\pi^2}} \\ &= \frac{1}{2\sqrt{r}} \operatorname{arcsinh}\left(\frac{2\sqrt{r}}{1-r}\right). \end{aligned}$$

It follows that $\lim_{r\to 1^-} \frac{F(r)}{-\ln(1-r)} \leq \frac{1}{2}$ and hence

$$\lim_{r\to 1^-} \sqrt{1-r}F(r) = 0. \qquad \square$$

Remark 8.8. With basically the same techniques, but with more work, one can prove the following beautiful theorem of Erdős and Fuchs:

Theorem 8.9. *Let A be a set of positive integers and let $r(n)$ be the number of pairs (a, b) with $a, b \in A$ such that $a + b = n$. Suppose that for some $\varepsilon > 0$ we have*

$$\frac{r(0) + r(1) + \cdots + r(n)}{n+1} = C + O\left(\frac{1}{n^{\frac{3}{4}+\varepsilon}}\right)$$

for some constant C. Then $C = 0$.

For the origin of this result and a complete proof, see [62].

8.5 Miscellaneous problems

The solution of the following problem is quite short, but the problem is actually quite tricky.

22. Is it possible to partition the set of all 12-digit numbers (leading zeroes are allowed) into groups of four numbers such that the numbers in each group have the same digits in 11 places and four consecutive digits in the remaining place?

St. Petersburg Olympiad

Proof. The answer is negative. Assume that we found such a partition and let A be the set of all 12-digit numbers. Consider $f(X) = \sum_{a \in A} X^{s(a)}$, where $s(a)$ is the sum of digits of a, and let $G_1, \dots, G_k$ be the groups appearing in the partition. By the condition imposed on the structure of each group we deduce that $\sum_{a \in G_i} X^{s(a)}$ is a multiple of $1 + X + X^2 + X^3$ for any i. Thus

$$1 + X + X^2 + X^3 | f(X) = \sum_{i=1}^{k} \sum_{a \in G_i} X^{s(a)}.$$

But we can actually compute f in a closed form, since

$$f(X) = (1 + X + \cdots + X^9)^{12} = \left(\frac{X^{10} - 1}{X - 1}\right)^{12}.$$

Since this is clearly not a multiple of $1 + X + X^2 + X^3$ (for instance, because it does not vanish at i), the problem is solved. □

The following problems are quite challenging. They establish congruences using generating functions. The first one is taken from [64]. See also [77], [78], [79] for other references and delicate congruences.

23. Let p be a prime and let $d \in \{0, 1, \ldots, p\}$. Prove that

$$\sum_{k=0}^{p-1} \binom{2k}{k+d} \equiv r \pmod{p},$$

where $r \equiv p - d \pmod{3}$, $r \in \{-1, 0, 1\}$.

H. Pan, Z.W. Sun

Proof. The key point is to observe that

$$\binom{2k}{k+d} = [X^{-2d}] \left(X + \frac{1}{X}\right)^{2k},$$

so

$$S = \sum_{k=0}^{p-1} \binom{2k}{k+d} = [X^{-2d}] \sum_{k=0}^{p-1} \left(X + \frac{1}{X}\right)^{2k} = [X^{-2d}] \frac{\left(X + \frac{1}{X}\right)^{2p} - 1}{X^2 + 1 + X^{-2}}.$$

Now, since we are only interested in $S \pmod{p}$, it is enough to understand the previous rational function taken mod p. Since in $\mathbb{F}_p((X))$ we have

$$\left(X + \frac{1}{X}\right)^{2p} = \left(X^p + \frac{1}{X^p}\right)^2,$$

we deduce that $S \pmod{p}$ is the coefficient of X^{-2d} in $\frac{X^{2p}+X^{-2p}+1}{X^2+X^{-2}+1} \in \mathbb{F}_p((X))$. So $S \pmod{p}$ is also the coefficient of X^{-d} in

$$\begin{aligned} F &= \frac{X^p + X^{-p} + 1}{X + X^{-1} + 1} \\ &= \frac{X(1-X)}{1-X^3}(X^p + X^{-p} + 1) \\ &= (X^p + X^{-p} + 1)\left(\sum_{n\geq 0} X^{3n+1} - \sum_{n\geq 0} X^{3n+2}\right). \end{aligned}$$

By inspection of the last product, the result follows. □

The following problem is similar in nature to problem 23, but more complicated.

24. Let $p > 3$ be a prime. Prove that

$$\sum_{k=1}^{p^2-1} \binom{2k}{k} \equiv 0 \pmod{p^2}.$$

David Callan, AMM 11292

Proof. Let $S = \sum_{k=0}^{p^2-1} \binom{2k}{k}$, so we need to prove that $S \equiv 1 \pmod{p^2}$. The first key point is to observe that

$$S = [X^0] \sum_{k=0}^{p^2-1} X^{-k}(1+X)^{2k} = [X^0] \frac{(1+X)^{2p^2} - X^{p^2}}{X^{p^2-1}(X^2+X+1)},$$

thus

$$\begin{aligned} S &= [X^{p^2-1}] \frac{(1-X)\left(\sum_{k=0}^{2p^2} \binom{2p^2}{k} X^k - X^{p^2}\right)}{1-X^3} \\ &= [X^{p^2-1}] \frac{1-X}{1-X^3} \sum_{k=0}^{p^2-1} \binom{2p^2}{k} X^k. \end{aligned}$$

The second key ingredient is the following nice

Lemma 8.10. *For all $1 \le k < p^2$ we have*

$$\binom{2p^2}{k} \equiv 2\binom{p^2}{k} \pmod{p^2}.$$

Proof. The left-hand side is the coefficient of X^k in $(1+X)^{2p^2}$. But since $(1+X)^{p^2} \equiv 1 + X^{p^2} \pmod{p}$ (this follows by raising to the p-th power the equality $(1+X)^p \equiv 1 + X^p \pmod{p}$), we can write

$$(1+X)^{p^2} = 1 + X^{p^2} + pA(X)$$

for some $A \in \mathbb{Z}[X]$. The lemma follows then by taking the square of the previous relation and comparing the coefficients of X^k. □

Coming back to the proof, we deduce that $S-1$ is the same (mod p^2) as twice the coefficient of X^{p^2-1} in

$$\frac{1-X}{1-X^3}\sum_{k=1}^{p^2-1}\binom{p^2}{k}X^k=(1-X)\sum_{p^2>k\geq 1, j\geq 0}\binom{p^2}{k}X^{k+3j},$$

which is precisely

$$S_1=\sum_{1\leq k<p^2, 3|k}\binom{p^2}{k}-\sum_{0\leq k<p^2-1, 3|k+1}\binom{p^2}{k}.$$

We need to prove that S_1 is a multiple of p^2. Well, the good news is that $S_1=0$, since if $z=e^{\frac{2i\pi}{3}}$, then

$$S_1=\sum_{k=0}^{p^2}\binom{p^2}{k}\frac{z^{k-1}-z^{1-k}}{z^{-1}-z}-1=\frac{z^{-1}(1+z)^{p^2}-z(1+z^{-1})^{p^2}}{z^{-1}-z}-1.$$

On the other hand, since $p^2\equiv 1 \pmod 6$, we have

$$z^{-1}(1+z)^{p^2}=1+z^{-1},\quad z(1+z^{-1})^{p^2}=1+z,$$

which combined with the previous relation shows that $S_1=0$. The result follows. $\square$

The following result is very difficult. It was conjectured by Rodriguez-Villegas and proved in a rather difficult way by E. Mortenson. The following beautiful proof is due to R. Tauraso, taken from [83].

25. Prove that for any $p>2$ we have

$$\sum_{k=0}^{p-1}\frac{1}{16^k}\binom{2k}{k}^2\equiv(-1)^{\frac{p-1}{2}} \pmod{p^2}.$$

Proof. Observe first of all that

$$\begin{aligned}\frac{1}{16^k}\binom{2k}{k} &= \frac{\prod_{j=1}^k (2j-1)^2}{4^k \cdot (2k)!}\\ &\equiv (-1)^k \cdot \frac{\prod_{j=1}^k (p^2-(2j-1)^2)}{4^k \cdot (2k)!}\\ &= (-1)^k \binom{n+k}{2k} \pmod{p^2},\end{aligned}$$

where $n = \frac{p-1}{2}$.

We will prove that for all nonnegative integers n we have

$$\sum_{k=0}^{n} (-1)^k \binom{2k}{k}\binom{n+k}{2k} = (-1)^n,$$

by computing the generating function of the left-hand side. We have

$$\begin{aligned}&\sum_{n\ge 0}\left(\sum_{k=0}^{n}(-1)^k\binom{2k}{k}\binom{n+k}{2k}\right)X^n = \sum_{k\ge 0}(-1)^k\binom{2k}{k}\sum_{n\ge k}X^n\binom{n+k}{2k}\\ =&\sum_{k\ge 0}(-1)^k\binom{2k}{k}X^k\sum_{n\ge 0}X^n\binom{n+2k}{2k} = \sum_{k\ge 0}(-1)^k\binom{2k}{k}X^k\cdot\frac{1}{(1-X)^{2k+1}}\\ =&\frac{1}{1-X}\sum_{k\ge 0}\binom{2k}{k}\left(-\frac{X}{(1-X)^2}\right)^k = \frac{1}{1-X}\cdot\left(1-4\cdot\frac{-X}{(1-X)^2}\right)^{-\frac{1}{2}}\\ =&\frac{1}{1+X}.\end{aligned}$$

Combining the previous two paragraphs yields the desired result. □

8.6 Notes

The following people provided solutions for the problems in this chapter: Robin Chapman (problem 24), Prasad Chebolu (problem 14), Darij Grinberg (problem 14), Xiangyi Huang (problems 18, 22), Mitchell Lee (problems 4, 10, 12), Palmer Mebane (problem 11), Greg Martin (problem 2), Fedja Nazarov (problem 21), Fedor Petrov (problem 23), Richard Stong (problems 8, 12, 18), Qiaochu Yuan (problems 1, 6, 16), Victor Wang (problem 12).

Addendum 8.A Lagrange's Inversion Theorem

In many counting problems, we start by finding a recursive relation for the number of objects we are trying to count, then we consider the generating function of that sequence (or the exponential generating function, according to the context) and establish a functional equation for it, based on the recursive relation. For instance, when counting the number of ways to put parantheses in a product of n terms, the associated generating function turns out to satisfy the easy functional equation $f(z) = 1 + zf(z)^2$. Of course, in this case one can solve this quadratic equation and find a formula for $f(z)$, which in turn allows us to find the desired number of ways (which is the classical nth Catalan number). But what if the equation was $f(z) = 1+zf(z)^5$? Then surely such a method would not work, simply because it is impossible to solve this equation using radicals. Of course, one might argue that such an equation is unlikely to come up in enumerative combinatorics, but this is also wrong!

The purpose of this addendum is to present a basic and very powerful tool for dealing with such problems (and many more), the Lagrange inversion formula. After discussing the very beautiful proof of this result, we will turn to applications, which will hopefully show its power.

8.A.1 Statement and proof of Lagrange's inversion formula

Before stating Lagrange's inversion formula, let us recall a very basic result on compositional inverses.

Proposition 8.A.1. *Let K be a field and let $F(T) = a_1T+a_2T^2+\cdots \in K[[T]]$ be a formal series with $a_1 \neq 0$. Then there exists a unique $f \in K[[T]]$ such that $F(f(T)) = T$ and it also satisfies $f(F(T)) = T$.*

Proof. Let us look for solutions of the equation $F(f(T)) = T$ having the form $f(T) = b_1T + b_2T^2 + \cdots$ (it is clear that if f is a solution, then f has zero constant term). By comparing the coefficient of T on both sides, we obtain $b_1 = \frac{1}{a_1}$. Doing the same for the coefficient of T^2 yields $a_1b_2 + a_2b_1^2 = 0$ and, in general, looking at the coefficient of T^n we obtain an equation of the form $a_1 b_n + G_n(a_1, a_2, \dots, a_n, b_1, \dots, b_{n-1}) = 0$ for some polynomial G_n. This already shows that if the solution exists, then it is unique. But it also shows that a solution exists, since the previous equations can be solved recursively.

In order to prove that $f(F(T)) = T$, observe that f has the same property as F (namely its constant term vanishes and its linear term is nonzero). So, by the previous paragraph there is a unique g such that $f(g(T)) = T$. But then $F(T) = F(f(g(T))) = g(T)$ and so $F = g$ and $f(F(T)) = T$. □

We are now ready to state and prove the main theoretical result of this addendum. The proof given here (due to Hofbauer) is an adaptation of an analytic proof that would work over complex numbers. However, using the purely algebraic notion of residue of a Laurent series, one can obtain a proof over any field of characteristic zero. Recall that if $f \in K[[T]]$, then $[T^n]f(T)$ denotes the coefficient of T^n in f.

Theorem 8.A.2. *(Lagrange's inversion formula) Let K be a field of characteristic 0 and let e be a series in $K[[T]]$ whose constant term is nonzero. If $f \in K[[T]]$ satisfies $f(T) = Te(f(T))$, then for any $g \in K[[T]]$ and any $n > 0$ we have*

$$[T^n]g(f(T)) = \frac{1}{n}[T^{n-1}](g'(T)e(T)^n).$$

Proof. Let $F(T) = \frac{T}{e(T)}$, so $F(T) = \alpha \cdot T + \cdots$ for some $\alpha \in K^*$, as the constant term of e is nonzero. The hypothesis $f(T) = Te(f(T))$ becomes $F(f(T)) = T$. Thus f is the (compositional) inverse of F and so $f(F(T)) = T$, too. If $h(T) = \sum_{n \in \mathbb{Z}} a_n T^n \in K[[T]][1/T]$ is a Laurent series with coefficients in K (so $a_n = 0$ for all n small enough), let $\operatorname{res}(h dT) = a_{-1}$. Obviously, for all h we have $\operatorname{res}(h' dT) = 0$. The key point is the following "change of variable formula"

Lemma 8.A.3. *If $G \in K[[T]][1/T]$, we have*

$$\operatorname{res}(G(T)dT) = \operatorname{res}(G(F(T))F'(T)dT).$$

Proof. By linearity, we may assume that $G(T) = T^k$, with $k \in \mathbb{Z}$. If $k \neq -1$, both $G(T)$ and $G(F(T))F'(T)$ are of the form g' with $g \in K[[T]][1/T]$ and so both terms of the equality we want to establish are zero. So, assume that $k = -1$. Then[2]

$$\frac{F'(T)}{F(T)} = \frac{1}{T} - \frac{e'(T)}{e(T)} = \frac{1}{T} - \left(\log\frac{e(T)}{e(0)}\right)',$$

so $\operatorname{res}\left(\frac{F'(T)}{F(T)}dT\right) = \operatorname{res}\left(\frac{dT}{T}\right)$ and we are done again. □

Coming back to the proof of the theorem and using the lemma, we can write

$$\begin{aligned}[T^n]g(f(T)) &= \operatorname{res}\left(\frac{g(f(T))}{T^{n+1}}dT\right)\\ &= \operatorname{res}\left(\frac{g(f(F(T)))}{F(T)^{n+1}}F'(T)dT\right)\\ &= -\frac{1}{n}\operatorname{res}(g(T)(F(T)^{-n})'dT).\end{aligned}$$

As $u'v+uv' = (uv)'$, we have $\operatorname{res}(u'v) = -\operatorname{res}(uv')$ for all u, v, which combined with the previous equality yields

$$\begin{aligned}[T^n]g(f(T)) &= \frac{1}{n}\operatorname{res}\left(\frac{g'(T)}{F(T)^n}dT\right)\\ &= \frac{1}{n}\operatorname{res}\left(\frac{g'(T)}{T^n}e(T)^n dT\right)\\ &= \frac{1}{n}[T^{n-1}](g'(T)e(T)^n dT).\end{aligned}$$

The result follows. □

Here is an easy application of the inversion formula, which is not so easy to prove by other means:

---

[2]Note that

$$\log\frac{e(T)}{e(0)} = \sum_{n\geq 1}\frac{(-1)^{n-1}}{n}\left(\frac{e(T)}{e(0)} - 1\right)^n$$

exists in $K[[T]]$, as $\frac{e(T)}{e(0)} - 1 \in TK[[T]]$.

Example 8.A.4. Suppose that two sequences $(a_n)_n, (b_n)_n$ of complex numbers satisfy

$$b_n = \sum_k \binom{k}{n-k} a_k$$

for all $n \geq 0$. Prove that

$$a_n = \frac{1}{n} \sum_k \binom{2n-k-1}{n-k} (-1)^{n-k} k b_k$$

for all $n \geq 1$.

Proof. Let $A(X) = \sum_{n \geq 0} a_n X^n$ and $B(X) = \sum_{n \geq 0} b_n X^n$ be the generating functions of the two sequences. An easy computation shows that

$$\begin{aligned} B(X) &= \sum_n X^n \cdot \left(\sum_k \binom{k}{n-k} a_k \right) = \sum_k a_k X^k \sum_n \binom{k}{n-k} X^{n-k} \\ &= \sum_k a_k X^k \sum_{n=0}^{k} \binom{k}{n} X^n = \sum_k a_k X^k (1+X)^k = A(X+X^2). \end{aligned}$$

Let $Y = X + X^2$, so that $X = f(Y)$ with $f(Y) = Y \cdot \frac{1}{1+f(Y)}$. Using Lagrange's inversion formula, we obtain

$$\begin{aligned} a_n &= [Y^n](A(Y)) = [Y^n](B(X)) = [Y^n](B(f(Y)) \\ &= \frac{1}{n}[Y^{n-1}](B'(Y)(1+Y)^{-n}) = \sum_k \frac{kb_k}{n}[Y^{n-1}](Y^{k-1}(1+Y)^{-n}) \\ &= \sum_k \frac{kb_k}{n}[Y^{n-k}]((1+Y)^{-n}) = \sum_k \frac{kb_k}{n} \binom{-n}{n-k} \end{aligned}$$

and the result follows from the easily checked equality

$$\binom{-n}{n-k} = (-1)^{n-k} \binom{2n-k-1}{n-k}.$$

□

8.A.2 Two variations and some applications

In applications, one also encounters the following versions of the inversion formula. For instance, we will use the following result to prove a very nice combinatorial identity due to Abel.

Theorem 8.A.5. *Let K be a field of characteristic 0 and let e be a series in $K[[T]]$ whose constant term is nonzero. Then for all $f \in K[[T]]$ we have*

$$f(T) = f(0) + \sum_{n\geq 1}\left(\frac{T}{e(T)}\right)^n \cdot \frac{1}{n}[T^{n-1}](f'(T)e^n(T)).$$

Proof. Let $X = X(T)$ be a formal series such that $X = Te(X)$ (it exists by proposition 8.A.1 applied to $F(T) = \frac{T}{e(T)}$). Then Lagrange's inversion formula yields

$$f(X(T)) = f(0) + \sum_{n\geq 1}[T^n](f(X(T))T^n = f(0) + \sum_{n\geq 1}\frac{1}{n}[T^{n-1}](f'(T)e(T)^n)\cdot T^n.$$

Now, substitute $T = Y/e(Y)$ to obtain the desired result. □

Here is the promised application, which is really not easy to prove by direct computational means.

Theorem 8.A.6. *(Abel's identity) For all complex numbers a, x, y and all positive integers n,*

$$(x+y)^n = \sum_{k=0}^{n}\binom{n}{k}x(x+ak)^{k-1}(y-ak)^{n-k}.$$

Proof. The desired equality is equivalent to

$$\frac{(x+y)^n}{n!} = \sum_{k=0}^{n}\frac{x(x+ak)^{k-1}}{k!}\frac{(y-ak)^{n-k}}{(n-k)!}.$$

Let us consider the generating functions of the two sides of the previous equality. The generating function of the left-hand side is $e^{(x+y)T}$. The generating function of the right-hand side is

$$\sum_n \sum_{0\le k\le n} \frac{x(x+ak)^{k-1}}{k!}\frac{(y-ak)^{n-k}}{(n-k)!}T^n = \sum_k T^k \frac{x(x+ak)^{k-1}}{k!}e^{T(y-ak)}.$$

Thus, by dividing by e^{yT}, it suffices to prove that

$$e^{xT} = 1 + \sum_{k\ge 1}\frac{1}{k!}x(x+ak)^{k-1}e^{-akT}T^k.$$

But this is a consequence of theorem 8.A.5 with

$$f(T) = e^{xT} \text{ and } e(T) = e^{aT}.$$ □

Let us apply this result to prove the following nice-looking identity.

Example 8.A.7. Prove that for all $n \ge 1$ we have

$$\sum_{i,j\ge 0, i+j=n}\binom{n}{i}(i+1)^{i-1}(j+1)^{j-1} = 2(n+2)^{n-1}.$$

AMM E. 2828

Proof. Take $a = 1$, $x = 1$ and $y = n+1$ in the previous theorem. We obtain

$$\sum_{i+j=n}\binom{n}{i}(i+1)^{i-1}(j+1)^{j} = (n+2)^n.$$

Unfortunately, this is not really what we want to prove, but it shows that we are on the right track. To get rid of the extra 1 in the exponent of $j+1$, differentiate with respect to y the equality in the previous theorem and then take $a = 1$, $x = 1$ and $y = n+1$. This time we end up with

$$n(n+2)^{n-1} = \sum_{i+j=n}\binom{n}{i}(i+1)^{i-1}j(j+1)^{j-1}.$$

Now, observing that $j = j+1-1$, we can write the last sum as

$$\sum_{i+j=n}\binom{n}{i}(i+1)^{i-1}(j+1)^{j} - \sum_{i+j=n}\binom{n}{i}(i+1)^{i-1}(j+1)^{j-1}.$$

Combining this with the first relation, the result follows. □

Here is a quite exotic problem: suppose that $e \in K[[T]]$ and consider the sequence $a_n = [T^n](e^n(T))$. What is the generating function of the sequence a_n? The following result answers this question in a more general context:

Theorem 8.A.8. *Let X, Y be variables satisfying $Y = Xe(Y)$, where $e \in K[[T]]$ has nonzero constant term. Then for any $F \in K[[T]]$ we have*

$$\sum_{n\geq 0}[T^n](F(T)e(T)^n)\cdot X^n = \frac{F(Y)}{1-Xe'(Y)}.$$

Proof. Apply theorem 8.A.5 to $f(T) = \int_0^T \frac{F(u)}{e(u)}du$ (recall that this is formal integration, so $f(T)$ is the unique formal series vanishing at 0 and such that $f'(u)e(u) = F(u)$). Using that $f' = \frac{F}{e}$, we obtain

$$\begin{aligned} f(Y) &= \sum_{n\geq 1}\frac{1}{n}(Y/e(Y))^n\cdot[T^{n-1}](F(T)e(T)^{n-1}) \\ &= \sum_{n\geq 1}\frac{1}{n}[T^{n-1}](F(T)e(T)^{n-1})X^n. \end{aligned}$$

Differentiating this equality with respect to Y we obtain

$$\frac{F(Y)}{e(Y)} = \frac{d}{dY}f(Y) = \sum_{n\geq 1}[T^{n-1}](F(T)e(T)^{n-1})X^{n-1}\frac{dX}{dY}.$$

Finally, differentiating $Y = Xe(Y)$ we obtain $dY = dX\cdot e(Y) + Xe'(Y)dY$, so

$$\frac{dX}{dY} = \frac{1-Xe'(Y)}{e(Y)}.$$

Replacing this in the previous equality yields the desired equality. □

Example 8.A.9. Find a closed form for the generating function of the sequence a_n, where a_n is the constant term of $\left(1+X+\frac{1}{X}\right)^n$.

Proof. Note that $a_n = [T^n](e^n(T))$, where $e(T) = T^2 + T + 1$. So, by the previous theorem

$$\sum_n a_n X^n = \frac{1}{1 - Xe'(Y)},$$

where $Y = Xe(Y)$. Solving the equation in Y yields

$$Y = \frac{1 - X - \sqrt{1 - 2X - 3X^2}}{2X}$$

and then an easy computation shows that

$$\sum_n a_n X^n = \frac{1}{\sqrt{1 - 2X - 3X^2}}.$$

□

8.A.3 Examples from enumerative combinatorics

In this section we consider applications of the inversion formula in counting problems. We start with an absolutely classical and beautiful theorem of Cayley, but we need a series of definitions before stating it. Recall that a tree is a connected graph with no cycle. A labeled tree on the set $\{1, 2, \ldots, n\}$ is a tree whose set of vertices is $\{1, 2, \ldots, n\}$. A rooted tree is a tree in which one of the vertices (called the root) is distinguished. There is a unique (non-backtracking) path between any two vertices of a tree. The parent of a vertex in a rooted tree is the vertex connected to it on the unique path to the root. A child of a vertex v is a vertex whose parent is v. A tree is called ordered if one is given an ordering of the children of each vertex.

Theorem 8.A.10. *There are n^{n-2} unordered labeled trees on the set $\{1, 2, \ldots, n\}$.*

Proof. Let a_n be the number of unordered rooted labeled trees on $\{1, 2, \ldots, n\}$, with the natural convention that $a_0 = 0$. It is enough to prove that $a_n = n^{n-1}$, as clearly the number of unordered labeled trees is a_n/n.

The point is that giving such a tree is the same as giving its root and a forest of subtrees whose roots are the children of the root. Suppose that the root has k children and that the corresponding subtrees have $n_1, n_2, \ldots, n_k$

vertices, so $n_1 + n_2 + \cdots + n_k = n - 1$. The number of ways to distribute the $n - 1$ vertices different from the root in these k subtrees is

$$\binom{n-1}{n_1} \cdot \binom{n-1-n_1}{n_2} \cdots = \frac{(n-1)!}{n_1! n_2! \cdots n_k!}.$$

Once such a distribution is made, we have $a_{n_1} a_{n_2} \cdots a_{n_k}$ ways to label the elements of the forest and so the contribution to the total number is

$$\frac{(n-1)!}{n_1! n_2! \cdots n_k!} a_{n_1} a_{n_2} \cdots a_{n_k}.$$

The total contribution coming from all partitions of $n - 1$ is

$$\sum_{n_1 + \cdots + n_k = n-1} \frac{(n-1)!}{n_1! n_2! \cdots n_k!} a_{n_1} a_{n_2} \cdots a_{n_k},$$

however each configuration is counted $k!$ times (since we do not care about the order of the children of the root) and the main root can be chosen in n ways. We finally deduce the very complicated recurrence relation

$$a_n = n \sum_k \frac{1}{k!} \sum_{n_1 + \cdots + n_k = n-1} \frac{(n-1)!}{n_1! n_2! \cdots n_k!} a_{n_1} a_{n_2} \cdots a_{n_k}.$$

This simplifies drastically if we consider the exponential generating function $T(X) = \sum_{n \geq 0} a_n \frac{X^n}{n!}$, since

$$[X^{n-1}] T(X)^k = \sum_{n_1 + \cdots + n_k = n-1} \frac{a_{n_1} \cdots a_{n_k}}{n_1! \cdots n_k!}.$$

Hence the recurrence relation can also be written

$$\frac{a_n}{n!} = [X^{n-1}] \sum_k \frac{T(X)^k}{k!} = [X^{n-1}] e^{T(X)},$$

that is $X e^{T(X)} = T(X)$. Using Lagrange's inversion formula, we obtain

$$\frac{a_n}{n!} = [X^n](T(X)) = \frac{1}{n} [X^{n-1}] e^{nX} = \frac{n^{n-1}}{n!},$$

from where the result follows. □

We end this addendum with two more difficult examples. The following beautiful result is taken from [65].

Example 8.A.11. An intransitive tree on the set of vertices $\{1, 2, \ldots, n\}$ is a tree such that for all $1 \leq i < j < k \leq n$, $\{i, j\}$ and $\{j, k\}$ are not simultaneously edges. Prove that the number of such trees is

$$\frac{1}{n \cdot 2^{n-1}} \sum_{k=1}^{n} \binom{n}{k} k^{n-1}.$$

Note that it is absolutely not clear that the above quantity is an integer!

Proof. Let F_n be the number of such trees and let

$$F(T) = \sum_{n \geq 0} F_{n+1} \frac{T^n}{n!}$$

be an associated exponential generating function. Call a vertex i left if all of its neighbors are greater than i. Let L_n be the number of rooted intransitive trees on the set of vertices $\{1, 2, \ldots, n\}$, whose root is a left vertex. Then $L_1 = 1$ and we clearly have $L_n = \frac{n}{2} F_n$ for $n \geq 2$ (n comes from the choice of the root, division by 2 takes into account the fact that the probability that the root is left is $1/2$). If

$$L(T) = \sum_{n \geq 1} L_n \frac{T^n}{n!}$$

is the exponential generating function associated to L_n, then $L_n = \frac{n}{2} F_n$ for $n \geq 2$ yields $L(T) = \frac{T}{2}(1 + F(T))$. But the exponential formula implies that $F(T) = e^{L(T)}$, as an intransitive tree on the set of vertices $\{1, 2, \ldots, n+1\}$ is obtained from a forest of left-rooted trees on $\{1, 2, \ldots, n\}$, by connecting $n+1$ to each root. Thus, we obtain $F(T) = e^{\frac{T}{2}(1+F(T))}$. If $f(T) = T(1 + F(T))$, we deduce that $f(T) = T(1 + e^{f(T)/2})$. An application of Lagrange's inversion formula yields

$$\frac{F_n}{(n-1)!} = [T^n] f(T) = \frac{1}{n}[T^{n-1}](1 + e^{T/2})^n = \frac{1}{n}[T^{n-1}] \sum_{k=0}^{n} \binom{n}{k} e^{kT/2}.$$

If we expand $e^{kT/2}$ and collect terms according to the exponent of T, we finally deduce that

$$F_n = \frac{1}{n \cdot 2^{n-1}} \sum_{k=1}^{n} \binom{n}{k} k^{n-1}. \qquad \square$$

Finally, a question by James Propp with a nice proof from [69].

Example 8.A.12. The vertices of a polygon P with $N+2$ vertices are labeled $1, 2, 1, 2, \ldots$ in order (stopping when the end is reached). Let a_N be the number of triangulations of P with no monochromatic triangle. Then $a_N = \frac{2^n}{2n+1}\binom{3n}{n}$ if $N = 2n$ and $a_N = \frac{2^{n+1}}{2n+2}\binom{3n+1}{n}$ if $N = 2n+1$.

Proof. Define $a_0 = 1$. Suppose that $N = 2n+1$ and call a triangulation proper if it contains no monochromatic triangle. Consider a proper triangulation π of P. Note that P has an edge labeled $1, 1$. This edge must be a side of a triangle with a vertex labeled 2. If this vertex is the ith vertex labeled 2, with $i \geq 0$, then the two sides of the triangle split P into a $2i+2$-gon and a $2n-2i+2$-gon and both of these polygons are properly triangulated by π. By adding over all i we obtain $a_{2n+1} = \sum_{i=0}^{n} a_{2i} a_{2n-2i}$ and a similar argument for N even yields $a_{2n} = \sum_{i=0}^{2n-1} a_i a_{2n-1-i}$. This suggests considering the two generating functions

$$A(X) = \sum_{n \geq 1} a_{2n} X^n, \quad B(X) = \sum_{n \geq 0} a_{2n+1} X^n.$$

Then the previous relations can be written in a compact form

$$A(X) = 2X(1 + A(X))B(X), \quad B(X) = (1 + A(X))^2.$$

Indeed, we have

$$\begin{aligned} B(X) &= \sum_{n \geq 0} a_{2n+1} X^n = \sum_{n \geq 0} \left(\sum_{i=0}^{n} a_{2i} a_{2(n-i)} \right) X^n \\ &= \left(\sum_{i \geq 0} a_{2i} X^i \right)^2 = (1 + A(X))^2 \end{aligned}$$

and

$$\begin{aligned}A(X) &= \sum_{n\geq 1}\left(\sum_{i=0}^{2n-1} a_i a_{2n-1-i}\right) X^n = X\cdot\sum_{n\geq 0}\left(\sum_{i=0}^{2n+1} a_i a_{2n+1-i}\right) X^n\\ &= X\cdot\sum_{n\geq 0}\left(\sum_{i=0}^{n} a_{2i} a_{2n-2i+1}\right) X^n + X\cdot\sum_{n\geq 0}\left(\sum_{i=0}^{n} a_{2i+1} a_{2n-2i}\right) X^n\\ &= 2X(1+A(X))B(X).\end{aligned}$$

We deduce that $A(X) = 2X(1+A(X))^3$ and an easy application of Lagrange's inversion formula finishes the proof. □

8.A.4 Composition of generating functions

One of the key points in the proof of Cayley's theorem 8.A.10 is to establish the functional equation $Xe^{T(X)} = T(X)$ for the exponential generating function of the number of labeled trees. We would like to give a more abstract and general context for this kind of argument, which appears very often in counting problems. We follow rather closely the wonderful book [76] and we strongly advise the reader to take a look at the first chapter of it, which contains an impressive number of examples and problems on this topic.

Let K be a field of characteristic 0 and let $f, g : \mathbb{N} \to K$ be two sequences of elements of K. Let $E_f = \sum_{n\geq 0} f(n)\frac{X^n}{n!}$ be the generating function of f. We would like to give a combinatorial interpretation of the generating functions $E_f \cdot E_g$ and $E_f \circ E_g$. Note that $E_f \cdot E_g = E_h$, where

$$h(n) = \sum_{k=0}^{n}\binom{n}{k} f(k)g(n-k).$$

We deduce that for any finite set X we have

$$h(|X|) = \sum_{S\subset X} f(|S|)g(|X-S|) = \sum_{(S_1,S_2)} f(|S_1|)g(|S_2|),$$

the second sum being taken over all ordered partitions (S_1, S_2) of X (and S_1, S_2 may be empty). By an obvious induction, we deduce that

$$E_{f_1}\cdot E_{f_2}\cdot \dots \cdot E_{f_s} = E_h,$$

where

$$h(|X|) = \sum_{(S_1,S_2,\ldots,S_s)} f_1(|S_1|) \cdot f_2(|S_2|) \cdot \cdots \cdot f_s(|S_s|),$$

the sum being taken again over ordered partitions of X. For instance, consider the problem of counting the number of partitions $(S_1, S_2, \ldots, S_k)$ of $\{1, 2, \ldots, n\}$, where each S_i is nonempty. By taking $f(n) = 1$ if $n \geq 1$ and $f(0) = 0$, we deduce that the exponential generating function for the number of such partitions is

$$E_f^k(X) = (e^X - 1)^k = \sum_{j=0}^{k} (-1)^{k-j} \binom{k}{j} \cdot e^{jX}$$

and expanding e^{jX} yields the desired number of partitions.

Suppose now that $f(0) = 0$. We would like to understand the generating function $E_g \circ E_f$.

Theorem 8.A.13. *Let $f, g : \mathbb{N} \to K$ be sequences such that $f(0) = 0$ and $g(0) = 1$. Then $E_g \circ E_f = E_h$, where $h : \mathbb{N} \to K$ is a sequence such that $h(0) = 1$ and for all finite sets X we have*

$$h(|X|) = \sum_{\pi} g(k) \cdot f(|S_1|) \cdot f(|S_2|) \cdot \cdots \cdot f(|S_k|),$$

the sum being taken over all unordered partitions $(S_1, S_2, \ldots, S_k)$ (with arbitrary k) of X into nonempty subsets.

Proof. We clearly have $h = \sum_{k \geq 0} g(k) h_k$, where

$$h_k(n) = \sum_{\pi} f(|S_1|) \cdot \cdots \cdot f(|S_k|),$$

the sum being taken over unordered partitions with k classes. Hence it is enough to prove that $E_{h_k} = \frac{E_f^k}{k!}$. But this follows from the previous discussion and the fact that we are only considering unordered partitions here (thus the k classes may be permuted in $k!$ ways and yield the same unordered partition). □

Example 8.A.14. Let us consider again the problem of finding the functional equation for the exponential generating function of the number of unordered rooted trees on $\{1, 2, \ldots, n\}$. Let T be this generating function. Then by the previous theorem $Xe^{T(X)}$ is the generating function for the number of pairs (r, F), where r is a root and F is a forest of unordered rooted trees starting from this root. But it is clear that any unordered rooted tree arises in this way, so we actually have $Xe^{T(X)} = T(X)$.

Example 8.A.15. Suppose that we want to count rooted unordered labeled trees such that the number of children of each node is in a fixed set S, containing 0. The argument used in the previous example yields the functional equation

$$f(X) = X \sum_{s \in S} \frac{f(X)^s}{s!}.$$

Using Lagrange's inversion formula, we obtain a formula for the number of such trees.

Example 8.A.16. Let E be the generating function for the number of connected graphs with vertices $1, 2, \ldots, n$. Giving a graph with vertices $1, 2, \ldots, n$ is the same as giving a family of disjoint connected graphs (its connected components), thus by theorem 8.A.13 the generating function for the graphs with vertices $1, 2, \ldots, n$ is e^E. But since there are $2^{\binom{n}{2}}$ such graphs, we deduce that

$$E = \log \left(\sum_{n \geq 0} 2^{\binom{n}{2}} \cdot \frac{X^n}{n!} \right).$$

Example 8.A.17. Consider two sequences f, g such that $f(0) = 0$, $g(0) = 1$ and define $h(0) = 1$ and

$$h(|X|) = \sum_{\sigma \in \mathrm{Sym}(X)} g(k) \cdot f(|C_1|) \cdot f(|C_2|) \cdot \cdots \cdot f(|C_k|),$$

where $\mathrm{Sym}(X)$ is the set of permutations of X and $C_1, C_2, \ldots, C_k$ are the cycles of σ. Then theorem 8.A.13 yields[3]

$$E_h(X) = E_g\left(\sum_{n\geq 1} f(n)\frac{X^n}{n}\right).$$

Example 8.A.18. For nonnegative integers $c_1, c_2, \ldots$, let $a_n(c_1, c_2, \ldots)$ be the number of permutations $\sigma \in S_n$ having c_i cycles of length i for all $i \leq n$. Consider indeterminates $X_1, X_2, \ldots$. Then the previous example yields the following cycle-index formula

$$\sum_{\substack{n\geq 0 \\ c_1,c_2,\ldots,c_n\geq 0}} a_n(c_1, c_2, \ldots)\cdot X_1^{c_1}\cdot X_2^{c_2}\cdot\cdots\cdot X_n^{c_n}\cdot\frac{T^n}{n!} = \exp\left(\sum_{n\geq 1} X_n\cdot\frac{T^n}{n}\right).$$

Example 8.A.19. Let us count the number of permutations of odd order, i.e. for which all cycles have odd length. By taking $X_i = 1$ when i is odd and 0 otherwise, we deduce that the exponential generating function for this counting problem is

$$\begin{aligned} E_f &= \exp\left(\sum_{\substack{n\geq 1 \\ n \text{ odd}}} \frac{T^n}{n}\right) \\ &= \exp\left(\frac{\log(1+T) - \log(1-T)}{2}\right) \\ &= \sqrt{\frac{1+T}{1-T}} \\ &= (1+T)(1-T^2)^{-\frac{1}{2}} \end{aligned}$$

[3] Note that we also have

$$h(|X|) = \sum_{\pi} g(k) f(|S_1|)(|S_1| - 1)! \cdot \cdots \cdot f(|S_k|)(|S_k| - 1)!,$$

the sum being taken over unordered partitions π of X, since the cycle decomposition of a permutation yields a partition of X and since one can cyclically permute in $(|S_i| - 1)!$ ways the elements of a class with $|S_i|$ elements.

and an easy application of the binomial formula yields the number of permutations $f(n) = (1\cdot 3\cdot\dots\cdot(n-1))^2$ when n is even and $f(n) = (1\cdot 3\cdot\dots\cdot(n-2))^2\cdot n$ when n is odd.

Example 8.A.20. Let k be a positive integer and let $f(n)$ be the number of permutations $\sigma \in S_n$ such that $\sigma^k = 1$. This is equivalent to the fact that the length of each cycle of σ divides k. Thus by the previous examples

$$E_f = \exp\left(\sum_{d|k} \frac{X^d}{d}\right).$$

8.A.5 More tree-counting problems

In this section we present another proof of Cayley's theorem as well as some similar counting problems, all related to trees. The following general result is quite useful in problems concerning trees.

Theorem 8.A.21. *Let $d_1, d_2, \dots, d_n$ be positive integers such that $d_1 + d_2 + \dots + d_n = 2n - 2$. Then the number of trees on the set $\{v_1, v_2, \dots, v_n\}$ such that vertex v_i has degree d_i is $\frac{(n-2)!}{(d_1-1)!\cdot(d_2-1)!\cdot\dots\cdot(d_n-1)!}$.*

Proof. We will prove the result by induction on n. We may assume that $d_n = 1$, by permuting the d_i's if necessary. Consider a tree on $\{v_1, v_2, \dots, v_n\}$ such that $\deg(v_i) = d_i$ and remove vertex v_n and the unique edge whose endpoint is v_n. We obtain a tree on $\{v_1, v_2, \dots, v_{n-1}\}$ whose degrees are $d_1, \dots, d_{j-1}, d_j - 1, d_{j+1}, \dots, d_{n-1}$ if v_n is connected to v_j. Conversely, any such tree on $\{v_1, \dots, v_{n-1}\}$ yields a tree on $\{v_1, \dots, v_n\}$ simply by connecting v_n with v_j. It follows that there are

$$\sum_{j=1}^{n-1} \frac{(n-3)!}{(d_1-1)!\cdot\dots\cdot(d_j-2)!\cdot\dots\cdot(d_{n-1}-1)!}$$
$$= \left(\sum_{k=1}^{n-1} d_k - n + 1\right)\cdot\frac{(n-3)!}{\prod(d_k-1)!} = \frac{(n-2)!}{\prod(d_k-1)!}$$

such trees and the result follows. □

Cayley's theorem is a direct consequence of the previous theorem and of the multinomial formula: the number of trees on $\{1, 2, \ldots, n\}$ is

$$\sum_{\substack{d_1+\cdots+d_n=2n-2\\ d_i\geq 1}} \frac{(n-2)!}{(d_1-1)!\cdot\cdots\cdot(d_n-1)!} = \sum_{i_1+\cdots+i_n=n-2} \frac{(n-2)!}{i_1!\cdot i_2!\cdot\cdots\cdot i_n!}$$
$$= (1+1+\cdots+1)^{n-2} = n^{n-2}.$$

Proposition 8.A.22. *There are $\binom{n-2}{k-1}\cdot(n-1)^{n-k-1}$ labeled trees with vertices $1, 2, \ldots, n$ in which vertex 1 has degree k.*

Proof. This is also an easy consequence of the previous theorem. The desired number of trees is

$$\sum_{k+d_2+\cdots+d_n=2(n-1)} \frac{(n-2)!}{(k-1)!\cdot(d_2-1)!\cdot\cdots\cdot(d_n-1)!} = \binom{n-2}{k-1}\cdot(n-1)^{n-k-1},$$

the second equality being a consequence of the multinomial formula. □

Let us introduce a very useful notion in graph theory.

Definition 8.A.23. Let G be a loopless graph. A spanning forest is a subgraph without cycles and having the same vertices as G. A spanning tree is a connected spanning forest.

Here is a nice application of Abel's identity.

Example 8.A.24. There are $(n-2)\cdot n^{n-3}$ spanning trees of K_n which do not contain a fixed edge of K_n.

Proof. Call $1, 2, \ldots, n$ the vertices of K_n and assume without loss of generality that the fixed edge is $e = 12$. Let $f(n)$ be the number of spanning trees that contain e. Such a tree appears uniquely as a result of the following process: consider two trees T_1, T_2 whose vertices form a partition of $1, 2, \ldots, n$ with 1 a vertex of T_1 and 2 a vertex of T_2, and then join these two trees by the edge e. If T_1 has k vertices different from 1, these vertices can be chosen in $\binom{n-2}{k}$

ways. Once these vertices are chosen, we have $(k+1)^{k-1}$ possibilities for T_1 and $(n-k-1)^{n-k-3}$ possibilities for T_2. Thus

$$f(n) = \sum_{k=0}^{n-2} \binom{n-2}{k} (k+1)^{k-1} \cdot (n-k-1)^{n-k-3} = 2 \cdot n^{n-3},$$

the last equality being an easy consequence of example 8.A.7. The result follows. □

Remark 8.A.25. Here is another approach, suggested by Richard Stong. Let X_e be the probability that a randomly chosen spanning subtree of K_n contains the edge e. Then by symmetry it is clear that $E[X_e]$ is the same for all edges e. Since any spanning subtree of K_n has exactly $n-1$ edges we have

$$\sum_e E[X_e] = n - 1.$$

Therefore since all $\binom{n}{2}$ terms in this sum are equal,

$$E[X_e] = \frac{n-1}{\binom{n}{2}} = \frac{2}{n}.$$

Hence by Cayley's formula, there are $2 \cdot n^{n-3}$ spanning subtrees containing e and $(n-2)n^{n-3}$ that do not contain e.

Chapter 9

A Little Introduction to Algebraic Number Theory

This rather long chapter is concerned with elementary algebraic number theory. The techniques are rather diverse: basic linear algebra, algebraic numbers and symmetric polynomials, cyclotomy and p-adic analysis are some of the topics discussed in this chapter. Since we will use the notion of algebraic number quite often in this chapter, we end this introduction with a few recollections. For more details and some proofs, the reader is referred to the addendum 9.B.

A complex number z is called algebraic if it is root of some nonzero polynomial with rational coefficients. In this case, there exists a unique monic polynomial with rational coefficients, called the minimal polynomial of z, which vanishes at z and has minimal degree. The roots of this polynomial are called the conjugates of z. The crucial property of the minimal polynomial is that it is irreducible over $\mathbb{Q}$ and divides any polynomial with rational coefficients that vanishes at z. A fundamental theorem in algebraic number theory states that the algebraic numbers form an algebraically closed subfield of $\mathbb{Q}$, thus an algebraic closure of $\mathbb{Q}$. If z is an algebraic number, we let $\mathbb{Q}(z)$ (or $\mathbb{Q}[z]$) be the subfield of $\mathbb{C}$ generated by z. It consists of all numbers of the form $f(z)$, with $f \in \mathbb{Q}[X]$ (or equivalently $f \in \mathbb{Q}(X)$). This is a finite extension of $\mathbb{Q}$, of degree equal to the degree of the minimal polynomial of z. The primitive element theorem ensures that all finite extensions of $\mathbb{Q}$ are of the form $\mathbb{Q}(z)$ for some algebraic number z. We call such extensions number fields. We will frequently use the notation $[L : K]$ to denote the dimension of L as K-vector space, as well as the fundamental tower relation $[M : K] = [M : L] \cdot [L : K]$ for any finite extensions $M/L/K$.

A more refined notion is that of algebraic integer. This is a complex number that is killed by some monic polynomial with integer coefficients. By Gauss' lemma, we can characterize algebraic integers as those algebraic numbers whose minimal polynomial has integer coefficients. An easy but fundamental result is that a rational number which is also an algebraic integer is necessarily a rational integer. Another important result is that the algebraic integers form a subring of the field of algebraic numbers.

9.1 Tools from linear algebra

In this section we consider a few applications of linear algebra to number theory. These concern especially divisibility issues and linear diophantine equations.

1. Let a, b, c be relatively prime nonzero integers. Prove that for any relatively prime integers u, v, w satisfying $au+bv+cw = 0$, there are integers m, n, p such that

$$a = nw - pv, \ \ b = pu - mw, \ \ c = mv - nu.$$

Octavian Stănăşilă, Romanian TST 1989

Proof. Consider the linear system in the variables m, n, p

$$a = nw - pv, \ \ b = pu - mw, \ \ c = mv - nu.$$

Trivially, the determinant of this system is 0 and the rank of its associated matrix is 2. It is thus enough to solve in integers the system $a = nw - pv$, $b = pu - mw$. This system has integer solutions if and only if there is an integer p such that $vp \equiv -a \pmod{w}$ and $up \equiv b \pmod{w}$. Now, the hypothesis implies the existence of integers A, B, C such that $Au + Bv + Cw = 1$. We deduce that $Aua + Bva \equiv a \pmod{w}$. Since $au + bv \equiv 0 \pmod{w}$, we deduce that $(Ab - Ba)v \equiv -a \pmod{w}$, so that we can take $p = Ab - Ba$ to get $up \equiv b \pmod{w}$. Also, we can immediately check that

$$u(Ab - Ba) \equiv Aub - Bua \equiv b(Au + Bv) \equiv b \pmod{w}. \qquad \square$$

Proof. We will actually prove a stronger result: for any integers a, b, c and any integers u, v, w such that $au + bv + cw = 0$ and $\gcd(u, v, w) = 1$, there exist integers A, B, C such that $a = Bw - Cv$, $b = Cu - Aw$ and $c = Av - Bu$.

Indeed, since $\gcd(u, v, w) = 1$, a standard application of Bézout's lemma yields the existence of integers X, Y, Z such that $Xu + Yv + Zw = 1$. Let us define

$$A = cY - bZ, \quad B = aZ - cX, \quad C = bX - aY.$$

Then,

$$\begin{aligned} Bw - Cv &= (aZ - cX)\,w - (bX - aY)\,v \\ &= a(Xu + Yv + Zw) - X(au + bv + cw) \\ &= a. \end{aligned}$$

Thus $a = Bw - Cv$ and similarly $b = Cu - Aw$ and $c = Av - Bu$. The result follows. $\square$

A very nice and classical result is that $\prod_{1 \le i < j \le n} \frac{a_i - a_j}{i - j}$ is an integer for any integers $a_1, a_2, \ldots, a_n$. There are many proofs of this result, at least two of them being presented in [3]. The following problem is a variation on this topic.

2. Prove that for any integers $a_1, a_2, \ldots, a_n$ the number

$$\frac{\operatorname{lcm}(a_1, a_2, \ldots, a_n)}{a_1 a_2 \cdots a_n} \prod_{1 \le i < j \le n} (a_j - a_i)$$

is an integer divisible by $1!2! \cdots (n-2)!$. Moreover, we cannot replace $1!2! \cdots (n-2)!$ by any other multiple of $1!2! \cdots (n-2)!$.

Proof. Consider the matrix $A = \{a_{i,j}\}_{1\le i,j\le n}$ with $a_{i,j} = \binom{a_i-1}{j-2}$ for $j \ge 2$ and $a_{i,1} = \frac{\operatorname{lcm}(a_1,a_2,\ldots,a_n)}{a_i}$. We will prove that

$$\det(A) = \frac{\operatorname{lcm}(a_1,a_2,\ldots,a_n)}{a_1a_2\cdots a_n}\cdot\frac{\prod_{1\le i<j\le n}(a_j-a_i)}{1!2!\cdots(n-2)!}.$$

Since the entries of A are integers, it is clear that its determinant is an integer, from which the first part of the problem will follow.

Factoring an $L = \operatorname{lcm}(a_1,a_2,\ldots,a_n)$ out of the first column and multiplying the i-th row by a_i, it follows that

$$\begin{vmatrix} \frac{L}{a_1} & 1 & \binom{a_1-1}{1} & \cdots & \binom{a_1-1}{n-1} \\ \frac{L}{a_2} & 1 & \binom{a_2-1}{1} & \cdots & \binom{a_2-1}{n-1} \\ \vdots & \vdots & \vdots & \ddots & \vdots \\ \frac{L}{a_n} & 1 & \binom{a_n-1}{1} & \cdots & \binom{a_n-1}{n-1} \end{vmatrix} = \frac{L}{a_1a_2\cdots a_n}\begin{vmatrix} 1 & a_1 & a_1(a_1-1) & \cdots & a_1\binom{a_1-1}{n-2} \\ 1 & a_2 & a_2(a_2-1) & \cdots & a_2\binom{a_2-1}{n-2} \\ \vdots & \vdots & \vdots & \ddots & \vdots \\ 1 & a_n & a_n(a_n-1) & \cdots & a_n\binom{a_n-1}{n-2} \end{vmatrix}.$$

Taking out the numbers $\frac{1}{j!}$ that appear at the denominators of the binomials in each column shows that

$$\det A = \frac{L}{a_1a_2\cdots a_n 1!2!\cdots(n-2)!}\cdot\begin{vmatrix} 1 & a_1 & a_1^2-a_1 & \cdots & a_1^{n-1}+\cdots \\ 1 & a_2 & a_2^2-a_2 & \cdots & a_2^{n-1}+\cdots \\ \vdots & \vdots & \vdots & \ddots & \vdots \\ 1 & a_n & a_n^2-a_n & \cdots & a_n^{n-1}+\cdots \end{vmatrix}.$$

Note that the (i,j) entry of this matrix can be written as $P_j(a_i)$, where P_j, $0 \le j \le n-1$ is a monic polynomial of degree j. For each column j add a suitable linear combination of the previous columns to reduce the previous determinant to

$$\frac{L}{a_1\cdots a_n 1!2!\cdots(n-2)!}\cdot\begin{vmatrix} 1 & a_1 & a_1^2 & \cdots & a_1^{n-1} \\ 1 & a_2 & a_2^2 & \cdots & a_2^{n-1} \\ \vdots & \vdots & \vdots & \ddots & \vdots \\ 1 & a_n & a_n^2 & \cdots & a_n^{n-1} \end{vmatrix}.$$

Since the last determinant is Vandermonde, the identity stated above is proved.

To see that the result is optimal, simply choose $a_n = (n!)^2$ and $a_i = (n!)^2 + i$ for $1 \leq i < n$. Then

$$\operatorname{lcm}(a_1, a_2, \ldots, a_n) = n!na_1 \cdots a_{n-1}$$

because the numbers $a_n, \frac{a_i}{i}, i = 1, 2, \ldots, n-1$ are pairwise relatively prime. The result follows easily from this. □

Remark 9.1. Quantities such as $\prod_{i<j} \frac{a_i - a_j}{i-j}$ and the one in the previous problem have natural combinatorial interpretations: they are the dimensions of some irreducible representations of special unitary groups. Of course, explaining this is beyond the scope of this modest book, but the reader should know that these are not "just some random problems."

Remark 9.2. A similar result is proved in the beautiful paper [6]: if $a_0, a_1, \ldots, a_n$ are integers, then $\prod_{i<j}(a_i^2 - a_j^2)$ is a multiple of $\frac{2! \cdot 4! \cdot \cdots \cdot (2n)!}{2^{n+1}}$ and this result is optimal. This is also related to the dimension of some irreducible representations of the symplectic group.

We continue with a nice application of linear-algebraic arguments. The ideas used in the following solutions are very useful in other contexts, too.

3. Let p be a prime and let $a_1, a_2, \ldots, a_{p+1}$ be real numbers such that no matter how we eliminate one of them, the remaining numbers can be divided into at least two nonempty pairwise disjoint subsets each having the same arithmetic mean. Prove that $a_1 = a_2 = \cdots = a_{p+1}$.

Marius Rădulescu, Romanian TST 1994

Proof. Subtracting from the a_i's their arithmetic mean (observe that the new numbers have the same property), we may assume that $a_1 + a_2 + \cdots + a_{p+1} = 0$. Fix some $1 \leq j \leq p+1$ and let $C_1, \ldots, C_r$ be the classes of a partition of $\{a_1, \ldots, a_{p+1}\} - \{a_j\}$, such that $\frac{1}{|C_l|} \sum_{x \in C_l} x$ does not depend on l. Since the sum of the a_i's is zero and since

$$\sum_l \sum_{x \in C_l} x = -a_j,$$

we deduce that we have a linear relation of the form

$$\frac{1}{|C_1|}a_{i_1} + \cdots + \frac{1}{|C_1|}a_{i_{|C_1|}} + \frac{a_j}{p} = 0,$$

with $|C_1| < p$.

Now, consider all such linear relations, obtained by making j run over all $1, 2, \ldots, p+1$. This gives us a linear system with $p+1$ equations and $p+1$ unknowns (the a_i's), whose matrix has $\frac{1}{p}$ on the main diagonal and numbers of the form $\frac{1}{k}$ with $k < p$ elsewhere. But then the determinant of the matrix will be of the form $\frac{m}{p^{p+1}}$ for some rational $m \equiv 1 \pmod{p}$. Thus, the determinant is nonzero and since the system is homogeneous, the only solution is the trivial one. This implies that all a_i's are zero and the conclusion follows. □

Proof. First, we will reduce the problem to the case when all a_i are integers. The following method is classical and very useful in a whole variety of situations: consider the vector space spanned over $\mathbb{Q}$ by the a_i's. This is a finite dimensional $\mathbb{Q}$-vector space and if we take a basis of it and write each a_i as a linear combination with rational coefficients of the elements of the basis, we easily see that the coordinates of the a_i's also satisfy the conditions of the problem (because by definition the elements of the basis are linearly independent over $\mathbb{Q}$). Working coordinate by coordinate reduces therefore the problem to the case when all a_i are rational. Multiplying all a_i's by $N!$ for some sufficiently large N reduces then the problem to the case when all a_i are integers.

Assume now that all a_i are integers and let us prove the result by induction on $\max|a_i|$. The base case is obvious, so let us focus on the inductive step. Removing every element a_i gives sets $S_{i,j} \subset \{a_1, a_2, \ldots a_{p+1}\} - \{a_i\} = A_i$ nonempty, pairwise disjoint so that $\frac{1}{|S_{i,j}|}\sum_{x \in S_{i,j}} x$ is independent of j, say equal to k. Then

$$\frac{1}{|S_{i,1}|}\left(\sum_{x \in S_{i,1}} x\right) = k = \frac{1}{\sum_{j \neq 1}|S_{i,j}|}\left(\sum_{x \in \cup_{j \neq 1} S_{i,j}} x\right)$$

Let $|S_{i,1}| = m$ and observe that $\sum_{j\neq 1} |S_{i,j}| = p - m$ and

$$(p-m)\left(\sum_{x\in S_{i,1}} x\right) = m\left(\sum_{x\in A_i - S_{i,1}} x\right) \implies \sum_{x\in A_i} x \equiv 0 \pmod p$$

Summing over all choices of i we obtain

$$\sum_{i=1}^{p+1} a_i \equiv r \pmod p \implies a_i \equiv r \pmod p\, \forall i \in \{1,2,\ldots,p+1\}$$

Thus, we can write $a_i = pb_i + r$ for some integers b_i. But clearly $\{b_1, b_2, \ldots, b_{p+1}\}$ also satisfy the conditions of the problem and moreover $\max|b_i| < \max|a_i|$. By the inductive hypothesis all b_i's are equal and so all a_i's are equal □

Proof. Here's a "no formula" proof, which uses the same kind of argument, but replaces the choice of a basis in a vector space with an approximation argument: first, we reduce to the case when all a_i are integers in the following way. By Dirichlet's approximation theorem there exists a large integer M such that all Ma_i are very close to some integer A_i. The linear equations deduced from the fact that the a_i's satisfy the conditions of the problem become approximate linear equations for the A_i's. But there are finitely many such equations and each has rational coefficients. Thus, if at the beginning we ensured that Ma_i are sufficiently close to the A_i's, the approximate equations in A_i are actually exact. Thus the A_i's are integers satisfying the conditions of the problem. If we solved the problem over the integers, it follows that all A_i are equal. But then any two a_i's are less than $2/M$ apart and since M is arbitrarily large, this implies that all a_i are equal.

Now, let us assume that the a_i's are integers. If we remove one number, the common rational arithmetic mean for the rest of the numbers cannot have p in the denominator, so the sum of all other numbers is rational with p in the numerator and, thereby, an integer divisible by p. Hence all numbers have the same remainder modulo p as their sum. Now continue as in the end of the previous solution. □

9.2 Cyclotomy

There are $\varphi(n)$ primitive nth roots of unity, namely $e^{\frac{2i\pi k}{n}}$, where k is relatively prime to n. Hence the nth cyclotomic polynomial

$$\phi_n(X) = \prod_{\substack{1\le k\le n \\ \gcd(k,n)=1}} (X - e^{\frac{2i\pi k}{n}})$$

has degree $\varphi(n)$. The splitting field $\mathbb{Q}(e^{\frac{2i\pi}{n}})$ of ϕ_n is called the nth cyclotomic extension of $\mathbb{Q}$. These polynomials and their splitting fields play a very important role in many areas of mathematics and gave rise to a whole series of very deep results. Their study would require a whole book by itself, so we decided to focus only on some very elementary and classical applications.

Since any nth root of unity in $\mathbb{C}$ is primitive of order d for a unique $d|n$, we get the:

Proposition 9.3. *(Fundamental identity) We have*

$$X^n - 1 = \prod_{d|n} \phi_d(X).$$

This easily implies (by strong induction) that $\phi_n(X) \in \mathbb{Z}[X]$ for all n. The following result is not trivial and plays an important role in many proofs concerning cyclotomic polynomials. We'll also see that a weak form of Dirichlet's theorem follows very easily from it. For a proof of Dirichlet's theorem in full generality, see addendum 7.A.

Theorem 9.4. *Let a be an integer and let p be a prime divisor of $\phi_n(a)$. Then either the order of a modulo p is n (and so $p \equiv 1 \pmod{n}$) or p divides n.*

Proof. By assumption, p divides $\phi_n(a)$, and so if a has order $k \pmod{p}$ then $k|n$. If $k < n$, then p divides both $a^k - 1$ and $\frac{a^n-1}{a^k-1}$ (the second because of the fundamental identity and the fact that p divides $\phi_n(a)$). As $\gcd\left(a^k - 1, \frac{a^n-1}{a^k-1}\right)$ divides $\frac{n}{k}$ by the Euclidean algorithm, $p|n$ and we're done. □

Remark 9.5. Note that the proof also works for prime powers p.

Theorem 9.6. *For all n there are infinitely many primes $p \equiv 1 \pmod{n}$.*

Proof. For $k > n$ large enough (it is actually enough to take $k > 2$) we have $\phi_n(k!) > 1$ and so we can choose some $p_k | \phi_n(k!)$. Since $\phi_n(0)$ is 1 or -1, we have $\phi_n(k!) \equiv 1, -1 \pmod{k!}$, which obviously implies that $\gcd(p_k, k!) = 1$. As $k > n$ we get $p_k > k > n$ and by the previous theorem we deduce that $p_k \equiv 1 \pmod{n}$. The result follows. $\square$

For the next three problems we will use a very useful rationality result: if r and $\cos(r\pi)$ are both rational numbers, then $\cos(r\pi) \in \{\pm 1, \pm \frac{1}{2}, 0\}$. Let us recall the argument: $2\cos(r\pi) = e^{ir\pi} + e^{-ir\pi}$ and the numbers $e^{ir\pi}, e^{-ir\pi}$ are algebraic integers (they are roots of unity), so $2\cos(r\pi)$ is an algebraic integer. Thus, if it is rational, it has to be a rational integer and the result follows. Before passing to the next problem, let us discuss a beautiful consequence of the previous observation. We will prove that the only regular n-gons all of whose vertices are lattice points are the squares. Indeed, let A, B, C be three consecutive vertices of the polygon and observe that

$$\frac{1 + \cos \frac{4\pi}{n}}{2} = \cos^2 \frac{2\pi}{n} = \frac{(AB^2 + BC^2 - AC^2)^2}{4AB^2 \cdot BC^2} \in \mathbb{Q}.$$

Using the previous observation, the result follows easily. We strongly advise the reader to look for a geometric proof in order to appreciate the power of algebraic numbers!

4. Let A, B, C be lattice points such that the angles of triangle ABC are rational multiples of π. Prove that triangle ABC is right and isosceles.

Proof. Note that any angle $\theta = \angle ABC$ with A, B, and C lattice points must have $\tan\theta$ rational or infinity. To see this note that all lines between lattice points have rational or infinite slopes and if $\tan\alpha$ and $\tan\beta$ are rational (or infinite) then so is $\tan(\alpha - \beta) = \frac{\tan\alpha - \tan\beta}{1 + \tan\alpha\tan\beta}$. This implies that

$$\tan^2\theta = \sec^2\theta - 1 = \frac{1 - \cos 2\theta}{1 + \cos 2\theta}$$

is rational and hence $\cos 2\theta$ is rational. Combining this with the discussion preceding the problem shows that $\cos 2A$, $\cos 2B$, $\cos 2C$ are all equal to ± 1,

$\pm\frac{1}{2}$ and 0. It is immediate to check that the condition $\tan A$, $\tan B$, $\tan C$ rational or infinite says A, B, C must be integer multiples of $\pi/4$. Hence the only possibility is when ABC is right and isosceles. □

5. Let α be a rational number with $0 < \alpha < 1$ and

$$\cos(3\pi\alpha) + 2\cos(2\pi\alpha) = 0.$$

Prove that $\alpha = \frac{2}{3}$.

IMO Shortlist 1991

Proof. Let $x = \cos\pi\alpha$ and observe that the equation satisfied by α can be written as

$$4x^3 + 4x^2 - 3x - 2 = 0 \implies (2x+1)(2x^2+x-2) = 0.$$

Of course, if $x = -\frac{1}{2}$, we must have $\alpha = \frac{2}{3}$ and we are done. The difficult point is to prove that we cannot have $2x^2 + x - 2 = 0$. If this is the case, then $x = \frac{-1+\sqrt{17}}{4}$, because $|x| \leq 1$. We will then prove that $\cos(2^n\pi\alpha)$ takes infinitely many values as n runs over the positive integers. This will clearly contradict the hypothesis that α is rational. But since $\cos(2^n\pi\alpha) = 2\cos^2(2^{n-1}\pi\alpha) - 1$, it is easy to prove that we can write

$$\cos(2^n\alpha\pi) = \frac{a_n + b_n\sqrt{17}}{4}, \quad b_{n+1} = a_nb_n, \quad a_{n+1} = \frac{a_n^2 + 17b_n^2 - 8}{2}.$$

The previous relations yield by induction that a_n, b_n are odd integers and that $a_{n+1} > a_n$. Thus $\cos(2^n\pi\alpha)$ takes infinitely many values. □

Remark 9.7. In general, let us choose relatively prime integers m, n with $n > 2$ and find the degree of the algebraic number $x = \cos\left(\frac{2\pi m}{n}\right)$. Define $z = e^{\frac{2i\pi m}{n}}$, a primitive n-th root of unity. The irreducibility of the cyclotomic polynomials (which is a very nontrivial theorem) implies that z has degree $\varphi(n)$ as an algebraic number. On the other hand, we have

$$[\mathbb{Q}(z) : \mathbb{Q}] = [\mathbb{Q}(z) : \mathbb{Q}(x)] \cdot [\mathbb{Q}(x) : \mathbb{Q}]$$

and we have $[\mathbb{Q}(z) : \mathbb{Q}(x)] = 2$. Indeed, $2x = z + z^{-1}$, which implies that z satisfies a quadratic equation with coefficients in $\mathbb{Q}(x)$, so $[\mathbb{Q}(z) : \mathbb{Q}(x)] \leq 2$. On the other hand, we cannot have $\mathbb{Q}(z) = \mathbb{Q}(x)$, because z is not a real number. Putting these observations together, we deduce that x has degree $\frac{\varphi(n)}{2}$. Using the previous result and the fact that $\cos\left(\frac{\pi}{2} - x\right) = \sin x$, we can compute the degree of $\sin\frac{2\pi}{n}$. The answer is a bit complicated: if $n \neq 4$, the degree of $\sin\frac{2\pi}{n}$ is $\frac{\varphi(n)}{2}$ if 8 divides n, $\frac{\varphi(n)}{4}$ if $\gcd(n, 8) = 4$ and $\varphi(n)$ if $\gcd(n, 8) < 4$.

6. Prove that none of the numbers $\sqrt{n+1} - \sqrt{n}$ for positive integers n can be the written in the form $2\cos\left(\frac{2k\pi}{m}\right)$ for some integers k, m.

Chinese Olympiad

Proof. First, we will find a polynomial with $x = \sqrt{n+1} - \sqrt{n}$ as a root. We have $x^2 = 2n + 1 - 2\sqrt{n^2 + n}$ and so $(x^2 - 2n - 1)^2 = 4n^2 + 4n$, from where we easily find that $x^4 - 2(2n+1)x^2 + 1 = 0$. Note that the other roots of the polynomial $f(X) = X^4 - 2(2n+1)X^2 + 1$ are $x = \pm\sqrt{n+1} \pm \sqrt{n}$. Next, we will find a polynomial with roots $x = 2\cos\frac{2k\pi}{m}$. Let T_m be the mth Chebyshev polynomial, defined by the equality $T_m(\cos x) = \cos mx$ for all x. Then $T_m\left(\frac{x}{2}\right) = \cos 2k\pi = 1$. Thus the numbers $2\cos\frac{2k\pi}{m}$ for $k = 0, \ldots, m-1$ are roots of $g(X) = T_m\left(\frac{X}{2}\right) - 1$. These m numbers are not distinct, but $2\cos\frac{2k\pi}{m} = 2\cos\frac{2(m-k)\pi}{m}$ for $1 \leq k < m/2$ are double roots of this polynomial since g achieves a local maximum at these points. Thus these are the only roots of $g(X)$.

If $\sqrt{n+1} - \sqrt{n} = 2\cos\frac{2k\pi}{m}$, then $f(X)$ and $g(X)$ have a common factor in $\mathbb{Z}[X]$. The only roots of f which lie in the interval $[-2, 2]$ (which contains all roots of g) are $\sqrt{n+1} - \sqrt{n}$ and $\sqrt{n} - \sqrt{n+1}$. Therefore this common factor is either $X - (\sqrt{n+1} - \sqrt{n})$ or $X^2 - (\sqrt{n+1} - \sqrt{n})^2$. In either case we see that $(\sqrt{n+1} - \sqrt{n})^2 = 2n + 1 - 2\sqrt{n(n+1)}$ is an integer and hence $n(n+1)$ is a square. But this would make $4n(n+1)$ and $(2n+1)^2$ consecutive positive squares, a contradiction. $\square$

We continue with a very beautiful and classical result ([50]) concerning linear equations in roots of unity.

7. a) Suppose that $a_1, a_2, \dots, a_k$ are rational numbers and $\zeta_1, \zeta_2, \dots, \zeta_k$ are roots of unity such that $a_1\zeta_1 + a_2\zeta_2 + \dots + a_k\zeta_k = 0$. Moreover, suppose that $\sum_{i\in I} a_i\zeta_i \neq 0$ for any proper subset I of $\{1, 2, \dots, k\}$. Prove that $\zeta_i^m = \zeta_j^m$ for all i, j, where m is the product of primes smaller than or equal to k.

b) Let z be a complex number. Prove that there are at most $2^{4k^2} \cdot k^k$ k-tuples $(\zeta_1, \zeta_2, \dots, \zeta_k)$ of roots of unity with the following property: there exist rational numbers $a_1, a_2, \dots, a_k$ such that $z = \sum_{i=1}^k a_i\zeta_i$ and $z \neq \sum_{i\in I} a_i\zeta_i$ for any proper subset I of $\{1, 2, \dots, k\}$.

Mann's theorem

Proof. a) We may assume that $a_1 = \zeta_1 = 1$. Let m be the least positive integer such that $\zeta_i^m = 1$ for all i and choose a prime factor p of m. If $m = p^j n$ with $\gcd(n, p) = 1$, we will prove that $j = 1$ and $p \leq k$. This will imply that m divides $\prod_{p\leq k} p$ and the first part of the theorem will follow. Proving this is however not a simple task.

We start with an observation: let $z = e^{\frac{2i\pi}{p^j}}$ and let ζ be an m-th root of unity. We claim that there exists $0 \leq r < p$ and x such that $x^{\frac{m}{p}} = 1$ and $\zeta = z^r \cdot x$. This is very easy: if $\zeta = e^{\frac{2i\pi l}{m}}$, simply choose $0 \leq r < p$ such that $rn \equiv l \pmod{p}$.

Applying this observation to each ζ_i, we can write $\zeta_i = z^{r_i} x_i$ with x_i, r_i as above. We have $x_1 = 1$ and $r_1 = 0$. The equation $\sum_{i=1}^k a_i\zeta_i = 0$ can be written $\sum_{l=0}^{p-1} b_l z^l$, where $b_l = \sum_{r_i = l} a_i x_i$. Note that $b_l \in \mathbb{Q}(e^{\frac{2i\pi p}{m}})$. On the other hand, we can compute the degree of z over $\mathbb{Q}(e^{\frac{2i\pi p}{m}})$. Indeed, observe that $\mathbb{Q}(z, e^{\frac{2i\pi p}{m}}) = \mathbb{Q}(e^{\frac{2i\pi}{m}})$, so that

$$[\mathbb{Q}(e^{\frac{2i\pi p}{m}})(z) : \mathbb{Q}(e^{\frac{2i\pi p}{m}})] = \frac{[\mathbb{Q}(e^{\frac{2i\pi}{m}}) : \mathbb{Q}]}{[\mathbb{Q}(e^{\frac{2i\pi p}{m}}) : \mathbb{Q}]} = \frac{\varphi(m)}{\varphi(m/p)} = \frac{\varphi(p^j)}{\varphi(p^{j-1})}$$

and the last quantity is $p - 1$ for $j = 1$ and p otherwise.

Note that $\sum_{l=0}^{p-1} b_l X^l$ is not the zero polynomial, since otherwise we obtain the relation $\sum_{r_i=l} a_i\zeta_i = 0$ for all $0 \leq l < p$. But the hypothesis yields then $\{i | r_i = l\} = \emptyset$ or $\{1, 2, \dots, k\}$ for all l. As $r_1 = 0$, this gives $r_i = 0$ for all i and so $\zeta_i^{\frac{m}{p}} = 1$ for all i, contradicting the minimality of m.

If we combine the results of the previous two paragraphs, we see that we must have $j = 1$, as z is killed by the nonzero polynomial $\sum_{l=0}^{p-1} b_l X^l$, of degree at most $p-1$. But then z has degree $p-1$ over $\mathbb{Q}(e^{\frac{2i\pi p}{m}})$ (as follows from the previous computation) and so $\sum_{l=0}^{p-1} b_l X^l$ is the minimal polynomial of z over $\mathbb{Q}(e^{\frac{2i\pi p}{m}})$. As z is also killed by $1 + X + \cdots + X^{p-1}$, we deduce that these two polynomials differ by a constant. In particular, all b_l are nonzero. So for all $0 \le l < p$ one can find i such that $r_i = l$. Clearly, this implies that $p \le k$ and the proof is finished.

b) Fix a solution $z = \sum_{i=1}^{k} a_i \zeta_i$ of the equation and consider another solution, say $z = \sum_{i=1}^{k} b_i z_i$. Thus

$$\sum_{i=1}^{k} a_i \zeta_i - \sum_{i=1}^{k} b_i z_i = 0,$$

but one has to be a little bit careful, as this relation does not necessarily satisfy the conditions of (a). However, if we fix $1 \le i \le k$, we can find a minimal sub-relation of the previous relation which contains z_i. By hypothesis, such a sub-relation must contain some ζ_j. As the length of this sub-relation is at most $2k$ and as it clearly satisfies the hypothesis of (a), we deduce that $z_i^m = \zeta_j^m$ for all ζ_j in this sub-relation. Here $m = \prod_{p \le 2k} p$. So, for any i, z_i can take at most km values and so the number of solutions of the equation in $z_1, z_2, \ldots, z_k$ is at most $(km)^k$. It remains to use Erdős's famous inequality (theorem 3.A.3) $\prod_{p \le n} p \le 4^n$ to conclude. □

Remark 9.8. Let $a_1, a_2, \ldots, a_n$ be nonzero complex numbers and consider the equation $a_1 z_1 + a_2 z_2 + \cdots + a_n z_n = 1$. A non-degenerate solution is an n-tuple $(z_1, z_2, \ldots, z_n)$ of roots of unity which satisfies the equation and such that $\sum_{i \in I} a_i z_i \neq 0$ for any nonempty subset I of $\{1, 2, \ldots, n\}$. Conway and Jones [20] improved Mann's theorem by proving that if $a_i \in \mathbb{Q}$, then for any non-degenerate solution we have $z_1^d = z_2^d = \cdots = z_n^d = 1$ where d is the product of primes $p_1, p_2, \ldots, p_s$ such that $\sum_{i=1}^{s} (p_i - 2) \le n - 1$. Also, in [30] the author proves using rather elementary and very beautiful arguments that there are at most $(n+1)^{3(n+1)^2}$ non-degenerate solutions of the equation.

9.3 The gcd trick

The division algorithm shows that if $K \subset L$ are fields and if $f, g \in K[X]$ are two polynomials, then their gcd is the same if we see f, g as polynomials with coefficients in K or with coefficients in L. That is, the greatest common divisor of two polynomials is not sensitive to the field in which the coefficients of these polynomials live. Combining this observation with Gauss' lemma, we also obtain that if f and g are monic polynomials with integer coefficients, then their gcd computed in $\mathbb{Q}[X]$ has integer coefficients. This gives a very indirect, but sometimes very useful way to prove the rationality or integrality of a real number x: it is enough to exhibit $X - x$ as the gcd of two polynomials with rational coefficients (respectively of two monic polynomials with integer coefficients). The next problems in this section illustrate this trick.

8. Let a, b be two positive rational numbers such that for some $n \geq 2$ the number $\sqrt[n]{a} + \sqrt[n]{b}$ is rational. Prove that $\sqrt[n]{a}$ is also rational.

Marius Cavachi, Gazeta Matematică

Proof. Let us write $\sqrt[n]{a} + \sqrt[n]{b} = c$ for some (positive) rational number c. Then $\sqrt[n]{a}$ is a root of $X^n - a$ and also of $(c - X)^n - b$. The key point is that it is the **unique** common root of these polynomials. Indeed, if z is a common root, then we can write $z = \sqrt[n]{a} z_1$ and $c - z = \sqrt[n]{b} z_2$ for some nth roots of unity z_1, z_2. We deduce that $\sqrt[n]{a} + \sqrt[n]{b} = \sqrt[n]{a} z_1 + \sqrt[n]{b} z_2$. Since $|z_i| = 1$, the real parts of z_1, z_2 are at most 1. Passing to real parts in the previous equality then implies that $z_1 = z_2 = 1$ and the claim is proved. Now, since the two polynomials don't have multiple roots, it follows that $\gcd(X^n - a, (c - X)^n - b) = X - \sqrt[n]{a}$. The result follows now from the gcd trick. □

9. Let m, n be relatively prime numbers and let $x > 1$ be a real number such that $x^m + \frac{1}{x^m}$ and $x^n + \frac{1}{x^n}$ are integers. Prove that $x + \frac{1}{x}$ is also an integer.

Proof. Let $a = x^m + \frac{1}{x^m}$ and $b = x^n + \frac{1}{x^n}$ and consider the polynomials

$$p(X) = X^{2m} - aX^m + 1 = (X^m - x^{-m})(X^m - x^m)$$

and

$$q(X) = X^{2n} - bX^n + 1 = (X^n - x^{-n})(X^n - x^n).$$

The crucial claim is that

$$\gcd(p, q) = X^2 - (x + x^{-1})X + 1 = (X - x)(X - x^{-1}).$$

Assuming this for a moment, we can conclude that $x + x^{-1}$ is an integer by the gcd trick.

It remains to establish the claim and for that it is enough to prove that x and x^{-1} are the only common zeros of p, q (since clearly p, q have no multiple root). But if z is a common zero, we have $z^m = x^m$ or $z^m = x^{-m}$ and similarly $z^n = x^n$ or $z^n = x^{-n}$. We may assume (by changing z and z^{-1}) that $z^m = x^m$, so that $|z| > 1$. Then clearly we must have $z^n = x^n$. But then z/x is a root of unity whose order divides both m and n. Since $\gcd(m, n) = 1$, it follows that $z = x$ and we are done. □

The following problem is very similar to the previous problem, but a bit more difficult.

10. Let $\theta \in (0, \pi/2)$ be an angle such that $\cos\theta$ is irrational. Suppose that $\cos k\theta$ and $\cos[(k+1)\theta]$ are rational for some positive integer k. Prove that $\theta = \pi/6$.

USA TST 2007

Proof. We will actually prove more: it is enough to replace $k+1$ by any integer l which is relatively prime to k. The key point is the following

Lemma 9.9. *If $\cos k\theta$ and $\cos l\theta$ are rational for relatively prime positive integers k, l, then either $\cos\theta$ is rational or θ is a rational multiple of π.*

Proof. If $\cos k\theta = p$ and $\cos l\theta = q$, then $e^{i\theta}$ is a common root of the polynomials

$$f(X) = X^{2k} - 2pX^k + 1, \quad g(X) = X^{2l} - 2qX^l + 1.$$

On the other hand, it is not difficult to check that if θ is not a rational multiple of π, then $e^{i\theta}$ and $e^{-i\theta}$ are the only common roots of f and g. Indeed, all roots of f are $e^{\pm i\theta+\frac{2\pi ij}{k}}$ for $0\leq j<k$ and all roots of g are $e^{\pm i\theta+\frac{2\pi ij}{l}}$ with $0\leq j<l$. On the other hand, since $\gcd(k,l)=1$, the only solution of the equation $e^{\pm i\theta+\frac{2\pi ij_1}{k}}=e^{\pm i\theta+\frac{2\pi ij_2}{l}}$ with $0\leq j_1<k$ and $0\leq j_2<l$ (for some choices of signs) is $j_1=j_2=0$. This proves that the greatest common divisor of f and g is precisely $(X-e^{i\theta})(X-e^{-i\theta})=X^2-2\cos\theta X+1$, thus $\cos\theta$ is rational. □

Coming back to the proof, the previous lemma shows that θ is a rational multiple of π. On the other hand, we saw in section 9.2 that the only rational numbers $r\in[0,1]$ such that $\cos r\pi$ is rational are $r=0,\frac{1}{3},\frac{1}{2},\frac{2}{3},1$. We deduce that $k\theta$ and $l\theta$ are integer multiples of $\frac{\pi}{6}$. Since $\gcd(k,l)=1$, Bézout's lemma implies that θ is an integer multiple of $\frac{\pi}{6}$. Since $\cos\theta$ is irrational, we deduce that $\theta=\frac{\pi}{6}$. □

9.4 The theorem of symmetric polynomials

The proof of the following result is quite elementary, but the result itself is incredibly powerful and useful. If R is a commutative ring and if $f\in R[X_1,X_2,\ldots,X_n]$ is a polynomial, we say that f is symmetric if for all permutations σ of $\{1,2,\ldots,n\}$ we have

$$f(X_1,\ldots,X_n)=f(X_{\sigma(1)},\ldots,X_{\sigma(n)}).$$

Recall that the fundamental symmetric polynomials are

$$\sigma_k=\sum_{1\leq i_1<i_2<\cdots<i_k\leq n}X_{i_1}X_{i_2}\cdot\dots\cdot X_{i_k},$$

for $1\leq k\leq n$. We have the equality

$$(t+X_1)(t+X_2)\cdot\dots\cdot(t+X_n)=t^n+\sigma_1\cdot t^{n-1}+\cdots+\sigma_n\in R[t,X_1,\ldots,X_n].$$

Theorem 9.10. *(Fundamental theorem of symmetric polynomials.) Let R be a commutative ring and let $f\in R[X_1,\ldots,X_n]$ be a symmetric polynomial. Then there is $g\in R[X_1,\ldots,X_n]$ such that $f(X_1,\ldots,X_n)=g(\sigma_1,\sigma_2,\ldots,\sigma_n)$.*

Proof. We will use induction on n and inside the induction step an induction on $\deg(f)$. For $n = 1$ everything is clear, so assume the result holds for $n - 1$. We now prove by induction on $\deg(f)$ the assertion of the theorem with n variables. If $\deg(f) = 0$ or 1, everything is clear. It is clear that the polynomial $g(X_1, \dots, X_{n-1}) = f(X_1, \dots, X_{n-1}, 0)$ is still symmetric, so by (the first) induction it is a polynomial of the form $h(X_1 + \dots + X_{n-1}, \dots, X_1 \cdots X_{n-1})$ for some $h \in R[X_1, \dots, X_{n-1}]$. Note that the difference

$$f(X_1, \dots, X_n) - h(X_1 + \cdots + X_n, \dots, X_2 \cdots X_n + \cdots + X_1 X_2 \cdots X_{n-1})$$

vanishes when $X_n = 0$ and is a symmetric polynomial. Therefore this polynomial is a multiple of $X_1 \cdots X_n$. Applying the inductive hypothesis to the quotient between this polynomial and $X_1 \cdots X_n$ (which has degree less than $\deg f$), the result follows. □

Remark 9.11. It is not difficult to prove that the polynomial g is unique. This means that there are no algebraic relations between the polynomials $\sigma_1, \sigma_2, \dots, \sigma_n$.

Remark 9.12. The theorem also implies that any symmetric rational function $f \in R(X_1, X_2, \dots, X_n)$ is a rational function in the σ_i's. Indeed, let

$$\sigma \cdot P(X_1, X_2, \dots, X_n) = P(X_{\sigma(1)}, X_{\sigma(2)}, \dots, X_{\sigma(n)})$$

for $P \in R[X_1, X_2, \dots, X_n]$. Then we can write

$$f = \frac{P}{Q} = \frac{P_1}{\prod_{\sigma \in S_n} \sigma \cdot Q}$$

for some polynomials P, Q, P_1. Since f is symmetric, so is P_1.

The result follows from the theorem of symmetric polynomials applied to P_1 and to $\prod_\sigma \sigma \cdot Q$.

Remark 9.13. We refer the reader to [66], chapter 5 for the proof of the following theorem of Lagrange: let K be a field of characteristic 0. If $f \in K(X_1, X_2, \dots, X_n)$, let G_f be the set (actually group) of those permutations $\sigma \in S_n$ such that $f(X_1, X_2, \dots, X_n) = f(X_{\sigma(1)}, X_{\sigma(2)}, \dots, X_{\sigma(n)})$. If

$f, g \in K(X_1, X_2, \ldots, X_n)$ satisfy $G_f \subset G_g$, then one can find a rational function h whose coefficients are symmetric polynomials in $X_1, X_2, \ldots, X_n$ such that $g = h(f)$.

A very important consequence of theorem 9.10 is the following result, that will be constantly used in this section.

Corollary 9.14. *a) Let $f \in \mathbb{Q}[X_1, X_2, \ldots, X_n]$ be a symmetric polynomial and let $g \in \mathbb{Q}[X]$ be a polynomial of degree n, with complex roots $z_1, z_2, \ldots, z_n$. Then $f(z_1, z_2, \ldots, z_n) \in \mathbb{Q}$.*

b) If f has integer coefficients and if g is monic with integer coefficients, then $f(z_1, z_2, \ldots, z_n)$ is an integer.

Proof. Using theorem 9.10, we can write

$$f(X_1, \ldots, X_n) = h(\sigma_1, \sigma_2, \ldots, \sigma_n)$$

for some $h \in \mathbb{Q}[X_1, \ldots, X_n]$ (resp $\mathbb{Z}[X_1, \ldots, X_n]$). The result follows from the fact that $\sigma_i(z_1, z_2, \ldots, z_n)$ are rational (respectively integers), because the coefficients of g are so. □

Another very useful result is the following generalization of Fermat's little theorem.

Corollary 9.15. *Let $f \in \mathbb{Z}[X]$ be a monic polynomial with complex roots $z_1, z_2, \ldots, z_n$ (multiplicities counted) and let p be a prime number. Then*

$$z_1^p + z_2^p + \cdots + z_n^p \equiv (z_1 + z_2 + \cdots + z_n)^p \pmod{p}.$$

Proof. Corollary 9.14 implies that both sides are integers. Consider the quotient by p of the difference between the left-hand side and the right-hand side. Using the multinomial formula, it is easy to see that this quotient is a symmetric polynomial with integer coefficients in $z_1, z_2, \ldots, z_n$, thus the result follows from corollary 9.14. □

Here is a nice application of the previous corollary. It was one of the difficult problems given in the Romanian IMO Team Selection Tests in 2004.

11. Let a, b, c be integers. Define the sequence $(x_n)_{n\geq 0}$ by $x_0 = 4$, $x_1 = 0$, $x_2 = 2c$, $x_3 = 3b$ and $x_{n+3} = ax_{n-1} + bx_n + cx_{n+1}$. Prove that for any prime p and any positive integer m, the number x_{p^m} is divisible by p.

Călin Popescu, Romanian TST 2004

Proof. Let r_1, r_2, r_3, r_4 be the roots of the characteristic polynomial of the recurrence relation, namely $X^4 - cX^2 - bX - a$. The crucial point and by far the hardest step in the proof is to realize that

$$x_n = r_1^n + r_2^n + r_3^n + r_4^n$$

for all n. This is suggested by $x_0 = 4$ and by the fact that problem creators tend to try to be sneaky.[1] Proving the previous formula is immediate by induction, once we prove it for $n = 0, 1, 2, 3$. For $n = 0, 1$ this is trivial, for $n = 2$ follows from the identity

$$\sum r_i^2 = \left(\sum r_i\right)^2 - 2\sum_{i<j} r_i r_j = 2c$$

and for $n = 3$ we can use the recursive relation (since it is easy to see that $y_n = r_1^n + r_2^n + r_3^n + r_4^n$ together with $y_{-1} = -b/a$ satisfies the same recursive relation as x_n). With this closed formula for the general term of the sequence, we need to prove that $\sum r_i^{p^m}$ is a multiple of p. Since $\sum r_i = 0$, the result follows from corollary 9.15 and by induction on m. □

Let us consider now a few more or less direct applications of theorem 9.10 and of corollary 9.14.

12. a) Let P, R be polynomials with rational coefficients such that $P \neq 0$. Prove that there exists a non-zero polynomial $Q \in \mathbb{Q}[X]$ such that $P(X)|Q(R(X))$

    b) Let P, R be polynomials with integer coefficients and suppose that P is monic. Prove that there exists a monic polynomial $Q \in \mathbb{Z}[X]$ such that $P(X)|Q(R(X))$

Iranian Olympiad 2006

[1] As Richard Stong kindly remarks...

Proof. The idea is very natural: the first condition that should be satisfied in order to have $P(X)|Q(R(X))$ is that for each root z of P we have $Q(R(z)) = 0$. Therefore, if $x_1, x_2, \ldots, x_n$ are the roots of P (some of the x_i's may be equal), then we would like to have $Q(R(x_i)) = 0$. The most natural choice is to take

$$Q(X) = \prod_{i=1}^{n}(X - R(x_i)).$$

Note that it satisfies $P(X)|Q(R(X))$, because $X - x_i$ divides $R(X) - R(x_i)$ for all i. It remains to check that Q has rational (respectively integer, for the second part of the problem) coefficients. This follows from corollary 9.14. □

13. a) Let $a_1, a_2, \ldots, a_m, b_1, b_2, \cdots, b_n \in \mathbb{C}$ be such that

$$f_1(X) = \prod_{i=1}^{m}(X - a_i), \quad f_2(X) = \prod_{i=1}^{n}(X - b_i) \in \mathbb{Z}[X].$$

Suppose that there exist $g_1, g_2 \in \mathbb{Z}[X]$ such that $f_1g_1 + f_2g_2 = 1$. Prove that:

$$\left|\prod_{i=1}^{m}\prod_{j=1}^{n}(a_i - b_j)\right| = 1.$$

b) If a_i, b_i are integers and

$$\left|\prod_{i=1}^{m}\prod_{j=1}^{n}(a_i - b_j)\right| = 1,$$

prove that there exist polynomials $g_1, g_2 \in \mathbb{Z}[X]$ such that $f_1g_1 + f_2g_2 = 1$.

Ibero-American Olympiad

Proof. a) Note that the relation to be proved can also be written as

$$\left|\prod_{i=1}^{m} f_2(a_i)\right| = 1.$$

Evaluating the relation $f_1g_1 + f_2g_2 = 1$ at a_i yields $f_2(a_i)g_2(a_i) = 1$. Thus

$$\left|\prod_{i=1}^{m} f_2(a_i)\right| \cdot \left|\prod_{i=1}^{m} g_2(a_i)\right| = 1.$$

On the other hand, $\prod_{i=1}^{m} f_2(a_i)$ and $\prod_{i=1}^{m} g_2(a_i)$ are integers, by corollary 9.14. These two observations are enough to conclude.

b) Note that $|a_i - b_j| = 1$ for all i, j, because a_i, b_j are integers. It is then immediate that we have only two possible cases:

1) A, B are singletons of the form $\{a\}, \{a+1\}$ or $\{a+1\}, \{a\}$.

2) $A = \{a\}$ and $B = \{a-1, a+1\}$ for some integer a or $B = \{a\}$ and $A = \{a-1, a+1\}$.

Thus, by symmetry in A and B and by making a translation of the variable $X \to X - a$, it is enough to consider the cases when $A = \{0\}, B = \{1\}$ and $A = \{0\}, B = \{-1, 1\}$. In each case f_1 divides some X^n and f_2 divides some $(X^2-1)^m$. Thus f_1 divides X^{2^k} and f_2 divides $(X^{2^k}-1)^n$ for k, n sufficiently large. It is thus enough to find Bézout relations with integral coefficients for the polynomials X^{2^k} and $(X^{2^k}-1)^n$. But this is immediate. □

Remark 9.16. The assumption that a_i and b_i are integers is useless. Here is a proof, due to Richard Stong. We advise the reader not familiar with the notion of resultant to read the discussion before problem 27 in chapter 12. It is not too difficult to check that the resultant of f_1 and f_2 is $\prod_{i=1}^{m}\prod_{j=1}^{n}(b_j - a_i) = \pm 1$. But then the map

$$\varphi : \mathbb{Z}[X]_{\deg(f_2)-1} \times \mathbb{Z}[X]_{\deg(f_1)-1} \to \mathbb{Z}[X]_{\deg(f_1)+\deg(f_2)-1}$$

defined by $\varphi(g_1, g_2) = g_1(X)f_1(X) + g_2(X)f_2(X)$ is invertible, thus we can find $g_1, g_2 \in \mathbb{Z}[X]$ such that $f_1g_1 + f_2g_2 = 1$.

A classical problem is to prove that

$$|a + b\sqrt{2}| \geq \frac{1}{\sqrt{2}(|a| + |b|)}$$

for any integers a, b, not both of them equal to 0. The idea is that it is not clear how to deal with $|a + b\sqrt{2}|$ directly, but it is very easy to say something about the product of this number and its conjugate $|a - b\sqrt{2}|$. Indeed, this product is a nonzero integer, thus at least 1. The result follows immediately. With a similar idea, it is not difficult to prove the following absolutely classical theorem of Liouville: if z is an algebraic irrational number of degree d, then there exists $c > 0$ such that for all integers p and q we have $\left|z - \frac{p}{q}\right| > \frac{c}{|q|^d}$. That is, irrational algebraic numbers are badly approximable with rational numbers. A much deeper result, for which Roth won the Fields medal, is that we can improve the previous inequality to $\left|z - \frac{p}{q}\right| > \frac{c(\varepsilon)}{|q|^{2+\varepsilon}}$ for all $\varepsilon > 0$. The following problems use this trick of multiplying by conjugates and estimating the conjugates, but they are much more challenging than the very simple example discussed above.

14. Let k, n be positive integers and let $P(X)$ be a polynomial of degree n with all coefficients in the set $\{-1, 0, 1\}$. Suppose that $(X-1)^k | P(X)$ and that there exists a prime q such that $\frac{q}{\ln q} < \frac{k}{\ln(n+1)}$. Prove that the primitive complex roots of unity of order q are roots of P.

IMC 2001

Proof. The problem looks rather complicated because of the strange inequality imposed on q. Let us forget first about that and consider the product of all values of P at the primitive qth roots of unity, $\prod_{i=1}^{q-1} P(z_i)$. This is an integer, by corollary 9.14. If it is not 0, then $\prod_{i=1}^{q-1} |P(z_i)| \geq 1$. However, by assumption there exists a polynomial Q (necessarily with integer coefficients) such that $P(X) = (1-X)^k Q(X)$. Therefore

$$\prod_{i=1}^{q-1} P(z_i) = \prod_{i=1}^{q-1} (1-z_i)^k \cdot \prod_{i=1}^{q-1} Q(z_i).$$

Since

$$X^{q-1} + X^{q-2} + \cdots + X + 1 = \prod_{i=1}^{q-1} (X - z_i),$$

we have $\prod_{i=1}^{q-1}(1-z_i) = q$ and so

$$\prod_{i=1}^{q-1} P(z_i) = q^k \prod_{i=1}^{q-1} Q(z_i).$$

But the same argument as before shows that $\prod_{i=1}^{q-1} Q(z_i)$ is a nonzero integer, therefore $\prod_{i=1}^{q-1} |P(z_i)| \geq q^k$. This is however impossible, since by assumption we have $|P(z_i)| \leq n+1$, therefore

$$\prod_{i=1}^{q-1} |P(z_i)| \leq (n+1)^{q-1} < q^k.$$

The previous arguments show that P vanishes at one of the primitive roots of unity of order q. But since the polynomial $1 + X + \cdots + X^{q-1}$ is irreducible over the rational numbers, if P vanishes at a primitive root of unity of order q, then it also vanishes at all the other roots. This ends the proof. □

One needs some gymnastics if one wants to avoid the use of Galois theory for the following problem.

15. Let p be a prime and let $n_1, n_2, \ldots, n_k$ be integers. Define

$$S = \left| \sum_{j=1}^{k} \cos \frac{2\pi n_j}{p} \right|.$$

Prove that either $S = 0$ or $S \geq k \left(\frac{1}{2k}\right)^{\frac{p-1}{2}}$.

Holden Lee

Proof. Let $z = e^{\frac{2i\pi}{p}}$. The crucial ingredient in the proof is the following:

Lemma 9.17. *The number*

$$N = \prod_{l=1}^{\frac{p-1}{2}} \left(\sum_{j=1}^{k} \left(z^{l \cdot n_j} + z^{-l \cdot n_j} \right) \right)$$

is an integer. Moreover, $N = 0$ if and only if $S = 0$.

Let us admit this for a moment and see how we can finish the proof. Assume that $S \neq 0$. Then $|N| \geq 1$. So

$$2^{\frac{1-p}{2}} \leq \frac{|N|}{2^{\frac{p-1}{2}}} = |S| \cdot \prod_{l=2}^{\frac{p-1}{2}} \left| \sum_j \cos \frac{2\pi l n_j}{p} \right| \leq |S| \cdot k^{\frac{p-1}{2}-1}$$

and the conclusion follows.

Now, let us prove the lemma. As is well-known (and easily proved by induction) there are polynomials $F_j \in \mathbb{Z}[X]$ of degree j such that $X^j + X^{-j} = F_j(X + X^{-1})$. Let $F = \sum_{j=1}^k F_{n_j}$. Then $N = \prod_{l=1}^{\frac{p-1}{2}} F(z^l + z^{-l})$. The lemma will be proved if we prove the following result:

Lemma 9.18. *The minimal polynomial of $z + z^{-1}$ is*

$$f(X) = \prod_{l=1}^{\frac{p-1}{2}} (X - (z^l + z^{-l})) = F_{\frac{p-1}{2}} + F_{\frac{p-3}{2}} + \cdots$$

Proof. Note that

$$\begin{aligned} f(X + X^{-1}) &= \prod_{l=1}^{\frac{p-1}{2}} \frac{(X - z^l)(X - z^{-l})}{X} \\ &= X^{-\frac{p-1}{2}} \prod_{l=1}^{\frac{p-1}{2}} (X - z^l)(X - z^{p-l}) \\ &= \frac{1 + X + \cdots + X^{p-1}}{X^{\frac{p-1}{2}}}. \end{aligned}$$

Thus $f = F_{\frac{p-1}{2}} + F_{\frac{p-3}{2}} + \cdots$ by definition of the polynomials F_j. In particular, f has integer coefficients and so N is an integer (by the fundamental theorem of symmetric polynomials). Moreover, f has degree $\frac{p-1}{2}$ and vanishes at $z + z^{-1}$. But

$$[\mathbb{Q}(z + z^{-1}) : \mathbb{Q}] = \frac{[\mathbb{Q}(z) : \mathbb{Q}]}{[\mathbb{Q}(z) : \mathbb{Q}(z + z^{-1})]} \geq \frac{p-1}{2},$$

so f must be the minimal polynomial of $z + z^{-1}$. □

Now, assume that $N = 0$. Thus, there exists $1 \leq l \leq \frac{p-1}{2}$ such that $F(z^l + z^{-l}) = 0$. By the lemma, F is a multiple of f and so F vanishes at $z + z^{-1}$. But this means that $S = 0$, a contradiction. Thus, we have proved the crucial claim and the result follows. □

The following result is certainly classical, but it is rather difficult to find an elementary proof in the literature. We follow one of the approaches proposed in the beautiful article [10].

16. Let $a_1, a_2, \ldots, a_n$ be positive rational numbers and let $k_1, k_2, \ldots, k_n$ be integers greater than 1. If $a_1^{1/k_1} + a_2^{1/k_2} + \cdots + a_n^{1/k_n}$ is a rational number, then any term of the previous sum is also a rational number.

Proof. It is clearly enough to prove the following result: let $k > 1$ and suppose that the positive rational numbers $a_1, \ldots, a_n, b_1, \ldots, b_n$ satisfy

$$a_1 \sqrt[k]{b_1} + \cdots + a_n \sqrt[k]{b_n} \in \mathbb{Q}.$$

Then $\sqrt[k]{b_i} \in \mathbb{Q}$ for all i.

Let

$$A_i = \{\text{roots of } X^k - a_i^k b_i\} = \{\omega^j a_i \sqrt[k]{b_i} \mid 1 \leq j \leq k\},$$

where ω is a primitive root of order k of 1. Also, let

$$S = \sum_{i=1}^{n} a_i \sqrt[k]{b_i}$$

and

$$P(X) = \prod_{x_2 \in A_2, \ldots, x_n \in A_n} (S - X - x_2 - \cdots - x_n).$$

By theorem 9.14 we have $P \in \mathbb{Q}[X]$. Note that $P(a_1 \sqrt[k]{b_1}) = 0$. Let d be the least positive divisor of k for which $\sqrt[k]{b_1^d} \in \mathbb{Q}$ (it exists, as $\sqrt[k]{b_1^k} \in \mathbb{Q}$). If we manage to prove that $d = 1$, it will follow that $\sqrt[k]{b_1} \in \mathbb{Q}$, so we can delete the first term of S and conclude by induction on n. So, let us prove that $d = 1$. By definition, we can write $a_1 \sqrt[k]{b_1} = \sqrt[d]{x}$ with $x \in \mathbb{Q}_+$. The crucial fact is the following:

Lemma 9.19. *$X^d - x$ is irreducible in $\mathbb{Q}[X]$.*

Proof. If F is a monic polynomial with rational coefficients of degree between 1 and $d-1$ that divides $X^d - x$, all roots of F have absolute value $\sqrt[d]{x}$ and so $|F(0)| = (\sqrt[d]{x})^{\deg(F)}$ is a rational number, that is $\sqrt[k]{b_1^{\deg(F)}} \in \mathbb{Q}$, contradicting the minimality of d. □

Since $P(\sqrt[d]{x}) = 0$, the previous lemma yields $X^d - x \mid P$ in $\mathbb{Q}[X]$. Thus, if z is a primitive root of order d of 1, we have $P(z\sqrt[d]{x}) = 0$ and so there are $(x_2, \ldots, x_n) \in A_2 \times A_3 \times \cdots \times A_n$ with $S - z\sqrt[d]{k} = x_2 + \cdots + x_n$. If $d \geq 2$, then $\mathrm{Re}(z) < 1$, so

$$\begin{aligned}\mathrm{Re}(S) &= S \\ &= \mathrm{Re}(z\sqrt[d]{x} + x_2 + \cdots + x_n) \\ &\leq \mathrm{Re}(z\sqrt[d]{x}) + \sum_{i=2}^{n} |x_i| \\ &= \mathrm{Re}(z\sqrt[d]{x}) + \sum_{i=2}^{n} a_i\sqrt[k]{b_i} \\ &< \sqrt[d]{x} + \sum_{i=2}^{n} a_i\sqrt[k]{b_i} \\ &= S,\end{aligned}$$

a contradiction. So $d = 1$ and $a_1\sqrt[k]{b_1} \in \mathbb{Q}$. □

9.5 Ideal theory and local methods

We strongly advise the reader not familiar with algebraic number theory to read the appendices on number fields and p-adic numbers before reading this section, which is short but rather challenging.

We start with a beautiful result of Polya concerning linear recurrence sequences. Recall that a sequence $(a_n)_n$ is called a linear recurrence sequence if one can find d and $x_1, \ldots, x_d$ such that $a_{n+d} + x_1 a_{n+d-1} + \cdots + x_d a_n = 0$ for all n.

17. Suppose that $(a_n)_{n\geq 1}$ is a linear recurrence sequence of integers such that n divides a_n for all positive integers n. Prove that $\left(\frac{a_n}{n}\right)$ is also a linear recurrence sequence.

Polya

Proof. By the general theory of linear recurrence sequences, we can find distinct nonzero algebraic numbers $z_1, z_2, \ldots, z_m$ and polynomials $f_1, f_2, \ldots, f_m$ with algebraic coefficients such that

$$a_n = f_1(n)z_1^n + f_2(n)z_2^n + \cdots + f_m(n)z_m^n$$

for all n. We will prove that if $n|a_n$ for all n, then $f_i(0) = 0$ for all i, from which the result follows easily.

Let K be the field obtained by adjoining to $\mathbb{Q}$ all z_i's and all coefficients of the polynomials f_i. Choose a prime p which does not divide any of the norms of the (nonzero) coefficients of f_i's or the norms of one of the z_i's. All sufficiently large primes satisfy this property. Fix such a prime p and consider a prime ideal I of K over p, with norm $N(I) = p^f$. Impose the condition that jp^f divides a_{jp^f}. Note that

$$a_{jp^f} \equiv \sum_{i=1}^{m} f_i(0)z_i^j \pmod{I},$$

since $z_i^{p^f} \equiv z_i \pmod{I}$ and since $p \in I$. Thus, we must have

$$\sum_{i=0}^{m} f_i(0)z_i^j \equiv 0 \pmod{I}$$

for all $j = 0, 1, \ldots, m-1$. Seeing this as a linear system in the $f_i(0)$'s, it follows that $f_i(0) \in I$ for all i, unless I divides the determinant of the matrix associated to this system. However, this is a Vandermonde determinant in the z_i's and so, if we ensure that I and $\prod_{i\neq j}(z_i - z_j)$ are relatively prime, we will be able to conclude that $f_i(0) \in I$. But to ensure the last property, it is enough to choose a prime p which does not divide the norm of the algebraic number $\prod_{i\neq j}(z_i - z_j)$. Again, all sufficiently large primes have this property.

The previous paragraph shows that we can find infinitely many primes p and for each such prime an ideal I over p such that $f_i(0) \in I$ for all i. But then p will divide the norm of $f_i(0)$ for infinitely many primes p and so $f_i(0) = 0$ for all i. It is then clear that $\frac{a_n}{n}$ is still a linearly recurrence sequence. □

We present two approaches for the following challenging problem: a rather exotic elementary one and a more advanced approach which uses standard facts about number fields and their p-adic completions.

18. Let $a_1, a_2, \ldots, a_n$ be complex numbers such that $a_1^m + a_2^m + \cdots + a_n^m$ is an integer for all positive integers m. Prove that $(X - a_1)(X - a_2) \cdots (X - a_n) \in \mathbb{Z}[X]$.

Michael Larsen, AMM E 2993

Proof. Let
$$\sigma_k = \sum_{1 \le i_1 < i_2 < \cdots < i_k \le n} a_{i_1} a_{i_2} \cdots a_{i_k}$$
and $P_k = a_1^k + a_2^k + \cdots + a_n^k$, so Newton's identities[2] can be written (for $1 \le k \le n$)
$$P_k - \sigma_1 P_{k-1} + \sigma_2 P_{k-2} - \cdots + (-1)^k k \sigma_k = 0.$$

It follows immediately from these relations that if $P_k \in \mathbb{Z}$ for all k, then $\sigma_k \in \frac{1}{n!}\mathbb{Z}$ for all $1 \le k \le n$. In particular, $(X - a_1)(X - a_2) \cdots (X - a_n) \in \frac{1}{n!}\mathbb{Z}[X]$ and so $n! \cdot a_i$ are algebraic integers. Observe that if $a_1, a_2, \ldots, a_n$ satisfy the conditions of the problem, then so do $a_1^r, a_2^r, \ldots, a_n^r$ for all $r \ge 1$. We deduce that $n! a_i^r$ is an algebraic integer for all r. The next lemma shows that all a_i's are algebraic integers, so the coefficients of $(X - a_1)(X - a_2) \cdots (X - a_n)$ are algebraic integers. Since these coefficients are rational numbers (this has already been established), they must be integers and the result follows.

Lemma 9.20. *Let n be a positive integer and a be an algebraic number. If na^k is an algebraic integer for all positive integers k, then a is an algebraic integer.*

---

[2]See the remark 9.22 for a proof of these.

Proof. If d_i is the degree of the algebraic number a^i, it is clear that $d_{2^k} \geq d_{2^{k+1}}$ (because $\mathbb{Q}(a^{2^{k+1}}) \subset \mathbb{Q}(a^{2^k})$). Thus there exists an integer j and a positive integer d such that $d_{2^k} = d$ for all $k \geq j$. Let $a_1 = a^{2^j}, a_2, \ldots, a_d$ denote the conjugates of a^{2^j} and choose a positive integer c such that $f_0 = c(X - a_1)\cdots(X - a_d) \in \mathbb{Z}[X]$ is primitive. Then $g_0 = c(X + a_1)\cdots(X + a_d) \in \mathbb{Z}[X]$ is also primitive, so by an easy application of Gauss' lemma $f_1 = c^2(X - a_1^2)\cdots(X - a_d^2) \in \mathbb{Z}[X]$ is also primitive. Since a_1^2 has degree d and since $\deg f_1 = d$, it follows that f_1 is irreducible over $\mathbb{Q}$. Repeating this argument, we obtain that $f_r = c^{2^r}(X - a_1^{2^r})\cdots(X - a_d^{2^r}) \in \mathbb{Z}[X]$ is primitive and irreducible. Next, since $na_1^{2^r}$ is an algebraic integer, we have $h_r = n^d(X - a_1^{2^r})\cdots(X - a_d^{2^r}) = (nX - na_1^{2^r})\cdots(nX - na_d^{2^r}) \in \mathbb{Z}[X]$, so we must have $\frac{n^d}{c^{2^r}} \in \mathbb{Z}$. Since this happens for all sufficiently large r, it follows that $c = \pm 1$ and so a^{2^j} is an algebraic integer. As a result, a is an algebraic integer and we're done. □

Proof. This proof uses rather heavy material, but it is much more conceptual than the previous one. Namely, we will use a local-global principle, stating that an algebraic number x is an algebraic integer if and only if $v(x) \geq 0$ for any valuation v on $\overline{\mathbb{Q}}$. This follows easily from the relations between a number field and its completions (see the addendum on number fields), but the result is not obvious at all. Anyway, once we have this, the lemma is immediate: if v is a valuation, then we know that $v(n) + kv(a) \geq 0$ for all k. Dividing by k and making $k \to \infty$ yields $v(a) \geq 0$, which is enough to ensure that a is an algebraic integer. □

The lemma is proven, and so we are done. □

Remark 9.21. The case when all a_i's are rational numbers is much easier and is a rather folklore problem. In this case, the problem reduces easily to the following: if p is a prime number and if $a_1, a_2, \ldots, a_k$ are integers such that p^n divides $a_1^n + a_2^n + \cdots + a_k^n$ for all n, then p divides all a_i's. This follows easily from Euler's theorem, by choosing $n = \varphi(p^N)$ with N sufficiently large. Actually, using ideal theory as in the previous problem and imitating the proof for rational numbers, one can give yet another solution of the problem. We leave this as a nice exercise for the reader.

Remark 9.22. Let us recall the proof of Newton's relations. Let a_i be elements of a field K of characteristic 0 and define

$$f(X) = \prod_{i=1}^{n}(1 - a_i X) = \sum_{i=0}^{n} b_i X^i.$$

Let $P_k = a_1^k + a_2^k + \cdots + a_n^k$. Observe that

$$\frac{f'(X)}{f(X)} = -\sum_{i=1}^{n} \frac{a_i}{1 - a_i X} = -\sum_{k\geq 1} P_k X^{k-1}.$$

Identifying coefficients in the equality

$$f'(X) = -f(X) \cdot \sum_{k\geq 1} P_k X^{k-1}$$

yields Newton's relations

$$mb_m + P_1 b_{m-1} + \cdots + P_m b_0 = 0$$

for $1 \leq m \leq n$.

The following is also a very tricky problem. We use a p-adic approach to solve it and we refer the reader to the appendices on p-adic numbers and number fields for more details.

19. Let p, q be prime numbers and let r be a positive integer such that $q|p-1$, q does not divide r and $p > r^{q-1}$. Let $a_1, a_2, \ldots, a_r$ be integers such that $a_1^{\frac{p-1}{q}} + a_2^{\frac{p-1}{q}} + \cdots + a_r^{\frac{p-1}{q}}$ is a multiple of p. Prove that at least one of the a_i's is a multiple of p.

J. Borosh, D.A. Hensley, J. Zinn, AMM 10748

Proof. Let $z = e^{\frac{2i\pi}{q}}$ and let $K = \mathbb{Q}(z)$. This is an extension of degree $q-1$ of $\mathbb{Q}$. By choosing a prime dividing p in the ring of algebraic integers of K and completing K with respect to this prime, we obtain an extension of the p-adic valuation v_p on K. Moreover, if x is an algebraic integer in K, then $v_p(x) \geq 0$.

Assume that no a_i is a multiple of p and let $z_i = z^{i-1}$ $(1 \leq i \leq q)$. Since

$$\sum_{j=1}^{q} v_p(a_i^{\frac{p-1}{q}} - z_j) = v_p(a_i^{p-1} - 1) > 0,$$

there exists $\sigma(i) \in \{1, 2, \ldots, q\}$ such that $v_p(a_i^{\frac{p-1}{q}} - z_{\sigma(i)}) > 0$. Since we have $v_p\left(\sum_i a_i^{\frac{p-1}{q}}\right) > 0$, it follows that $v_p\left(\sum z_{\sigma(i)}\right) > 0$. So, if $f(X) = \sum_{i=1}^{r} X^{\sigma(i)-1}$, we have $v_p(f(z)) > 0$. Since $f(z_i)$ is an algebraic integer for all i, it follows that $v_p(f(1)\prod_{i=2}^{q} f(z_i)) > 0$. Let $N = \prod_{i=2}^{q} f(z_i)$. N is an integer, because it is a symmetric polynomial expression with integer coefficients in the roots of the polynomial $X^{q-1} + \cdots + X + 1$. We claim that N is nonzero. Otherwise, there exists $2 \leq i \leq q$ such that $f(z_i) = 0$. The irreducibility of the polynomial $1 + X + \cdots + X^{q-1}$ over the rational numbers implies that f is a multiple of $1 + X + \cdots + X^{q-1}$. But then $r = f(1)$ is a multiple of q, a contradiction with the hypothesis.

So N is a nonzero integer and $v_p(rN) > 0$, so that $v_p(N) > 0$ (clearly p does not divide r). Thus $|N| \geq p$. On the other hand, we have $|f(z_i)| \leq r$, so that $|N| \leq r^{q-1} < p$. This contradiction finishes the proof. □

Proof. Here is a more elementary, but still very tricky solution, based on the theorem of symmetric polynomials. Suppose that none of the a_i's is a multiple of p and let $h = g^{\frac{p-1}{q}}$, where g is a primitive root mod p. We can therefore find positive integers m_i such that $a_i^{\frac{p-1}{q}} \equiv h^{m_i} \pmod{p}$. Let $f = X^{m_1} + X^{m_2} + \cdots + X^{m_r}$ and let $g \in \mathbb{Z}[X]$ satisfy

$$f(X_1) \cdot f(X_2) \cdot \cdots \cdot f(X_{q-1}) = g(\sigma_1(X_1, X_2, \ldots, X_{q-1}), \ldots, \sigma_r(X_1 X_2 \cdots X_{q-1})),$$

where σ_i are the symmetric fundamental sums. Let $z_1, z_2, \ldots, z_{q-1}$ be the complex roots of $\frac{X^q-1}{X-1}$. Note that

$$\frac{X^q - 1}{X - 1} = (X - h)(X - h^2) \cdot \cdots \cdot (X - h^{q-1}) \in \mathbb{F}_p[X],$$

as $h, h^2, \ldots, h^{q-1}$ are distinct qth roots of unity in $\mathbb{F}_p$. This implies that

$$\sigma_i(z_1, z_2, \ldots, z_{q-1}) \equiv \sigma_i(h, h^2, \ldots, h^{q-1}) \pmod p$$

for all i. Therefore

$$\begin{aligned}\prod_{i=1}^{q-1} f(z_i) &\equiv g(\sigma_1(h, h^2, \ldots, h^{q-1}), \ldots, \sigma_{q-1}(h, h^2, \ldots, h^{q-1}))\\ &= \prod_{i=1}^{q-1} f(h^i) \equiv 0 \pmod p,\end{aligned}$$

that is p divides the integer $N = f(z_1) \cdot \cdots \cdot f(z_{q-1})$. As in the previous solution we obtain $|N| \le r^{q-1} < p$ and so $N = 0$. We conclude as in the previous solution. $\square$

9.6 Miscellaneous problems

It is really not easy to solve the following problem without the use of minimal polynomials. However, once the yoga of minimal polynomials is understood, the argument is rather standard.

20. Find the least positive integer n such that $\cos\frac{\pi}{n}$ cannot be written in the form $p + \sqrt{q} + \sqrt[3]{r}$ with $p, q, r \in \mathbb{Q}$.

O. Mushkarov, N. Nikolov, Bulgaria

Proof. For $n \le 6$, explicit computations show that $\cos\frac{\pi}{n}$ can be written in the desired form (the argument is a bit tricky for $n = 5$, but note that $z = e^{\frac{i\pi}{5}}$ is a solution of the equation $z^4 - z^3 + z^2 - z + 1 = 0$, which can also be written as $(z + z^{-1})^2 - (z + z^{-1}) - 1 = 0$.) The question is whether we can write $\cos\frac{\pi}{7}$ in the form $p + \sqrt{q} + \sqrt[3]{r}$ with $p, q, r \in \mathbb{Q}$ and the answer turns out to be negative, implying that the answer to the problem is $n = 7$.

Let us assume that

$$\cos\frac{\pi}{7} = p + \sqrt{q} + \sqrt[3]{r}$$

and first compute the minimal polynomial of $\cos\frac{\pi}{7}$. In order to do this, we will first find a rational equation of low degree satisfied by $\cos\frac{\pi}{7}$. Let $z = e^{\frac{i\pi}{7}}$, so that $z^7 = -1$ and

$$z^6 - z^5 + z^4 - z^3 + z^2 - z + 1 = 0.$$

Dividing this by z^3 and rearranging terms yields

$$z^3 + \frac{1}{z^3} - \left(z^2 + \frac{1}{z^2}\right) + z + \frac{1}{z} - 1 = 0.$$

Thus, if $x = \cos\frac{\pi}{7} = \frac{z+\frac{1}{z}}{2}$, then the previous relation gives

$$8x^3 - 6x - (4x^2 - 2) + 2x - 1 = 0,$$

that is $8x^3 - 4x^2 - 4x + 1 = 0$. Since the polynomial

$$f(X) = X^3 - \frac{1}{2}X^2 - \frac{1}{2}X + \frac{1}{8}$$

is trivially irreducible over the rational numbers (it has degree 3, so we only have to look for rational roots), this is the minimal polynomial of x. Therefore x and $x - p$ have degree 3 over $\mathbb{Q}$.

But then (observe that the identity $((\sqrt{q} + \sqrt[3]{r}) - \sqrt{q})^3 = r$ easily implies that $\sqrt{q} \in \mathbb{Q}(\sqrt{q} + \sqrt[3]{r})$)

$$[\mathbb{Q}(\sqrt{q} + \sqrt[3]{r}) : \mathbb{Q}(\sqrt{q})] \cdot [\mathbb{Q}(\sqrt{q}) : \mathbb{Q}] = [\mathbb{Q}(\sqrt{q} + \sqrt[3]{r}) : \mathbb{Q}] = 3$$

and since $[\mathbb{Q}(\sqrt{q}) : \mathbb{Q}]$ is 1 or 2, it follows that $\sqrt{q}$ is a rational number. Thus $x - p - \sqrt{q} = \sqrt[3]{r}$ and $u = p + \sqrt{q}$ is a rational number. Now, since x is irrational, we must have $\sqrt[3]{r}$ irrational and so $X^3 - r$ is irreducible over the rational numbers. Since $f(u + \sqrt[3]{r}) = f(x) = 0$, it follows that $X^3 - r$ divides $f(u + X)$ and so (for degree reasons) we must have $f(u + X) = X^3 - r$. It is trivial now, by identifying coefficients, to see that this is not possible. The result follows. $\square$

Proof. As in the previous solution it is enough to show that we cannot have

$$\cos\frac{\pi}{7} = p + \sqrt{q} + \sqrt[3]{r}.$$

As before we compute that $\cos\frac{\pi}{7}$ satisfies $8x^3 - 4x^2 - 4x + 1$ and that this polynomial is irreducible since it has no rational roots. Also either by noting that the other two roots are $\cos\frac{3\pi}{7}$ and $\cos\frac{5\pi}{7}$ or by plugging in a few values, we see that this polynomial has three real roots.

Now let $z = e^{2\pi i/3}$ and suppose $x = p \pm \sqrt{q} + z^k\sqrt[3]{r}$. Then

$$r = (x - p \mp \sqrt{q})^3 = (x-p)^3 + 3q(x-p) \mp (3(x-p)^2 + q)\sqrt{q}.$$

So

$$[(x-p)^3 + 3q(x-p) - r]^2 = (3(x-p)^2 + q)^2 q.$$

Thus

$$g(X) = [(X-p)^3 + 3q(X-p) - r]^2 - (3(X-p)^2 + q)^2 q \in \mathbb{Q}[X]$$

is a sixth degree polynomial with roots $p \pm \sqrt{q} + z^k\sqrt[3]{r}$. If the equality above holds, then this polynomial must be a multiple of $f(X)$, the minimal polynomial of $\cos\frac{\pi}{7} = p + \sqrt{q} + \sqrt[3]{r}$. However $f(X)$ has three real roots and $g(X)$ has only two real roots (the ones with $k = 0$). Thus this cannot occur. □

We continue with a very beautiful problem and a very elegant solution.

21. Let $s_1, s_2, \ldots$ and $t_1, t_2, \ldots$ be two infinite nonconstant sequences of rational numbers such that $(s_i - s_j)(t_i - t_j)$ is an integer for all $i, j \geq 1$. Prove that there exists a rational number r such that $(s_i - s_j)r$ and $\frac{t_i - t_j}{r}$ are integers for all i, j.

USAMO 2009

Proof. We start with some useful reductions: first of all, by working with the sequences $(s_i - s_1)_i$ and $(t_i - t_1)_i$, we may assume that $s_1 = t_1 = 0$. Secondly, there is u such that $s_u \neq 0$ and, by working with the sequences $\left(\frac{s_n}{s_u}\right)_n$ and $(s_u \cdot t_n)_n$, we may assume that $s_u = 1$.

Now, by assumption $s_n t_n$ is an integer for all n. But then

$$s_i t_j + s_j t_i = s_i t_i + s_j t_j - (s_i - s_j)(t_i - t_j)$$

is also an integer for all i, j. Since $s_i t_j + s_j t_i$ and $(s_i t_j) \cdot (s_j t_i) = (s_i t_i)(s_j t_j)$ are integers, $s_i t_j$ and $s_j t_i$ are algebraic integers. Since they are also rational numbers, they must be rational integers. Thus $s_i t_j$ is an integer for all i, j. For $i = u$, we obtain that all t_j are integers. Let d be their greatest common divisor. Then clearly $\frac{t_i}{d}$ is an integer for all i. We claim that ds_i is also an integer for all i, which will solve the problem. But since d is a linear combination with integer coefficients of some t_j's (by Bézout's lemma) and since $s_i t_j \in \mathbb{Z}$ for all i, j, it is clear that $ds_i \in \mathbb{Z}$ for all i. The conclusion follows. □

In order to motivate the next problem, we will discuss first a very classical and nontrivial result in elementary number theory. The reader is advised to read the addendum 9.A before reading the proof.

Theorem 9.23. *(Lucas-Lehmer) Define a sequence by $a_0 = 4$ and $a_{n+1} = a_n^2 - 2$ for $n \geq 0$. Let m be an odd positive integer and let $n = 2^m - 1$. Then n is a prime if and only if $n | a_{m-2}$.*

Proof. The first difficulty is to actually find a manageable formula for the general term of the sequence. We use the identity $x^2 + x^{-2} = \left(x + x^{-1}\right)^2 - 2$ and set $a_n = x_n + x_n^{-1}$ for a sequence $x_n > 1$ (note that $a_n > 2$, so x_n exists). Then $x_{n+1} = x_n^2$, so $x_n = x_0^{2^n}$ and we easily conclude that

$$a_n = (2 + \sqrt{3})^{2^n} + (2 - \sqrt{3})^{2^n}.$$

Suppose that $n = p$ is a prime and $m \geq 3$. Since $p \equiv 1 \pmod 3$ and $p \equiv -1 \pmod 8$, the quadratic reciprocity law implies that $\left(\frac{2}{p}\right) = 1$ and $\left(\frac{3}{p}\right) = -1$. Pick some α in an algebraic closure[3] of $\mathbb{F}_p$ such that $\alpha^2 = 3$.

[3]One does not need the existence of an algebraic closure to prove the existence of α: if 3 is a quadratic residue mod p, it is clear what we have to do; otherwise, it is easy to check that $\mathbb{F}_p[X]/(X^2 - 3)$ is a field with p^2 elements and we can take for α the image of X in this field.

Note that α is actually an element of $\mathbb{F}_{p^2}$ and that we can define a map $f : \mathbb{Z}[\sqrt{3}] \to \mathbb{F}_{p^2}$ by $f(a+b\sqrt{3}) = \bar{a} + \bar{b}\alpha$, where $\bar{a} = a \pmod{p}$ (seen as an element of $\mathbb{F}_{p^2}$). Since $\alpha^2 = 3$, it is immediate to check that this is a ring homomorphism. Trivially, f vanishes on $p\mathbb{Z}$. Let $x = f(2+\sqrt{3}) = 2+\alpha$ and $y = f(2-\sqrt{3}) = 2-\alpha$. Thus $x, y \in \mathbb{F}_{p^2}$ and they are nonzero, since $xy = f(1) = 1$. We want to prove that $f(a_{m-2}) = 0$ or equivalently that $x^{2^{m-2}} + x^{-2^{m-2}} = 0$, i.e. $x^{\frac{p+1}{2}} = -1$. Since $2x = (1+\alpha)^2$, we obtain the following equality in $\overline{\mathbb{F}_p}$:

$$2x^{\frac{p+1}{2}} = 2\left(\frac{2}{p}\right) \cdot x^{\frac{p+1}{2}} = (2x)^{\frac{p+1}{2}} = (1+\alpha)^{p+1} = (1+\alpha)(1+\alpha^p).$$

Since $\alpha^2 = 3$, we have $\alpha^p = \left(\frac{3}{p}\right) \cdot \alpha = -\alpha$, which combined with the previous equality yields the desired result.

Let us prove the converse now. Suppose that $n|a_{m-2}$, we need that n is a prime. It is enough to check that for all $p|n$ we have $p > \sqrt{n}$. Since p divides a_{m-2}, the previous arguments yield the equality $(2+\alpha)^{\frac{n+1}{2}} = -1$ in $\mathbb{F}_{p^2}^*$. Thus $2+\alpha \in \mathbb{F}_{p^2}^*$ has order $n+1$ and Lagrange's theorem yields $n+1|p^2-1$. The result follows. □

22. The sequence $a_0, a_1, a_2, \ldots$ is defined by $a_0 = 2$ and $a_{k+1} = 2a_k^2 - 1$ for $k \geq 0$. Prove that if an odd prime p divides a_n, then 2^{n+3} divides $p^2 - 1$.

IMO Shortlist 2003

Proof. Note that $2a_n$ is precisely the sequence studied in theorem 9.23, so

$$a_n = \frac{(2+\sqrt{3})^{2^n} + (2-\sqrt{3})^{2^n}}{2}.$$

Let now $p > 2$ be a prime factor of a_n and let $\alpha \in \overline{\mathbb{F}_p}$ be such that $\alpha^2 = 3$. Define f, x, y as in the proof of the previous theorem. Since $p|a_n$, we have $x^{2^n} + y^{2^n} = 0$, thus $x^{2^{n+1}} = -1$. Hence x has order 2^{n+2} in the group $\mathbb{F}_{p^2}^*$ and so by Lagrange's theorem 2^{n+2} divides $p^2 - 1$. Unfortunately, this is not enough, but we are close.

If $x \in \mathbb{F}_p^*$, everything is easy, since then Lagrange's theorem for this subgroup yields $2^{n+2}|p-1$ and so trivially $2^{n+3}|p^2-1$. So, assume that x is not in $\mathbb{F}_p^*$. Then x, y are roots of the irreducible polynomial $X^2 - 4X + 1 \in \mathbb{F}_p[X]$, so that we must have $x^p = y$. Indeed, since $x^2 - 4x + 1 = 0$, we also have (by raising the previous equality to the p-th power and by using the formula $(x+y)^p = x^p + y^p$, valid in fields of characteristic p) $x^{2p} - 4x^p + 1 = 0$, so that x^p is also a root of $X^2 - 4X + 1$, which cannot be x (because otherwise $x^p = x$ and $x \in \mathbb{F}_p^*$). Thus $x^p = y$ and so $x^{p+1} = 1$. But then 2^{n+2}, which is the order of x, must divide $p+1$ and we are done again. □

There is really no obvious approach to the following rather exotic problem.

23. Let k be a positive integer and let $a_1, a_2, \ldots, a_k$ and $b_1, b_2, \ldots, b_k$ be two sequences of rational numbers with the property: for any irrational numbers $x_1, x_2, \ldots, x_k > 1$ there exist positive integers $n_1, n_2, \ldots, n_k$ and $m_1, m_2, \ldots, m_k$ such that

$$a_1[x_1^{n_1}] + a_2[x_2^{n_2}] + \cdots + a_k[x_k^{n_k}] = b_1[x_1^{m_1}] + b_2[x_2^{m_2}] + \cdots + b_k[x_k^{m_k}].$$

Prove that $a_i = b_i$ for all i.

Gabriel Dospinescu, Mathlinks Contest

Proof. The key point is the following result:

Lemma 9.24. *For any integer $N \geq 2$ we can find irrational numbers $a, b > 1$ such that for every positive integer m we have $[a^m] \equiv -1 \pmod{N}$ and $[b^m] \equiv 0 \pmod{N}$.*

Proof. We will choose a, b to be algebraic integers of degree 2. Let us show how to construct a and leave to the reader the details for the construction of b. We want to find a polynomial with integer coefficients

$$f(X) = X^2 + uX + v = (X-a)(X-c)$$

for some irrational numbers $a > 1, 0 < c < 1$. In this case, since $a^m + c^m$ is an integer for all positive integers m, it follows that $[a^m] = a^m + c^m - 1$ for all m. Thus, we need to ensure that $a^m + c^m \equiv 0 \pmod{N}$ for all m. Since

$$a^{m+1} + c^{m+1} = -u(a^m + c^m) - v(a^{m-1} + c^{m-1})$$

for all m, it is enough to ensure that N divides u, v. Also, to ensure that $0 < c < 1$ we will choose $v > 0$ and $1 + u + v < 0$. For instance, we can take $u = -2N, v = N$, yielding $a = N + \sqrt{N^2 - N}$.

Similarly for b, we will choose $u = -(2N + 1)$ and $v = N$, so

$$b = \frac{2N + 1 + \sqrt{4N^2 + 1}}{2}. \qquad \square$$

Coming back to the proof, choose a positive integer N and a, b irrational numbers as in the lemma. Set $x_1 = a$ and $x_2 = \cdots = x_k = b$. By hypothesis, we can find positive integers $n_1, n_2, \ldots, n_k$ and $m_1, m_2, \ldots, m_k$ such that

$$a_1[x_1^{n_1}] + a_2[x_2^{n_2}] + \cdots + a_k[x_k^{n_k}] = b_1[x_1^{m_1}] + b_2[x_2^{m_2}] + \cdots + b_k[x_k^{m_k}].$$

By the properties of a and b we deduce that $a_1 \equiv b_1 \pmod{N}$. Since N was arbitrary, it follows that $a_1 = b_1$. Since we can do the same with the other pairs (a_i, b_i), the result follows. $\square$

The following result is really a mathematical gem, taken from [5]. It is quite difficult and has a very elementary proof.

24. Prove that if $p_1, p_2, \ldots, p_n$ are distinct primes and if

$$a_1\sqrt{p_1} + a_2\sqrt{p_2} + \cdots + a_n\sqrt{p_n} = 0$$

for some rational numbers $a_1, a_2, \ldots, a_n$, then $a_i = 0$ for all i.

Besicovitch's theorem

Proof. We will prove by induction on n the following statement: for any $m \geq 1$ and any distinct primes $q_1, q_2, \ldots, q_m, p_1, p_2, \ldots, p_n$ we have[4]

$$\sqrt{q_1 q_2 \cdots q_m} \notin \mathbb{Q}(\sqrt{p_1}, \sqrt{p_2}, \ldots, \sqrt{p_n}).$$

[4]If K/F is an extension of fields and if $x_1, x_2, \ldots, x_n \in K$, we let $F(x_1, x_2, \ldots, x_n)$ be the smallest subfield of K which contains F and $x_1, x_2, \ldots, x_n$. It is also the set of elements of the form $f(x_1, x_2, \ldots, x_n)$, where f is a rational function in n variables with coefficients in F.

Let us prove the basc case: assume that $n = 1$ and that

$$\sqrt{q_1 q_2 \cdots q_m} = a + b\sqrt{p_1}$$

for some rational numbers a, b. Squaring this relation and using that $\sqrt{p_1}$ is irrational, we deduce that $ab = 0$. But then either $q_1 q_2 \cdots q_m$ or $q_1 q_2 \cdots q_m p_1$ is a perfect square, which is clearly not possible. Now, assume that the result holds for n and let us prove it for $n+1$. Let $F = \mathbb{Q}(\sqrt{p_1}, \sqrt{p_2}, \dots, \sqrt{p_n})$ and assume that $\sqrt{q_1 q_2 \cdots q_m} = a + b\sqrt{p_{n+1}}$ for some $a, b \in F$. Again, we square this relation to deduce that

$$2ab\sqrt{p_{n+1}} = q_1 q_2 \cdots q_m - a^2 - p_{n+1} b^2 \in F.$$

However, by the inductive hypothesis we have $\sqrt{p_{n+1}} \notin F$, so we must have $ab = 0$. If $a = 0$, we obtain that $\sqrt{p_{n+1} q_1 q_2 \cdots q_m} \in F$, contradicting the inductive hypothesis. If $b = 0$, we get a similar contradiction. In all cases, the inductive step is proved and the conclusion follows. □

Remark 9.25. In [58], Mordell proved the following generalization:

Theorem 9.26. *Let $K \subset L$ be fields of characteristic 0 and let $x_1, x_2, \dots, x_r$ be elements of L such that for all i there exists a least positive integer n_i such that $x_i^{n_i} \in K$. Suppose that for all integers $e_1, e_2, \dots, e_r$, if $x_1^{e_1} \cdot x_2^{e_2} \cdot \dots \cdot x_r^{e_r} \in K$, then n_i divides e_i for all i. Finally, suppose that $L \subset \mathbb{R}$ or that K contains all n_ith roots of unity, for all i. Then $(x_1^{i_1} \cdot x_2^{i_2} \cdot \dots \cdot x_r^{i_r})_{0 \le i_j < n_i}$ is a linearly independent set. In particular, $[K(x_1, x_2, \dots, x_r) : K] = n_1 \cdot \dots \cdot n_r$.*

9.7 Notes

We thank the following people for providing solutions: Amol Aggarwal (problem 18), Darij Grinberg (problem 1), Daniel Harrer (problem 9), Holden Lee (problems 8, 10), Thanasin Nampaisarn (problem 13), Fedja Nazarov (problem 3), Richard Stong (problems 4, 6, 20), Qiaochu Yuan (problems 4, 17), Victor Wang (problems 9, 19), Gjergji Zaimi (problems 2, 3).

Addendum 9.A Equations over Finite Fields

This addendum is a modest introduction to finite fields and polynomial equations over finite fields. There are some very beautiful and extremely deep results on the subject, which are far beyond the scope of this book. But the fact that their proofs are very difficult should not be a reason for not presenting them. We highly recommend the introductory text [43] for more details.

To avoid spending too much time on preliminaries, we will fix a prime number p and an algebraic closure $\overline{\mathbb{F}_p}$ of the field $\mathbb{F}_p = \mathbb{Z}/p\mathbb{Z}$. Recall that this means that any $x \in \overline{\mathbb{F}_p}$ is a root of some nonzero polynomial $f \in \mathbb{F}_p[X]$ and that any $f \in \mathbb{F}_p[X]$ has at least one root in $\overline{\mathbb{F}_p}$ (which actually implies that it splits into linear factors over $\overline{\mathbb{F}_p}$). It is a rather nontrivial theorem of Steinitz that any field has an algebraic closure and any two algebraic closures are isomorphic. We take this approach when introducing finite fields since it is pretty rapid, though not very elegant...

Before proving the first fundamental result, let us glorify the following easy result, which will be constantly used in this chapter:

Proposition 9.A.1. *Let p be a prime and let A be a ring such that*[5] $pa = 0$ *for all $a \in A$. Then for all powers q of p and for all $a_1, a_2, \ldots, a_n \in A$ we have*

$$(a_1 + a_2 + \cdots + a_n)^q = a_1^q + a_2^q + \cdots + a_n^q.$$

Proof. By induction on n, we may assume that $n = 2$. Then everything follows from the usual binomial formula, the hypothesis on A and the fact that $\binom{q}{i} \equiv 0 \pmod{p}$ for any $1 \leq i < q$. □

If q is a power of p, let

$$\mathbb{F}_q = \{x \in \overline{\mathbb{F}_p} | x^q = x\}.$$

We have the following easy, but crucial result:

Theorem 9.A.2. $\mathbb{F}_q$ *is the unique field with q elements contained in* $\overline{\mathbb{F}_p}$.

[5] We say that A has characteristic p.

Proof. First, let us check that $\mathbb{F}_q$ is a field. It is clearly stable by multiplication and stability under addition follows from the previous proposition. $\mathbb{F}_q$ has q elements since $X^q - X$ splits into linear factors over $\overline{\mathbb{F}_p}$ (because $\overline{\mathbb{F}_p}$ is algebraically closed) and all of these linear factors are distinct (because $X^q - X$ is prime to its derivative -1).

Let us consider now a subfield L of $\overline{\mathbb{F}_p}$ with q elements. As L^* is a group with $q-1$ elements, Lagrange's theorem yields $x^{q-1} = 1$ for all $x \in L^*$. Thus $x^q = x$ for all $x \in L$ and so $L \subset \mathbb{F}_q$. A cardinality argument finishes the proof. □

A more subtle result is the following generalization of Gauss' classical theorem on primitive roots modulo prime numbers.

Theorem 9.A.3. *$\mathbb{F}_q^*$ is a cyclic group of order $q-1$. More generally, if K is any field and G is a finite subgroup of K^*, then G is cyclic.*

Proof. Let d be the maximal order of the elements of G. It is a general property of finite abelian groups that if $x, y \in G$ have orders m, n, then one can find $z \in G$ with order $\operatorname{lcm}(m, n)$ (the reader can take this as an easy exercise). Using this, we deduce that the order of any element of G divides d. Thus for all $g \in G$ we have $g^d = 1$. But the polynomial $X^d - 1 \in K[X]$ vanishes at all elements of G, so $d \geq |G|$. On the other hand, d is the order of some element of G, so $d \mid |G|$ by Lagrange's theorem. Therefore $d = |G|$ and G is cyclic. □

There is a trap concerning finite fields: it is not true that if $n \geq m$, then $\mathbb{F}_{p^m} \subset \mathbb{F}_{p^n}$. Actually, this inclusion takes place if and only if $X^{p^m-1} - 1$ divides $X^{p^n-1} - 1$ (this follows immediately from the definition and the fact that the roots of $X^q - X$ are simple) and this happens if and only if $p^m - 1$ divides $p^n - 1$, which in turns happens if and only if m divides n.

A fundamental object in the theory of finite fields is the Frobenius map

$$\operatorname{Fr}_q : \mathbb{F}_{q^n} \to \mathbb{F}_{q^n}, \quad \operatorname{Fr}_q(x) = x^q,$$

an automorphism of $\mathbb{F}_{q^n}$ which acts as identity on $\mathbb{F}_q$. Moreover, any such automorphism is an iterate of the Frobenius map and there are precisely n such automorphisms.[6] All these results would be pretty hard to prove without theorem 9.A.3, but they become easy exercises once we have it. The following result is fundamental. It says that if you know a root of an irreducible polynomial over $\mathbb{F}_q$, then the other roots are obtained by successively applying the Frobenius map to that root.

Theorem 9.A.4. *Let $f \in \mathbb{F}_q[X]$ be a monic irreducible polynomial of degree n and let $x \in \overline{\mathbb{F}_p}$ be a root of f. Then the roots of f are $x, x^q, x^{q^2}, \ldots, x^{q^{n-1}}$. In other words, $f(X) = \prod_{i=0}^{n-1}(X - x^{q^i})$.*

Proof. The key point is that $x^{q^n} = x$. Indeed, the field generated by x over $\mathbb{F}_q$ (inside $\overline{\mathbb{F}}_p$) has q^n elements, because x has degree n over $\mathbb{F}_q$, so this field is $\mathbb{F}_{q^n}$. But in $\mathbb{F}_{q^n}$ all elements are roots of $X^{q^n} - X$. Having done this, define the polynomial $G(X) = \prod_{i=0}^{n-1}(X - x^{q^i})$. The key point and proposition 9.A.1 yield

$$G(X)^q = \prod_{i=0}^{n-1}(X^q - x^{q^{i+1}}) = \prod_{i=0}^{n-1}(X^q - x^{q^i}) = G(X^q).$$

Thus, if we write $G(X) = g_0 + g_1 X + \cdots + g_l X^l$, then again by proposition 9.A.1

$$g_0^q + g_1^q X^q + \cdots + g_l^q X^{ql} = g_0 + g_1 X^q + \cdots + g_l X^{ql},$$

which implies that $g_i^q = g_i$ for all i and so $g_i \in \mathbb{F}_q$. Thus $G \in \mathbb{F}_q[X]$. Since G vanishes at x and f is irreducible, we deduce that f divides G. A degree argument finishes the proof. □

9.A.1 Norm and trace maps

Consider a finite field $\mathbb{F}_q$ and a finite extension $\mathbb{F}_{q^n}$. Define the norm and trace maps by

$$N_{\mathbb{F}_{q^n}/\mathbb{F}_q} : \mathbb{F}_{q^n} \to \mathbb{F}_{q^n}, \quad x \mapsto \prod_{j=0}^{n-1} x^{q^j}, \quad \mathrm{Tr}_{\mathbb{F}_{q^n}/\mathbb{F}_q} : \mathbb{F}_{q^n} \to \mathbb{F}_{q^n}, \quad x \mapsto \sum_{j=0}^{n-1} x^{q^j}.$$

[6]In fancy terms, the Galois group of the extension $\mathbb{F}_{q^n}/\mathbb{F}_q$ is cyclic of order n and generated by Fr_q.

The following result summarizes the basic properties of these maps, that will be used in future sections.

Proposition 9.A.5. *The norm and trace maps are surjective maps from $\mathbb{F}_{q^n}$ to $\mathbb{F}_q$. The norm map is multiplicative and the trace map is additive.*

Proof. To avoid complicated notations, write N and T for the norm, respectively trace map. First, let us check that $N(x), T(x) \in \mathbb{F}_q$ for all $x \in \mathbb{F}_{q^n}$. It is enough to see that $N(x)^q = N(x)$ and $T(x)^q = T(x)$. For $N(x)$, this is clear since $x^{q^n} = x$, while for $T(x)$, this follows from proposition 9.A.1 and the equality $x^{q^n} = x$. It is clear that N is multiplicative and proposition 9.A.1 shows that T is additive. It remains to prove the surjectivity of these maps.

Let ξ be a generator of $\mathbb{F}_q^*$ and let u be a generator of $\mathbb{F}_{q^n}^*$. There exists $a \in \mathbb{Z}$ such that $\xi = u^a$. As $\xi^{q-1} = 1$, we have $u^{a(q-1)} = 1$ and so there is an integer b such that $a = b \cdot \frac{q^n-1}{q-1}$. But then $\xi = N(u^b)$ and the surjectivity of N follows. For the trace map, this argument does not work, however we note that $T(ax) = aT(x)$ for any $a \in \mathbb{F}_q$ and any $x \in \mathbb{F}_{q^n}$. Thus, it is enough to prove that there exists x such that $T(x) \neq 0$. But this is clear, as the polynomial $X + X^q + \cdots + X^{q^{n-1}}$ has at most q^{n-1} roots and so it cannot vanish on all of $\mathbb{F}_{q^n}$. □

9.A.2 Characters of finite fields

As $\mathbb{F}_{p^n}$ is an n-dimensional vector space over $\mathbb{F}_p$, the choice of a basis yields a group isomorphism $\mathbb{F}_{p^n} \simeq \mathbb{F}_p \times \cdots \times \mathbb{F}_p$. Now, basic properties of the dual of a group discussed in section 7.A.1 yield the following result.

Proposition 9.A.6. *Let q be a power of p. There is an isomorphism of groups $a \to \psi_a$ between $\mathbb{F}_q$ and its character group, where*

$$\psi_a(x) = e^{\frac{2i\pi}{p}\mathrm{Tr}_{\mathbb{F}_q/\mathbb{F}_p}(ax)}.$$

Also, $\mathbb{F}_q^*$ is a cyclic group, so its group of characters is also cyclic of order $q-1$. The following result will play an important role in the following sections, when we will compute the zeta function of a diagonal hypersurface.

Proposition 9.A.7. *Let d be a divisor of $q-1$. The map $\chi \to \chi_n$, where $\chi_n(x) = \chi(N_{\mathbb{F}_{q^n}/\mathbb{F}_q}(x))$ induces a bijection between characters of order d of $\mathbb{F}_q^*$ and characters of order d of $\mathbb{F}_{q^n}^*$.*

Proof. The fact that χ_n is a character of order dividing d is a consequence of the multiplicativity of the norm map. The fact that it has order exactly d and that $\chi \to \chi_n$ is injective is a consequence of the surjectivity of the norm map (proposition 9.A.5). It remains to check the surjectivity of $\chi \to \chi_n$. Let $\tilde{\chi}$ be a character of order d of $\mathbb{F}_{q^n}^*$ and let u be a generator of $\mathbb{F}_{q^n}^*$. Then $\xi = u^{\frac{q^n-1}{q-1}}$ is a generator of $\mathbb{F}_q^*$ and since $\tilde{\chi}(u)^{q-1} = 1$ (because $\tilde{\chi}^d = 1$ and $d|q-1$), there is a unique character χ of $\mathbb{F}_q^*$ such that $\chi(\xi) = \tilde{\chi}(u)$. By construction, $\tilde{\chi} = \chi_n$ and the result follows. □

Just as for Dirichlet characters, it is convenient to extend the definition of a multiplicative character χ of $\mathbb{F}_q^*$ to $\mathbb{F}_q$, by defining $\chi(0) = 0$ if χ is nontrivial and $\chi(0) = 1$ if χ is trivial. The following innocent-looking identity will play a crucial role in future arguments and is constantly used when dealing with equations over finite fields:

Proposition 9.A.8. *Let d be a divisor of $q-1$ and let $x \in \mathbb{F}_q$. The number of solutions of the equation $y^d = x$ with $y \in \mathbb{F}_q$, denoted $N(y^d = x)$ is $\sum_{\chi^d=1} \chi(x)$, the sum being taken over all multiplicative characters whose order divides d.*

Proof. If $x = 0$, this is clear, as both sides are equal to 1. Assume that $x \neq 0$. If the equation $y^d = x$ has a solution in $\mathbb{F}_q$, then it has exactly d such solutions, as the equation $y^d = 1$ has precisely d solutions in $\mathbb{F}_q^*$ (because $d|q-1$ and $\mathbb{F}_q^*$ is cyclic of order $q-1$). On the other hand, the dual group of $\mathbb{F}_q^*$ is also cyclic of order $q-1$, so the equation $\chi^d = 1$ has d solutions and for each of them $\chi(x) = \chi(y^d) = \chi(y)^d = 1$, so both sides of the equality we want to prove are equal to d and we are done. Finally, if the equation has no solution, the result is a consequence of the orthogonality relations (theorem 7.A.5) for the abelian group $\mathbb{F}_q^*/\{x^d | x \in \mathbb{F}_q^*\}$, whose dual group is precisely the subgroup of those multiplicative characters χ such that $\chi^d = 1$ (actually, this argument also covers the previous case...). □

Finally, the following result will be used in the proof of the Davenport-Hasse relation, to which the next section is devoted. It is by no means specific to finite fields, but the short proof we are going to give uses properties of finite fields developed in the previous sections.

Proposition 9.A.9. *Let $x \in \mathbb{F}_{q^n}$ and let*

$$f = X^d - a_1X^{d-1} + \cdots + (-1)^d a_d \in \mathbb{F}_q[X]$$

be its minimal polynomial over $\mathbb{F}_q$. Then $d|n$ and $\prod_{j=0}^{n-1}(X - x^{q^j}) = f^{\frac{n}{d}}$. In particular, $N_{\mathbb{F}_{q^n}/\mathbb{F}_q}(x) = a_d^{\frac{n}{d}}$ and $\mathrm{Tr}_{\mathbb{F}_{q^n}/\mathbb{F}_q}(x) = \frac{n}{d}a_1$.

Proof. Since $[\mathbb{F}_q(x) : \mathbb{F}_q] = \deg(f) = d$, we have $\mathbb{F}_q(x) = \mathbb{F}_{q^d}$ (this uses theorem 9.A.2). But then $\mathbb{F}_{q^d} \subset \mathbb{F}_{q^n}$ and, as we have already remarked, this implies that $d|n$. Next, for degree reasons it is enough to prove that $g = \prod_{j=0}^{n-1}(X - x^{q^j})$ has only one irreducible monic factor, namely f. But if h is such a factor, then h has some root x^{q^j}. But proposition 9.A.4 implies that f also vanishes at x^{q^j} (note that the cited proposition applies only for $j < d$, but we have $x^{q^d} = x$ anyway, since we have seen that $\mathbb{F}_q(x) = \mathbb{F}_{q^d}$). Thus $\gcd(f, h)$ is nonconstant and by irreducibility $f = h$. This finishes the proof. □

9.A.3 Gauss and Jacobi sums, the Davenport-Hasse relation

Gauss and Jacobi sums play a fundamental role in the theory of equations over finite fields and in number theory, in general. We give here their basic properties, that we will need in the following sections. But before doing that, it is convenient to define them...

Definition 9.A.10. 1) If ψ and χ are characters of $\mathbb{F}_q$, respectively $\mathbb{F}_q^*$, the associated Gauss sum is

$$g(\chi, \psi) = \sum_{x \in \mathbb{F}_q^*} \chi(x)\psi(x).$$

2) If χ_1 and χ_2 are characters of $\mathbb{F}_q^*$, the associated Jacobi sum is

$$J(\chi_1, \chi_2) = \sum_{x,y \in \mathbb{F}_q, x+y=1} \chi_1(x)\chi_2(y).$$

Theorem 9.A.11. *If χ and ψ are nontrivial, then $|g(\chi,\psi)| = \sqrt{q}$.*

Proof. The orthogonality relations (theorem 7.A.5) yield (using also the substitution $\frac{x}{y} = t$)

$$\begin{aligned}|g(\chi,\psi)|^2 &= \sum_{x,y\in\mathbb{F}_q^*} \chi(x/y)\psi(x-y) = \sum_{t,y\in\mathbb{F}_q^*} \chi(t)\psi(y(t-1))\\ &= \sum_{t\in\mathbb{F}_q^*} \chi(t)\left(\sum_{y\in\mathbb{F}_q} \psi(y(t-1)) - 1\right) = \sum_{t\in\mathbb{F}_q^*} \chi(t)(q\cdot 1_{t=1} - 1)\\ &= q - 1 - \sum_{t\neq 0,1} \chi(t) = q - \sum_{t\in\mathbb{F}_q^*} \chi(t) = q.\end{aligned}$$

□

Corollary 9.A.12. *If χ and ψ are nontrivial, then*

$$g(\chi,\psi)\cdot g(\chi^{-1},\psi) = \chi(-1)q.$$

Proof. This is just a long string of obvious computations, using the previous theorem and the fact that $g(\chi,\psi(-\cdot)) = \chi(-1)g(\chi,\psi)$ (which is immediate by definition and the fact that $x \to -x$ is a permutation of $\mathbb{F}_q^*$). More precisely, we have

$$\begin{aligned}g(\chi^{-1},\psi) &= g(\overline{\chi},\psi) = \overline{g(\chi,\overline{\psi})} = \overline{g(\chi,\psi(-\cdot))}\\ &= \chi(-1)\overline{g(\chi,\psi)} = \chi(-1)\frac{q}{g(\chi,\psi)}.\end{aligned}$$

□

One has the following beautiful result which connects Gauss and Jacobi sums. Note the striking similarity with Euler's famous formula

$$B(x,y) = \frac{\Gamma(x)\Gamma(y)}{\Gamma(x+y)},$$

where

$$\Gamma(x) = \int_0^\infty e^{-t}t^{x-1}dt, \quad B(x,y) = \int_0^1 t^{x-1}(1-t)^{y-1}dt,$$

the integrals being convergent for $\mathrm{Re}(x), \mathrm{Re}(y) > 0$.

Theorem 9.A.13. *If χ_1, χ_2 are nontrivial characters of $\mathbb{F}_q^*$ such that $\chi_1 \cdot \chi_2$ is nontrivial, then for all nontrivial characters ψ of $\mathbb{F}_q$ we have*

$$J(\chi_1, \chi_2) = \frac{g(\chi_1, \psi) \cdot g(\chi_2, \psi)}{g(\chi_1\chi_2, \psi)}.$$

Proof. This is a rather tricky computation:

$$J(\chi_1, \chi_2)g(\chi_1\chi_2, \psi) = \sum_{x \in \mathbb{F}_q - \{0,1\}} \sum_{y \in \mathbb{F}_q^*} \chi_1(x)\chi_1(y)\chi_2(1-x)\chi_2(y)\psi(y).$$

Using the substitution $a = xy$ and $b = y(1-x)$, this becomes

$$\sum_{a,b \in \mathbb{F}_q^*, a+b \neq 0} \chi_1(a)\chi_2(b)\psi(a+b) = g(\chi_1, \psi)g(\chi_2, \psi) - \sum_{a \in \mathbb{F}_q^*} \chi_1(a)\chi_2(-a).$$

As $\chi_1\chi_2$ is nontrivial, the orthogonality relations (theorem 7.A.5) yield the desired result. □

Here is a striking application. Assume that $p \equiv 1 \pmod 4$ is a prime. As $\mathbb{F}_p^*$ is cyclic of order $p-1$, there exists a nontrivial character χ_1 of order 4 of $\mathbb{F}_p^*$. Let $\chi_2(x) = \left(\frac{x}{p}\right)$ be Legendre's symbol. The previous two theorems imply that $|J(\chi_1, \chi_2)|^2 = p$. On the other hand, it is clear that χ_1 takes only the values $0, \pm 1, \pm i$, thus $J(\chi_1, \chi_2) \in \mathbb{Z}[i]$. In particular, $|J(\chi_1, \chi_2)|^2$ is the sum of the squares of two integers. We recovered thus Fermat's celebrated theorem that any prime of the form $4k+1$ is the sum of the squares of two integers.

We end this section with a much deeper result, the famous Davenport-Hasse relation. This is a quite strong identity between Gauss sums, which is crucial for the proof of Weil's theorem 9.A.24 that we will see a bit later on. It also has relations to the Langlands program, but that is really beyond the scope of this book. The proof is very ingenious.

Theorem 9.A.14. *(Davenport-Hasse) Let χ and ψ be nontrivial characters of $\mathbb{F}_q^*$, respectively $\mathbb{F}_q$. Then $-g_n(\chi,\psi) = (-g(\chi,\psi))^n$, where*

$$g_n(\chi,\psi) = \sum_{x\in\mathbb{F}_{q^n}} \chi(N_{\mathbb{F}_{q^n}/\mathbb{F}_q}(x))\psi(\mathrm{Tr}_{\mathbb{F}_{q^n}/\mathbb{F}_q}(x)).$$

Proof. As $N_{\mathbb{F}_{q^n}/\mathbb{F}_q}(x)$ and $\mathrm{Tr}_{\mathbb{F}_{q^n}/\mathbb{F}_q}(x)$ only depend on the minimal polynomial of x and not on its roots, we will partition $\mathbb{F}_{q^n}$ into collections of conjugate elements over $\mathbb{F}_q$. Define

$$\lambda(X^d - b_1X^{d-1} + \cdots + (-1)^d b_d) = \chi(b_d)\psi(b_1)$$

for any $b_i \in \mathbb{F}_q$. It is easy to check that $\lambda(fg) = \lambda(f)\cdot\lambda(g)$ for all monic polynomials $f, g \in \mathbb{F}_q[X]$. Combining this with proposition 9.A.9 shows that if $x \in \mathbb{F}_{q^n}$ has minimal polynomial $P = X^d - a_1X^{d-1} + \cdots + (-1)^d a_d$ over $\mathbb{F}_q$, then

$$\chi\left(N_{\mathbb{F}_{q^n}/\mathbb{F}_q}(x)\right)\psi\left(\mathrm{Tr}_{\mathbb{F}_{q^n}/\mathbb{F}_q}(x)\right) = \chi(a_d)^{n/d}\psi\left(\frac{n}{d}a_1\right) = \lambda(P)^{n/d}.$$

Summing over all conjugates of x and then over collections of conjugate elements in $\mathbb{F}_{q^n}$ yields the crucial identity

$$g_n(\chi,\psi) = \sum_{d=\deg P|n} d\lambda(P)^{n/d},$$

the sum being taken over all irreducible monic polynomials $P \in \mathbb{F}_q[X]$ whose degree divides n.

To exploit this relation, consider the L-function

$$L(T) = \prod_P \frac{1}{1-\lambda(P)T^{\deg P}} = \sum_f \lambda(f)T^{\deg f}.$$

Here the product is taken over all $P \in \mathbb{F}_q[X]$ monic irreducible, while the sum is over all $f \in \mathbb{F}_q[X]$ monic. The second equality follows from multiplicativity of λ and the unique factorization theorem in $\mathbb{F}_q[X]$. Taking log and differentiating, we obtain

$$\begin{aligned} T \cdot \frac{L'(T)}{L(T)} &= T \cdot \frac{d}{dT} \log L(T) \\ &= \sum_P \sum_{n \geq 1} \deg(P) \lambda(P)^n T^{n \deg(P)} \\ &= \sum_n \left(\sum_{d = \deg(P) \mid n} d\lambda(P)^{n/d} \right) T^n. \end{aligned}$$

Combining this with the key relation of the previous paragraph, we conclude that

$$T \cdot \frac{L'(T)}{L(T)} = \sum_n g_n(\chi, \psi) T^n.$$

Finally, we will show that $L(T)$ has a very simple expression. Note that

$$L(T) = 1 + \sum_{n \geq 1} \left(\sum_{\deg f = n} \lambda(f) \right) T^n.$$

On the other hand,

$$\sum_{\deg f = 1} \lambda(f) = \sum_a \lambda(X - a) = \sum_a \chi(a)\psi(a) = g(\chi, \psi),$$

while for $n \geq 2$

$$\sum_{\deg f = n} \lambda(f) = \sum_{a_1, \ldots, a_n} \chi(a_n)\psi(a_1) = q^{n-2} \sum_{a_1} \psi(a_1) \cdot \sum_{a_n} \chi(a_n) = 0,$$

by the orthogonality relations (theorem 7.A.5). We deduce that

$$L(T) = 1 + g(\chi, \psi)T$$

and the result follows immediately from this and the last equality of the previous paragraph. □

We end this section by stating a very deep result of Dwork, a consequence of his proof of the rationality of the zeta function of algebraic varieties. It is a vast and very difficult generalization of the Davenport-Hasse relation.

Theorem 9.A.15. *Let $f, g \in \mathbb{F}_q[X]$ and let χ, ψ be multiplicative, respectively additive characters of $\mathbb{F}_q$. If*

$$S_n = \sum_{x \in \mathbb{F}_{q^n}} \chi\left(N_{\mathbb{F}_{q^n}/\mathbb{F}_q}(f(x))\right) \cdot \psi\left(\mathrm{Tr}_{\mathbb{F}_{q^n}/\mathbb{F}_q}(g(x))\right),$$

then there exist polynomials $P, Q \in \mathbb{C}[T]$ such that $P(0) = Q(0) = 1$ and

$$\exp\left(\sum_{n\geq 1} \frac{S_n}{n} T^n\right) = \frac{P(T)}{Q(T)}.$$

9.A.4 Diagonal equations and a theorem of Weil

Using almost everything we have done so far, we can prove the following beautiful theorem of Weil. The true beauty of the result will be revealed in a next section, when we will use this result and the Davenport-Hasse relation to compute the zeta function of a diagonal hypersurface. Before stating the theorem, we need some notation. Let $a_0, a_1, \ldots, a_l \in \mathbb{F}_q^*$ and let $\chi_0, \chi_1, \ldots, \chi_l$ be multiplicative characters of $\mathbb{F}_q^*$. Consider the additive character $\psi(x) = e^{\frac{2i\pi}{p}\mathrm{Tr}_{\mathbb{F}_q/\mathbb{F}_p}(x)}$ and let $g(\chi_i) = g(\chi_i, \psi)$. Finally, define

$$W_q(\chi_0, \chi_1, \ldots, \chi_l) = \frac{g(\chi_0)}{\chi_0(a_0)} \cdot \frac{g(\chi_1)}{\chi_1(a_1)} \cdot \cdots \cdot \frac{g(\chi_l)}{\chi_l(a_l)}.$$

Note that this quantity depends on the a_i's, but we suppress the dependence from the notation, as we will consider the a_i's as fixed elements, while the characters χ_i will vary. We are now ready to state and prove:

Theorem 9.A.16. *(Weil) Let $a_0, a_1, \ldots, a_l \in \mathbb{F}_q^*$ and let X be the projective variety defined by $a_0 x_0^m + a_1 x_1^m + \cdots + a_l x_l^m = 0$. Then*

$$|X(\mathbb{F}_q)| = \sum_{j=0}^{l-1} q^j + \frac{1}{q} \sum_{\chi_0, \chi_1, \ldots, \chi_l} W_q(\chi_0, \chi_1, \ldots, \chi_l),$$

the sum being taken over all nontrivial characters χ_i of $\mathbb{F}_q^*$ such that $\chi_i^m = 1$ and $\chi_0\chi_1\cdots\chi_l = 1$.

Proof. Let M be the number of solutions $(x_0, x_1, \ldots, x_l) \in \mathbb{F}_q^{l+1}$ of the equation $a_0x_0^m + \cdots + a_lx_l^m = 0$. Since we work with a projective variety, we have $|X(\mathbb{F}_q)| = \frac{M-1}{q-1}$, so it remains to find M. Note that

$$\begin{aligned} M &= \sum_{x_0,\ldots,x_l\in\mathbb{F}_q} 1_{a_0x_0^m+\cdots+a_lx_l^m=0} \\ &= \sum_{u_0,\ldots,u_l\in\mathbb{F}_q} 1_{a\cdot u=0} \sum_{x_0,\ldots,x_l} 1_{x_0^m=u_0}\cdot\cdots\cdot 1_{x_l^m=u_l} \\ &= \sum_{a\cdot u=0} N(x_0^m = u_0)\cdot\cdots\cdot N(x_l^m = u_l), \end{aligned}$$

where we wrote for simplicity $a \cdot u = 0$ for $a_0u_0 + a_1u_1 + \cdots + a_lu_l = 0$. Using proposition 9.A.8, then expanding the product and re-arranging terms, we obtain

$$M = \sum_{a\cdot u=0}\prod_{j=0}^{l}\left(\sum_{\chi_j^m=1}\chi_j(u_j)\right) = \sum_{\chi_j^m=1,\forall 0\le j\le l}\left(\sum_{a\cdot u=0}\chi_0(u_0)\chi_1(u_1)\cdots\chi_l(u_l)\right).$$

Next, note that

$$\sum_{a\cdot u=0}\chi_0(u_0)\chi_1(u_1)\cdots\chi_l(u_l) = \prod_{j=0}^{l}\chi_j(a_j)^{-1}J_0(\chi_0,\ldots,\chi_l),$$

where

$$J_0(\chi_0,\chi_1,\ldots,\chi_l) = \sum_{u_0+u_1+\cdots+u_l=0}\chi_0(u_0)\chi_1(u_1)\cdots\chi_l(u_l).$$

So, we end up with the pretty complicated formula

$$M = \sum_{\chi_0^m=\cdots=\chi_l^m=1}\prod_{j=0}^{l}\chi_j(a_j)^{-1}J_0(\chi_0,\ldots,\chi_l).$$

It is convenient to study in more detail these sums $J_0(\chi_0, \chi_1, \ldots, \chi_l)$, which are generalizations of the Jacobi sums discussed in section 9.A.3. The following lemma deals with those terms in the sum defining M for which some character is trivial.

Lemma 9.A.17. *Suppose that the trivial character appears in the list $\chi_0, \chi_1, \ldots, \chi_l$. Then either $\chi_j = 1$ for all j, in which case*

$$J_0(\chi_0, \chi_1, \ldots, \chi_l) = q^l, \quad \textit{or } J_0(\chi_0, \chi_1, \ldots, \chi_l) = 0.$$

Proof. If all $\chi_j = 1$, it is clear that $J_0(\chi_0, \chi_1, \ldots, \chi_l) = q^l$ (don't forget that the trivial character evaluated at 0 yields 1 by convention). Assume that not all χ_i are trivial, say (without loss of generality) $\chi_0 = \chi_1 = \cdots = \chi_k = 1$ and $\chi_j \neq 1$ for $j > k$. Then

$$\begin{aligned} &J_0(\chi_0, \chi_1, \ldots, \chi_l) \\ &\quad = \sum_{u_0+u_1+\cdots+u_l=0} \chi_{k+1}(u_{k+1}) \cdots \chi_l(u_l) \\ &\quad = q^k \sum_{u_{k+1},\ldots,u_l} \chi_{k+1}(u_{k+1}) \cdots \chi_l(u_l) \\ &\quad = q^k \prod_{j=k+1}^{l} \left(\sum_{u_j} \chi_j(u_j) \right) \\ &\quad = 0, \end{aligned}$$

the last equality being a consequence of the orthogonality relations (theorem 7.A.5). □

Using this, we obtain

$$M = q^l + \sum_{\chi_i^m=1, \chi_i \neq 1} \prod_{j=0}^{l} \chi_j(a_j)^{-1} J_0(\chi_0, \chi_1, \ldots, \chi_l).$$

The crucial step is the following lemma, whose proof uses a generalization of theorem 9.A.13.

Lemma 9.A.18. *If $\chi_0, \chi_1, \chi_2, \dots, \chi_l$ are nontrivial characters and*

$$\chi_0 \cdot \chi_1 \cdot \dots \cdot \chi_l \neq 1,$$

then $J_0(\chi_0, \chi_1, \dots, \chi_l) = 0$. If $\chi_0 \cdot \chi_1 \cdot \dots \cdot \chi_l = 1$, then

$$J_0(\chi_0, \chi_1, \dots, \chi_l) = \frac{q-1}{q} g(\chi_0) g(\chi_1) \cdots g(\chi_l).$$

Proof. Let $J_0 = J_0(\chi_0, \chi_1, \dots, \chi_l)$ and

$$J_1 = J_1(\chi_1, \chi_2, \dots, \chi_l) = \sum_{u_1+u_2+\dots+u_l=1} \chi_1(u_1)\chi_2(u_2)\cdots\chi_l(u_l).$$

Then

$$\begin{aligned} J_0 &= \sum_{t\in\mathbb{F}_q^*} \sum_{x_1+\dots+x_l=-t} \chi_0(t)\chi_1(x_1)\cdots\chi_l(x_l) \\ &= \sum_{t\in\mathbb{F}_q^*} \chi_0(t)(\chi_1\cdots\chi_l)(-t)J_1 \\ &= (\chi_1\cdots\chi_l)(-1)\left(\sum_{t\in\mathbb{F}_q^*} \chi_0\chi_1\cdots\chi_l(t)\right) J_1. \end{aligned}$$

If $\chi_0\chi_1\cdots\chi_l \neq 1$, we are done since $\sum_{t\in\mathbb{F}_q^*} \chi_0\chi_1\cdots\chi_l(t) = 0$ in this case (orthogonality relations). If $\chi_0\chi_1\cdots\chi_l = 1$, the previous equality becomes $J_0 = \chi_0(-1)(q-1)J_1$. Since $g(\chi_0)g(\chi_0^{-1}) = \chi_0(-1)q$ (by corollary 9.A.12), it remains to prove that

$$J_1 = \frac{g(\chi_1)g(\chi_2)\cdots g(\chi_l)}{g(\chi_1\chi_2\cdots\chi_l)}.$$

Now, by definition we have

$$\begin{aligned} g(\chi_1)g(\chi_2)\cdots g(\chi_l) &= \sum_{x_1,x_2,\dots,x_l} \chi_1(x_1)\chi_2(x_2)\cdots\chi_l(x_l)\psi(x_1+x_2+\dots+x_l) \\ &= J_0(\chi_1,\chi_2,\dots,\chi_l) + \left(\sum_{t\neq 0} \psi(t)\chi_1\chi_2\cdots\chi_l(t)\right) J_1 \\ &= J_0(\chi_1,\chi_2,\dots,\chi_l) + g(\chi_1\chi_2\cdots\chi_l)J_1. \end{aligned}$$

As $\chi_1\chi_2\cdots\chi_l \neq 1$, we have $J_0(\chi_1,\chi_2,\ldots,\chi_l) = 0$ by the first part of the proposition and the conclusion follows. □

Finally, using the previous lemma, we obtain

$$M = q^l + \frac{q-1}{q} \sum_{\chi_0,\ldots,\chi_l} \prod_{j=0}^{l} (g(\chi_j)\chi_j(a_j)^{-1}),$$

the sum being taken over all nontrivial characters χ_j such that $\chi_j^m = 1$ and $\chi_0\cdots\chi_l = 1$. The result follows. □

9.A.5 The zeta function of an algebraic variety

In essence, an affine variety over a field k is the locus in k^N of a bunch of polynomial equations in N variables with coefficients in k. A projective variety is the locus in the projective space $\mathbb{P}^n(k)$ of a bunch of homogeneous polynomial equations in $n+1$ variables and coefficients in k. If X is an algebraic variety over $\mathbb{F}_q$, it is natural to consider the number of points of X over the various finite extensions $\mathbb{F}_{q^n}$. The zeta function of the variety X is (up to a convenient normalization) the generating function of the sequence obtained in this way, i.e.

$$Z_X(T) = \exp\left(\sum_{n\geq 1} \frac{|X(\mathbb{F}_{q^n})|}{n} T^n\right).$$

Clearly, $Z_X(T)$ is a formal series in T with rational coefficients.

Example 9.A.19. Consider finitely many polynomials $f_1, f_2, \ldots, f_s$ in k variables with coefficients in $\mathbb{F}_p$, and let a_n be the number of solutions in $\mathbb{F}_{p^n}$ of the system of equations

$$f_1(x_1,\ldots,x_k) = f_2(x_1,\ldots,x_k) = \cdots = f_s(x_1,\ldots,x_k) = 0.$$

The zeta function of the affine variety defined by the polynomials $f_1,\ldots,f_s$ is $\exp\left(\sum_{n\geq 1} \frac{a_n}{n} T^n\right)$. On the other hand, if $f_1,\ldots,f_s$ are homogeneous polynomials , the zeta function of the projective variety defined by these polynomials is $\exp\left(\sum_{n\geq 1}\frac{a_n-1}{n(p^n-1)}T^n\right)$, as this time two solutions that differ by a nonzero element of $\mathbb{F}_{p^n}$ are the same element of the projective space.

Remark 9.A.20. Suppose that

$$|X(\mathbb{F}_{q^n})| = z_1^n + z_2^n + \cdots + z_s^n - u_1^n - \cdots - u_t^n$$

for all $n \geq 1$ and some complex numbers z_i, u_j (as we will see, this always happens, but this is very difficult to prove). Then

$$Z_X(T) = \frac{(1-Tu_1)(1-Tu_2)\cdots(1-Tu_t)}{(1-Tz_1)(1-Tz_2)\cdots(1-Tz_s)},$$

essentially by definition of $Z_X(T)$ and by the equality of formal series

$$\sum_{n\geq 1}\frac{\alpha^n}{n}T^n = -\log(1-\alpha T).$$

For instance, if $X = \mathbb{P}^r$, the projective space, then

$$|X(\mathbb{F}_{q^n})| = q^{nr} + q^{(r-1)n} + \cdots + q^n + 1,$$

so

$$Z_X(T) = \prod_{j=0}^{r}\frac{1}{1-q^jT}.$$

Another trivial example is the variety defined by the equation $xyz = 1$ in three-dimensional affine space. Then clearly $|X(\mathbb{F}_{q^n})| = (q^n-1)^2$, so that

$$Z_X(T) = \frac{(1-qT)^2}{(1-T)(1-q^2T)}.$$

Let $\mathbb{Z}[[T]]$ be the set of formal series in T with integer coefficients. Note that in all previous examples we have $Z_X(T) \in \mathbb{Z}[[T]]$. It turns out that this is always the case. The following result discusses more generally when $\exp\left(\sum_n \frac{a_n}{n}T^n\right) \in \mathbb{Z}[[T]]$.

Proposition 9.A.21. *Let a_n be a sequence of integers. There exist unique sequences of rational numbers b_n and c_n such that*

$$\exp\left(\sum_{n\geq 1}\frac{a_n}{n}T^n\right) = 1+\sum_{n\geq 1} b_nT^n = \prod_{n\geq 1}(1-T^n)^{c_n}.$$

Moreover, all b_n are integers if and only if all c_n are integers, if and only if n divides $\sum_{d|n}\mu\left(\frac{n}{d}\right)a_d$ for all n (where μ is Möbius' function).

Proof. The existence and uniqueness of b_n is clear. As for c_n, the key point is to consider

$$\begin{aligned}\log\prod_{n\geq 1}(1-T^n)^{c_n} &= \sum_{n\geq 1} c_n\log(1-T^n)\\ &= -\sum_{n} c_n\sum_{j\geq 1}\frac{T^{nj}}{j}\\ &= -\sum_{n\geq 1}\frac{T^n}{n}\left(\sum_{d|n} dc_d\right).\end{aligned}$$

Thus we need to find the sequence c_n for which $\sum_{d|n} dc_d = -a_n$ for all n. But the Möbius inversion formula yields the explicit form

$$c_n = \frac{-1}{n}\sum_{d|n} a_d\mu\left(\frac{n}{d}\right),$$

showing the existence and uniqueness of c_n.

It is clear that if $c_n\in\mathbb{Z}$, then $b_n\in\mathbb{Z}$ for all n. The converse is proved by induction. Since $c_1=-b_1$, we have $c_1\in\mathbb{Z}$. Suppose that $c_1,\ldots,c_{n-1}\in\mathbb{Z}$ and observe that $f=\prod_{i<n}(1-T^i)^{c_i}$ is invertible in $\mathbb{Z}[[T]]$, as its constant term is 1. Thus

$$(1-T^n)^{c_n}(1-T^{n+1})^{c_{n+1}}\cdots = \frac{1+\sum_j b_jT^j}{f}\in\mathbb{Z}[[T]].$$

Considering the coefficient of T^n in the left-hand side of the previous equality, we obtain that $c_n \in \mathbb{Z}$. Using also the explicit formula of the c_n's in terms of the a_n's yields the last statement of the proposition and finishes the proof. $\square$

Let X be an algebraic variety over $\mathbb{F}_q$, say defined by some polynomials $f_1, f_2, \ldots, f_d$ in n variables with coefficients in $\mathbb{F}_q$. If

$$x = (x_1, \ldots, x_n) \in X(\mathbb{F}_{q^m}),$$

define

$$\mathrm{Fr}(x) = (x_1^q, x_2^q, \ldots, x_n^q).$$

It is easy to see that this is again an element of $X(\mathbb{F}_{q^m})$. Let f be the smallest positive integer such that $x \in X(\mathbb{F}_{q^f})$ and associate to x the cycle $(x, \mathrm{Fr}(x), \ldots, \mathrm{Fr}^{f-1}(x))$, of length f. Then $X(\mathbb{F}_{q^m})$ is the disjoint union of the cycles of length dividing m, so if a_n is the number of cycles of length n, then $|X(\mathbb{F}_{q^m})| = \sum_{d|m} d \cdot a_d$. Combining this and the previous proposition yields

$$Z_X(T) = \prod_{n\geq 1}(1-T^n)^{-a_n} \in \mathbb{Z}[[T]].$$

Remark 9.A.22. The previous proposition is powerful in other contexts, too. For instance, it immediately yields the equality of formal series

$$\exp(X) = \prod_{n\geq 1}(1-X^n)^{-\frac{\mu(n)}{n}},$$

where μ is Möbius' function. The proposition also easily implies the following equality

$$\exp\left(\sum_{n\geq 0}\frac{X^{p^n}}{p^n}\right) = \prod_{\substack{n\geq 1\\ \gcd(n,p)=1}}(1-X^n)^{-\frac{\mu(n)}{n}},$$

showing that the Artin-Hasse exponential $\exp\left(\sum_{n\geq 0}\frac{X^{p^n}}{p^n}\right)$ has coefficients in $\mathbb{Z}_p$. This is absolutely not clear from the definition and plays a major role in p-adic analysis and also in Dwork's proof of the rationality of the zeta functions attached to algebraic varieties over finite fields.

In general, it is a very deep problem to compute the zeta function of a given variety. Yet even without computing the zeta function, one can say a great deal of things about it! This was conjectured by Weil in the wonderful paper [84] and proved after a gigantic work by Deligne and Grothendieck. The following theorem is really one of the most difficult and beautiful results of modern mathematics:

Theorem 9.A.23. *(Deligne-Grothendieck) Let X be a non-singular projective variety of dimension n over $\mathbb{F}_q$. Then the zeta function of X is a rational function. More precisely, there are polynomials $P_0, P_1, \dots, P_{2n} \in \mathbb{Z}[T]$ such that*

$$Z_X(T) = \frac{P_1(T)P_3(T)\cdots P_{2n-1}(T)}{P_0(T)P_2(T)\cdots P_{2n}(T)}$$

and

a) $P_0(T) = 1 - T$ and $P_{2n}(T) = 1 - q^n T$.

b) We can write $P_i = \prod_{j=1}^{b_i}(1 - \omega_{ij}T)$, where ω_{ij} are algebraic integers such that $|\omega_{ij}| = q^{i/2}$ for all i, j.

c) If $\chi = \sum_{i=0}^{2n}(-1)^i b_i$, then the zeta function satisfies the functional equation

$$Z_X\left(\frac{1}{q^n T}\right) = \pm(q^{\frac{n}{2}}T)^{\chi} Z_X(T)$$

for some sign $\pm$.

If X is a smooth projective curve of genus g (this is an important invariant attached to curves; in the notations of the theorem, we have $b_1 = 2g$) over $\mathbb{F}_q$, Weil proved in 1940 that its zeta function can be written in the form $\frac{P(t)}{(1-t)(1-qt)}$ for some polynomial $P \in 1 + t\mathbb{Z}[t]$. Moreover, the roots of P have absolute value $1/\sqrt{q}$. So in this case $b_0 = b_2 = 1$, $b_1 = 2g$ and moreover we can write

$$|X(\mathbb{F}_{q^n})| = 1 + q^n - \left(\omega_1^n + \omega_2^n + \cdots + \omega_{2g}^n\right)$$

with $|\omega_i| = \sqrt{q}$. In particular, we obtain the very nontrivial estimate

$$|X(\mathbb{F}_{q^n}) - (1 + q^n)| \le 2gq^{n/2}.$$

This last estimate was obtained by Lang and Weil in 1954 for arbitrary varieties, before the proof of the previous deep theorem. For an even more concrete example, consider integers a, b and a prime $p > 3$. The condition that the curve $y^2 = f(x)$ be non-singular is that p does not divide $4a^3 + 27b^2$. In this case the curve $y^2 = f(x)$ has genus 1 and one point at infinity. Hence

$$|X(\mathbb{F}_p)| = 1 + |\{(x, y) \in \mathbb{F}_p \times \mathbb{F}_p | y^2 = x^3 + ax + b\}|$$

and the bound above becomes

$$\Big| |\{(x, y) \in \mathbb{F}_p \times \mathbb{F}_p | y^2 = x^3 + ax + b\}| - p \Big| \leq 2\sqrt{p}.$$

This reproves a famous theorem of Hasse (there are however easier proofs, but they require a good knowledge of the theory of elliptic curves and quite a lot of algebraic geometry).

Dwork's p-adic proof (1960) of the rationality of zeta functions works for affine or projective varieties, be they non-singular or not. For a non-singular projective hypersurface defined by a polynomial $f \in \mathbb{F}_q[X_1, \ldots, X_n]$, homogeneous of degree d, Dwork proved that its zeta function is of the form $\frac{P(t)^{(-1)^{n+1}}}{(1-t)\cdots(1-q^{n-2}t)}$ for some $P \in 1+t\mathbb{Z}[t]$ of degree $\frac{(d-1)^n+(-1)^n(d-1)}{d}$. We will prove this in the next section, in the much easier case of diagonal hypersurfaces (a famous theorem of Weil).

9.A.6 Zeta function of diagonal hypersurfaces

In this part, we show how to compute the zeta function of a diagonal hypersurface. This beautiful result, due to Weil was also the starting point of the famous Weil conjectures.

Theorem 9.A.24. *(Weil) Let $l \geq 1$, $m | q-1$ and let X be the projective hypersurface of equation $a_0 x_0^m + a_1 x_1^m + \cdots + a_l x_l^m = 0$. There exists $P \in \mathbb{Z}[T]$ of degree $d = \frac{(m-1)^{l+1}+(-1)^{l+1}(m-1)}{m}$ such that*

a) $Z_X(T) = \frac{P(T)^{(-1)^l}}{(1-T)(1-qT)\cdots(1-q^{l-1}T)}$.

b) If $P(z) = 0$, then $1/z$ is an algebraic integer of absolute value $q^{\frac{l-1}{2}}$.

c) There exists an explicit integer χ *such that*

$$Z_X\left(\frac{1}{q^{l-1}T}\right) = \pm\left(q^{\frac{l-1}{2}}T\right)^{\chi} Z_X(T).$$

Proof. Recall that by theorem 9.A.16 we have for any q an equality

$$|X(\mathbb{F}_q)| = \sum_{j=0}^{l-1} q^j + \frac{1}{q} \sum_{\chi_0,\chi_1,\ldots,\chi_l} W_q(\chi_0, \chi_1, \ldots, \chi_l),$$

the sum being taken over all nontrivial characters χ_i of $\mathbb{F}_q^*$ such that $\chi_i^m = 1$ and $\chi_0\chi_1\cdots\chi_l = 1$. The main point is to study how the numbers $W_{q^n}(\chi_0, \ldots, \chi_l)$ vary and this is accomplished by the Davenport-Hasse relation. More precisely, recall that for a character χ of $\mathbb{F}_q^*$ we have a character $\chi_n(x) = \chi(N_{\mathbb{F}_{q^n}/\mathbb{F}_q}(x))$ and that $\chi \to \chi_n$ induces a bijection between characters of $\mathbb{F}_q^*$ of a given order and characters of $\mathbb{F}_{q^n}^*$ of the same order (proposition 9.A.7). So, if S is the set of $l+1$-tuples $(\chi_0, \ldots, \chi_l)$ of nontrivial characters of $\mathbb{F}_q^*$ such that $\chi_i^m = 1$ and $\chi_0\chi_1\cdots\chi_l = 1$, then

$$|X(\mathbb{F}_{q^n})| = \sum_{j=0}^{l-1} q^{nj} + \frac{1}{q^n} \sum_{(\chi_0,\chi_1,\ldots,\chi_l)\in S} W_{q^n}(\chi_{0,n}, \chi_{1,n}, \ldots, \chi_{l,n}).$$

On the other hand, by the Davenport-Hasse relation (theorem 9.A.14) we can write

$$W_{q^n}(\chi_{0,n}, \chi_{1,n}, \ldots, \chi_{l,n}) = (-1)^{l+1}(-1)^{n(l+1)}(W_q(\chi_0, \ldots, \chi_l))^n.$$

We deduce that

$$|X(\mathbb{F}_{q^n})| = \sum_{j=0}^{l-1} q^{nj} - (-1)^l \sum_{(\chi_0,\chi_1,\ldots,\chi_l)\in S} \left(\frac{(-1)^{l+1}}{q} W_q(\chi_0, \ldots, \chi_l)\right)^n$$

and so by the previous paragraph we finally obtain the first part of the theorem, with

$$P(T) = \prod_{(\chi_0,\ldots,\chi_l)\in S} \left(1 - \frac{(-1)^{l+1}}{q} W_q(\chi_0, \ldots, \chi_l)T\right).$$

Note that $W_q(\chi_0, \ldots, \chi_l) \neq 0$ for all $(\chi_0, \ldots, \chi_l) \in S$, as the Gauss sum associated to a nontrivial character is nonzero. Thus $\deg(P) = |S|$. It remains to find this number. See $|S|$ as a function $f(l)$ of l. Let $g(l)$ be the number of $l+1$-tuples $(\chi_0, \ldots, \chi_l)$ of nontrivial characters such that $\chi_i^m = 1$ and $\chi_0 \cdots \chi_l \neq 1$. Clearly, $g(l) = f(l+1)$. But $f(l) + g(l)$ is just the number of tuples of nontrivial characters such that $\chi_i^m = 1$. As there are $m-1$ nontrivial characters of order dividing m, we deduce that $f(l) + g(l) = (m-1)^{l+1}$. One immediately deduces that

$$f(l) = \frac{(m-1)^{l+1} + (-1)^{l+1}(m-1)}{m},$$

finishing therefore the computation of $\deg(P)$.

By definition of $W_q(\chi_0, \ldots, \chi_l)$ and by the fact that $|g(\chi_i)| = \sqrt{q}$, we deduce that $|W_q(\chi_0, \ldots, \chi_l)| = q^{\frac{l+1}{2}}$. This yields part 2) of the theorem.

Next, the fact that $g(\chi_i)g(\chi_i^{-1}) = \chi_i(-1)q$ and $\chi_0 \cdots \chi_l = 1$ implies that

$$W_q(\chi_0^{-1}, \ldots, \chi_l^{-1}) = \frac{q^{l+1}}{W_q(\chi_0, \ldots, \chi_l)}.$$

As clearly the set S is stable by inversion, we deduce that the map $z \to \frac{1}{q^{l-1}z}$ is a permutation of the roots of P. From here, it is an easy but tedious exercise to deduce the third part of the theorem.

Finally, it remains to prove that $P \in \mathbb{Z}[T]$. As $Z_X(T) \in \mathbb{Q}[[T]]$, we must have $P \in \mathbb{Q}[T]$. We will prove that the coefficients of P are algebraic integers, which will be enough to conclude. It is enough (taking into account the definition of P) to check that $W_q(\chi_0, \ldots, \chi_l)/q$ is an algebraic integer. As $\chi_i(a_i)$ are roots of unity, it will therefore suffice to check that $\frac{g(\chi_0)\cdots g(\chi_l)}{q}$ is an algebraic integer. But this is an obvious consequence of lemma 9.A.18. This finally proves the theorem! □

Addendum 9.B A Glimpse of Algebraic Number Theory

This addendum recalls the basic properties of number fields. Of course, one would need a whole book (and actually much more...) to properly develop the theory of number fields, as even proving the basic properties requires a lot of commutative algebra. We will try to stay as elementary as possible, while still giving some proofs. We warn the reader that a long part of this addendum is very abstract. To see the power of the notions and theorems discussed, we advise the reader to start with the last part of the addendum, which discusses applications to problems with very elementary statements and very non-elementary solutions...

9.B.1 Ideals and quotient rings

Let R be a commutative ring. An ideal of R is a nonempty subset I of R which is stable under addition and such that $ax \in I$ for all $a \in R$ and $x \in I$. Note that this is far stronger than the stability of I under multiplication. It is fairly easy to construct ideals of R: if $x_1, x_2, \ldots, x_n \in R$ then

$$(x_1, x_2, \ldots, x_n) = \{a_1x_1 + \cdots + a_nx_n | a_i \in R\}$$

is obviously an ideal, called the ideal generated by $x_1, x_2, \ldots, x_n$.

Once we have an ideal I in a ring R, we can naturally construct a quotient ring, whose elements are coset classes $\overline{a} = a + I$ with $a \in R$ and addition, multiplication are defined by $\overline{a} + \overline{b} = \overline{a+b}$ and $\overline{a} \cdot \overline{b} = \overline{ab}$. It is an easy exercise to check that it is well-defined (the issue is that we may have $\overline{a} = \overline{a'}$ even if $a \neq a'$ and one needs to check that if $\overline{a} = \overline{a'}$ and $\overline{b} = \overline{b'}$, then $\overline{a+b} = \overline{a'+b'}$, similarly for multiplication).

Definition 9.B.1. 1) An ideal I of R is called maximal if $I \neq R$ and if I is not contained in any ideal different from I and R.

2) An ideal I of R is called prime if $I \neq R$ and $ab \in R - I$ for any $a, b \in R - I$.

There is a very nice characterization of prime and maximal ideals of a ring in terms of quotient rings. The proof is essentially trivial unwinding of definitions, but the result is crucial:

Proposition 9.B.2. *1) An ideal I of R is maximal if and only if R/I is a field.*

2) An ideal I of R is prime if and only if R/I has no zero divisors.

As a field has no zero divisors, this proposition implies that any maximal ideal is a prime ideal. There are however prime ideals which are not maximal: the ideal (2) in $\mathbb{Z}[X]$ is prime and not maximal, as the quotient ring is $\mathbb{F}_2[X]$, which has no nonzero zero divisors but is not a field.

There are natural operations on ideals: if I, J are ideals of a ring R, one defines their sum $I + J = \{i + j | i \in I, j \in J\}$. It is easy to check that this is an ideal. The analogous definition for multiplication would fail (in general) to yield an ideal, so one defines the product of ideals I, J as the ideal generated by all products $i \cdot j$ with $(i, j) \in I \times J$.

9.B.2 Field extensions

We say that L is a field extension of K if both K, L are fields and $K \subset L$. The extension[7] L/K is called finite if L is a finite dimensional K-vector space. In this case, we define the degree of the extension to be

$$[L : K] = \dim_K(L).$$

For instance, the extension $\mathbb{C}/\mathbb{R}$ has degree 2, as $1, i$ is a basis of $\mathbb{C}$ over $\mathbb{R}$. On the other hand, the extension $\mathbb{C}/\mathbb{Q}$ is infinite (for example, because $\mathbb{C}$ is uncountable and $\mathbb{Q}$ is countable). We will mostly be interested in finite extensions, for which the following result is of constant use:

Proposition 9.B.3. *Let L/K and M/L be finite extensions of fields. Then M/K is finite and*

$$[M : K] = [M : L] \cdot [L : K].$$

[7]This notation should not be confused with the quotient ring previously discussed, simply because K is not an ideal in L unless $K = L$.

Proof. One can easily check that if $(x_i)_i$ is a basis of M as L-vector space and $(y_i)_i$ is a basis of L as K-vector space, then $(x_i y_j)_{i,j}$ is a basis of M as K-vector space. □

9.B.3 Algebraic numbers and algebraic integers

If L/K is an extension of fields and if $l \in L$, we say that l is algebraic over K if there is a nonzero polynomial $f \in K[X]$ such that $f(l) = 0$. In this case, there is a unique monic polynomial $\pi_l \in K[X]$ of least degree which vanishes at l. It is called the minimal polynomial of l and it is irreducible in $K[X]$, by minimality. The division algorithm shows that the only polynomials $f \in K[X]$ such that $f(l) = 0$ are the multiples of π_l. Recall that $K(l)$ is the smallest field containing K and l and it can also be described as

$$K(l) = \left\{ \frac{f(l)}{g(l)} | f, g \in K[X], g(l) \neq 0 \right\} = K[l] := \{f(l) | f \in K[X]\}.$$

To prove this equality, it suffices to show that if $f \in K[X]$ does not vanish at l, then $\frac{1}{f(l)}$ is of the form $A(l)$ for some $A \in K[X]$. But since $f(l) \neq 0$ and π_l is irreducible, f and π_l are relatively prime, so there are polynomials $A, B \in K[X]$ such that $Af + B\pi_l = 1$. Evaluation at l yields the result. The following proposition is easy, but fundamental.

Proposition 9.B.4. *Let L/K be any extension of fields.*

1) *Let $l \in L$ be algebraic over K. Then there is an isomorphism of K-algebras*[8] *between $K(l)$ and $K[X]/(\pi_l)$. Moreover, $[K(l) : K] = \deg \pi_l < \infty$.*

2) *Conversely, if $l \in L$ and $[K(l) : K] < \infty$, then l is algebraic over K.*

Proof. 1) Consider the map sending $f \in K[X]$ to $f(l)$. It is a map of K-algebras, vanishing precisely on the ideal (π_l), by definition of π_l. It is easy to check that it induces an isomorphism between $K[X]/(\pi_l)$ and $K[l]$, obtained by sending $f + (\pi_l)$ to $f(l)$. Next, if $d = \deg \pi_l$,

[8]This means an isomorphism of rings which is K-linear.

then (the classes of) $1, X, \dots, X^{d-1}$ form a K-basis of $K[X]/(\pi_l)$ and so $[K[X]/(\pi_l) : K] = d$. Combining this with the previous isomorphism finishes the proof.

2) This is clear: if $d = [K(l) : K]$, then $1, l, l^2, \dots, l^d \in K(l)$ cannot be linearly independent over K. This forces a nonzero polynomial equation with coefficients in K satisfied by l and so l is algebraic over K.

□

Combining the previous results, we can now prove the following nontrivial result (which can also be obtained using the theorem on symmetric polynomials):

Theorem 9.B.5. *Let L/K be any field extension. The set of elements of L which are algebraic over K forms a subfield of L. This subfield is equal to L if L/K is finite.*

Proof. If $l_1, l_2 \in L$ are algebraic over K, then by proposition 9.B.4 $K(l_1)/K$ and $K(l_1)(l_2)/K(l_1)$ are finite extensions (note that l_2 is also algebraic over $K(l_1)$). Thus by proposition 9.B.3, $K(l_1)(l_2)/K$ is finite. But $K(l_1)(l_2)$ contains $K(l_1+l_2)$, $K(l_1l_2)$ and $K(l_1/l_2)$. We deduce that if $x \in \{l_1+l_2, l_1l_2, l_1/l_2\}$, then $K(x)/K$ is finite and the result follows by proposition 9.B.4. The second part is also a trivial consequence of proposition 9.B.4. □

Definition 9.B.6. 1) A number $z \in \mathbb{C}$ is called algebraic if it is algebraic over $\mathbb{Q}$. It is called an algebraic integer if its minimal polynomial over $\mathbb{Q}$ has integer coefficients. By Gauss' lemma, this is equivalent to the fact that z is root of some monic polynomial with integer coefficients.

2) We denote by $\overline{\mathbb{Q}}$ (respectively $\overline{\mathbb{Z}}$) the set of algebraic numbers (respectively algebraic integers).

The following result is an easy consequence of the theorem of symmetric polynomials 9.10.

Theorem 9.B.7. *$\overline{\mathbb{Q}}$ is an algebraically closed field and $\overline{\mathbb{Z}}$ is a ring. For any $x \in \overline{\mathbb{Q}}$ there exists $n \geq 1$ such that $nx \in \overline{\mathbb{Z}}$.*

Proof. The previous theorem shows that $\overline{\mathbb{Q}}$ is a field. Suppose that $x \in \mathbb{C}$ satisfies $x^n + a_{n-1}x^{n-1} + \cdots + a_0 = 0$ for some $a_i \in \overline{\mathbb{Q}}$. We want to prove that $x \in \overline{\mathbb{Q}}$. Let $a_k^{(j)}$ be the conjugates of a_k (i.e. the roots of the minimal polynomial of a_k, including a_k). The theorem of symmetric polynomials easily implies that

$$f = \prod_{k_0,\ldots,k_{n-1}} (X^n + a_{n-1}^{(k_{n-1})} X^{n-1} + \cdots + a_0^{(k_0)})$$

has rational coefficients and vanishes at x, from where the result follows. To prove that $\overline{\mathbb{Z}}$ is a ring, consider for instance $x, y \in \overline{\mathbb{Z}}$ and let $x_1, \ldots, x_n$ and $y_1, y_2, \ldots, y_m$ be all roots of the minimal polynomials of x and y. Another application of the theorem of symmetric polynomials shows that $\prod_{i,j}(X - x_i - y_j)$, respectively $\prod_{i,j}(X - x_i \cdot y_j)$ have integer coefficients and vanish at $x + y$, respectively $x \cdot y$. Finally, to prove the last statement, take $x \in \overline{\mathbb{Q}}$ and choose integers $a_0, a_1, \ldots, a_n$ such that $a_n x^n + a_{n-1}x^{n-1} + \cdots + a_0 = 0$ and $a_n \neq 0$. Then

$$(a_n x)^n + a_{n-1}(a_n x)^{n-1} + \cdots + a_0 a_n^{n-1} = 0,$$

so $a_n \cdot x \in \overline{\mathbb{Z}}$. □

Let us introduce now the main object of this addendum.

Definition 9.B.8. A number field is a finite extension of $\mathbb{Q}$. If K is a number field, then we let $O_K = K \cap \overline{\mathbb{Z}}$.

Example 9.B.9. 1) Let $d \neq \pm 1$ be a squarefree integer and consider $K = \mathbb{Q}(\sqrt{d})$ (as usual, if $d < 0$, $\sqrt{d} = i\sqrt{-d}$). Then K has degree 2 over $\mathbb{Q}$, but the structure of O_K depends on the residue class modulo 4 of d: if $d \equiv 1 \pmod 4$, then $O_K = \mathbb{Z}\left[\frac{1+\sqrt{d}}{2}\right]$, while if $d \equiv 2$ or 3 mod 4, then $O_K = \mathbb{Z}[\sqrt{d}]$. This is not difficult to prove: imagine that $x \in O_K$ and write $x = a + b\sqrt{d}$. If $b = 0$, then a must be an integer (since it is rational and algebraic integer), so we are done. Otherwise, the conjugates of x are x and $y = a - b\sqrt{d}$. Thus $2a$ and $a^2 - db^2$ must be integers, which easily implies the desired result.

2) Let ζ_n be a primitive nth root of unity in $\mathbb{C}$ and consider $K = \mathbb{Q}(\zeta_n)$. The irreducibility of cyclotomic polynomials (a fairly nontrivial theorem) implies that the nth cyclotomic polynomial is the minimal polynomial of ζ_n, so that K has degree $\varphi(n)$ over $\mathbb{Q}$. One can prove with quite a lot of effort that $O_K = \mathbb{Z}[\zeta_n]$, i.e. there are no algebraic integers in K except for the obvious ones.

9.B.4 Factorization in O_K, the fundamental theorems

Since the proofs of the theorems stated in this section are rather long and technical, we will simply state them without proof, referring the reader to any basic number theory book. We prefer to focus on their arithmetic applications. Let K be a number field of degree d over $\mathbb{Q}$ and let O_K be the subring of K consisting of algebraic integers.

Theorem 9.B.10. *If I is a nonzero ideal of O_K, then O_K/I is a finite ring.*

Remark 9.B.11. Actually, one can prove that if $[K : \mathbb{Q}] = d$, then there exist $x_1, x_2, \dots, x_d \in O_K$ such that the map $\mathbb{Z}^d \to O_K$ sending $(n_1, n_2, \dots, n_d)$ to $n_1x_1 + \cdots + n_dx_d$ is a bijection. This easily implies the previous theorem: let $x \in I$ be nonzero, then the norm n of x is again in I and is nonzero. Hence I contains nO_K. But the previous result shows that O_K/nO_K is in bijection with $\mathbb{Z}^d/n\mathbb{Z}^d$, which is finite, with $|n|^d$ elements.

Corollary 9.B.12. *If $\wp$ is a nonzero prime ideal, then $O_K/\wp$ is a finite field and so $\wp$ is a maximal ideal.*

Proof. The ring $R = O_K/\wp$ is finite with no zero divisors. Let $x \neq 0$ be any element of R. As R is finite, there must be $i < j$ such that $x^i = x^j$. Then $x^i(x^{j-i} - 1) = 0$ and as there are no zero divisors in R, we obtain $x^{j-i} = 1$. Thus x is a unit. This proves that R is a field and the result follows. $\square$

Definition 9.B.13. If I is a nonzero ideal of I, we define its norm

$$N(I) = |O_K/I|.$$

This is an integer which lies in I, by Lagrange's theorem in the group O_K/I.

We cannot emphasize enough the importance of the following result for the theory of algebraic number fields. Suffice it to say that it plays exactly the same role as the fundamental theorem of arithmetic (i.e. the unique factorization theorem) and we will leave the reader recall that basically all results of elementary number theory follow from it.

Theorem 9.B.14. *(Kummer, Dedekind) Let K be a number field.*

a) Any ideal of O_K, different from 0 and O_K, can be uniquely (up to permutation) written in the form $\wp_1 \cdot \wp_2 \cdot \dots \cdot \wp_n$ for some $n \geq 1$ and some prime ideals $\wp_i$, not necessarily distinct.

b) If I and J are two nonzero ideals of O_K, then $N(I \cdot J) = N(I) \cdot N(J)$. Moreover, $I \subset J$ if and only if there exists an ideal J' such that $I = J \cdot J'$.

Remark 9.B.15. One can also prove that for all $x \in O_K$ we have

$$N(xO_K) = |N(x)|.$$

Remark 9.B.16. Suppose that K has degree d over $\mathbb{Q}$ and that p is a rational prime. Let $pO_K = \wp_1^{e_1} \cdot \wp_2^{e_2} \cdot \dots \cdot \wp_g^{e_g}$ be the factorization of the ideal pO_K. As pO_K has norm p^d (by the first part of the remark), theorem 9.B.14 yields $p^d = N(\wp_1)^{e_1} \cdot \dots \cdot N(\wp_g)^{e_g}$ and so there are nonnegative integers f_i such that $N(\wp_i) = p^{f_i}$. We have the fundamental relation

$$e_1 f_1 + e_2 f_2 + \dots + e_g f_g = [K : \mathbb{Q}].$$

Remark 9.B.17. Let K be a number field, let $\wp$ be a nonzero prime ideal of O_K and let $x \in O_K$ be prime to $\wp$ (i.e. $\wp$ does not appear in the prime factorization of xO_K or, equivalently, $xO_K + \wp = O_K$). Since $O_K/\wp$ is a field with $N(\wp)$ elements, we obtain from Lagrange's theorem the following useful analogue of Fermat's little theorem: $x^{N(\wp)-1} \equiv 1 \pmod{\wp}$.

Consider now a finite extension L/K of number fields and let $\wp$ be a nonzero prime ideal of O_K. Let

$$\wp O_L = \{a_1 x_1 + a_2 x_2 + \dots + a_n x_n | n \geq 1, a_i \in \wp, x_i \in O_L\},$$

which is easily seen to be a nonzero ideal of O_L. Moreover, $\wp O_L \neq O_L$, as otherwise we would have $1 = a_1x_1 + \cdots + a_nx_n$ for some $a_i \in \wp$ and some $x_i \in O_L$ and by taking norms, we would get $1 \in \wp$. Hence, by the previous theorem, there exists a prime β of O_L dividing $\wp O_L$. By the same theorem, this simply means that $\wp \subset \beta$. Note that β is not unique, but there are only finitely many possibilities for it. On the other hand, given a nonzero prime ideal β of O_L, there is a unique prime $\wp$ of O_K such that $\wp \subset \beta$. Indeed, the existence is obtained by taking $\wp = \beta \cap O_K$ (it is an easy exercise for the reader to check that this is indeed a nonzero prime ideal of O_K) and the uniqueness follows from the fact that any nonzero prime ideal of O_K is maximal (hence, if $\wp_1 \subset \beta$ and $\wp_2 \subset \beta$ for some different primes $\wp_j$ of O_K, then $1 \in \wp_1 + \wp_2 \subset \beta$, a contradiction).

Definition 9.B.18. Let L/K be a finite extension of number fields and let $\wp$ be a nonzero prime ideal of O_K. A prime ideal β of O_L is said to lie above $\wp$ if $\wp \subset \beta$. We also write $\beta|\wp$ in this case.

We can resume the previous discussion by saying that any prime β of O_L has a unique prime $\wp$ of O_K below it, namely $\beta \cap O_K$ and any prime $\wp$ of O_K has at least one (but finitely many) prime β above it. Note that if $\beta|\wp$, then the inclusion $O_K \subset O_L$ induces an injective morphism $O_K/\wp \to O_L/\beta$, realizing therefore O_L/β as a finite extension of $O_K/\wp$.

Definition 9.B.19. Let $\beta|\wp$ be as above.

1) The residual degree of $\beta/\wp$ is defined by $f(\beta/\wp) = [O_L/\beta : O_K/\wp]$.

2) The ramification index of $\beta/\wp$ is the exponent $e(\beta/\wp)$ of β in the prime factorization of $\wp O_L$.

Note that by definition we have

$$\wp O_L = \prod_{\beta|\wp} \beta^{e(\beta/\wp)}.$$

Using this and proposition 9.B.3, we easily obtain the following useful property of the ramification index and residual degree in a tower of number fields.

Proposition 9.B.20. *Let $M/L/K$ be a tower of finite extensions of number fields and let $\rho|\beta|\wp$ be primes of O_M, O_L and O_K. Then*

$$e(\rho/\beta)\cdot e(\beta/\wp)=e(\rho/\wp),\quad f(\rho/\beta)\cdot f(\beta/\wp)=f(\rho/\wp).$$

Definition 9.B.21. Let L/K be a finite extension of number fields. A prime $\wp$ of O_K is called

a) unramified in L if $e(\beta/\wp)=1$ for all $\beta|\wp$ in L.

b) totally split in L if $e(\beta/\wp)=1$ and $f(\beta/\wp)=1$ for all $\beta|\wp$ in L. This is equivalent[9] to: $\wp O_L$ is the product of $[L:K]$ different prime ideals of O_L.

The following theorem gives a practical way to factor a prime in a number field. The precise statement is a bit complicated, but the message is very simple: for most primes p, factoring pO_K in O_K comes down to factoring $\overline{f}\in\mathbb{F}_p[X]$, where f is the minimal polynomial of any primitive element of K that lives in O_K.

Theorem 9.B.22. *(Dedekind, Kummer) Let $K=\mathbb{Q}(x)$ be a number field, where $x\in O_K$ has minimal polynomial f. Let p be a prime which does not divide*[10] *$|O_K/\mathbb{Z}[x]|$ and let*

$$\overline{f}=\prod_{i=1}^{g}\overline{f_i}^{e_i}$$

be the prime factorization of $\overline{f}\in\mathbb{F}_p[X]$. Then

$$pO_K=\prod_{i=1}^{g}\wp_i^{e_i},$$

where $\wp_i=pO_K+f_i(x)O_K$ are different prime ideals and $N(\wp_i)=p^{\deg\overline{f_i}}$.

[9] We have $\sum_{\beta|\wp}e(\beta/\wp)\cdot f(\beta/\wp)=[L:K]$, by an argument similar to that used in remark 9.B.16.

[10] Note that the result stated in remark 9.B.11 shows that $O_K/\mathbb{Z}[x]$ is a finite set.

Proof. Lift arbitrarily $\overline{f_i}$ to some monic polynomials $f_i \in \mathbb{Z}[X]$. The key point is to prove that the natural map $\mathbb{Z}[X]/(p, f_i) \to \mathbb{Z}[x]/(p, f_i(x))$ sending (the class of) h to (the class of) $h(x)$ is an isomorphism of rings. Assume for a moment that we proved this. Let $a = |O_K/\mathbb{Z}[x]|$, so $\gcd(p, a) = 1$ and $aO_K \subset \mathbb{Z}[x]$. A standard argument using Bézout's lemma shows that $O_K = pO_K + \mathbb{Z}[x]$, thus $O_K/\wp_i$ is naturally isomorphic to $\mathbb{Z}[x]/(p, f_i(x))$. Since $\mathbb{Z}[X]/(p, f_i(X))$ is clearly isomorphic to $\mathbb{F}_p[X]/(\overline{f_i})$, which is a field with $p^{\deg \overline{f_i}}$ elements, it follows from the previous discussion that the $\wp_i$ are different prime ideals with $N(\wp_i) = p^{\deg \overline{f_i}}$. Since $\wp_i^{e_i} \subset pO_K + f_i(x)^{e_i} O_K$, $f(x) = 0$ and $f \equiv \prod_{i=1}^{g} f_i^{e_i} \pmod{p\mathbb{Z}[X]}$, it follows that $\prod_{i=1}^{g} \wp_i^{e_i} \subset pO_K$ and so by theorem 9.B.14 we have $pO_K | \prod_{i=1}^{g} \wp_i^{e_i}$. So, if $pO_K = \prod_{i=1}^{g} \wp_i^{s_i}$, we must have $s_i \le e_i$. We conclude by noting that both sums $\sum_i e_i \cdot \deg \overline{f_i}$ and $\sum_i s_i \cdot \deg \overline{f_i}$ are equal to $[K : \mathbb{Q}]$ (see remark 9.B.16), so we must have $s_i = e_i$ for all i.

Let us prove now that $\mathbb{Z}[X]/(p, f_i) \to \mathbb{Z}[x]/(p, f_i(x))$ is bijective. Since surjectivity is clear, it remains to prove the following assertion: if $h \in \mathbb{Z}[X]$ satisfies $h(x) \in p\mathbb{Z}[x] + f_i(x)\mathbb{Z}[x]$, then $h \in p\mathbb{Z}[X] + f_i(X) \cdot \mathbb{Z}[X]$. Write $h(x) = pA(x) + f_i(x)B(x)$ for some $A, B \in \mathbb{Z}[X]$. Since f is the minimal polynomial of x, there is $r \in \mathbb{Z}[X]$ such that $h = pA + f_iB + rf$. It suffices to use again that $f \in p\mathbb{Z}[X] + f_i \cdot \mathbb{Z}[X]$ to finish the proof. □

Remark 9.B.23. It is not very difficult to prove that if p does not divide the discriminant of f, then p does not divide $|O_K/\mathbb{Z}[x]|$, so the theorem can be applied.

9.B.5 Two classical examples

Consider a squarefree integer $d \neq \pm 1$ and let $K = \mathbb{Q}(\sqrt{d})$. We saw that $O_K = \mathbb{Z}[x]$, where $x = \frac{1+\sqrt{d}}{2}$ if $d \equiv 1 \pmod 4$ and $x = \sqrt{d}$ otherwise. The minimal polynomial of x is $X^2 - X + \frac{1-d}{4}$ in the first case and $X^2 - d$ in the second case. Theorem 9.B.22 shows that in order to understand the prime factorization of pO_K, we need to understand the prime factorization of these polynomials modulo p. Since a quadratic polynomial modulo p is irreducible if and only if it has a root in $\mathbb{F}_p$, we easily deduce the following

Proposition 9.B.24. *Let $d \neq \pm 1$ be a squarefree integer and let $K = \mathbb{Q}(\sqrt{d})$. Let p be a prime.*

a) If $p > 2$, then pO_K is a prime ideal if $\left(\frac{d}{p}\right) = -1$, a product of two different prime ideals if $\left(\frac{d}{p}\right) = 1$ and the square of a prime ideal if $p|d$.

b) If $p = 2$, then pO_K is the square of a prime ideal if $d \equiv 2, 6, 3, 7 \pmod 8$, a prime ideal if $d \equiv 5 \pmod 8$ and a product of two different prime ideals if $d \equiv 1 \pmod 8$.

Consider now $n > 1$ and let $K = \mathbb{Q}(\zeta_n)$, where ζ_n is a primitive nth root of unity. As we have already said, it is rather difficult to prove that $O_K = \mathbb{Z}[\zeta_n]$ and we will take this for granted (actually, it is easier to use remark 9.B.23). In this case, theorem 9.B.22 reduces the prime factorization of pO_K to that of $\phi_n \in \mathbb{F}_p[X]$, where ϕ_n is the nth cyclotomic polynomial (whose roots are precisely the primitive nth roots of unity). Assume that p does not divide n and let g be an irreducible factor of degree f of $\phi_n \in \mathbb{F}_p[X]$. Let x be a root of g in an algebraic closure of $\mathbb{F}_p$. Theorem 9.A.4 shows that $g(X) = \prod_{i=0}^{f-1}(X - x^{p^i})$. Let d be the order of p modulo n. Since n is the least positive power of x equal to 1 (because x is a root of ϕ_n), it follows that the sequence x^{p^i} is periodic with period d, so we must have $f = d$. That is, ϕ_n factors mod p as a product of irreducible polynomials of degree d, the order of p modulo n. Since $X^n - 1$ is squarefree modulo p (because the hypothesis that p does not divide n implies that $X^n - 1$ is relatively prime to its derivative), we deduce that in theorem 9.B.22 we have $e_1 = e_2 = \cdots = e_g = 1$ and $f_i = d$ for all i, so $g = \frac{\varphi(n)}{d}$. Assume now that $p|n$ and let $n = p^k \cdot m$ for some m relatively prime to p. It is easy to see that $\phi_n(X) = \phi_m(X^{p^k})/\phi_m(X^{p^{k-1}})$, so modulo p we have $\phi_n = \phi_m^{p^k - p^{k-1}}$. This reduces the problem of factoring pO_K to the previous case. All in all, we have the following useful result:

Proposition 9.B.25. *Let n be an integer greater than 1, let $K = \mathbb{Q}(\zeta_n)$ and let p be a prime.*

a) If $\gcd(n,p) = 1$, then pO_K is a product of $\frac{\varphi(n)}{\operatorname{ord}(p \mod n)}$ different prime ideals, each of degree $\operatorname{ord}(p \mod n)$.

b) If $p|n$ and $n = p^k \cdot m$ with $\gcd(m, p) = 1$, *then* $pO_K = (\wp_1 \cdot \dots \cdot \wp_s)^{p^k - p^{k-1}}$, *where* $s = \frac{\varphi(m)}{\mathrm{ord}(p \mod m)}$ *and each* $\wp_i$ *is of degree* $\mathrm{ord}(p \mod m)$.

9.B.6 The primitive element theorem and embeddings of number fields

When working with subfields of $\mathbb{C}$, the following result is very handy. We will use it constantly to shorten proofs of results which actually hold in much greater generality.

Theorem 9.B.26. *(primitive element theorem) Let L/K be a finite extension of subfields of $\mathbb{C}$. Then there exists $l \in L$ such that $L = K(l)$.*

Proof. As L/K is finite, there are elements $x_1, \dots, x_n \in L$ such that $L = K(x_1, x_2, \dots, x_n)$ (for instance, the elements of a K-basis of L over K). Thus, by induction on n it is enough to prove that if x, y are algebraic over K, then there exists $l \in L = K(x, y)$ such that $L = K(l)$.

Let $f, g \in K[X]$ be the minimal polynomials of x, y respectively and let $x_1 = x, x_2, \dots, x_n$ and $y_1 = y, y_2, \dots, y_m$ be their roots. Clearly, there exists $c \in K$ such that $x_i + cy_j \neq x + cy$ for all $(i, j) \neq (1, 1)$ (each of the previous linear equations in c has at most one solution in K and K is infinite). We claim that $l = x + cy$ works. Clearly $K(l) \subset L$, so it is enough to check that $x, y \in K(l)$. As $l = x + cy$, it is enough to do it for y. Now, we know $f(l - cy) = 0$, so the polynomial $f(l - cX) \in K(l)[X]$ has a common root y with g. But by construction this is the only common root of these polynomials. Moreover, it is a simple root, as g is irreducible over $\mathbb{Q}$, so it has simple roots. Finally, we conclude that the greatest common divisor of these two polynomials is $X - y$. As these two polynomials have coefficients in $K(l)$, so does their greatest common divisor and so $y \in K(l)$. The result follows. $\square$

We will use the primitive element theorem to prove some basic results on the structure of number fields. The first one concerns the embeddings of a number field in $\mathbb{C}$. Note that $\mathbb{Q}(\sqrt{2})$ has two such embeddings, namely the identity map and $a + b\sqrt{2} \mapsto a - b\sqrt{2}$. It turns out that a number field with degree d has exactly d embeddings in $\mathbb{C}$.

Theorem 9.B.27. *Let L/K be an extension of number fields (L/K is automatically finite). Then any embedding $K \to \mathbb{C}$ extends to exactly $[L : K]$ embeddings $L \to \mathbb{C}$.*

Proof. Fix an embedding $\sigma : K \to \mathbb{C}$ and use the primitive element theorem to write $L = K(l)$ for some $l \in L$. Then l is algebraic over K, of degree $d = [L : K]$, with minimal polynomial $f \in K[X]$. Suppose that $\sigma' : L \to \mathbb{C}$ is an embedding that extends σ, so $\sigma'(x) = \sigma(x)$ for all $x \in K$. Thus, if $g \in K[X]$, then $\sigma'(g(l)) = g^{\sigma}(\sigma'(l))$ (where g^{σ} is the polynomial obtained from g by applying σ to its coefficients) and so σ' is determined by $\sigma'(l)$, which has to be a root of f^{σ} (use the previous equality with $g = f$). Conversely, if l' is a root of f^{σ}, we can define an embedding $\sigma'(g(l)) = g^{\sigma}(l')$ for all $g \in K[X]$, well-defined because any two polynomials g and h with $g(l) = h(l)$ differ by a multiple of f. Thus the embeddings of L into $\mathbb{C}$ that extend σ are in bijection with roots of f^{σ} and the result follows. □

Taking $K = \mathbb{Q}$ in the previous theorem, we deduce that any number field L of degree d over $\mathbb{Q}$ embeds in exactly d ways in $\mathbb{C}$.

9.B.7 A bit of Galois theory

Let L, M be two extensions of a field K and suppose for simplicity that they are contained in $\mathbb{C}$. A K-morphism $L \to M$ is a K-linear map from L to M which is also a ring homomorphism. Stated differently (but equivalently), it is an additive and multiplicative map $f : L \to M$ such that $f(x) = x$ for all $x \in K$.

Definition 9.B.28. Let $f \in K[X]$. The splitting field of f is the field $K(x_1, x_2, \ldots, x_n)$, where $x_i \in \mathbb{C}$ are the roots of f.

Note that a K-morphism $f : K(x_1, \ldots, x_n) \to M$ (where M/K is an extension) is uniquely determined by the values $f(x_i)$, as any element of $K(x_1, \ldots, x_n)$ is a polynomial with coefficients in K in the x_i's. However, $f(x_i)$ cannot be any element of M, since if $P \in K[X]$ kills x_i, then P also kills $f(x_i)$: note that $P(f(x_i)) = f(P(x_i)) = 0$, as f is K-linear and multiplicative. Moreover, there might be algebraic relations with coefficients in K

between the x_i's and the same argument shows that the numbers $f(x_i)$ must satisfy these relations. Thus, it is a fairly delicate issue to understand these K-morphisms. This is the content of Galois theory. Let us make an important definition first:

Definition 9.B.29. Let L be an extension of a field K. The Galois group of L over K, denoted $\mathrm{Gal}(L/K)$ is the set of bijective K-morphisms $f : L \to L$.

We can now prove one basic result of Galois theory, which is also of constant use:

Theorem 9.B.30. *Let L/K be an extension of number fields. Then*

$$|\mathrm{Gal}(L/K)| \leq [L : K],$$

with equality if and only if L is the splitting field of some polynomial $f \in K[X]$.

Proof. Let $L = K(l)$ for some $l \in L$. Each element $\sigma \in \mathrm{Gal}(L/K)$ is uniquely determined by $\sigma(l)$, which must be a conjugate of l. Thus we have a natural injection of $\mathrm{Gal}(L/K)$ in the set of conjugates of l, which has $[L : K]$ elements. The inequality follows.

Suppose that we have equality and let $f \in K[X]$ be the minimal polynomial of l. We will prove that L is the splitting field of f. It is enough to prove that if x is a conjugate of l, then $x \in L$. But the first paragraph and the equality case implies that if $l_1, \dots, l_n$ are the conjugates of l, then $\{\sigma(l) | \sigma \in \mathrm{Gal}(L/K)\} = \{l_1, \dots, l_n\}$. Thus there exists $\sigma \in \mathrm{Gal}(L/K)$ such that $x = \sigma(l)$. Since $\sigma(L) \subset L$, we have $x \in L$ and we are done.

Conversely, suppose that L is the splitting field of some $f \in K[X]$, with roots $x_1, x_2, \dots, x_n$. Theorem 9.B.27 applied to the natural inclusion map $K \to \mathbb{C}$ shows that there are $[L : K]$ K-linear morphisms of rings $L \to \mathbb{C}$. But for any such morphism $\sigma : L \to \mathbb{C}$ we have $\sigma(x_i) \in L$, as $\sigma(x_i)$ is just some root of f and L contains all these roots. Thus the image of any such morphism is a subset of L, i.e. any such morphism is an element of $\mathrm{Gal}(L/K)$. The result follows. □

Definition 9.B.31. We say that a finite extension L/K of number fields is Galois if $|\mathrm{Gal}(L/K)| = [L : K]$, or, equivalently, if L is the splitting field of a polynomial with coefficients in K.

We are now able to prove the main theorem of Galois theory for number fields. The result holds in much greater generality, but we will not need it and we prefer to use all the extra data in order to shorten the proof rather dramatically.

Theorem 9.B.32. *Let L/K be a finite Galois extension of number fields. Sending a subfield M of L containing K to $\mathrm{Gal}(L/M)$ yields a bijection between subfields of L which contain K and subgroups of $\mathrm{Gal}(L/K)$. The inverse of this bijection is the map sending the subgroup H of $\mathrm{Gal}(L/K)$ to*

$$L^H := \{x \in L | \sigma(x) = x, \forall \sigma \in H\}.$$

Moreover, we have $H_1 \subset H_2$ if and only if L^{H_1} contains L^{H_2}.

Proof. Let us prove first that $L^{\mathrm{Gal}(L/M)} = M$ for any intermediate field M between L and K. Note that L/M is again finite and Galois (as L is the splitting field of a polynomial with coefficients in K, thus also in M). Write $L = M(l)$ for some primitive element $l \in L$, with conjugates $l_1, l_2, \ldots, l_n$ over M. Suppose that $x \in L^{\mathrm{Gal}(L/M)}$ is not in M. Thus, we can find $f \in M[X]$ nonconstant of degree less than n and such that $x = f(l)$. We saw in the proof of the previous theorem that $\{\sigma(l) | \sigma \in \mathrm{Gal}(L/M)\} = \{l_1, \ldots, l_n\}$. So, for any i we can find $\sigma_i \in \mathrm{Gal}(L/M)$ such that $\sigma_i(l) = l_i$. Then $x = \sigma_i(x) = f(l_i)$. Hence $f(l_1) = f(l_2) = \cdots = f(l_n)$, contradicting the fact that f is nonconstant of degree less than n. The result follows.

Next, we need to prove that if H is any subgroup of $\mathrm{Gal}(L/K)$, then $\mathrm{Gal}(L/L^H) = H$. By the very definition of L^H we have an inclusion $H \subset \mathrm{Gal}(L/L^H)$. It is thus enough to check that $|\mathrm{Gal}(L/L^H)| \leq |H|$ and, using the previous theorem, it is enough to check that $[L : L^H] \leq |H|$. Write $L = L^H(l)$ for some $l \in L$. Consider $f(X) = \prod_{\sigma \in H}(X - \sigma(l))$. This is a polynomial of degree $|H|$ whose coefficients are clearly in L^H. Hence l has degree at most $|H|$ over L^H and the result follows.

It remains to check that $H_1 \subset H_2$ if and only if L^{H_1} contains L^{H_2}. It is clear that if $H_1 \subset H_2$, then L^{H_1} contains L^{H_2} (any element of L fixed by H_2 is also fixed by H_1). Assume that L^{H_1} contains L^{H_2}, so $L^{H_1} \cap L^{H_2} = L^{H_2}$. But it is clear that the left-hand side of this equality is simply L^H, where H

is the subgroup of $\mathrm{Gal}(L/K)$ generated by H_1 and H_2. Hence $L^H = L^{H_2}$ and, as we have seen, this forces $H = H_2$ and so $H_1 \subset H_2$. $\square$

Remark 9.B.33. Using similar arguments, it is not difficult to show that H is normal in $\mathrm{Gal}(L/K)$ (i.e. $gHg^{-1} = H$ for any $g \in \mathrm{Gal}(L/K)$) if and only if L^H/K is Galois.

9.B.8 Prime factorization in a Galois extension

The results in this section will be crucially used in the applications that will be presented at the end of this addendum. They are absolutely fundamental in algebraic number theory.

Theorem 9.B.34. *Let L/K be a Galois extension of number fields, with $G = \mathrm{Gal}(L/K)$. Let $\wp$ be a nonzero prime ideal of O_K and let β_1 and β_2 be two prime ideals of O_L which lie over $\wp$. Then there exists $\sigma \in G$ such that*[11] *$\beta_2 = \sigma(\beta_1)$.*

Proof. Suppose that $\beta_2 \neq \sigma(\beta_1)$ for all $\sigma \in G$. Using the general version[12] of the Chinese Remainder Theorem, we obtain the existence of $a \in \beta_2$ such that $a \notin \sigma(\beta_1)$ for all $\sigma \in G$. Then $\prod_{\sigma \in G} \sigma(a) \in O_K \cap \beta_2 = \wp \subset \beta_1$ and this contradicts the fact that β_1 is a prime ideal and $a \notin \sigma^{-1}(\beta_1)$ for any $\sigma \in G$. The result follows. $\square$

Corollary 9.B.35. *Let L/K be a Galois extension of number fields and let $\wp$ be a nonzero prime ideal of O_K. Then all primes above $\wp$ have the same ramification index and the same residual degree.*

Proof. For the residual degrees, note that any $\sigma \in G$ induces a bijection between O_L/β and $O_L/\sigma(\beta)$, so these two sets have the same number of elements. We conclude by the previous theorem. Next, if e is a positive integer such that

[11] Note that by definition $\sigma(\beta_1)$ is the set of all $\sigma(x)$, with $x \in \beta_1$. It is easy to check that this is again a prime ideal of O_L and that $\sigma(\beta_1) \cap O_K = \wp$.

[12] This is stated as follows: let A be a commutative ring and let $I_1, I_2, \ldots, I_n$ be ideals such that $I_i + I_j = A$ for all $i \neq j$ (which is satisfied if I_i are different maximal ideals of A, for instance). Then for any $x_1, \ldots, x_n \in A$ there exists $x \in A$ such that $x - x_i \in I_i$ for all i.

β^e divides $\wp O_L$, then $\sigma(\beta)^e$ divides $\sigma(\wp O_L) = \wp O_L$. The result follows again easily from the previous theorem. □

Definition 9.B.36. Let L/K be a Galois extension of number fields, with Galois group G. Let β be a prime of O_L. The decomposition group of β is

$$D_\beta = \{\sigma \in G | \sigma(\beta) = \beta\}.$$

Note that D_β is indeed a group and that any $\sigma \in D_\beta$ induces an automorphism $\sigma : O_L/\beta \to O_L/\beta$, which is trivial on $O_K/\wp$ for $\wp = \beta \cap O_K$. Hence, we have a natural map $D_\beta \to \mathrm{Gal}((O_L/\beta)/(O_K/\wp))$. Note that $\mathrm{Gal}((O_L/\beta)/(O_K/\wp))$ is a cyclic group, generated by the automorphism $x \to x^{N(\wp)}$, since $(O_L/\beta)/(O_K/\wp)$ is a finite extension of finite fields (see the addendum 9.A for the structure of finite fields). The following result is rather tricky, but fundamental.

Theorem 9.B.37. *With the previous notations, the map*

$$D_\beta \to \mathrm{Gal}((O_L/\beta)/(O_K/\wp))$$

is surjective.

Proof. We may assume that $O_L/\beta \neq O_K/\wp$, as otherwise the statement is trivial. Let α be a generator of the cyclic group $(O_L/\beta)^*$, so that clearly $O_L/\beta = (O_K/\wp)[\alpha]$. Using the general form of the Chinese Remainder Theorem (see the proof of theorem 9.B.34), we can find $a \in O_L$ such that $a \pmod{\beta} = \alpha$ and $a \in \beta'$ for any $\beta' \neq \beta$ above $\wp$. Let F be the minimal polynomial of a over K. Then $F \in O_K[X]$ and $F(a) = 0$, hence $\overline{F}(\alpha) = 0$ in O_L/β. But then[13] $\overline{F}(\alpha^{N(\wp)}) = 0$, so that $F(a^{N(\wp)}) \in \beta$. So, we can find a conjugate b of a so that $b - a^{N(\wp)} \in \beta$. It is not difficult[14] to see that there

[13] Recall that if $\mathbb{F}_q$ is the field with q elements, then for all $f \in \mathbb{F}_q[X]$ we have $f(X^q) = f(X)^q$, so if α is a root of f in some extension of $\mathbb{F}_q$, then α^q is also a root of f.

[14] Since L/K is Galois, L contains all conjugates of a over K, i.e. all roots of F. Let M be the field generated over K by these conjugates, i.e. the splitting field of K. Then M/K is a Galois subextension of L/K, thus any element of $\mathrm{Gal}(M/K)$ extends to an element of $\mathrm{Gal}(L/K)$. But the isomorphisms $K[X]/(F) \to M$ sending X to a and b respectively yield $\sigma \in \mathrm{Gal}(M/K)$ such that $\sigma(a) = b$.

exists $\sigma \in \mathrm{Gal}(L/K)$ such that $\sigma(a) = b$. We claim that σ is in D_β and that it induces the automorphism $x \to x^{N(\wp)}$ of O_L/β. First, if $\sigma(\beta) \neq \beta$, then $\sigma^{-1}(\beta) \neq \beta$ and so by our choice of a we have $a \in \sigma^{-1}(\beta)$, i.e. $\sigma(a) \in \beta$. But $\sigma(a) - a^{N(\wp)} \in \beta$, forcing $a^{N(\wp)} \in \beta$ and then $a \in \beta$, a contradiction with $a \pmod{\beta} = \alpha \neq 0$. Thus $\sigma \in D_\beta$. The automorphism induced by σ on O_L/β is uniquely determined by its action on α (generator of $(O_L/\beta)^*$) and by our choice this action is $\alpha^{N(\wp)}$. The result follows. □

Let us fix a prime β above $\wp$ in L. By theorem 9.B.34, all primes above $\wp$ are of the form $\sigma(\beta)$ with $\sigma \in \mathrm{Gal}(L/K)$, i.e. $\mathrm{Gal}(L/K)$ permutes transitively the different primes $\beta_1, \ldots, \beta_g$ over $\wp$. Since there are precisely $|D_\beta|$ elements of $\mathrm{Gal}(L/K)$ which fix β, we obtain $[L : K] = |\mathrm{Gal}(L/K)| = g \cdot |D_\beta|$. But we also have $[L : K] = e(\beta/\wp) \cdot f(\beta/\wp) \cdot g$, since all ramification indices and residual degrees of the β_i's are the same. Hence $|D_\beta| = e(\beta/\wp) \cdot f(\beta/\wp)$. In particular, if $\wp$ is unramified in L, i.e. if $e(\beta/\wp) = 1$, then $|D_\beta| = f(\beta/\wp)$, which is the same as the cardinality of $\mathrm{Gal}((O_L/\beta)/(O_K/\wp))$. We obtain therefore the following crucial result:

Theorem 9.B.38. *Let L/K be a Galois extension of number fields and let $\wp$ be a prime of O_K unramified in O_L. Let $\beta|\wp$ be a prime of O_L over $\wp$. Then the map $D_\beta \to \mathrm{Gal}((O_L/\beta)/(O_K/\wp))$ is a bijection. In particular, there exists a unique $(\beta, L/K) \in D_\beta$ such that $(\beta, L/K)(x) \equiv x^{N(\wp)}$ for all $x \in O_L$. We call $(\beta, L/K)$ the Frobenius substitution of β in L/K.*

Remark 9.B.39. Assume that we are only given $\wp$. The choice of some β over $\wp$ yields a Frobenius substitution $(\beta, L/K)$, but this depends on the choice of β. However, any other prime above $\wp$ is of the form $\sigma(\beta)$ for some $\sigma \in \mathrm{Gal}(L/K)$ and it is immediate to check that $(\sigma(\beta), L/K) = \sigma \circ (\beta, L/K) \circ \sigma^{-1}$ (indeed, the right-hand side is in $D_{\sigma(\beta)}$, which is trivially equal to $\sigma D_\beta \sigma^{-1}$, and satisfies the congruence that uniquely characterizes it). Hence, the conjugacy class of $(\beta, L/K)$ in $\mathrm{Gal}(L/K)$ does not depend on the choice of β and is called the Frobenius conjugacy class of $\wp$. Note that if $\mathrm{Gal}(L/K)$ is abelian, then this conjugacy class is reduced to an element, which we denote $(\wp, L/K)$: it is $(\beta, L/K)$ for any β over $\wp$.

We leave to the reader to check the following easy results:

Proposition 9.B.40. *Let $M/L/K$ be a tower of Galois extensions of number fields and let $\rho|\beta|\wp$ be a tower of prime ideals. Assume that $\wp$ is unramified in M (thus β is unramified in M and $\wp$ is unramified in L). Then we have the following relations.*

a) $(\rho, M/L) = (\rho, M/K)^{f(\beta/\wp)}$. *Actually, this relation still holds if M/K is not necessarily Galois.*

b) $(\rho, M/K)|_L = (\beta, L/K)$.

9.B.9 Bauer's theorem and Chebotarev's density theorem

Before stating the following fundamental result, we need one more definition. If S is a set of primes, its Dirichlet density is

$$d(S) = \lim_{s\to 1} \frac{\sum_{p\in S} p^{-s}}{\sum_p p^{-s}},$$

if the limit exists. The next very deep theorem was conjectured (and proved in some special cases) by Frobenius. It is a vast generalization of Dirichlet's theorem.

Theorem 9.B.41. *(Chebotarev) Let K be a finite Galois extension of $\mathbb{Q}$ of degree n and let $g \in Gal(K/\mathbb{Q})$. Let S be the set of primes p such that $(\wp, K/\mathbb{Q})$ is conjugate to g for all prime divisors $\wp$ of p. Then $d(S) = \frac{1}{n}\cdot|\{hgh^{-1}|h \in G\}|$.*

Here is a typical application of this deep theorem.

Example 9.B.42. Let l be a prime and consider the set S of those primes $p \equiv 1 \pmod{l}$ such that $2^{\frac{p-1}{l}} \equiv 1 \pmod{p}$. These two conditions are equivalent to the statement that $X^l - 2$ splits into distinct linear factors in $\mathbb{F}_p[X]$ (as $2^{\frac{p-1}{l}} \equiv 1 \pmod{p}$ is equivalent to the existence of $y \in \mathbb{F}_p$ such that $y^l = 2$). This is also equivalent to p being a product of different primes in O_K, where $K = \mathbb{Q}(e^{\frac{2i\pi}{l}}, \sqrt[l]{2})$ is the splitting field of $X^l - 2$. Using again Chebotarev's theorem, we deduce that $d(S) = \frac{1}{[K:\mathbb{Q}]}$. But K contains $\mathbb{Q}(e^{\frac{2i\pi}{l}})$ and $\mathbb{Q}(\sqrt[l]{2})$, which have degrees $l-1$, respectively l. Thus $[K:\mathbb{Q}]$ is a multiple of $l(l-1)$ and so necessarily $[K:\mathbb{Q}] = l(l-1)$ and $d(S) = \frac{1}{l(l-1)}$.

We will also need the following easy consequence of Chebotarev's density theorem:

Proposition 9.B.43. *Let $\sigma \in \mathrm{Gal}(L/K)$. There are infinitely many prime ideals $\wp$ of degree 1 of K such that one can find $\beta|\wp$ in L with $(\beta, L/K) = \sigma$.*

Proof. Take a Galois extension E over $\mathbb{Q}$ which contains L. Since L/K is Galois, we can extend σ to an automorphism again denoted σ of $\mathrm{Gal}(E/K)$. By Chebotarev, we can find infinitely many primes p such that there is $\rho|p$ in E with $(\rho, E/\mathbb{Q}) = \sigma$. Then $D_\rho \subset \mathrm{Gal}(E/K)$, so $\wp = \rho \cap K$ has degree 1 (lemma 9.B.49). Let $\beta = \rho \cap O_L$. Then

$$(\beta, L/K) = (\rho, E/K)|_L = (\rho, E/\mathbb{Q})|_L = \sigma$$

and we are done. □

Definition 9.B.44. If K is a number field, let $P_1(K)$ be the set of rational primes p for which pO_K has at least one prime ideal factor $\wp$ of residual degree 1 (i.e. such that $O_K/\wp$ is the field with p elements).

The following result explains why $P_1(K)$ is an interesting object:

Proposition 9.B.45. *Let $f \in \mathbb{Z}[X]$ be a monic irreducible polynomial, let θ be a root of f and let $K = \mathbb{Q}(\theta)$. Then there is c such that*

$$P_1(K) \cap [c, \infty) = \{p \geq c | \exists x \in \mathbb{Z} \textit{such that} \quad p|f(x)\}.$$

Proof. If p is a sufficiently large prime, then the residual degrees of the prime divisors of pO_K are given by the degrees of the irreducible factors of $\overline{f} \in \mathbb{F}_p[X]$, by theorem 9.B.14. Hence, for such p we have $p \in P_1(K)$ if and only if $\overline{f}$ has a linear factor in $\mathbb{F}_p[X]$, i.e. if and only if there exists $x \in \mathbb{Z}$ such that $p|f(x)$. □

Here is a rather nice application of this proposition:

Theorem 9.B.46. *(Nagell) If $f \in \mathbb{Z}[X]$, let $P(f)$ be the set of primes p for which the congruence $f(x) \equiv 0 \pmod{p}$ has at least one solution. If $f_1, f_2, \ldots, f_k$ are nonconstant polynomials with integer coefficients, then $P(f_1) \cap P(f_2) \cap \cdots \cap P(f_k)$ is infinite.*

Proof. The case $k = 1$ is a classical result due to Schur, but for the reader's convenience let us recall the proof. If $f_1(0) = 0$, everything is clear. Consider the numbers $x_n = \frac{f((2n)!\cdot f(0))}{f(0)}$. We have $|x_n| > 1$ and $x_n \equiv 1 \pmod{n!}$ if n is large enough, so there exists $p_n | x_n$ with $p_n > n$ and the result follows.

The general case is more difficult. We may assume that f_i are irreducible monic. Let z_i be roots of f_i and let $K_i = \mathbb{Q}(z_i)$. Let K be the least number field containing all K_i's and write $K = \mathbb{Q}(x)$ for some $x \in O_K$ with minimal polynomial f. Applying the case $k = 1$ to f and using the previous proposition, we see that we can find infinitely many p which have a prime $\wp$ of degree 1 in K. Let $\beta_i = \wp \cap O_{K_i}$. Since $f(\wp/p) = 1$, we also have $f(\beta_i/p) = 1$. Applying once more theorem 9.B.22 (this time for each K_i) we see that all but finitely many primes p (among the infinitely many we have just found) belong to $\cap_{i=1}^{k} P(f_i)$, which is therefore infinite. □

The following beautiful theorem due to Bauer was discovered before Chebotarev proved his theorem. It is however more conceptual nowadays to see Bauer's theorem as a consequence of Chebotarev's density theorem, which is what we will do.

Theorem 9.B.47. *(Bauer) Let K_1 be a number field and let K_2 be a number field which is Galois over $\mathbb{Q}$. Suppose that there exists a set of primes S of Dirichlet density 0, with the following property: if $p \notin S$ is in $P_1(K_1)$, then p is completely split in K_2. Then $K_2 \subset K_1$.*

Proof. We start with two easy lemmas, which express the condition $p \in P_1(K)$ in a group-theoretic way. They are also results of independent interest and will be used in other applications.

Lemma 9.B.48. *Let L be a number field which is Galois over $\mathbb{Q}$, with group G. Let $\mathbb{Q} \subset K \subset L$ be a subfield and let $H = \mathrm{Gal}(L/K)$. Suppose that $\beta|\wp|p$ is a chain of prime ideals in $L, K, \mathbb{Q}$ such that p is unramified in L. Then $f(\wp/p) = 1$ if and only if $D_\beta \subset H$.*

Proof. Note that the decomposition group of β with respect to the extension L/K is simply $D_\beta \cap H$. But since p is unramified in L, $\wp$ is unramified in L/K and so the decomposition group of β with respect to L/K has $f(\beta/\wp)$

elements. That is, $|H \cap D_\beta| = f(\beta/\wp)$. But this equals also $\frac{f(\beta/p)}{f(\wp/p)} = \frac{|D_\beta|}{f(\wp/p)}$. The result follows. $\square$

Lemma 9.B.49. *Assume that L and K are as in the previous lemma and let p be a prime which is unramified in L. Then*

1) p has a prime factor $\wp$ in K with $f(\wp/p) = 1$ if and only if there is $\beta|p$ in L such that $D_\beta \subset H$.

2) p is completely split in K if and only if for all $\beta|p$ in L, we have $D_\beta \subset H$.

Proof. 1) is an immediate consequence of the previous lemma. For 2), it is clear that if p is completely split in K, then for any $\beta|p$ we have $D_\beta \subset H$ (as if $\wp = \beta \cap O_K$, then necessarily $f(\wp/p) = 1$). Conversely, if $D_\beta \subset H$ for any $\beta|p$, then the previous lemma implies that $f(\wp/p) = 1$ for any $\wp|p$ in K. But note that p is unramified in K, as it is already in L. Thus p must be completely split in K. $\square$

Let us prove now the theorem. Choose[15] a finite Galois extension L of $\mathbb{Q}$ containing K_1 and K_2 and let $H_j = \mathrm{Gal}(L/K_j)$. By the main theorem of Galois theory (theorem 9.B.32), it is enough to prove that $H_1 \subset H_2$. Choose any $\sigma \in H_1$. By Chebotarev's theorem, there is $p \notin S$ such that p has a prime factor β in L with $D_\beta = \langle \sigma \rangle$ (cyclic subgroup of H_1 generated by σ). By lemma 9.B.49, we have $p \in P_1(K)$ and so p is totally split in K_2, which forces, by the same lemma, $D_\beta \subset H_2$. Hence $\sigma \in H_2$ and we are done. $\square$

Remark 9.B.50. Here is another useful consequence of lemma 9.B.49. Let $n > 1$ be an integer and let H be a subgroup of $(\mathbb{Z}/n\mathbb{Z})^*$. We claim that a prime p not dividing n is totally split in $\mathbb{Q}(\zeta_n)^H$ if and only if $p \pmod n \in H$. Indeed, such a prime is unramified in $\mathbb{Q}(\zeta_n)$ and by lemma 9.B.49 it is totally split in $\mathbb{Q}(\zeta_n)^H$ if and only if for any $\beta|p$ in $\mathbb{Q}(\zeta_n)$ we have $D_\beta \subset H$. Since $\mathrm{Gal}(\mathbb{Q}(\zeta_n)/\mathbb{Q})$ is abelian, this is also equivalent to $(p, \mathbb{Q}(\zeta_n)/\mathbb{Q}) \in H$, which is saying that $p \pmod n \in H$ (we naturally identified H with a subgroup of $\mathrm{Gal}(\mathbb{Q}(\zeta_n)/\mathbb{Q})$).

[15]For instance, if $K_1 = \mathbb{Q}(x)$ and $K_2 = \mathbb{Q}(y)$, consider the extension generated over $\mathbb{Q}$ by all conjugates of x and y.

9.B.10 Finally, a reward: applications to "elementary-looking" problems

In this section we show how the deep results in the previous sections can be combined to yield some fairly nontrivial theorems with "elementary" aspects. The first one uses a generalization of Bertrand's postulate to obtain the following difficult irreducibility result.

Theorem 9.B.51. *(Schur) If $n \geq 2$ and $a_1, a_2, \dots, a_{n-1}$ are integers, then the polynomial*

$$\frac{X^n}{n!} + a_{n-1} \cdot \frac{X^{n-1}}{(n-1)!} + \dots + a_2 \cdot \frac{X^2}{2!} + a_1 X + 1$$

is irreducible in $\mathbb{Q}$.

Proof. It is enough to prove that $f = X^n + na_{n-1}X^{n-1} + \dots + n!$ is irreducible in $\mathbb{Z}[X]$. Suppose that this is not the case and let g be a monic irreducible factor of f of degree m, with $m \leq \frac{n}{2}$. Choose any prime divisor p of $n(n-1) \cdot \dots \cdot (n-m+1)$ and consider the reduction $\overline{f}$ of mod p. The choice of p implies that X^{n-m+1} divides $\overline{f}$. If $f = g \cdot h$, it follows that X^{n-m+1} divides $\overline{g} \cdot \overline{h}$ and since $\deg \overline{h} = n - m$, we must have $X|\overline{g}$ and so $p|g(0)$.

Let z be a root of g and let $K = \mathbb{Q}[z]$, a number field of degree m. Since $p|g(0)$, it follows that p divides $N(z \cdot O_K)$ and so we can choose a prime $\wp$ of O_K dividing zO_K and such that $\wp|p$. Let $e = e(\wp/p) \leq [K : \mathbb{Q}] = m$ be the ramification index of $\wp$. Since $f(z) = 0$, we obtain (set $a_n = 1$)

$$v_p(n!) \geq \min_{1 \leq i \leq n} v_p\left(a_i \cdot z^i \cdot \frac{n!}{i!}\right),$$

so we can find i such that

$$v_p(i!) \geq iv_p(z) = i \cdot \frac{v_\wp(z)}{e} \geq \frac{i}{e}.$$

Using the inequalities $v_p(i!) < \frac{i}{p-1}$ and $e \leq m$, we deduce that $p \leq m$.

We have therefore proved that all prime factors of $n(n-1) \cdot \dots \cdot (n-m+1)$ are less than or equal to m. This contradicts a famous theorem of Sylvester and Schur, which generalizes Bertrand's postulate and for a proof of which we refer the reader to [26]. □

Theorem 9.B.52. *If $n \geq 2k$, then $\binom{n}{k}$ has a prime divisor greater than k.*

Let us consider now the second application.

Theorem 9.B.53. *(Davenport, Lewis, Schinzel) Let $f \in \mathbb{Q}[X]$ be a polynomial with the following property: any arithmetic progression $\mathcal{P} \subset \mathbb{Z}$ contains an integer x such that $f(x)$ is the sum of squares of two rational numbers. Then there exist polynomials $g, h \in \mathbb{Q}[X]$ such that $f = g^2 + h^2$.*

Proof. We may assume that f is not constant. Using Gauss' lemma, we can write $f = cf_1^{e_1} \cdot \dots \cdot f_m^{e_m}$ for some $c \in \mathbb{Q}^*$, some primitive, distinct, irreducible polynomials $f_i \in \mathbb{Z}[X]$ and some $e_i \in \mathbb{N}^*$. Fix an index j for which e_j is an odd number and let θ be a root of f_j. We claim that $L = \mathbb{Q}(\theta)$ contains $\mathbb{Q}(i)$. Using part a) of Bauer's theorem 9.B.47, it is enough to prove that if q is a sufficiently large prime in $P_1(\mathbb{Q}(\theta))$, then $q \in P_1(\mathbb{Q}(i))$. We will need the following standard argument:

Lemma 9.B.54. *There exists a prime q_0 with the following property: for all $q \in P_1(L) \cap [q_0, \infty)$, one can find an integer x such that $v_q(f_j(x)) = 1$ and $v_q(f_k(x)) = 0$ for all $k \neq j$.*

Proof. First, choose a prime $q_1 > [O_L : \mathbb{Z}[\theta]]$. Take a prime number q_2 greater than all prime factors of (the numerator or denominator of) c and such that q does not divide $\gcd(f_j(x), f_j'(x) \cdot \prod_{k \neq j} f_k(x))$ for any $x \in \mathbb{Z}$ and any $q \geq q_2$. To prove the existence of q_2, note that f_j is relatively prime to $f_j' \cdot \prod_{k \neq j} f_k$, write a Bézout relation over $\mathbb{Q}$ for f_j and $f_j' \cdot \prod_{k \neq j} f_k$ and clear denominators. We claim that $q_0 = \max(q_1, q_2)$ works. Indeed, let $q \in P_1(L) \cap [q_0, \infty]$. By Dedekind-Kummer's theorem 9.B.22, we can find $y \in \mathbb{Z}$ such that q divides $f_j(y)$. Then q does not divide $f_j'(y)$ or $\prod_{k \neq j} f_k(y)$, so one of y or $y + q$ satisfies the desired conditions. □

Now, fix q_0 as in the previous lemma and let $q \in P_1(L) \cap [q_0, \infty)$ and x as in the lemma. By hypothesis we can find $y \equiv x \pmod{q^2}$ such that $f(y) = a^2 + b^2$ for some rational numbers a, b. Note that $v_q(f_j(y)) = 1$ and

$v_q(f_k(y)) = 0$ for $k \neq j$. Then $e_j = v_q(f(y)) = v_q(a^2 + b^2)$ is odd by our choice of y. We deduce that $q \equiv 1 \pmod 4$ and so q is split in $\mathbb{Q}(i)$, hence $q \in P_1(\mathbb{Q}(i))$. This proves the claim made in the previous paragraph and we conclude that $i = \sqrt{-1} \in L$. Hence we can write $i = h(\theta)$ for some $h \in \mathbb{Q}[X]$. Then f_j must divide $h^2 + 1$. If $G = \gcd(h - i, f_j) \in \mathbb{Q}(i)[X]$, it is easy to see that $N_{\mathbb{Q}(i)/\mathbb{Q}}(G) = \gcd(h - i, f_j) \cdot \gcd(h + i, f_j)$ is a polynomial with rational coefficients dividing f_j. We deduce that f_j is a constant times $N_{\mathbb{Q}(i)/\mathbb{Q}}(G)$, and this last polynomial is obviously the sum of squares of two polynomials with rational coefficients.

Applying the previous arguments for each f_j such that e_j is odd, we deduce that f is of the form $c(u^2 + v^2)$ for a constant c and some $u, v \in \mathbb{Q}[X]$. But using once more the hypothesis on c we deduce that c is the sum of squares of two rational numbers and the result follows. □

Remark 9.B.55. Actually, Davenport, Lewis and Schinzel prove a more general result: let $K/\mathbb{Q}$ be a Galois extension of degree n and let $v_1, v_2, \ldots, v_n$ be such that $O_K = \mathbb{Z}v_1 + \mathbb{Z}v_2 + \cdots + \mathbb{Z}v_n$. Suppose that $f \in \mathbb{Q}[X]$ has the property that any arithmetic progression of integers contains an integer x for which the equation $f(x) = N_{K/\mathbb{Q}}(u_1v_1 + u_2v_2 + \cdots + u_nv_n)$ is solvable in $(u_1, u_2, \ldots, u_n) \in \mathbb{Q}^n$. If either $\mathrm{Gal}(K/\mathbb{Q})$ is a cyclic group or the multiplicity of any zero of f is prime to n, then one can find $u_1, u_2, \ldots, u_n \in \mathbb{Q}[X]$ such that

$$f(X) = N_{K/\mathbb{Q}}(u_1(X), u_2(X), \ldots, u_n(X)).$$

The proof follows precisely the same ideas, but is a bit more technical.

Theorem 9.B.56. *(Schinzel) Let $f \in \mathbb{Z}[X]$ be an irreducible polynomial over $\mathbb{Q}$ and let n be an integer greater than 1. Suppose that there is c with the following property: if $p \geq c$ divides $f(x)$ for some $x \in \mathbb{Z}$, then $p \equiv 1 \pmod n$. Then there are $a \in \mathbb{Q}$ and a polynomial $g \in \mathbb{Q}(\zeta_n)[X]$ such that*

$$f(X) = a \cdot N_{\mathbb{Q}(\zeta_n)/\mathbb{Q}}(g(X)).$$

Proof. Let θ be a root of f an let $K = \mathbb{Q}(\theta)$. We claim that $\mathbb{Q}(\zeta_n) \subset K$. By Bauer's theorem, it is enough to prove that if p is a large prime having a factor of degree 1 in K, then p is completely split in $\mathbb{Q}(\zeta_n)$. But if p is such a prime, greater than c and n, then p divides some $f(x)$, so $p \equiv 1 \pmod{n}$ and p is completely split in $\mathbb{Q}(\zeta_n)$.

Now, write $\zeta_n = u(\theta)$ for some $u \in \mathbb{Q}[X]$. Then, since f is irreducible and $\phi_n(u(\theta)) = 0$, the polynomial f divides $\phi_n \circ u$ in $\mathbb{Q}[X]$ (recall that ϕ_n is the nth cyclotomic polynomial, minimal polynomial of ζ_n over $\mathbb{Q}$). For j relatively prime to n, define $f_j = \gcd(f, u - \zeta_n^j) \in \mathbb{Q}(\zeta_n)[X]$. Let $\sigma_j \in \mathrm{Gal}(\mathbb{Q}(\zeta_n)/\mathbb{Q})$ be the automorphism sending ζ_n to ζ_n^j. Then clearly $f_j = \sigma_j(f_1)$, so

$$\prod_{\gcd(j,n)=1} f_j = N_{\mathbb{Q}(\zeta_n)/\mathbb{Q}}(f_1).$$

Clearly, the polynomials f_j are pairwise relatively prime, so their product divides f. But, as we have seen, their product has rational coefficients, hence by irreducibility of f we must have $f = a \cdot N_{\mathbb{Q}(\zeta_n)/\mathbb{Q}}(f_1)$ for some constant a. The result follows. □

Remark 9.B.57. It is much easier to check that the converse of the theorem also holds.

Theorem 9.B.58. *(Murty) Let $f \in \mathbb{Z}[X]$ be a polynomial and let l, n be relatively prime positive integers. Let S be the set of primes p for which the congruence $f(x) \equiv 0 \pmod{p}$ has solutions in $\mathbb{Z}$. Suppose that there exists c such that any $p \in S$ greater than c satisfies $p \equiv 1 \pmod{n}$ or $p \equiv l \pmod{n}$. Also, suppose that there are infinitely many primes $p \equiv l \pmod{n}$ in S. Then $l^2 \equiv 1 \pmod{n}$.*

Proof. Let θ be a root of f and let $K = \mathbb{Q}(\theta)$ and $L = K(\zeta_n)$, so L/K is a finite Galois extension (splitting field of $X^n - 1$). Let H be the subgroup of $\mathrm{Gal}(\mathbb{Q}(\zeta_n)/\mathbb{Q})$ generated by the automorphism σ_l, which sends ζ_n to ζ_n^l. Note that the hypothesis combined with Bauer's theorem and with remark 9.B.50 yield an inclusion $M \subset K$, where $M = \mathbb{Q}(\zeta_n)^H$. Here is the crucial technical result:

Lemma 9.B.59. *Restriction to $\mathbb{Q}(\zeta_n)$ yields an isomorphism between $\mathrm{Gal}(L/K)$ and H.*

Proof. Since $M \subset K$, the restriction of any $\sigma \in \mathrm{Gal}(L/K)$ yields an element of $H = \mathrm{Gal}(\mathbb{Q}(\zeta_n)/M)$. It is clear that if the restriction of σ is trivial, then σ is trivial, as $L = K(\zeta_n)$. Surjectivity is more delicate. Choose a prime $p \equiv l \pmod{n}$ which is unramified in L (this excludes only finitely many primes) and for which K has a prime $\wp | p$ of degree 1. The existence of p follows from the second hypothesis. Let β be a prime of L lying over $\wp$ and let $\rho = \beta \cap \mathbb{Q}(\zeta_n)$, a prime of $\mathbb{Q}(\zeta_n)$ lying over p. Let $\sigma = (\beta, L/K)$ be a Frobenius automorphism associated to β. Since $\wp$ has degree 1, we have $\sigma(x) \equiv x^p \pmod{\beta}$ for all $x \in O_L$. We claim that the restriction of σ to $\mathbb{Q}(\zeta_n)$ is σ_l, which will prove the surjectivity and end the proof of the lemma. But this restriction, call it τ, is an element of $\mathrm{Gal}(\mathbb{Q}(\zeta_n)/\mathbb{Q})$ which preserves ρ and such that $\tau(x) \equiv x^p \pmod{\rho}$ for all $x \in \mathbb{Z}[\zeta_n]$, thus it has to be $(\rho, \mathbb{Q}(\zeta_n)/\mathbb{Q}) = \sigma_p$ (it is easy to see that σ_p satisfies the properties which uniquely characterize $(\rho, \mathbb{Q}(\zeta_n)/\mathbb{Q})$). It remains to observe that $\sigma_p = \sigma_l$, as $p \equiv l \pmod{n}$. □

It is now easy to finish the proof of the theorem: consider $\sigma_{l^2} = \sigma_l^2 \in H$. By lemma 9.B.59 there is $\sigma \in \mathrm{Gal}(L/K)$ restricting to σ_{l^2}. By proposition 9.B.43 we can find infinitely many primes $\wp$ of degree 1 of K such that $(\beta, L/K) = \sigma$ for some $\beta | \wp$. These primes $\wp$ correspond therefore to primes p such that $p \equiv l^2 \pmod{n}$ and for which $p | f(x)$ for some $x \in \mathbb{Z}$. The first hypothesis yields therefore $l^2 \equiv 1 \pmod{n}$ or $l^2 \equiv l \pmod{n}$. The result follows. □

Using the tools developed in this addendum, we are also able to prove theorem 2.A.6. For the reader's convenience, let us recall the statement. The proof is adapted from the beautiful paper [32], where a much more general result is proved.

Theorem 9.B.60. *Let $n > 1$ and a be integers such that a is an nth power modulo p for all sufficiently large prime numbers p. Then either a is the nth power of an integer or $8|n$ and $a = 2^{\frac{n}{2}} b^n$ for some integer b.*

Proof. We will use without comment the following well-known, but nontrivial result: if an integer a is a perfect square mod p for all sufficiently large primes

p, then a is a perfect square. While elementary proofs exist,[16] this statement is also an immediate consequence of Chebotarev's density theorem 9.B.41 applied to the splitting field of $X^2 - a$.

So, if n is even, we already know that a is a perfect square, in particular positive (the case $a = 0$ is trivial and excluded in this proof). Hence, by replacing a by $-a$ if n is odd, we may and will assume that $a > 0$. The crucial ingredient of the proof is the following application of Bauer's theorem 9.B.47:

Lemma 9.B.61. *If z is a complex root of $X^n - a$, then $\mathbb{Q}(z) \subset \mathbb{Q}(\zeta_n)$, where $\zeta_n = e^{\frac{2i\pi}{n}}$.*

Proof. Let $K_1 = \mathbb{Q}(\zeta_n)$ and let K_2 be the splitting field of $X^n - a$ over $\mathbb{Q}$. It is enough to prove that $K_2 \subset K_1$. Using Bauer's theorem, it suffices to prove that for all sufficiently large primes $p \in P_1(K_1)$, p is completely split in K_2. If p does not divide n (which certainly happens if p is large enough), the condition $p \in P_1(K_1)$ is equivalent to $n|p-1$ by proposition 9.B.25. Also, if p is large enough, then a is an nth power modulo p. These two conditions imply that $X^n - a$ is a product of n distinct linear factors in $\mathbb{F}_p[X]$. Using the Dedekind-Kummer theorem 9.B.22,[17] we deduce that p is totally split in K_2, which finishes the proof of the lemma. $\square$

Coming back to the proof of the theorem, let d be the least positive divisor of n for which $\sqrt[n]{a^d} \in \mathbb{Q}$. Write $\sqrt[n]{a} = \sqrt[d]{c}$ for some $c \in \mathbb{Q}_+^*$. By lemma 9.19, the polynomial $X^d - c$ is irreducible over $\mathbb{Q}$ and by lemma 9.B.61 we have $\mathbb{Q}(\sqrt[d]{c}) \subset \mathbb{Q}(\zeta_n)$. Since $\mathbb{Q}(\zeta_n)/\mathbb{Q}$ is Galois with commutative Galois group, any subfield of $\mathbb{Q}(\zeta_n)$ is again Galois over $\mathbb{Q}$ and so $\mathbb{Q}(\sqrt[d]{c})/\mathbb{Q}$ is Galois. In particular, $\zeta_d \cdot \sqrt[d]{c}$ (which is a conjugate of $\sqrt[d]{c}$) is again in $\mathbb{Q}(\sqrt[d]{c})$ and so it is a real number. We deduce that $\zeta_d \in \mathbb{R}$, which implies that $d = 1$ or $d = 2$. If

[16] The Jacobi symbol $\left(\frac{a}{m}\right)$ for odd $m \geq 3$, defined as the product over the prime factors of m with multiplicity of the corresponding Legendre symbols, detects quadratic nonresidues mod m and is easily seen to obey quadratic reciprocity. The Chinese Remainder Theorem then permits the construction of a "bad" value of m.

[17] To be fair, we need the obvious version of this theorem in which $\mathbb{Q}$ is replaced by K_1 and K by $K_2 = K_1(\sqrt[n]{a})$. We leave to the reader to convince himself that the statement and proof of this general version are exactly the same.

$d = 1$, then a is the nth power of a rational number and we are done (as a is an integer, this rational number is actually an integer).

Suppose now that $d = 2$ and write $n = 2m$ so that a is an mth power. Note that a is also a perfect square by the discussion in the first paragraph. If m is odd, then a is an mth power and a perfect square, hence it is an $n = 2m$th power and we are done. Suppose that m is even, so $4|n$. Then a is a 4th power modulo all sufficiently large primes and, by what we have already seen, we can write $a = c^2$, with $\mathbb{Q}(\sqrt{c}) \subset \mathbb{Q}(\zeta_4) = \mathbb{Q}(i)$. Since $c > 0$, we deduce that $\mathbb{Q}(\sqrt{c}) \subset \mathbb{R} \cap \mathbb{Q}(i) = \mathbb{Q}$, so c is a square and a is a fourth power. If $m = 2k$ with k odd, this implies that a is a $\operatorname{lcm}(m, 4) = n$th power and we are done again. Finally, assume that $n = 2^i \cdot u$, with $i \geq 3$ and u odd. Since a is an 2^ith power modulo all sufficiently large primes, lemma 9.B.61 implies that $\mathbb{Q}(\sqrt{c}) \subset \mathbb{Q}(\zeta_{2^i})$. The extension $\mathbb{Q}(\zeta_{2^i})/\mathbb{Q}$ is Galois, with Galois group $(\mathbb{Z}/2^i\mathbb{Z})^*$. Its sub-extensions which are quadratic over $\mathbb{Q}$ are classified by the subgroups with two elements of $(\mathbb{Z}/2^i\mathbb{Z})^*$, which in turn are classified by the nontrivial solutions to the equation $x^2 \equiv 1 \pmod{2^i}$. There are three such solutions, giving rise to the quadratic sub-extensions $\mathbb{Q}(\sqrt{-1}), \mathbb{Q}(\sqrt{2}), \mathbb{Q}(\sqrt{-2})$. Since $\sqrt{c}$ is real, we deduce that $\mathbb{Q}(\sqrt{c}) \subset \mathbb{Q}(\sqrt{2})$ and so c is either a perfect square or twice a perfect square. As $a = c^{\frac{n}{2}}$, this finally yields the desired result. □

Chapter 10

Arithmetic Properties of Polynomials

This chapter deals with elementary, but rather challenging problems concerning the arithmetic of polynomials with integer coefficients. The techniques are very diverse, going from the very basic fact that $a-b$ divides $f(a)-f(b)$ for $f \in \mathbb{Z}[X]$ and $a, b \in \mathbb{Z}$ to Mahler expansions, finite differences, p-adic analysis, finite fields, etc.

10.1 The $a-b|f(a)-f(b)$ trick

One can extract a lot of arithmetic information from the fact that $a-b$ divides $f(a)-f(b)$ for all polynomials with integer coefficients f, but often one needs a lot of inventiveness to do this. The next problems are all quite tricky, even though they hide very simple ideas. The first problem is a version of a very classical result, stating that there are no nonconstant polynomials all of whose values are prime numbers.

1. Let $f_1, f_2, \ldots, f_k$ be nonconstant polynomials with integer coefficients. Prove that for infinitely many n all numbers $f_1(n), f_2(n), \ldots, f_k(n)$ are composite.

Proof. Let $g = f_1 f_2 \cdot \cdots \cdot f_k$ and fix a positive integer n_0. Since none of the polynomials g and f_i is constant, there are $a, b > n_0$ such that

$$\min(|g(a)|, |f_1(a)|, \ldots, |f_k(a)|) > 1 \text{ and } |f_i(a + |g(a)|b)| > |g(a)|$$

for all i. Since $|f_i(a)|$ divides $f_i(a+|g(a)|b)$ and is less than $|f_i(a+|g(a)|b)|$, it follows that all numbers $f_i(a + |g(a)|b)$ are composite. The result follows. □

The following two problems deal with periodic points of polynomials with integer coefficients.

2. Let $f \in \mathbb{Z}[X]$ and let $n \geq 3$. Prove that one cannot find distinct integers $x_1, x_2, \ldots, x_n$ such that $f(x_i) = x_{i-1}$, $i = 1, 2, \ldots, n$, indices being taken mod n.

Proof. If such a polynomial f and integers x_i existed,

$$x_i - x_{i-1} = f(x_{i+1}) - f(x_i)$$

would be divisible by $x_{i+1} - x_i$. In particular, $|x_i - x_{i-1}| \geq |x_{i+1} - x_i|$ for all i, which forces the numbers $|x_i - x_{i-1}|$ to be equal. Since

$$\sum_{i=1}^{n} (x_{i+1} - x_i) = 0,$$

there exists i such that $x_{i+1} - x_i$ and $x_{i+2} - x_{i+1}$ have different signs. But then $x_i = x_{i+2}$, which contradicts the hypothesis that x_i are all distinct. The result follows. □

3. Let $f \in \mathbb{Z}[X]$ be a polynomial of degree $n \geq 2$. Prove that the polynomial $f(f(X)) - X$ has at most n integer zeros.

Gh. Eckstein, Romanian TST

Proof. Suppose that $x_1 < x_2 < \cdots < x_{n+1}$ are distinct integers such that $f(f(x_i)) = x_i$. Then $f(x_i)$ are also pairwise distinct and $f(x_i) - f(x_j)$ is a multiple of $x_i - x_j$. But $x_i - x_j = f(f(x_i)) - f(f(x_j))$ is also a multiple of $f(x_i) - f(x_j)$. Thus, we must have $|f(x_i) - f(x_j)| = |x_i - x_j|$. But then

$$\begin{aligned}\sum_{i=1}^{n}|f(x_{i+1}) - f(x_i)| &= \sum_{i=1}^{n}|x_{i+1} - x_i| \\ &= x_{n+1} - x_1 \\ &= |f(x_{n+1}) - f(x_1)| \\ &= \left|\sum_{i=1}^{n}(f(x_{i+1}) - f(x_i))\right|,\end{aligned}$$

which implies that all numbers $f(x_{i+1}) - f(x_i)$ have the same sign. Combined with the equality $|f(x_i) - f(x_{i+1})| = |x_i - x_{i+1}|$ this shows that either all numbers $f(x_i) + x_i$ are equal or that $f(x_i) - x_i$ are all equal. This is however impossible, as f has degree n. □

Remark 10.1. A very similar problem was given at the IMO 2006: let $P(X)$ be a polynomial of degree $n > 1$ with integer coefficients and let k be a positive integer. Consider the polynomial

$$Q(X) = \underbrace{P(P(\cdots P(P}_{k \text{ times}}(X)))).$$

Prove that there are at most n integers such that $Q(t) = t$. It follows from problem 2 that any integral solution of the equation $Q(t) = t$ is a solution of the equation $P(P(t)) = t$, so this new problem is actually equivalent to the previously discussed one.

The following two problems are a bit trickier than the previous ones.

4. Find all integers $n > 1$ for which there is a polynomial $f \in \mathbb{Z}[X]$ with the property: for any integer k one has $f(k) \equiv 0 \pmod{n}$ or $f(k) \equiv 1 \pmod{n}$ and both these equations have solutions.

Proof. The answer is: exactly the powers of a prime number. If n is a prime power, one can see from Euler's theorem that the polynomial $X^{\varphi(n)}$ is a solution. Conversely, suppose there exists a polynomial f with the properties in the statement and that n has at least two different prime factors p, q. Changing f to $f(X-a)$ for a suitable a, we may assume that $f(0) \equiv 0 \pmod{n}$. Thus $f(0) \equiv 0 \pmod{q}$, which implies that for all $b \in \mathbb{Z}$ we also have $f(bq) \equiv 0 \pmod{q}$. Thus, we cannot have $f(bq) \equiv 1 \pmod{n}$ and so $f(bq) \equiv 0 \pmod{n}$. In particular, $f(bq) \equiv 0 \pmod{p}$. But then for all $a \in \mathbb{Z}$ we also have $f(ap+bq) \equiv f(bq) \equiv 0 \pmod{p}$ and so $f(ap+bq) \equiv 0 \pmod{n}$. Since any integer $x \in \mathbb{Z}$ can be written $ap+bq$ for suitable a, b, it follows that the equation $f(x) \equiv 1 \pmod{n}$ has no solutions, a contradiction. $\square$

5. Find all polynomials f with integer coefficients such that $f(n) | 2^n - 1$ for all positive integers n.

Polish Olympiad

Proof. Of course, if f is constant, then f has to be either 1 or -1 and these are solutions. So, assume that f is nonconstant. We may assume that the leading coefficient of f is positive. Take n such that $f(n) > 1$ and let p be a prime factor of $f(n)$. Then $p | 2^n - 1$ and also $p | f(n+p) | 2^{n+p} - 1$. But then $p | 2^p - 1$, which is certainly impossible. So any such f is constant and the solutions are $1, -1$. $\square$

6. Let $f \in \mathbb{Z}[X]$ be a nonconstant polynomial. Prove that the sequence $f(3^n) \pmod{n}$ is not bounded.

Proof. Changing f with its opposite, we may assume that the leading coefficient of f is positive. So, given N, there exists m such that $f(3^m) > N$. Choose a prime $p > 2f(3^m)$ and observe that $f(3^{pm}) \equiv f(3^m) \pmod{p}$ by Fermat's little theorem. In particular, if $r = f(3^{mp}) \pmod{mp}$, then $r \equiv f(3^m) \pmod{p}$ and so $r \geq f(3^m) > N$, finishing the proof. $\square$

7. Is there a nonconstant polynomial $f \in \mathbb{Z}[X]$ and an integer $a > 1$ such that the numbers $f(a), f(a^2), f(a^3), \ldots$ are pairwise relatively prime?

Saint Petersburg 1998

Proof. Assume that f is such a polynomial and a is as in the statement. Let $g = \gcd(a, f(a^k)) = \gcd(a, f(0))$. If $g > 1$, then g is a common factor of all $f(a^j)$. Thus $g = 1$, so a and $f(a^i)$ are relatively prime for all

i. Choose an i with $|f(a^i)| > 1$ and choose $j = i + \varphi(|f(a^i)|)$, where φ is Euler's totient function. Then $f(a^i)$ divides $a^j - a^i$ by Euler's Theorem, so $f(a^i)$ divides $f(a^j) - f(a^i)$ and $f(a^i)|f(a^j)$. But this contradicts the fact that $\gcd(f(a^i), f(a^j)) = 1$ and shows that the answer to the problem is negative. $\square$

The next problem is a variation on the following very classical (but non-trivial) problem: what are the polynomials $f \in \mathbb{Z}[X]$ such that $f(p)$ is a prime for all prime numbers p?

8. Find all polynomials f with integer coefficients, having the following property: there exists k such that for all primes p, $f(p)$ has at most k prime factors.

Proof. All polynomials of the form aX^m work, for obvious reasons. Let f be a solution of the problem and suppose f is not of this form. So we can write $f(X) = X^m g(X)$ for some polynomial g with integer coefficients such that $g(0) \neq 0$ and g is not constant. By Schur's theorem, there are infinitely many primes dividing at least one of the numbers $g(1), g(2), \dots$. In particular, we can choose $p_1, p_2, \dots, p_{k+1}$ distinct primes greater than $|g(0)|$ and we can choose positive integers $a_1, a_2, \dots, a_{k+1}$ such that $g(a_i) \equiv 0 \pmod{p_i}$. Using the Chinese Remainder Theorem, we can find an integer a such that $a \equiv a_i \pmod{p_i}$ for all i. Note that a is relatively prime to $p_1p_2\cdots p_{k+1}$ (otherwise some p_i would divide a, so $p_i|a_i$ and then $p_i|g(0)$, a contradiction). Thus, by Dirichlet's theorem, there exist infinitely many primes $p \equiv a \pmod{p_1p_2\cdots p_{k+1}}$. In particular, we can choose such a prime p with $f(p) \neq 0$. Since by construction $f(p)$ is a multiple of $p_1p_2\cdots p_{k+1}$, it follows that f is not a solution of the problem. The result follows. $\square$

Proof. Note that all polynomials of the form $f(X) = aX^m$ work. Suppose $f(X) = X^m g(X)$ is such a polynomial which is not of this form. Then $g(X)$ is also such a polynomial. Thus we may assume that f is nonconstant and $f(0) \neq 0$. Let p be a prime not dividing $f(0)$, hence p does not divide $f(p)$. By Dirichlet's Theorem there are infinitely many primes q of the form $q = p + kf(p)^2$. Choose such a q with $|f(q)| > |f(p)|$. For such a q we have

$f(q) \equiv f(p) \pmod{f(p)^2}$. Thus $f(q)$ is divisible by every prime factor of $f(p)$, with exactly the same multiplicities, and at least one additional prime factor. Iterating this construction, we can find a sequence of primes (p_n) for which $f(p_n)$ has at least n prime factors. Thus the only solutions are the ones already given. □

9. Find all polynomials $f \in \mathbb{Z}[X]$ with the property that for any relatively prime integers m, n, the numbers $f(m), f(n)$ are also relatively prime.

Iranian TST

Proof. The solutions are the polynomials $\pm X^d$ for $d \geq 0$. Trivially, these are solutions of the problem. Consider now any nonconstant solution f. Suppose that p is a large prime for which p does not divide $f(p)$. Then p and $p+f(p)$ are relatively prime and thus $f(p)$ and $f(p+f(p))$ are relatively prime. However, $f(p)$ divides $f(p+f(p))$. Thus necessarily $|f(p)| = 1$ or $f(p) = 0$. Since p was large, none of this happens (f is nonconstant). Thus, for p large enough we have $p|f(p)$, forcing $p|f(0)$. This implies that $f(0) = 0$ and so we can write $f(X) = Xg(X)$ for some polynomial g with integral coefficients. Of course, g satisfies the same property as f and has smaller degree. By an easy induction on the degree of f, we deduce that f is indeed of the form $\pm X^d$ for some d. □

Proof. We prove first the following easy lemma:

Lemma 10.2. *If $f \in \mathbb{Z}[X]$ is not of the form $\pm X^n$, then there exist different primes p, q such that $q|f(p)$.*

Proof. Assuming the contrary, we may assume that the leading coefficient of f is positive. So, for all large enough p, $f(p) = p^{k_p}$ for some integer k_p. Since $k_p = \frac{\log f(p)}{\log p}$ converges to $\deg(f)$, it follows that $k_p = \deg(f)$ for all sufficiently large p and so $f(p) = p^{\deg(f)}$ for all sufficiently large p. The result follows. □

Coming back to the proof, choose a solution f of the problem and suppose that f is not of the form $\pm X^n$. Pick primes p, q as in the lemma. Then q divides $f(p)$ and $f(p+q)$ (since q divides $f(p+q) - f(p)$). But this is impossible, since $p, p+q$ are relatively prime and so $f(p)$ and $f(p+q)$ are relatively prime. The result follows. □

It is a classical fact that for any integer a and any positive integers m and n we have

$$\gcd(a^m-1, a^n-1) = a^{\gcd(m,n)} - 1.$$

A sequence $(a_n)_n$ with the property that $\gcd(a_m, a_n) = a_{\gcd(m,n)}$ is called a Mersenne sequence. Thus $(a^n-1)_n$ is a Mersenne sequence. In chapter 3, problem 3 we proved that for any Mersenne sequence a_n there exists a sequence of integers b_n such that $a_n = \prod_{d|n} b_d$. The next problem shows how to construct a Mersenne sequence by iterating a polynomial with integer coefficients.

10. Let f be a polynomial with integer coefficients and let $a_0 = 0$ and

$$a_n = f(a_{n-1})$$

for all $n \geq 1$. Prove that $(a_n)_{n\geq 0}$ is a Mersenne sequence.

Romanian TST

Proof. Let $g = f^d$, the composite of f with itself d times. If $m = du$ and $n = dv$, with $\gcd(u,v) = 1$, are any positive integers, note that $a_n = g^v(0)$, $a_m = g^u(0)$ and $a_d = g(0)$. So, it is enough to prove that for any g with integer coefficients and any $\gcd(u,v) = 1$ we have $\gcd(g^u(0), g^v(0)) = g(0)$. Let us remark that for any polynomial h with integer coefficients and any k we have $h(0)|h^k(0)$. This is obvious. Applying this to g already shows that $g(0)$ divides $\gcd(g^u(0), g^v(0))$. Conversely, let $x = \gcd(g^u(0), g^v(0))$. Applying the previous remark to $h = g^u$ and $h = g^v$, we obtain that for all $A, B \geq 1$ we have $x|g^{Au}(0)$ and $x|g^{Bv}(0)$. Taking A, B such that $Bv = Au + 1$, we obtain that $x|g(g^{Au}(0))$ and $x|g^{Au}(0)$. It is then clear that x divides $g(0)$ and the conclusion follows. □

11. Find all positive integers k such that if a polynomial with integer coefficients f satisfies

$$0 \leq f(0), f(1), \ldots, f(k+1) \leq k,$$

then $f(0) = f(1) = \cdots = f(k) = f(k+1)$.

IMO 1997 Shortlist

Proof. If $k \leq 2$ we can easily find counterexamples, for instance $f(X) = X(2-X)$ for $k=1$ and $f(X) = X(3-X)$ for $k=2$. So, assume that $k \geq 3$ and let f be a polynomial with integral coefficients such that

$$0 \leq f(0), f(1), \ldots, f(k+1) \leq k.$$

As $f(k+1) - f(0)$ is between $-k$ and k and since it is a multiple of $k+1$, we must have $f(0) = f(k+1)$. Thus, we can write

$$f(X) - f(0) = X(X-(k+1))g(X)$$

for some polynomial g, which clearly has integer coefficients (note that X divides $f(X) - f(0)$ and X is relatively prime to $X - (k+1)$).

Thus, $f(0) + i(i-k-1)g(i)$ is between 0 and k for all $0 \leq i \leq k+1$. In particular, we have $k \geq i(k+1-i)|g(i)|$ for i in this range. However, $i(k+1-i) > k$ for $2 \leq i \leq k-1$, so that $g(i) = 0$ for $2 \leq i \leq k-1$. In particular, we can write

$$g(X) = (X-2)\cdots(X-k+1)h(X)$$

for some polynomial h, again with integer coefficients. But then

$$f(k) - f(0) = -k(k-2)!h(k), \quad f(1) - f(0) = (-1)^{k-1}k(k-2)!h(1).$$

This implies that $k(k-2)!|h(x)| \leq k$ for $x \in \{1, k\}$, which clearly implies that $h(1) = h(k) = 0$, unless $k = 3$. Therefore, unless $k = 3$ we can definitely conclude that $f(0) = f(1) = \cdots = f(k+1)$ and so all $k \geq 4$ are solutions of the problem. It remains to study the case $k = 3$. But we note that we can actually have equality in all previous inequalities, yielding without much difficulty the polynomial $f(X) = X(X-2)^2(4-X)$. This proves that $k = 3$ is not a solution and the problem is solved. □

Fermat's little theorem implies that p divides $2^p - 2$ for any prime p. The following question is then rather natural to ask, but not easy to answer.

12. Find all polynomials f with integer coefficients such that $f(p) | 2^p - 2$ for any odd prime number p.

Gabriel Dospinescu, Peter Scholze

Proof. We will use the fact that $f(p)$ divides $f(p+kf(p))$ for any k and choose k such that $p+kf(p)$ is again a prime. Of course, this is not possible unless p is relatively prime to $f(p)$.

Since $f(X)=X$ is a solution (by Fermat's little theorem), a reasonable first step will be to prove that any solution satisfies $f(0)=0$. This is the hardest step. Assume for a moment that we proved that any nonconstant solution satisfies $f(0)=0$, then we can write $f(X)=Xg(X)$ and of course g satisfies the same conditions. So either g is constant or it is a multiple of X. Continuing like this we obtain that $f(X)=\pm 2^a X^b$ for some $a, b \geq 0$. Clearly $a \leq 1$ and by taking $p=3$ we obtain that $b \leq 1$. Thus, all solutions are of the form $f(X)=\pm 2^a X^b$ with $a, b \leq 1$ (these are clearly solutions of the problem).

Now, let us prove that if f is nonconstant, then $f(0)=0$. Assume that $f(0) \neq 0$ and, by changing f with $-f$, that the leading coefficient of f is positive. Then for large primes p, p is relatively prime to $f(p)$ (as if p divides $f(p)$, then p divides $f(0)$). By Dirichlet's theorem for such a prime p we can choose infinitely many $k \geq 1$ (depending on p, of course) such that $q=p+kf(p)$ is again a prime. Then $f(p)$ divides 2^p-2 and also $2^{p+kf(p)}-2$. But then $f(p)$ also divides $2(2^{\gcd(p-1,p-1+kf(p))}-1)$. As $f(p) \equiv f(1) \pmod{p-1}$, we deduce that $f(p)$ divides $2(2^{\gcd(p-1,kf(1))}-1)$. Now, if $f(1)=0$, all this work will accomplish nothing, but the good news is that $f(1)$ cannot be 0: otherwise, $f(p)$ would be a multiple of $p-1$ and so $p-1$ would divide 2^p-2 for any prime p. This is of course false (take $p=5$). Thus $f(1) \neq 0$. However, there is a big obstacle in the previous approach: k depends on p and we have absolutely no control on it. The crucial remark is that we can however control $\gcd(p-1, kf(1))$ by choosing more carefully k. This is the object of the next paragraph.

We will choose k of the form $k=2+r(p-1)$ for suitable integers r. In order to do that, we need to ensure that $p+f(p)(2+r(p-1))$ is a prime. This is realized for infinitely many r if $p+2f(p)$ is relatively prime to $(p-1)f(p)$. Now, if l is a prime factor of $p+2f(p)$ that divides $(p-1)f(p)$, then we cannot have $l|f(p)$ (as otherwise l divides p and so $l=p$ divides $f(p)$, a contradiction), so l divides $p-1$ and also $1+2f(1)$ (since $f(p) \equiv f(1) \pmod{p-1}$). But we can now do the following: choose from the beginning large primes p such that $p-1$ is relatively prime to $1+2f(1)$, for instance primes $p \equiv 2 \pmod{1+2f(1)}$.

There are infinitely many such primes by Dirichlet's theorem. For such p, we saw that $p+2f(p)$ is relatively prime to $(p-1)f(p)$, so we can indeed choose k of the form $2+r(p-1)$ such that $q=p+kf(p)$ is a prime. The previous paragraph shows that $f(p)$ divides $2(2^{\gcd(p-1,kf(1))}-1)$. But $\gcd(k,p-1)=2$, so that $f(p)$ actually divides $2^{2f(1)}-1$. But since $f(p)$ can take arbitrarily large values and since $f(1)\neq 0$, this is a (so desired) contradiction. This shows that $f(0)=0$ and then the problem is solved by the first paragraph. □

10.2 Derivatives and p-adic Taylor expansions

Most of the time when working with congruences mod p it is enough to know that $f(a)\equiv f(b) \pmod p$ whenever $a\equiv b \pmod p$. However, when dealing with congruences with prime power moduli, one is sometimes obliged to use more precise estimates. Assume that $f\in\mathbb{Z}[X]$ has degree n and consider its Taylor expansion

$$f(x+h)=\sum_{k=0}^{n}\frac{f^{(k)}(x)}{k!}h^k,$$

which is valid for any complex numbers x,h. Note that $\frac{f^{(k)}(x)}{k!}$ is an integer. Indeed, by linearity it suffices to prove this when $f(X)=X^n$ and in this case

$$\frac{f^{(k)}(x)}{k!}=\binom{n}{k}x^{n-k}.$$

It follows easily from this that we have an equality in $\mathbb{Z}[X]$ (thus in $R[X]$ for any commutative ring R)

$$f(X)=\sum_{k=0}^{n}\frac{f^{(k)}(Y)}{k!}(X-Y)^k.$$

In particular, we have the very useful congruence $f(a+b)\equiv f(a)+bf'(a) \pmod{b^2}$ whenever $f\in\mathbb{Z}[X]$ and $a,b\in\mathbb{Z}$. We give a few applications of these ideas in this section.

13. Let p be a prime and let $f \in \mathbb{Z}[X]$ be a polynomial. If $f(0)$, $f(1), \dots$, $f(p^2-1)$ give distinct remainders when divided by p^2, prove that the numbers $f(0), f(1), \dots, f(p^3-1)$ give distinct remainders when divided by p^3.

Putnam 2008

Proof. Assume that $f(i) \equiv f(j) \pmod{p^3}$ for some i, j. Since $f(i) \equiv f(j) \pmod{p^2}$ and since f is injective mod p^2, we deduce that $i \equiv j \pmod{p^2}$, say $j = i + p^2k$. It is enough to prove that $k \equiv 0 \pmod{p}$. Assume that this is not the case. We have

$$f(i) \equiv f(j) \equiv f(i + kp^2) \equiv f(i) + kp^2 f'(i) \pmod{p^3},$$

so p divides $kf'(i)$, hence p divides $f'(i)$. But then

$$f(i + kp) \equiv f(i) + kpf'(i) \equiv f(i) \pmod{p^2},$$

which, combined with the hypothesis, yields $i + kp \equiv i \pmod{p^2}$, a contradiction. Thus $k \equiv 0 \pmod{p}$ and $i \equiv j \pmod{p^3}$. The result follows. $\square$

14. Let P be a polynomial with integer coefficients such that $P(0) = 0$ and

$$\gcd(P(0), P(1), P(2), \dots) = 1.$$

Show that there are infinitely many n such that

$$\gcd(P(n) - P(0), P(n+1) - P(1), P(n+2) - P(2), \dots) = n.$$

USA TST 2010

Proof. Let us try to study first

$$d_n = \gcd(P(n) - P(0), P(n+1) - P(1), \dots)$$

for any polynomial P with integer coefficients. Let q be a prime factor of d_n, so that $P(n+k) \equiv P(k) \pmod{q}$ for all k, i.e. P is n-periodic modulo q. But P is also q-periodic modulo q. Thus, if $\gcd(q, n) = 1$, then P is 1-periodic modulo q (by Bézout's lemma) and so q divides $P(n+1) - P(n)$ for all n. Then q divides $P(n) - P(0)$ for all n, so if $P(0) = 0$, then q must divide $\gcd(P(0), P(1), \ldots)$. In particular, for our polynomial we must have $q|n$ for any prime factor q of d_n.

The previous paragraph suggests taking for n a power of a prime, say $n = p^N$. Then we saw that d_n is also a power of p. Note that d_n is a multiple of n, since n divides $P(n+k) - P(k)$ for all k. It remains to see if we can have $p^{N+1}|P(k+p^N) - P(k)$ for all k. Since

$$P(k+p^N) \equiv P(k) + p^N P'(k) \pmod{p^{N+1}},$$

this would imply that p divides $P'(k)$ for all k. Now we see how to choose our numbers n: pick and fix once and for all a value k such that $P'(k) \neq 0$. For all sufficiently large p, p does not divide $P'(k)$. For any such p, the previous arguments show that $d_n = n$ for all $n = p^N$. The conclusion follows. $\square$

Before tackling the next problem, we need to recall a very standard result, known as Lagrange's theorem. Consider $f \in \mathbb{Z}[X]$ and a prime p. Let $\overline{f} \in \mathbb{F}_p[X]$ be the reduction of f mod p and assume that $\overline{f} \neq 0$. As $\mathbb{F}_p$ is a field, it follows that $\overline{f}$ has at most $\deg \overline{f} \leq \deg f$ distinct roots in $\mathbb{F}_p$. Therefore, if $\overline{f} \neq 0$ (which is equivalent to the fact that at least one coefficient of f is not a multiple of p), then the congruence $f(x) \equiv 0 \pmod{p}$ has at most $\deg f$ pairwise distinct solutions. The next result is a generalization of this classical theorem; it will also be used in the solution of problem 20. This was one of the problems proposed in the Iranian TST 2011, but the result is much older, see for instance theorem 27, chapter II of the beautiful book [21].

15. Let p be a prime, k a positive integer and $f \in \mathbb{Z}[X]$ such that p^k divides $f(x)$ for all $x \in \mathbb{Z}$. If $k \leq p$, prove that there are polynomials $g_0, g_1, \ldots, g_k \in \mathbb{Z}[X]$ such that

$$f(X) = \sum_{i=0}^{k} p^{k-i}(X^p - X)^i \cdot g_i(X).$$

Proof. The proof is by induction on k. If $k = 1$, perform the division algorithm in $\mathbb{Z}[X]$ for the polynomials f and $X^p - X$ (which we can do, as $X^p - X$ is monic) to find $q, r \in \mathbb{Z}[X]$ such that $f(X) = (X^p - X)q(X) + r(X)$ and $\deg r < p$. Then p divides $r(x)$ for all integers x (by Fermat's little theorem and the hypothesis) and the result follows from Lagrange's theorem. Assume that the result holds for k and that $k+1 \leq p$. If p^{k+1} divides $f(x)$ for all x, then by the inductive hypothesis we can write $f(X) = \sum_{i=0}^{k} p^{k-i}(X^p - X)^i \cdot g_i(X)$ for some $g_i \in \mathbb{Z}[X]$. Pick any integers x and z and write $x^p - x = py$ for some integer y. Then $(x+pz)^p - (x+pz) \equiv p(y-z) \pmod{p^2}$, thus

$$f(x+pz) \equiv \sum_{i=0}^{k} p^k (y-z)^i g_i(x+pz) \equiv p^k \sum_{i=0}^{k} (y-z)^i g_i(x) \pmod{p^{k+1}}.$$

Therefore the hypothesis on f implies that p divides $\sum_{i=0}^{k} z^i g_i(x)$ for all integers z. Using the fact that $k+1 \leq p$ and Lagrange's theorem, it follows that p divides $g_i(x)$ for all i and all x. By the case $k=1$ we can write $g_i(X) = (X^p - X)h_i(X) + pr_i(X)$ for some $h_i, r_i \in \mathbb{Z}[X]$. Replacing these expressions in $f(X) = \sum_{i=0}^{k} p^{k-i}(X^p - X)^i \cdot g_i(X)$ yields the desired result. □

10.3 Hilbert polynomials and Mahler expansions

Consider the polynomial

$$\binom{X}{n} = \frac{X(X-1)(X-2)\cdots(X-n+1)}{n!},$$

known as the nth Hilbert polynomial. It is clear that this polynomial has rational (not necessarily integer) coefficients and it is not difficult to check that it takes integer values at all integers (note that $x(x-1)\cdots(x-n+1) = n!\binom{x}{n}$ if $x \geq 1$, while

$$x(x-1)\cdots(x-n+1) = (-1)^n \binom{-x+n-1}{n}$$

if $x < 0$). These polynomials play a fundamental role in the theory of integer-valued polynomials because of the following beautiful and classical result of Polya:

Theorem 10.3. *Let f be a polynomial of degree n, with real coefficients. Then $f(\mathbb{Z}) \subset \mathbb{Z}$ if and only if there exist integers $a_0, a_1, a_2, \dots, a_n$ such that*

$$f(X) = a_0 + a_1\binom{X}{1} + a_2\binom{X}{2} + \cdots + a_n\binom{X}{n}.$$

Proof. One implication follows from the previous discussion, so assume that $f(\mathbb{Z}) \subset \mathbb{Z}$. The polynomials $\binom{X}{0}, \binom{X}{1}, \dots, \binom{X}{n}$ have degrees $0, 1, \dots, n$, thus they form a basis of the $\mathbb{R}$-vector space of real polynomials with degree at most n. Thus there exist unique real numbers $a_0, a_1, \dots, a_n$ such that

$$f(X) = \sum_{k=0}^{n} a_k \cdot \binom{X}{k}.$$

Consider the operator $\Delta f(X) = f(X+1) - f(X)$ and observe that

$$\Delta\binom{X}{n} = \binom{X}{n-1}.$$

Applying Δ successively to the relation

$$f(X) = \sum_{k=0}^{n} a_k \cdot \binom{X}{k},$$

we deduce that $a_j = \Delta^j f(0)$ for all j. On the other hand, an immediate induction shows that

$$\Delta^k f(X) = \sum_{j=0}^{k} (-1)^{k-j}\binom{k}{j} f(X+j).$$

Thus, if $f(0), f(1), \dots, f(n)$ are integers, then so are $a_0, a_1, \dots, a_n$. The result follows. □

The proof of the previous theorem shows that

$$a_k = \sum_{j=0}^{k} (-1)^{k-j}\binom{k}{j} f(j).$$

We call a_j the Mahler coefficients of f and we refer the reader to the addendum 3.B for more details on these coefficients, which play an absolutely fundamental role in p-adic analysis. Another consequence of the proof of theorem 10.3 is the following very useful result.

Proposition 10.4. *If f is a polynomial of degree n with coefficients in an arbitrary commutative ring, then*

$$\sum_{k=0}^{n+1}(-1)^{n+1-k}\binom{n+1}{k}f(X+k)=0.$$

Proof. The left-hand side is $\Delta^{n+1}f(X)$. However, we have $\deg \Delta g < \deg g$ for any nonzero polynomial g. We deduce that $\Delta^{n+1}f$ is the zero polynomial and the result follows. □

The next problem is an immediate application of theorem 10.3. The reader might try to solve it in a different way and he will realize that the problem is actually quite tricky.

16. Let n be a positive integer. What is the least degree of a monic polynomial f with integer coefficients such that $n|f(k)$ for any integer k?

Proof. Let $d = \deg f$. By theorem 10.3 we can write $\frac{f(X)}{n} = \sum_{k=0}^{d} a_k\binom{X}{k}$ for some integers a_i. Considering the leading coefficients in this equality shows that $d!$ is a multiple of n. On the other hand, if $d!$ is a multiple of n, then we can take $f(X) = X(X+1)\cdots(X+d-1)$. Thus the answer is: d is the smallest integer such that $d!$ is a multiple of n. □

17. Let f be a polynomial such that $f(n) \in \mathbb{Z}$ for all $n \in \mathbb{Z}$. Prove that for any integers m, n the number $\operatorname{lcm}[1, 2, \ldots, \deg(f)] \cdot \frac{f(m)-f(n)}{m-n}$ is an integer.

MOSP 2001

Proof. Fix m, n distinct integers and let $d = m-n$ and $g(X) = f(n+X)$. Then g has rational coefficients, sends integers to integers and $\deg(f) = \deg(g)$. So, we need to prove that

$$\operatorname{lcm}[1,2,\ldots,\deg(g)]\cdot\frac{g(d)-g(0)}{d}$$

is an integer. Let $D=\deg g$, so, using theorem 10.3, there exist integers $a_0,a_1,\ldots,a_D$ such that

$$g(X)=\sum_{i=0}^{D}a_i\binom{X}{i}.$$

It is thus enough to prove that for any $1\le i\le D$ we have

$$\operatorname{lcm}[1,2,\ldots,D]\cdot\frac{1}{d}\binom{d}{i}\in\mathbb{Z}.$$

But the left-hand side is equal to $\frac{\operatorname{lcm}[1,2,\ldots,D]}{i}\binom{d-1}{i-1}$, which is clearly an integer. The result follows. $\square$

18. Let f be a polynomial of degree d such that $f(\mathbb{Z})\subset\mathbb{Z}$ and for which $f(m)-f(n)$ is a multiple of $m-n$ for all $0\le m,n\le d$. Prove that $f(m)-f(n)$ is a multiple of $m-n$ for all integers m,n with $m\neq n$.

Holden Lee

Proof. Since $f(\mathbb{Z})\subset\mathbb{Z}$, theorem 10.3 yields the existence of integers a_0, $a_1,\ldots$, a_d such that

$$f(X)=\sum_{k=0}^{d}a_k\binom{X}{k}.$$

The result follows now from the previous problem and the following general result. $\square$

Lemma 10.5. *Let $a_i\in\mathbb{Z}$ and let*

$$f(X)=\sum_{k=0}^{d}a_k\binom{X}{k}\quad\text{and}\quad L_k=\operatorname{lcm}(1,2,\ldots,k)$$

(by convention $L_0=1$). Then the following assertions are equivalent:

a) $m - n$ divides $f(m) - f(n)$ for all $0 \le m \ne n \le d = \deg f$.

b) L_k divides a_k for all $0 \le k \le d$.

c) $m - n$ divides $f(m) - f(n)$ for all $m \ne n \in \mathbb{Z}$.

Proof. Suppose that a) holds. We will prove by induction on i that L_i divides a_i. This is clear for $i = 0$, so assume that $a_0, \dots, a_{i-1}$ are multiples of $L_0, L_1, \dots, L_{i-1}$ and fix $0 \le j < i$. Then $j - i$ divides

$$f(i) - f(j) = \sum_{k=0}^{i} a_k \left(\binom{i}{k} - \binom{j}{k} \right) = a_i + \sum_{0 \le k < i} a_k \left(\binom{i}{k} - \binom{j}{k} \right).$$

By the previous problem and the inductive hypothesis, each of the numbers $a_k \left(\binom{i}{k} - \binom{j}{k} \right)$ with $0 \le k < i$ is a multiple of $i - j$. We deduce that $i - j$ divides a_i and since $j < i$ was arbitrary, it follows that L_i divides a_i. Hence a) implies b). The previous problem shows that b) implies c) and since it is trivial that c) implies a), the result follows. We are done. □

Remark 10.6. The fact that the polynomial $f(X) = \sum_{k=0}^{n} a_k \cdot \binom{X}{k}$ satisfies $\frac{f(a)-f(b)}{a-b} \in \mathbb{Z}$ for all $a \ne b \in \mathbb{Z}$ if and only if $\operatorname{lcm}(1, 2, \dots, k) | a_k$ for all k seems to have been first noticed by R.R.Hall and I.Z.Ruzsa.

The following problem strongly suggests using Lagrange's interpolation theorem, but there are some difficulties in making the argument work, since the polynomials appearing in Lagrange's formula don't have integer coefficients. The problem is quite tricky and makes again use of Hilbert's polynomials.

19. a) Prove that for all positive integers n there is a polynomial $f \in \mathbb{Z}[X]$ such that all numbers $f(1) < f(2) < \cdots < f(n)$ are powers of 2.

    b) Let $a > 1$ be an integer and let n be a positive integer. Prove that there exists a polynomial f of degree n, having integer coefficients, such that $f(0), f(1), \dots, f(n)$ are pairwise distinct positive integers, all of the form $2a^k + 3$ for some integer k.

Chinese TST 2004

Proof. a) We will choose f of the form

$$f(X) = A \cdot \sum_{i=0}^{n} \binom{X}{i} B^i$$

for some suitable integers A, B, to be chosen later. By the binomial formula, for any $0 \le i \le n$ we have $f(i) = A(1+B)^i$. Now, of course we want f to have integral coefficients and unfortunately $\binom{X}{i}$ does not have integral coefficients. However, $n!\binom{X}{i}$ has integral coefficients for all $1 \le i \le n$. Note however that we cannot simply take for B a multiple of $n!$, since then $A(1+B)^i$ has no chance of being a power of 2. However, we can profit from the presence of A: take A to be $2^{v_2(n!)}$ and B an odd multiple of the greatest odd divisor of $n!$. Then the previous remarks show that f has integral coefficients. Finally, we want $1+B$ to be a power of 2. Thus, we can choose (for example) $1 + B = 2^{\varphi(d)}$, where d is the largest odd divisor of $n!$. With these choices, $f(1) < f(2) < \cdots < f(n)$ are powers of 2, as needed.

b) This uses the same idea as a). Write $n! = m \cdot q$, where all prime factors of m are among those of a and where $\gcd(q, a) = 1$. Let $b = a^{\varphi(q)} - 1$, so q divides b. Finally, define

$$f(X) = 2a^m \sum_{i=0}^{n} \binom{X}{i} b^i + 3.$$

It has integer coefficients because $i!|n!|a^m \cdot b$ for all $0 \le i \le n$. Moreover, for $1 \le k \le n$ we have

$$P(i) = 2a^m \cdot (b+1)^i + 3 = 2a^{m+\varphi(q)i} + 3. \qquad \square$$

Remark 10.7. It is also true that for all positive integers n there is a polynomial $f \in \mathbb{Z}[X]$ such that all numbers $f(1) < f(2) < \cdots < f(n)$ are prime numbers. We will try to find such a polynomial of the form

$$f(X) = a_1(X-2)(X-3)\cdots(X-n) + a_2(X-1)(X-3)\cdots(X-n) \\ + \cdots + a_n(X-1)\cdots(X-n+1) + 1,$$

for some suitable integers a_i. The reason is that $f(i)$ only depends on a_i, so that it will be rather easy to adjust a_i in order to make $f(i)$ prime. Indeed, note that

$$f(i) = (-1)^{n-i}(n-i)!(i-1)!a_i + 1.$$

Now, we will choose the a_i's inductively so that $f(1) < \cdots < f(n)$ are primes. We will heavily use theorem 9.6, according to which for all n there are infinitely many primes of the form $1 + kn$. Thus, there exists an integer a_1 such that $1 + (-1)^{n-1}(n-1)!a_1 = f(1)$ is a prime. Fix such a_1 and choose (again by the cited result) an integer a_2 such that $1 + a_2(n-2)!(-1)^{n-2} = f(2)$ is a prime greater than $f(1)$. Continuing like this, we find $a_1, a_2, \ldots, a_n$ such that $f(1) < f(2) < \cdots < f(n)$ are primes and the problem is solved.

20. Suppose that n is a positive integer not divisible by the cube of a prime number. Consider all sequences $(x_1, x_2, \ldots, x_n)$ with $x_i \in \mathbb{Z}/n\mathbb{Z}$. For how many of these can we find a polynomial f with integer coefficients such that $f(i) \pmod{n} = x_i$ for all i?

USA TST 2008

Proof. Let A_n be the additive group of those sequences $(x_1, x_2, \ldots, x_n) \in (\mathbb{Z}/n\mathbb{Z})^n$ associated to integer polynomials, as in the problem. We will show that the map

$$\Phi(\overline{a_0}, \overline{a_1}, \ldots, \overline{a_{n-1}}) = (\overline{f(1)}, \overline{f(2)}, \ldots, \overline{f(n)}),$$

where

$$f(X) = a_0 + \sum_{i=1}^{n-1} a_i \prod_{j=1}^{i} (X - j),$$

is an isomorphism of abelian groups $\Phi : \prod_{k=0}^{n-1} \left(\mathbb{Z}/\frac{n}{\gcd(n,k!)}\mathbb{Z}\right) \simeq A_n$. First, note that Φ is well-defined: indeed, since a product of d consecutive integers is a multiple of $d!$, it is clear that the sequence $(\overline{f(i)})_{1\le i\le n}$ does not depend on the choice of representatives a_i for $\overline{a_i}$.

Let us prove that Φ is surjective. Repeated division algorithm shows that any polynomial with integer coefficients of degree at most d can be written in the form

$$f(X) = a_0 + \sum_{i=1}^{d} a_i \prod_{j=1}^{i} (X - j)$$

for some integers a_i. We may restrict to $d < n$ since all a_k with $k \geq n$ do not matter when considering $f(i) \pmod{n}$. This yields the surjectivity.

It is clear that Φ is a group homomorphism. It remains to prove that Φ is injective, so suppose f satisfies $f(i) \equiv 0 \pmod{n}$ for $1 \leq i \leq n$. We want to show that a_k is a multiple of $\frac{n}{\gcd(n,k!)}$ for $0 \leq k < n$. Assuming the contrary, there is some least k for which this does not hold. Then we may assume $a_j = 0$ for $j < k$ (since replacing them by 0 does not change the values of f mod n). But then plugging in $X = k+1$ gives $f(k+1) = k!a_k \equiv 0 \pmod{n}$. and a_k is a multiple of $\frac{n}{\gcd(n,k!)}$, contrary to our assumption.

Thus the number of polynomial sequences $(x_1, \ldots, x_n)$ is

$$N = \prod_{k=0}^{n-1} \frac{n}{\gcd(n, k!)}.$$

If a prime p has $v_p(n) = 1$, then there are p factors in this product (those with $k = 0, 1, \ldots, p-1$) which are multiples of p. If a prime p has $v_p(n) = 2$, then there are p factors which are multiples of p^2 $(k = 0, \ldots, p-1)$ and p more factors which are multiples of just p $(k = p, \ldots, 2p-1)$. Therefore for n which are not divisible by the cube of a prime, we have

$$N = \prod_{p:v_p(n)=1} p^p \cdot \prod_{p:v_p(n)=2} p^{3p}. \qquad \square$$

Proof. For a polynomial $f \in \mathbb{Z}[X]$ and $n \geq 1$, let

$$S_n(f) = (f(1) \pmod{n}, f(2) \pmod{n}, \ldots, f(n) \pmod{n}) \in (\mathbb{Z}/n\mathbb{Z})^n$$

and let $A_n = \{S_n(f) | f \in \mathbb{Z}[X]\}$. We want to find $|A_n|$, at least when n is cube-free. First, we will prove the following

Lemma 10.8. *$|A_n|$ is multiplicative, i.e.*

$$|A_{mn}| = |A_m| \cdot |A_n|$$

for all $\gcd(m, n) = 1$.

Proof. We will exhibit a bijection between A_{mn} and $A_m \times A_n$. Consider the map

$$\varphi : A_{mn} \to A_m \times A_n, \quad \varphi(S_{mn}(f)) = (S_m(f), S_n(f)).$$

Note that it is well-defined, for if $S_{mn}(f) = S_{mn}(g)$, we have $f(x) \equiv g(x) \pmod{mn}$ for all $1 \leq x \leq mn$ and so $S_m(f) = S_m(g)$ and $S_n(f) = S_n(g)$. We claim that φ is bijective. Injectivity is very easy, for if $\varphi(S_{mn}(f)) = \varphi(S_{mn}(g))$, then m divides $(f-g)(i)$ for all $1 \leq i \leq m$ and n divides $(f-g)(i)$ for all $1 \leq i \leq n$. By the division algorithm, it follows that m divides $(f-g)(x)$ for all integers x and the same for n. Since $\gcd(m,n) = 1$, we have $mn | f(x) - g(x)$ for all x and we are done. For surjectivity, we need to prove that if $f, g \in \mathbb{Z}[X]$ are arbitrary, then there exists $h \in \mathbb{Z}[X]$ such that $h(i) \equiv f(i) \pmod{n}$ for all $1 \leq i \leq n$ and $h(i) \equiv g(i) \pmod{m}$ for all $1 \leq i \leq m$. Simply choose $h = Amf + Bng$, where A, B are integers such that $Am \equiv 1 \pmod{n}$ and $Bn \equiv 1 \pmod{m}$. The lemma is proved. □

Thanks to this lemma and the hypothesis that n is cube free, it remains to find $|A_p|$ and $|A_{p^2}|$. The first task is trivial: for any p-tuple $(a_1, \dots, a_p)$ of remainders mod p there exists a polynomial f such that $f(i) = a_i \pmod{p}$ for all $1 \leq i \leq p$, by Lagrange's interpolation formula. Thus $|A_p| = p^p$. Finding $|A_{p^2}|$ is more delicate, since $\mathbb{Z}/p^2\mathbb{Z}$ is not a field and so Lagrange's interpolation formula is useless. Moreover, there are nontrivial relationships between the numbers $f(i) \pmod{p^2}$, which implies that $|A_{p^2}|$ is definitely smaller than p^{2p^2}.

The first point is to note that

$$A_{p^2} = \{S_{p^2}(f) | f \in \mathbb{Z}[X], \deg(f) < 2p\},$$

as for any $f \in \mathbb{Z}[X]$ one can find $g \in \mathbb{Z}[X]$ of degree smaller than $2p$ and such that $f(x) \equiv g(x) \pmod{p^2}$ for all x. Indeed, simply take for g the remainder of f when divided by $(X^p - X)^2$. Let $G = \{f \in \mathbb{Z}/p^2\mathbb{Z}[X] | \deg(f) < 2p\}$, an abelian group of order $(p^2)^{2p} = p^{4p}$ and let $S : G \to \left(\mathbb{Z}/p^2\mathbb{Z}\right)^{p^2}$ be defined by $S(f) = (f(1), f(2), \dots, f(p^2))$. The previous observation shows that A_{p^2} is the image of the map $S : G \to \left(\mathbb{Z}/p^2\mathbb{Z}\right)^{p^2}$. As $|\mathrm{Im}(S)| = \frac{|G|}{|\mathrm{Ker}(S)|}$, it remains to find $|\mathrm{Ker}(S)|$. We can actually describe $\mathrm{Ker}(S)$ rather explicitly, thanks to problem 15. Indeed, this problem (for $k = 2$) implies that $\mathrm{Ker}(S)$ is in bijection with the set of polynomials $u \in \mathbb{F}_p[X]$ of degree smaller than p. As this set has p^p elements, it follows that $|A_{p^2}| = p^{3p}$. The problem is finally solved. $\square$

10.4 p-adic estimates

In this section we consider a few problems dealing with p-adic properties of polynomials.

21. Let $(a_n)_{n\geq 1}$ be an increasing sequence of positive integers such that for some polynomial $f \in \mathbb{Z}[X]$ we have $a_n \leq f(n)$ for all n. Suppose also that $m - n | a_m - a_n$ for all distinct positive integers m, n. Prove that there exists a polynomial $g \in \mathbb{Q}[X]$ such that $a_n = g(n)$ for all n.

USAMO 1995

Proof. Let d be the degree of f and choose a polynomial P of degree at most d with rational coefficients and such that $P(i) = a_i$ for $1 \leq i \leq d+1$. This is possible by Lagrange's interpolation formula. Choose (and fix) $N \geq 1$ such that $h = NP$ has integral coefficients. Then $h(i) = Na_i$ and h has degree d. Fix any integer $n > d + 1$ and observe that $m - n$ divides $Na_m - Na_n$ and $m-n$ divides $h(m)-h(n)$. Thus, if $m \leq d+1$, then $m-n$ divides $Na_n - h(n)$. Consequently, $Na_n - h(n)$ is a multiple of $\mathrm{lcm}(n-1, \ldots, n-(d+1))$. Note that $|Na_n - h(n)| \leq Cn^d$ for some constant C, because a_n is bounded by f and because h has degree at most d. On the other hand, we have the following result:

Lemma 10.9. *For any positive integers* $x_1, x_2, \ldots, x_n$, $\mathrm{lcm}(x_1, x_2, \ldots, x_n)$ *is a multiple of* $\frac{x_1 x_2 \cdots x_n}{\prod_{1\leq i<j\leq n} \gcd(x_i, x_j)}$.

Proof. It is enough to prove that for any prime p, the p-adic valuation of $\mathrm{lcm}(x_1, x_2, \ldots, x_n)$ is at least that of $\frac{x_1 x_2 \cdots x_n}{\prod_{1\leq i<j\leq n} \gcd(x_i, x_j)}$. Writing $y_i = v_p(x_i)$, this comes down to the inequality

$$\max(y_i) \geq \sum y_i - \sum_{i<j} \min(y_i, y_j),$$

which is clear (simply order the y_i's). $\square$

Coming back to the problem, we infer that

$$\operatorname{lcm}(n-1, n-2, \ldots, n-(d+1)) \geq \frac{(n-d-1)^{d+1}}{\prod_{1 \leq i<j \leq d+1} \gcd(n-i, n-j)},$$

which is greater than $C_1 n^{d+1}$ for some constant $C_1 > 0$ and all large n (this is because $\gcd(n-i, n-j)$ divides $j-i$). Thus, if n is large enough, then necessarily

$$\operatorname{lcm}(n-1, n-2, \ldots, n-(d+1)) > |Na_n - h(n)|.$$

Combining this with the result of the previous paragraph, we infer that there is n_0 such that for all $n \geq n_0$ we have $Na_n = h(n)$.

Finally, pick any $m \geq 1$ and observe that for all $n \geq n_0$ we have $m - n | Na_m - Na_n$ and $m - n | h(m) - h(n)$. Since $Na_n = h(n)$, we deduce that $m-n$ divides $Na_m - h(m)$ for all $n \geq n_0$, forcing $Na_m = h(m)$. Thus, we proved the existence of a polynomial $g = \frac{1}{N}h$, with rational coefficients, such that $a_n = g(n)$ for all n. $\square$

Remark 10.10. There are a lot of non-polynomial maps f such that $m-n$ divides $f(m) - f(n)$ for all $m \neq n$. For instance, pick any sequence of integers a_n, infinitely many of which are nonzero and define

$$f(n) = \sum_{j \geq 0} a_j \cdot (n+j)(n+j-1) \cdot \cdots \cdot (n-j).$$

It is an easy exercise to check that f satisfies the desired congruences and that f is not polynomial.

Remark 10.11. It is not true that if $a_n = g(n)$ for all n, then $g \in \mathbb{Z}[X]$. For instance, one can easily check that if $g(n) = \frac{n^4-n^2}{2}$, then $\frac{g(m)-g(n)}{m-n}$ is an integer for all different integers m and n.

22. Consider all sequences $(f(1) \pmod{n}, f(3) \pmod{n}, \ldots, f(1023) \pmod{n})$, where $n = 1024$ and f is an arbitrary polynomial with integer

coefficients. Prove that at most 2^{35} of these sequences are permutations of $1, 3, 5, \ldots, 1023 \pmod{n}$.

USA TST 2007

Proof. Define $P_0(X) = 1$ and $P_i(X) = \prod_{k=1}^{i}(X-(2k-1))$ for $i \geq 1$. Repeated euclidean division (taking into account that P_i is monic) shows that each polynomial $f \in \mathbb{Z}[X]$ can be written $f = c_0P_0 + c_1P_1 + \cdots + c_nP_n$ for integers c_i, where $n = \deg f$.

Note that $P_i(2n-1)$ is a multiple of $2^i \cdot i!$, so by Legendre's formula we have $v_2(P_i(x)) \geq a_i$ for all odd numbers x, where $a_i = \sum_{k=0}^{\infty} \left\lfloor \frac{i}{2^k} \right\rfloor$. Note that $a_0 = 0, a_1 = 1, a_2 = 3, a_3 = 4, a_4 = 7, a_5 = 8$, and $a_i \geq 10$ for $i \geq 6$.

Say a polynomial f is good if its associated sequence (as in the statement of the problem) is a permutation of $1, 3, 5, \ldots, 1023$. Let

$$f(X) = \sum_{0 \leq i \leq n} c_iP_i(X)$$

be a good polynomial. If we delete the terms with P_i for $i \geq 6$ (where $a_i \geq 10$), we get a polynomial with the same associated sequence as f, so we are only interested in $c_0, c_1, \ldots, c_5$. Note that c_0 is odd and that $f(1) \not\equiv f(3) \pmod 4$, as $f(1) \equiv f(4k+1) \pmod 4$ and $f(3) \equiv f(4k+3) \pmod 4$ for all k. But since

$$f(3) - f(1) \equiv c_1(P_1(3) - P_1(1)) \equiv 2c_1 \pmod 4,$$

it follows that c_1 is odd. Finally, note that if we mod out c_i by 2^{10-a_i}, we get a polynomial with the same associated sequence as f, so we are only interested in the values of $c_i \pmod{2^{10-a_i}}$. But since c_0 is odd, there are at most 2^9 choices for it; for the same reason there are at most 2^8 choices for $c_1 \pmod{2^9}$ and for $2 \leq i \leq 5$ there are at most 2^{10-a_i} choices for $c_i \pmod{2^{10-a_i}}$. Hence the number of complete remainder sequences is at most

$$2^9 \cdot 2^8 \cdot \prod_{i=2}^{5} 2^{10-\alpha_i} = 2^9 \cdot 2^8 \cdot 2^7 \cdot 2^6 \cdot 2^3 \cdot 2^2 = 2^{35}. \qquad \square$$

It is a classical result of Mahler that continuous functions on the ring of p-adic integers $\mathbb{Z}_p$ can be uniformly approximated by polynomials (this is the

p-adic analogue of a classical theorem of Weierstrass, stating that continuous functions on a compact interval can be uniformly approximated by polynomials). It is not difficult to see that the characteristic function of $2\mathbb{Z}_2 + 1$ is continuous, so it can be uniformly approximated with polynomials. The next problem gives a more precise estimate.

23. Prove that for all n there exists a polynomial f with integer coefficients and degree not exceeding n such that 2^n divides $f(x)$ for all even integers x and 2^n divides $f(x) - 1$ for all odd integers x.

P. Hajnal, KöMaL

Proof. Define $O(x) = 1$ if x is odd and 0 otherwise. Then we compute that

$$O(m) = \frac{1}{2}(1 - (-1)^m) = \frac{1}{2}(1 - (-2+1)^m) = \sum_{k=1}^{m} \binom{m}{k} (-2)^{k-1}.$$

Since the binomial coefficients are all integers it follows that truncating after the nth term gives a polynomial

$$g(X) = \sum_{k=1}^{n} \binom{X}{k} (-2)^{k-1} \in \mathbb{Q}[X]$$

of degree n such that for all integers m, $g(m)$ is an integer with $g(m) \equiv O(m) \pmod{2^n}$. Recall that $v_2(k!) = k - s_2(k) \le k - 1$. Therefore every coefficient of g has an odd denominator. Hence we can choose a polynomial $f(X) \in \mathbb{Z}[X]$ of degree at most n such that every coefficient of f is congruent to the corresponding coefficient of g mod 2^n. Then $f(m) \equiv O(m) \pmod{2^n}$ for all m and we are done. □

Proof. Define $O(x) = 1$ if x is odd and 0 otherwise. We want to find a polynomial f of degree at most n and such that $f(x) \equiv O(x) \pmod{2^n}$ for all integers x. We will prove the existence by induction, but in order to prove the inductive step, we will need the following key result:

Lemma 10.12. *Let f be a polynomial of degree at most n and let*

$$F(x) = f(x) - O(x).$$

Then for all integers x we have

$$F(x+n+1) \equiv \sum_{k=0}^{n} (-1)^{n-k} \binom{n+1}{k} F(x+k) \pmod{2^n}.$$

Proof. Since f has degree at most n, proposition 10.4 shows that

$$\sum_{k=0}^{n+1} (-1)^k \binom{n+1}{k} f(x+k) = 0.$$

Thus, to prove the lemma is remains to prove that 2^n divides

$$\sum_{k=0}^{n+1} (-1)^k \binom{n+1}{k} O(x+k).$$

However, if x is even this is equal to -2^n, because

$$\sum_{k} \binom{n+1}{2k+1} = 2^n.$$

Similarly, if x is odd, then it is equal to 2^n, because

$$\sum_{k} \binom{n+1}{2k} = 2^n.$$

This ends the proof of the lemma. □

The crucial point that we deduce from this lemma is the following: if $f(x) \equiv O(x) \pmod{2^n}$ for all $0 \le x \le n$, then $f(x) \equiv O(x) \pmod{2^n}$ for all integers x. This follows trivially by induction, using the congruence in the lemma.

We can now finally prove by induction the existence of f. For $n = 1$, everything is clear, so assume that we constructed a polynomial f of degree at most $n-1$ such that $f(x) \equiv O(x) \pmod{2^{n-1}}$ for all integers x. We will choose our polynomial of the form

$$g(X) = f(X) + \sum_{k=0}^{n} a_k \cdot \prod_{j \neq k, 0 \leq j \leq n} (X - j).$$

It clearly has degree at most n and for all $0 \leq k \leq n$ we have

$$g(k) = f(k) + (-1)^{n-k} a_k \cdot k!(n-k)!.$$

By hypothesis, there are integers b_k such that $f(k) = O(k) + b_k \cdot 2^{n-1}$ for all $0 \leq k \leq n$. Now, since

$$v_2(k!(n-k)!) \leq v_2(n!) \leq n-1,$$

it is clear that we can choose integers a_k such that 2^n divides

$$2^{n-1} b_k + (-1)^{n-k} a_k \cdot k!(n-k)!.$$

By construction, 2^n will divide $g(k) - O(k)$ for all $0 \leq k \leq n$ and thus for all x, by the previous results. The inductive step is thus proved and the problem is solved. □

We end this section with a challenging problem having a very elementary but delicate solution. The main ideas are the pigeonhole principle and p-adic estimates, but the way in which they are combined is rather subtle. A special case of this problem (whose proof is very similar) is discussed in chapter 3, problem 13.

24. Let $f \in \mathbb{Z}[X]$ be a polynomial and let a be an integer. Consider the sequence $a_0 = a, a_{n+1} = f(a_n)$. If $a_n \to \infty$ and the set of prime divisors of $(a_n)_n$ is finite, prove that $f(X) = AX^d$ for some A, d.

Tuymaada Olympiad, 2003

Proof. Let f^d be the composition of f with itself d times. Assume that $p_1, p_2, \ldots, p_N$ comprise all prime factors of all numbers $a_0, a_1, \ldots$. First, we will prove that one of the numbers $f(0), f^2(0), \ldots, f^N(0)$ is 0. To do this, fix any positive integer r and n_0 such that for all $n \geq n_0$ we have $a_n > (p_1p_2\cdots p_N)^r$. Such n_0 exists by assumption. Thus, for all $n \geq n_0$ there exists i such that $v_{p_i}(a_n) > r$. Considering $N+1$ consecutive values $n = n_0, n_0+1, \ldots, n_0+N$, the i's that we find take only N possible values, so two of them will be equal. Thus, we can find $u \in \{1, 2, \ldots, N\}$, $i \geq n_0$ and $k \in \{1, 2, \ldots, N\}$ such that $v_{p_k}(a_i) > r$ and $v_{p_k}(a_{i+u}) > r$. Since $a_{i+u} = f^u(a_i) \equiv f^u(0) \pmod{a_i}$, it follows that p_k^r divides $f^u(0)$. Thus, we proved that for any r we can find $u \in \{1, 2, \ldots, N\}$ and k such that p_k^r divides $f^u(0)$. It follows that one of $f(0), f^2(0), \ldots, f^N(0)$ has infinitely many divisors and the claim is proved.

If $f^i(0) = 0$, then working with f^i instead of f (the iterates of a under f^i form a subsequence of the sequence $(a_n)_n$, and so will still have only finitely many prime divisors), we may assume that $f(0) = 0$ (it is easy to see that if $f^i(X)$ is of the form AX^d for some $i \geq 1$, then f itself is of this form). Write then $f(X) = X^d(b_0 + b_1X + \cdots + b_eX^e)$ for some integers $b_0, \ldots, b_e$ and some $d \geq 1$ such that $b_0b_e \neq 0$. Assume that the conclusion of the problem does not hold, so that $e \geq 1$. Note that a_n divides a_{n+1}. Since for each i there is n_i such that a_{n_i} is a multiple of p_i, the number $a_{n_1+\cdots+n_N}$ is a multiple of $p_1\cdots p_N$ and so there is n_0 such that for all $n \geq n_0$ we have $p_1\cdots p_N | a_n$. It follows easily from this that the prime factors of b_0 are all amongst the p_i.

Consider now some i such that $p_i|b_0$. Then

$$\begin{aligned} v_{p_i}(a_{n+1}) &= dv_{p_i}(a_n) + v_{p_i}(b_0 + b_1a_n + \cdots + b_ea_n^e) \\ &\geq v_{p_i}(a_n) + \min(1, v_{p_i}(a_n)), \end{aligned}$$

from which is trivially follows that $v_{p_i}(a_n) \to \infty$ for $n \to \infty$ (we know that for large enough n we have $v_{p_i}(a_n) > 0$ by the previous paragraph). Now, we are finally ready to conclude: let $p_{i_1}, \ldots, p_{i_k}$ be those primes among $p_1, \ldots, p_N$ that divide b_0. By the previous paragraph, eventually by increasing n_0, we may assume that that $v_{p_{i_j}}(a_n) > v_{p_{i_j}}(b_0)$ for all $1 \leq j \leq k$ and all $n \geq n_0$. If q is a prime factor of $b_0 + \cdots + b_ea_n^e$ with $n \geq n_0$, then q divides a_{n+1}, so q is one of the p_i's. But all p_i's divide a_n, so q must divide b_0 and so $q \in \{p_{i_1}, \ldots, p_{i_k}\}$. Moreover, since $v_{p_{i_j}}(a_n) > v_{p_{i_j}}(b_0)$ for all $1 \leq j \leq k$, we have $v_{p_{i_j}}(b_0 + \cdots + b_ea_n^e) = v_{p_{i_j}}(b_0)$. Putting together these observations shows that $b_0 + b_1 a_n + \cdots + b_e a_n^e \leq b_0$ for all sufficiently large n, contradicting the fact that $a_n \to \infty$. The result finally follows. $\square$

10.5 Miscellaneous problems

The following problem is a version of a very classical topic, known as the Prouhet-Tarry-Escott problem. It concerns disjoint sets of n integers having the same sum of dth powers for all $1 \leq d \leq k$ (k depending on n).

25. Let d, r be positive integers with $d \geq 2$. Prove that for any nonconstant polynomial f with real coefficients and of degree smaller than r, the numbers $f(0), f(1), \ldots, f(d^r - 1)$ can be divided into d disjoint groups such that the sum of the elements of each group is the same.

J. O. Shallit, AMM E 3032

Proof. Define the sets

$$A_i(r) = \{x \in [0, d^r - 1] | s_d(x) \equiv i \pmod{d}\},$$

where $s_d(x)$ is the sum of digits of x when written in base d. We will prove that $\sum_{x \in A_i(r)} f(x)$ is independent of i, for any polynomial f of degree smaller than r. The proof will be by induction on r, the case $r = 1$ being clear (since then each $A_i(r)$ has exactly one element and f must be constant). Assuming that the result holds for r, observe that (by linearity) it is enough to prove that $\sum_{x \in A_i(r+1)} x^r$ is independent of i in order to prove the inductive step. But, by definition (and considering the last digit in base d) we have

$$A_i(r+1) = \cup_{j=0}^{d-1} \{dx + j | x \in A_{i-j}(r)\},$$

the union being disjoint (and the sets A_i being numbered mod d). Thus

$$\begin{aligned}\sum_{x \in A_i(r+1)} x^r &= \sum_{j=0}^{d-1} \sum_{x \in A_{i-j}(r)} (dx + j)^r \\ &= \sum_{j=0}^{d-1} \left(d^r \sum_{x \in A_{i-j}(r)} x^r + r d^{r-1} j \sum_{x \in A_{i-j}(r)} x^{r-1} + \cdots + j^r |A_{i-j}(r)| \right).\end{aligned}$$

Separate in the last sum the term corresponding to x^r and observe that it does not depend on i, since it is just

$$d^r \sum_{x \in A_0(r) \cup \cdots \cup A_{d-1}(r)} x^r.$$

The other terms are independent of i by the induction hypothesis. Thus $\sum_{x \in A_i(r+1)} x^r$ is also independent of i, finishing the proof of the inductive step. □

Proof. This will use the same sets A_i, but the proof that they satisfy the desired conditions is different. Let z be any dth root of unity different from 1 and let

$$\Delta_a f(X) = f(X) + z f(X+a) + z^2 f(X+2a) + \cdots + z^{d-1} f(X+(d-1)a).$$

Since $1 + z + \cdots + z^{d-1} = 0$, we have $\deg(\Delta_a f) < \deg(f)$.

Let our disjoint sets be $A_0, A_1, \ldots A_{d-1}$ as in the previous solution and consider

$$A(z) = \sum_{k=0}^{d-1} z^k \left(\sum_{X \in A_k} f(X) \right).$$

By an immediate induction, we obtain that

$$A(z) = \Delta_{d^{r-1}} \Delta_{d^{r-2}} \cdots \Delta_d \Delta_1 f(X).$$

But $\deg(f) < r$, so that $\deg(\Delta_{d^{r-1}} \Delta_{d^{r-2}} \cdots \Delta_d \Delta_1 f(X)) < 0$ and so $A(z) = 0$. Since this holds for all such z, it follows that the polynomial $A(t)$ is a multiple of $1 + t + \cdots + t^{d-1}$ and hence $\sum_{x \in A_k} f(x)$ is independent of k. □

The following problem is rather technical, but the idea is very simple: after some algebraic manipulations, everything comes down to the fact that a nonzero polynomial of degree at most d cannot vanish at more than d distinct points.

26. Let $p \geq 5$ be a prime and let a, b, c be integers such that p does not divide any of the numbers $a - b, b - c, c - a$. Let i, j, k be nonnegative integers such that $i + j + k$ is divisible by $p - 1$ and such that for all integers x, the number

$$(x-a)(x-b)(x-c)[(x-a)^i(x-b)^j(x-c)^k - 1]$$

is divisible by p. Prove that each of i, j, k is divisible by $p - 1$.

Kiran Kedlaya and Peter Shor, USA TST 2009

Proof. We start with some formal reductions. First, note that we may assume that $0 \leq i, j, k < p - 1$, as we can replace i, j, k with their remainders mod $p - 1$, without affecting the hypothesis or the conclusion (this uses Fermat's little theorem). We want to prove that $i = j = k = 0$, so assume the contrary. By hypothesis, $i + j + k = p - 1$ or $2(p - 1)$. In the second case, replace each $x \in \{i, j, k\}$ with $p - 1 - x$. As this does not change the hypothesis or the conclusion, we can assume from now on that $i + j + k = p - 1$. Finally, we can clearly assume that i is the largest among i, j, k.

Multiplying the congruence

$$(x-a)(x-b)(x-c)[(x-a)^i(x-b)^j(x-c)^k - 1] \equiv 0 \pmod{p}$$

by $(x - a)^{j+k}$ and using Fermat's little theorem, we deduce that

$$f(x) = (x-a)(x-b)(x-c)[(x-b)^j(x-c)^k - (x-a)^{j+k}] \equiv 0 \pmod{p}.$$

for all integers x. On the other hand, f has degree at most

$$3 + j + k - 1 \leq 2 + \frac{2(p-1)}{3} < p$$

(for $p \geq 5$) and p different roots mod p. Thus f vanishes in $\mathbb{F}_p[X]$ and we deduce the equality $(X - b)^j(X - c)^k = (X - a)^{j+k}$ in $\mathbb{F}_p[X]$. Note that $j + k \neq 0$, as $i < p - 1$ and $i + j + k = p - 1$. Thus $(X - b)^j(X - c)^k$ vanishes at b or c. But this is impossible, as by hypothesis $(X - a)^{j+k}$ does not vanish at either b or c. □

The next problem is very exotic and tricky.

27. Prove the existence of a number $c > 0$ with the following property: for any prime p, there are at most $cp^{2/3}$ positive integers n such that p divides $n! + 1$.

Chinese TST 2009

Proof. Of course, if $p|n! + 1$, then $n \leq p - 1$. Let $p > 2$ and let $1 < n_1 < n_2 < \cdots < n_m < p$ be all solutions of the equation $n! \equiv -1 \pmod{p}$. Assume that $m > 1$ (otherwise everything is clear). The congruences $n_i! \equiv -1 \pmod{p}$ and $n_{i+1}! \equiv -1 \pmod{p}$ imply that

$$(n_i + 1)(n_i + 2)\cdots(n_i + n_{i+1} - n_i) \equiv 1 \pmod{p}.$$

Letting $k = n_{i+1} - n_i$, we see that $x = n_i$ is a solution to

$$(x + 1)(x + 2)\cdots(x + k) \equiv 1 \pmod{p}.$$

Since the polynomial $(x + 1)(x + 2)\cdots(x + k) - 1 \in (\mathbb{Z}/p\mathbb{Z})[x]$ has at most k distinct roots modulo p, it follows that for each $1 < k < p$ there are at most k indices i such that $n_{i+1} - n_i = k$. We will prove that this is enough to force $m < cp^{2/3}$.

Choose a positive integer j such that

$$\frac{(j + 1)(j + 2)}{2} \geq m \geq \frac{j(j + 1)}{2}.$$

Since $m \geq \frac{j(j+1)}{2} = \sum_{i=1}^{j} j$, when the differences $n_{i+1} - n_i$ are written in ascending order, the first is at least 1, the next two are at least 2, and so on, each time the next i differences are at least i (this is because for a fixed k, $1 \leq k < p$, $n_{i+1} - n_i = k$ has at most k solutions i). Thus

$$\sum_{i=1}^{m-1}(n_{i+1} - n_i) \geq 1^2 + 2^2 + \cdots + j^2 = \frac{j(j + 1)(2j + 1)}{6}.$$

We deduce that

$$p > n_m - n_1 > \frac{j(j + 1)(2j + 1)}{6}.$$

In particular, $p > \frac{j^3}{3}$ and so $j < (3p)^{1/3}$. Since $m \leq (j+1)^2$, the result follows. $\square$

Before passing to the next problem, let us recall that Bézout's lemma does not hold in $\mathbb{Z}[X]$. However, if f and g are polynomials with integer coefficients and with no common complex root, then they are relatively prime and by Bézout's lemma in $\mathbb{Q}[X]$ there is a nonzero integer c and polynomials $A, B \in \mathbb{Z}[X]$ such that $Af + Bg = c$. Usually $|c| > 1$ and it is not at all clear what is the smallest possible value of $|c|$, given f and g. The following problem discusses the case $f = (X+1)^n$ and $g = X^n + 1$. It is a challenging problem, combining several ideas in a very tricky way.

28. Let n be an even positive integer. Find the least positive integer k for which one can find polynomials with integer coefficients f, g such that

$$f(X)(X+1)^n + g(X)(X^n+1) = k.$$

IMO Shortlist 1996

Proof. Let us write $n = 2^r \cdot m$ for some odd integer m and assume that we have

$$f(X)(X+1)^n + g(X)(X^n+1) = k$$

for some $f, g \in \mathbb{Z}[X]$ and some positive integer k. Taking for X a root z_i of the polynomial $X^{2^r} + 1$, we deduce that $f(z_i)(z_i+1)^n = k$. Multiplying all these relations and taking into account to $\prod_{i=1}^{2^r}(1+z_i) = 2$, it follows that $\prod_{i=1}^{2^r} f(z_i) \cdot 2^n = k^{2^r}$. Since $\prod_{i=1}^{2^r} f(z_i)$ is an integer (by theorem 9.10), 2^n divides k^{2^r} and so k must be a multiple of 2^m. In particular, $k \geq 2^m$.

We will prove now that $k = 2^m$ works. Let us see what happens when $m = 1$ first. We need to find polynomials f, g with integer coefficients such that

$$f(X)(X+1)^{2^r} + g(X)(X^{2^r}+1) = 2.$$

The idea is to find f such that

$$f(z)(z+1)^{2^r} = 2$$

for some root z of $X^{2^r}+1$. Indeed, since $X^{2^r}+1$ is irreducible over the rational numbers (because $(X+1)^{2^r}+1$ is Eisenstein for the prime 2), this would imply that $X^{2^r}+1$ divides $f(X)(X+1)^{2^r}-2$, which would give us g. The key point is to take $z=e^{\frac{i\pi}{2^r}}$, because all the other roots z_i of $X^{2^r}+1$ are of the form z^j, with **odd** j. Thus, if $z_1=z,\dots,z_{2^r}$ are the roots of $X^{2^r}+1$, then we can write $z_i+1=(z+1)Q_i(z)$ for some polynomials Q_i with integer coefficients. And since $\prod_{i=1}^{2^r}(1+z_i)=2$, it follows that $(1+z)^{2^r}\prod_{i=1}^{2^r}Q_i(z)=2$ which gives us the polynomial f and finishes the proof in the case $m=1$.

Finally, it is rather formal to deduce the general case from the case $m=1$. Namely, pick polynomials with integer coefficients f,g such that

$$f(X)(X+1)^{2^r}+g(X)(X^{2^r}+1)=2.$$

Then

$$f(X)^m(X+1)^n=(2-g(X)(X^{2^r}+1))^m=2^m+(X^{2^r}+1)h(X)$$

for some $h\in\mathbb{Z}[X]$. The last equality follows from the binomial formula. Now, replace X by X^m in the previous equality, to get

$$f(X^m)^m(X^m+1)^n=2^m+(X^n+1)h(X^m)$$

and observe that $(X^m+1)^n=(X+1)^nA(X)$ for some $A\in\mathbb{Z}[X]$ (because m is odd). The conclusion is now clear. □

The idea of the following problem is very natural, but there are a lot of technical details one has to deal with, which makes the proof rather long.

29. Suppose that f is a polynomial of degree at least 2, with positive leading coefficient and integer coefficients. Show that there are infinitely many n such that $f(n!)$ is composite.

IMO Shortlist 2005

Proof. We will try first to find prime numbers p and positive integers n such that $p|f(n!)$. Then, we will ensure that n is large enough and finally we will have to get rid of the cases $f(n!)=0,p,-p$. Write

$$f(X)=a_dX^d+a_{d-1}X^{d-1}+\cdots+a_0,$$

with $a_d > 0$ and $d \geq 2$. Note that we may assume that $a_0 \neq 0$, otherwise the problem is trivial.

First, let us consider the equation $f(n!) \equiv 0 \pmod{p}$. Unless p divides a_0, this forces $n < p$. So, let us look for $n = p - k$ with $k > 0$. We have to compute first $(p-k)! \pmod{p}$, which is very easy by Wilson's theorem:

$$\begin{aligned} -1 &\equiv (p-1)! \\ &\equiv (p-k)!(p-k+1)\cdots(p-1) \\ &\equiv (p-k)!(-1)^{k-1}(k-1)! \pmod{p}. \end{aligned}$$

Thus, we have $f(n!) \equiv 0 \pmod{p}$ if and only if $p|x_k$, where

$$x_k = a_0(k-1)!^d + a_1(k-1)!^{d-1}(-1)^k + \cdots + a_d(-1)^{kd}.$$

We will prove first the existence of large prime factors of x_k, more precisely such that $p \geq k$. This is the content of the following

Lemma 10.13. *There exists k_0 such that for all $k > k_0$, there exists a prime factor p_k of x_k such that $p_k \geq k$.*

Proof. This is easy: choose k_1 such that $v_p((k_1-1)!) > v_p(a_d)$ for all primes $p \leq |a_d|$. If all prime factors p of x_k are less than k for some $k \geq k_1$, they divide $(k-1)!$ and x_k, so they divide a_d. But for such a prime p, since $v_p((k-1)!) > v_p(a_d)$, we must have $v_p(x_k) = v_p(a_d)$. We deduce that $|x_k| \leq |a_d|$. Now, choose $k_0 > k_1$ such that $|x_k| > |a_d|$ for all $k \geq k_0$, which is possible as $a_0 \neq 0$. □

Fix now k_0 and p_k as in the lemma. Fix also a positive integer N and assume that none of the numbers $f(n!)$ with $n \geq N$ is composite. By increasing N, we may assume that $x \to f(x!) - x$ is increasing on $[N, \infty)$. By construction, p_k divides $f((p_k-k)!)$. Thus, if $p_k - k \geq N$, then we must have $f((p_k-k)!) = p_k$ and this will happen if we ensure that $k, k+1, \ldots, k+N-1$ are composite. To have this, we can choose $k = k_a = a(N+1)! + 2$ for $a \geq 1$. Denoting $x_a = p_{k_a} - k_a$, we deduce that $f(x_a!) = x_a + a(N+1)! + 2$ for all sufficiently large a (so that $k_a > k_0$). In particular, the last relation shows that $x_a \to \infty$, because the map $a \mapsto x_a$ is injective. In particular, for infinitely many a we have $x_{a+1} \geq x_a + 1$ and so

$$f(x_a!) - x_a + (N+1)! = f(x_{a+1}!) - x_{a+1} \geq f((x_a+1)!) - (x_a+1).$$

This implies that

$$f((x_a+1)!) - f(x_a!) \leq 1 + (N+1)!,$$

which is certainly impossible because $\frac{f((x_a+1)!)}{f(x_a!)} \to \infty$ for $a \to \infty$. Thus our assumption was wrong and at least one of the numbers $f(n!)$ with $n \geq N$ is composite. Since N was arbitrary, the conclusion follows. $\square$

Remark 10.14. The hypothesis that $d \geq 2$ was not used in the proof, so the result still holds for linear polynomials with positive leading coefficient. For instance, this also solves the following Chinese TST 2011 problem: prove that for all positive integers d there are infinitely many n such that $d \cdot n! - 1$ is composite.

10.6 Notes

The following people helped us with solutions and we would like to thank them: Alexandru Chirvăsitu (problem 24), Xiangyi Huang (problems 6, 12, 27), Holden Lee (problem 22), Mitchell Lee (problem 2), Fedja Nazarov (problem 16), Hunter Spink (problem 10), Richard Stong (problems 7, 8, 20, 23, 25), Qiaochu Yuan (problem 1), Victor Wang (problem 14), Gjergji Zaimi (problems 3, 5), Alex Zhu (problems 14).

Chapter 11

Lagrange Interpolation Formula

It is a standard fact that a nonzero polynomial f with coefficients in a field K has at most $\deg f$ roots in K. Lagrange's interpolation formula is in some sense a refinement of this standard result: it makes precise the way in which $\deg f + 1$ points of K determine the polynomial f.

Theorem 11.1. *Let K be a field and let $x_0, x_1, \ldots, x_n$ be distinct elements of K. Then for any polynomial $f \in K[X]$ of degree at most n we have*

$$f(X) = \sum_{k=0}^{n} f(x_k) \cdot \prod_{j \neq k} \frac{X - x_j}{x_k - x_j}.$$

Proof. Simply note that the difference between the polynomials in both sides of the desired equality is a polynomial of degree at most n which vanishes at $x_0, x_1, \ldots, x_n$, thus it is the zero polynomial. □

The following corollary is an immediate consequence of the previous theorem, but it is rather useful in practice, especially when proving complicated identities.

Corollary 11.2. *Let $f \in K[X]$ be a polynomial of degree at most n and let $x_0, x_1, \ldots, x_n \in K$ be distinct elements. Then*

$$\sum_{k=0}^{n} \frac{f(x_k)}{\prod_{j \neq k}(x_k - x_j)}$$

is 0 *if* $\deg f < n$ *and equals the leading coefficient of f otherwise.*

Proof. Simply identify the coefficients of X^n in Lagrange's identity. □

The first two problems are direct applications of Lagrange's interpolation formula.

1. A polynomial p of degree n satisfies $p(k) = 2^k$ for all $0 \leq k \leq n$. Find its value at $n + 1$.

Murray Klamkin

Proof. Using theorem 11.1, we obtain

$$\begin{aligned} p(n+1) &= \sum_{k=0}^{n} p(k) \cdot \prod_{j \neq k} \frac{n+1-j}{k-j} \\ &= \sum_{k=0}^{n} \binom{n+1}{k} (-1)^{n-k} 2^k \\ &= 2^{n+1} - 1. \end{aligned}$$

□

Remark 11.3. There is also a neat solution without use of the interpolation formula: consider the polynomial

$$f(X) = \binom{X}{0} + \binom{X}{1} + \cdots + \binom{X}{n}.$$

It has degree n and satisfies $f(k) = 2^k$ for all $0 \leq k \leq n$ by the binomial formula. Thus we must have $f = p$ and then clearly $p(n+1) = 2^{n+1} - 1$. Even though very neat, this solution is not so conceptual.

2. A polynomial f of degree n satisfies

$$f(k) = \frac{1}{\binom{n+1}{k}}$$

for all $0 \le k \le n$. Find $f(n+1)$.

Titu Andreescu, IMO Shortlist 1981

Proof. This is also immediate using theorem 11.1:

$$\begin{aligned} f(n+1) &= \sum_{k=0}^{n} f(k) \prod_{j \neq k} \frac{n+1-j}{k-j} \\ &= \sum_{k=0}^{n} \frac{1}{\binom{n+1}{k}} \cdot (-1)^{n-k} \binom{n+1}{k} \\ &= \sum_{k=0}^{n} (-1)^{n-k} \\ &= \frac{1-(-1)^{n+1}}{2}. \end{aligned}$$

□

3. Prove that for any real number a we have the following identity

$$\sum_{k=0}^{n} (-1)^k \binom{n}{k} (a-k)^n = n!.$$

Tepper's identity

Proof. Use corollary 11.2 for the polynomial $P(X) = (a-X)^n$ and the points $x_k = k$. □

4. Let

$$S_p = \sum_{k=0}^{n} \frac{x_k^{n+p}}{\prod_{j \neq k} (x_k - x_j)}.$$

Prove that $S_1 = x_0 + x_1 + \cdots + x_n$ and compute S_2.

Proof. Actually, the following method shows how to compute S_p in general. The idea is to consider the remainder $f(X)$ of X^{n+p} when divided by

$$\prod_{j=0}^{n}(X - x_j)$$

and to apply Lagrange's interpolation formula to it. Identifying leading coefficients and using the fact that $f(x_j) = x_j^{n+p}$, we deduce that S_p is precisely the leading coefficient of $f(X)$. Now, for $p = 1$ we clearly have

$$f(X) = X^{n+1} - \prod_{j=0}^{n}(X - x_j),$$

so that its leading coefficient is $\sum_{i=0}^{n} x_i$. On the other hand, for $p = 2$ we must have

$$X^{n+2} = Q(X)\prod_{j=0}^{n}(X - x_j) + f(X)$$

for some polynomial Q. Comparing degrees and leading coefficients shows that $Q(X) = X + c$ for some constant c. To determine c, we impose the condition that $\deg(f) \leq n$. This implies that $c = \sum_{j=0}^{n} x_j$. Then, it is easy to find the coefficient of X^n in $f(X)$ and the answer is $\left(\sum_{j=0}^{n} x_j\right)^2 - \sum_{0\leq i<j\leq n} x_i x_j$. Actually, we leave as a nice exercise for the reader to prove that for all p we have

$$S_p = \sum_{a_1+a_2+\cdots+a_n=p} x_1^{a_1} x_2^{a_2} \cdots x_n^{a_n},$$

where in the sum above the a_i are nonnegative integers. □

The following problems deal with extremal properties of polynomials. The underlying philosophy is that imposing conditions on the values of a polynomial at sufficiently many points of an interval automatically imposes conditions at all other values. Lagrange's interpolation formula is a very handy tool in such situations, but there is an extra ingredient which appears all the time, namely the Chebyshev polynomials. Recall that the nth Chebyshev polynomial T_n is the unique polynomial of degree n such that $\cos(nx) = T_n(\cos x)$

for all $x \in \mathbb{R}$. It is not really obvious that T_n exists, but the reader can easily check the existence inductively, by establishing the recurrence relation $T_{n+1}(X) + T_{n-1}(X) = 2XT_n(X)$. This also implies that the leading coefficient of T_n is 2^{n-1}. A fundamental theorem of Chebyshev states that for any monic polynomial $f \in \mathbb{R}[X]$ of degree n we have $\max_{x\in[-1,1]} |f(x)| \geq \frac{1}{2^{n-1}}$, with equality if and only if $f = \pm\frac{1}{2^{n-1}}T_n$. This explains why Chebyshev polynomials are so important. For a proof of Chebyshev's theorem, see problem 14 in this chapter.

5. Let a, b, c be real numbers and let $f(x) = ax^2 + bx + c$ be such that

$$\max\{|f(\pm 1)|, |f(0)|\} \leq 1.$$

Prove that if $|x| \leq 1$ then $|f(x)| \leq \dfrac{5}{4}$ and $\left|x^2 f\left(\dfrac{1}{x}\right)\right| \leq 2.$

Spain 1996

Proof. Using Lagrange interpolation, we can write

$$f(X) = f(0)(1 - X^2) + f(1)\frac{X^2 + X}{2} + f(-1)\frac{X^2 - X}{2}.$$

We deduce that for all $|x| \leq 1$ we have

$$|f(x)| \leq 1 - x^2 + \frac{|x^2 + x|}{2} + \frac{|x^2 - x|}{2} = 1 - x^2 + |x| \leq \frac{5}{4},$$

the last inequality being equivalent to $(|x| - \frac{1}{2})^2 \geq 0$. Similarly, we find that

$$|x^2 f(1/x)| \leq 1 - x^2 + \frac{1 + x}{2} + \frac{1 - x}{2} = 2 - x^2 \leq 2. \qquad \square$$

6. Find the maximal value of the expression $a^2 + b^2 + c^2$ if $|ax^2 + bx + c| \leq 1$ for all $x \in [-1, 1]$.

Laurenţiu Panaitopol

Proof. Define

$$A = f(1), \quad B = f(0), \quad C = f(-1).$$

Then we easily obtain

$$a = \frac{A+C}{2} - B, \quad b = \frac{A-C}{2}, \quad c = B.$$

Therefore, an immediate computation gives

$$a^2 + b^2 + c^2 = \frac{A^2 + C^2}{2} + 2B^2 - B(A+C).$$

Since $|A|, |B|, |C| \leq 1$, the last expression is clearly bounded above by 5 (use the obvious estimate for each term of it). Thus, we always have $a^2+b^2+c^2 \leq 5$. To see that it is optimal, simply take Chebyshev's polynomial $2X^2 - 1$. □

7. Define $F(a,b,c) = \max\limits_{x \in [0,3]} |x^3 - ax^2 - bx - c|$. What is the least possible value of this function over $\mathbb{R}^3$?

Chinese TST 2001

Proof. Let $P_{a,b,c}(X) = X^3 - aX^2 - bX - c$. The idea is to map the interval $[0,3]$ to $[-1,1]$ via an affine map and then to use Chebyshev's least deviation theorem in order to bound from below $\max_{x\in[0,3]} |P_{a,b,c}(x)|$. Note that

$$\max_{x \in [0,3]} |P_{a,b,c}(x)| = \max_{x \in [-1,1]} \left| P_{a,b,c}\left(\frac{3(x+1)}{2}\right) \right|.$$

Since $P_{a,b,c}\left(\frac{3(X+1)}{2}\right)$ is a polynomial of third degree with leading coefficient 27/8, Chebyshev's theorem gives us the estimate

$$\max_{x \in [-1,1]} \left| P_{a,b,c}\left(\frac{3(x+1)}{2}\right) \right| \geq \frac{27}{32}.$$

Thus $F(a,b,c) \geq \frac{27}{32}$ and this is optimal, since equality holds for

$$P_{a,b,c} = \frac{27}{32} T_3(2X/3 - 1),$$

where $T_3(X) = 4X^3 - 3X$. □

Remark 11.4. For third degree polynomials, Chebyshev's theorem is very easy to prove using Lagrange's interpolation formula: if f is a monic polynomial of third degree, identifying leading coefficients in Lagrange's formula yields the equality

$$1 = \frac{2f(1)}{3} - \frac{4f(1/2)}{3} + \frac{4f(-1/2)}{3} - \frac{2f(-1)}{3}.$$

Of course, this can also be trivially checked by hand. This equality and the triangle inequality imply that $\max_{x\in[-1,1]} |f(x)| \geq \frac{1}{4}$ for any such f, with equality when $f(X) = X^3 - \frac{3}{4}X$.

8. Let $a, b, c, d \in \mathbb{R}$ such that $|ax^3 + bx^2 + cx + d| \leq 1$ for all $x \in [-1, 1]$. What is the maximal value of $|c|$? For which polynomials is the maximum attained?

Gabriel Dospinescu

Proof. Choose distinct numbers x_0, x_1, x_2, x_3 and identify the coefficients of X in Lagrange's formula

$$aX^3 + bX^2 + cX + d = \sum_{k=0}^{3} f(x_k) \prod_{j\neq k} \frac{X - x_j}{x_k - x_j}.$$

We deduce that

$$\begin{aligned} |c| &= \left| \sum f(x_0) \frac{x_1x_2 + x_2x_3 + x_3x_1}{(x_0 - x_1)(x_0 - x_2)(x_0 - x_3)} \right| \\ &\leq \sum \left| \frac{x_1x_2 + x_2x_3 + x_3x_1}{(x_0 - x_1)(x_0 - x_2)(x_0 - x_3)} \right|. \end{aligned}$$

The problem is to find a 4-tuple (x_0, x_1, x_2, x_3) which minimizes the last expression. As usual in this kind of problems, a good idea is to choose the points where $|T_3(x)|$ takes its maximal value 1 on the interval $[-1, 1]$. These are the points $x_0 = -1, x_1 = -\frac{1}{2}, x_2 = \frac{1}{2}$ and $x_3 = 1$. It is easy to compute the last sum in this case and we find that $|c| \leq 3$. Since this value is attained for the polynomial $T_3(X) = 4X^3 - 3X$, this is the maximal value. Also, it is not difficult to check that equality appears in the above chain of inequalities only for T_3 and $-T_3$. □

9. If a polynomial $f \in \mathbb{R}[X]$ of degree n satisfies $|f(x)| \le 1$ for all $x \in [0,1]$, then
$$\left| f\left(-\frac{1}{n}\right) \right| \le 2^{n+1} - 1.$$

Proof. We will use Lagrange's interpolation formula at the points k/n for $0 \le k \le n$. We have
$$f\left(-\frac{1}{n}\right) = \sum_{k=0}^{n} f\left(\frac{k}{n}\right) \prod_{j \ne k} \frac{-(j+1)}{k-j},$$
which shows that
$$\left| f\left(-\frac{1}{n}\right) \right| \le \sum_{k=0}^{n} \prod_{j \ne k} \frac{j+1}{|k-j|} = \sum_{k=0}^{n} \binom{n+1}{k+1} = 2^{n+1} - 1. \qquad \square$$

10. Let $a \ge 3$ be a real number and let p be a real polynomial of degree n. Prove that
$$\max_{i=0,1,\ldots,n+1} |a^i - p(i)| \ge 1.$$

Proof. The crucial observation is that we have
$$\sum_{k=0}^{n+1} (-1)^{n-k} \binom{n+1}{k} p(k) = 0.$$
This is simply Lagrange's interpolation formula written in a slightly changed form (though a more conceptual way of seeing this is using the finite differences theory). We deduce from this and the binomial formula that
$$(a-1)^{n+1} = \sum_{k=0}^{n+1} (-1)^{n-k} \binom{n+1}{k} (p(k) - a^k).$$
Thus, if $|p(k) - a^k| < 1$ for all $0 \le k \le n+1$, we must have
$$|a-1|^{n+1} < \sum_{k=0}^{n+1} \binom{n+1}{k} = 2^{n+1},$$
contradicting the fact that $a \ge 3$. This finishes the solution. $\square$

Remark 11.5. The identity

$$\sum_{k=0}^{n+1}(-1)^{n-k}\binom{n+1}{k}p(k)=0$$

for polynomials of degree at most n is extremely useful. We proved it here as a consequence of Lagrange's interpolation formula, but it also follows from the theory of finite differences, for a glimpse of which we refer the reader to section 10.3.

11. Let $a, b, c, d \in \mathbb{R}$ such that $|ax^3 + bx^2 + cx + d| \leq 1$ for all $x \in [-1, 1]$. Prove that

$$|a| + |b| + |c| + |d| \leq 7.$$

IMO Shortlist 1996

Proof. Let us look at the values of $P(X) = aX^3 + bX^2 + cX + d$ at $-1, -1/2, 1/2, 1$, which are the classical interpolation points in Chebyshev's theorem (for $n = 3$). Writing

$$A = f(1), \quad B = f(1/2), \quad C = f(-1/2), \quad D = f(-1),$$

we can easily express a, b, c, d in terms of A, B, C, D. An easy computation gives

$$a = \frac{2}{3}(A - D) - \frac{4}{3}(B - C), \quad b = \frac{2}{3}(A + D) - \frac{2}{3}(B + C),$$

and

$$c = -\frac{1}{6}(A - D) + \frac{4}{3}(B - C), \quad d = -\frac{1}{6}(A + D) + \frac{2}{3}(B + C).$$

This shows that $f(a, b, c, d) = |a| + |b| + |c| + |d|$ is actually a convex function of $A, B, C, D \in [-1, 1]$. Thus it attains its maximum when all A, B, C, D are equal to 1 or -1. Now, it is a tedious matter to check that in all cases the expression is at most 7. We have equality for the Chebyshev polynomial (or its opposite) of degree 3. $\square$

12. Let $A = \left\{ p \in \mathbb{R}[X] | \ \deg p \leq 3, \ |p(\pm 1)| \leq 1, \ \left| p\left(\pm \frac{1}{2}\right) \right| \leq 1 \right\}$.

    Find $\sup\limits_{p \in A} \max\limits_{|x| \leq 1} |p''(x)|$.

IMC 1998

Proof. Write $p(X) = aX^3 + bX^2 + cX + d$. Since $p''(x)$ is an affine function on $[-1, 1]$, the maximum of its absolute value is obtained for $x = 1$ or $x = -1$. Thus, we need to bound in an optimal way the number

$$\max(|6a + 2b|, |-6a + 2b|).$$

Taking the Chebyshev polynomial shows that this can be already as large as 24. But it is always at most 24 (under the given hypothesis on p), since for instance using the formulas in the solution of problem 11 yields

$$6a + 2b = \frac{16}{3}p(1) - \frac{8}{3}p(-1) - \frac{28}{3}p(1/2) + \frac{20}{3}p(-1/2).$$

Thus $|6a + 2b| \leq 24$ by the triangle inequality. We proceed in a similar way to bound $|-6a + 2b|$ or we can observe that if p satisfies the conditions in the problem, so does $p(-x)$. □

13. Let $n \geq 3$ and let $f, g \in \mathbb{R}[X]$ be polynomials such that the points $(f(1), g(1))$, $(f(2), g(2)), \ldots, (f(n), g(n))$ are the vertices of a regular n-gon in counterclockwise order. Prove that $\max(\deg f, \deg g) \geq n - 1$.

Putnam 2008

Proof. It is enough to prove that $P(X) = f(X) + ig(X)$ has degree at least $n-1$. Clearly, we may assume that the regular n-gon has vertices $z, z^2, \ldots, z^n$ for $z = e^{\frac{2i\pi}{n}}$ (simply apply a translation and a homothety to reduce the general case to this one). So, it remains to prove that if a polynomial P satisfies $P(i) = z^i$ for all $1 \leq i \leq n$, then $\deg P \geq n - 1$. Assume that this is not the case. Using Lagrange's interpolation formula, we obtain

$$z = P(1) = \sum_{i=2}^{n} z^i \cdot \prod_{j \neq i} \frac{1-j}{i-j} = \sum_{i=2}^{n} \binom{n-1}{i-1} (-1)^i z^i.$$

Dividing both sides by z and using the binomial formula yields $(1-z)^{n-1}=0$, a contradiction. $\square$

Remark 11.6. There are many other approaches. The most efficient is the method of finite differences, which yields

$$\Delta^{n-1}P(1)=\sum_{j=0}^{n-1}(-1)^j\binom{n-1}{j}P(n-j)=z(z-1)^{n-1}\neq 0.$$

Another very neat proof considers the polynomial $P(X+1)-zP(X)$. It has degree at most $n-2$, it is clearly nonzero and vanishes at $1,2,\dots,n-1$, a contradiction.

We leave the land of routine problems in the Lagrange interpolation formula and tackle some more delicate results. The following is taken from [66].

14. Let f be a complex polynomial of degree n such that $|f(x)|\leq 1$ for all $x\in[-1,1]$. Prove that $|f^{(k)}(x)|\leq |T_n^{(k)}(x)|$ for all k and all real numbers x such that $|x|\geq 1$. Deduce Chebyshev's theorem: if f is monic of degree n, then

$$\max_{x\in[-1,1]}|f(x)|\geq \frac{1}{2^{n-1}}.$$

W.W. Rogosinski

Proof. Pick for the moment $n+1$ points $x_0,\dots,x_n$ in $[-1,1]$ and apply Lagrange's interpolation formula:

$$f(X)=\sum_{i=0}^{n}f(x_i)\prod_{j\neq i}\frac{X-x_j}{x_i-x_j}.$$

Next, differentiate this equality k times, so if we set $P_i(X)=\prod_{j\neq i}(X-x_j)$, then

$$f^{(k)}(x)=\sum_{i=0}^{n}\frac{f(x_i)}{P_i(x_i)}P_i^{(k)}(x)$$

for all x. Choose any $|x| \geq 1$ and apply the triangle inequality to the previous identity, together with the hypothesis that $|f(x_i)| \leq 1$. We infer that

$$|f^{(k)}(x)| \leq \sum_{i=0}^{k} \frac{|P_i^{(k)}(x)|}{|P_i(x_i)|}.$$

But we also have

$$T_n^{(k)}(x) = \sum_{i=0}^{n} T_n(x_i) \frac{P_i^{(k)}(x)}{P_i(x_i)},$$

so it would be really nice if we could ensure that

$$\left| \sum_{i=0}^{n} T_n(x_i) \frac{P_i^{(k)}(x)}{P_i(x_i)} \right| = \sum_{i=0}^{k} \frac{|P_i^{(k)}(x)|}{|P_i(x_i)|}.$$

In order to have this, we already need $|T_n(x_i)| = 1$ for all i, which suggests taking the old friends $x_i = \cos\left(\frac{i\pi}{n}\right)$ for $0 \leq i \leq n$. Then $T_n(x_i) = (-1)^i$ and we would like to prove that the numbers $(-1)^i \cdot \frac{P_i^{(k)}(x)}{P_i(x_i)}$ have the same sign. If this were true, we would be done by the previous arguments. Note that $x_0 > x_1 > \cdots > x_i$, so that

$$P_i(x_i) = (x_i - x_0)(x_i - x_1) \cdots (x_i - x_{i-1})(x_i - x_{i+1}) \cdots (x_i - x_n)$$

has the sign $(-1)^i$. So, it remains to see if the signs of the numbers $P_i^{(k)}(x)$ are independent of i as long as $|x| \geq 1$. The crucial remark is that $P_i^{(k)}$ have all their roots in $[-1, 1]$. Indeed, this follows from Rolle's theorem and the fact that P_i have all their roots in $[-1, 1]$. But then it is clear that the sign of $P_i^{(k)}(x)$ only depends on the degree of $P_i^{(k)}$ (which is independent of i) and on the position of x with respect to 1 (ie whether $x \geq 1$ or $x \leq -1$). So, it is independent of i and we are done.

Finally, let us prove Chebyshev's theorem using this result. Suppose that $|f(x)| < \frac{1}{2^{n-1}}$ for all $x \in [-1, 1]$. By compactness there exists $c > 2^{n-1}$ such that $|cf(x)| \leq 1$ for all $x \in [-1, 1]$. Using the result established above, we deduce that for $|x| \geq 1$ we have $\left|cf^{(n)}(x)\right| \leq \left|T_n^{(n)}(x)\right|$. As f is monic of degree n and since T_n has degree n and leading coefficient 2^{n-1}, this yields $c \leq 2^{n-1}$, a contradiction. □

15. The polynomials $f, g \in \mathbb{R}[X]$ and $a \in \mathbb{R}[X, Y]$ satisfy

$$f(x) - f(y) = a(x, y)(g(x) - g(y))$$

for all x, y. Prove that we can find a polynomial h such that

$$f(x) = h(g(x)).$$

Russia 2004

Proof. Fix an integer $N > \deg f$. If g is constant, so is f and everything is clear. So, assume that g is not constant, thus we can choose $N + 1$ distinct numbers $x_0, x_1, \ldots, x_N$ at which g takes different values. By Lagrange's interpolation formula, there exists a polynomial h of degree at most N such that $h(g(x_i)) = f(x_i)$ for all $0 \le i \le N$. Note that by hypothesis

$$f(X) - f(x_i) = a(X, x_i)(g(X) - g(x_i)),$$

so that $g(X) - g(x_i)$ divides $f(X) - f(x_i)$. On the other hand, $g(X) - g(x_i)$ also divides $h(g(X)) - h(g(x_i)) = h(g(X)) - f(x_i)$ (this follows by linearity and the fact that $g(X) - g(x_i)$ divides $g(X)^k - g(x_i)^k$ for all $k \ge 1$). Thus $h(g(X)) - f(X)$ is a multiple of $g(X) - g(x_i)$. Now, the polynomials $g(X) - g(x_i)$ are clearly pairwise relatively prime, since the numbers $g(x_i)$ are pairwise distinct. Thus $h(g(X)) - f(X)$ is a multiple of $\prod_{i=0}^{N}(g(X) - g(x_i))$, but has degree smaller than that of $\prod_{i=0}^{N}(g(X) - g(x_i))$. Therefore $f(X) = h(g(X))$ and we are done. □

The next problem presents P.J. O'Hara's beautiful proof [63] of a famous inequality of Bernstein.

16. a) Prove that for all complex polynomials f having degree at most n and for all $x \in \mathbb{C}$

$$xf'(x) = \frac{n}{2} f(x) + \frac{1}{n} \sum_{k=1}^{n} f(xz_k) \frac{2z_k}{(1 - z_k)^2},$$

where z_k are the roots of the polynomial $X^n + 1$.

b) Deduce Bernstein's inequality: for all complex polynomials f we have $\|f'\| \leq n\|f\|$, where $\|f\| = \max_{|z|=1} |f(z)|$.

P.J. O'Hara

Proof. a) The polynomial

$$g_x(X) = \frac{f(xX) - f(x)}{X - 1}$$

has degree at most $n-1$ and it is easy to see that $g_x(1) = xf'(x)$. Lagrange's interpolation formula for the points z_i yields

$$g_x(X) = \sum_{i=1}^{n} g_x(z_i) \prod_{j \neq i} \frac{X - z_j}{z_i - z_j},$$

which combined with the equality[1]

$$\prod_{j \neq i} (z_i - z_j) = nz_i^{n-1} = -\frac{n}{z_i}$$

gives

$$g_x(X) = \frac{1}{n} \sum_{i=1}^{n} z_i g_x(z_i) \frac{X^n + 1}{z_i - X}.$$

Taking $X = 1$ and remembering that $g_x(1) = xf'(x)$, we obtain

$$xf'(x) = \frac{1}{n} \sum_{i=1}^{n} (f(xz_i) - f(x)) \frac{2z_i}{(1 - z_i)^2},$$

so we only need to prove that

$$\sum_{i=1}^{n} \frac{2z_i}{(1 - z_i)^2} = -\frac{n^2}{2}.$$

[1]Obtained by differentiating the identity $X^n + 1 = \prod_{i=1}^{n} (X - z_i)$ and taking $X = z_i$.

Simply choose $f(X) = X^n$ in the equality

$$xf'(x) = \frac{1}{n}\sum_{i=1}^{n}(f(xz_i) - f(x))\frac{2z_i}{(1-z_i)^2}.$$

b) By scaling f, we may assume that $||f|| = 1$. Choose any $|z| = 1$ and use the identity in (a) to deduce that

$$|f'(z)| \leq \frac{n}{2} + \frac{1}{n}\sum_{k=1}^{n}\left|\frac{2z_k}{(1-z_k)^2}\right|.$$

The miracle is that all numbers

$$\frac{2z_k}{(1-z_k)^2} = \frac{1}{\frac{1}{2}\left(z_k + \frac{1}{z_k}\right) - 1} = \frac{1}{\mathrm{Re}(z_k) - 1}$$

are actually negative, so the previous inequality becomes

$$|f'(z)| \leq \frac{n}{2} - \frac{1}{n}\sum_{k=1}^{n}\frac{2z_k}{(1-z_k)^2}.$$

Since we've already seen that

$$\sum_{i=1}^{n}\frac{2z_i}{(1-z_i)^2} = -\frac{n^2}{2},$$

we can now safely conclude that $|f'(z)| \leq n$, finishing the proof of this hard problem. $\square$

We end this chapter with a beautiful inequality due to Gelfond, who used it in his solution of Hilbert's seventh problem. This famous problem asked to prove that a^b is transcendental whenever b is an algebraic irrational number and a is an algebraic number different from 0 and 1.

17. Let f be a complex polynomial of degree at most n and let $z_0, z_1, \ldots, z_d$ be the zeros of the polynomial $X^{d+1} - 1$, where $d > n$. Define

$$||f|| = \max_{|z|=1} |f(z)|.$$

a) Prove that if there exist $n+1$ pairwise distinct numbers $x_0, x_1, \ldots, x_n$ among $z_0, z_1, \ldots, z_d$ such that $|f(x_i)| \leq \frac{1}{2^d}$ then $||f|| < 1$.

b) Deduce that

$$||f|| \cdot ||g|| \leq 4^{\deg(f)+\deg(g)} \cdot ||fg||.$$

A.O. Gelfond

Proof. a) The first step is rather clear: Lagrange's interpolation formula for the points $x_0, x_1, \ldots, x_n$ yields

$$f(z) = \sum_{i=0}^{n} f(x_i) \prod_{j \neq i} \frac{z - x_j}{x_i - x_j}.$$

Combining this with the triangle inequality and the hypothesis, we obtain for $|z| = 1$

$$|f(z)| \leq \frac{1}{2^d} \sum_{i=0}^{n} \prod_{j \neq i} \frac{|z - x_j|}{|x_i - x_j|}.$$

We brutally bound each $|z - x_j|$ by 2, so that $\prod_{j \neq i} |z - x_j| \leq 2^n$. The subtle part is to deal with $\prod_{j \neq i} |x_i - x_j|$. Write

$$\{y_1, \ldots, y_{d-n}\} = \{z_0, \ldots, z_d\} - \{x_0, \ldots, x_n\}$$

and observe that

$$\prod_{j \neq i} |x_i - x_j| \cdot \prod_{k=1}^{d-n} |x_i - y_k| = \left| \prod_{z_j \neq x_i} (x_i - z_j) \right| = d + 1,$$

because $\prod_{z_j \neq x_i} (x_i - z_j)$ is exactly the derivative of $X^{d+1} - 1$ at x_i, which has absolute value $d + 1$. Since $|x_i - y_k| \leq 2$, we infer that

$$\prod_{j \neq i} |x_i - x_j| \geq \frac{d + 1}{2^{d-n}}.$$

All in all, each term $\prod_{j\neq i}\frac{|z-x_j|}{|x_i-x_j|}$ can be bounded from above by

$$\frac{2^n}{\frac{d+1}{2^{d-n}}}=\frac{2^d}{d+1}.$$

Since there are $n+1$ terms and $n+1<d+1$, we deduce that for all $|z|=1$ we have $|f(z)|<1$ and so $||f||<1$.

b) This part is an easy consequence of the first one. Scaling f and g allows us to assume that $||f||=||g||=1$. Assuming that $||fg||<\frac{1}{4^d}$, we deduce that for any $0\leq i\leq d$ we have $|f(z_i)|<\frac{1}{2^d}$ or $|g(z_i)|<\frac{1}{2^d}$. With $d=\deg(f)+\deg(g)$, it follows that either we can find at least $\deg(f)+1$ z_i's such that $|f(z_i)|<\frac{1}{2^d}$ or we can find at least $\deg(g)+1$ z_i's for which $|g(z_i)|<\frac{1}{2^d}$. Both cases are excluded by part a), and the result follows. $\square$

11.1 Notes

We would like to thank the following people for providing solutions to the following problems: Xiangyi Huang (problems 6, 7, 8, 9, 12), Qiaochu Yuan (problems 1, 3, 10).

Chapter 12

Higher Algebra in Combinatorics

This last chapter deals with applications of linear algebra in combinatorics. There is a huge literature on this subject, which is still very active, so all we can do in this chapter is to barely scratch the surface and present the reader with some interesting elementary examples. Of course, this does not mean that there aren't a lot of difficult problems in this chapter.

Before passing to problems, let us recall some basic results of linear algebra. Let K be a field. A K-vector space V is an abelian group endowed with an external multiplication $K \times V \to V$, denoted $(a, v) \to a \cdot v$, which is compatible with the additive operation of V and with the operations in K. A subspace of V is an additive subgroup which is stable under multiplication by elements of K. A family $(v_i)_{i \in I}$ of elements of V is called linearly independent if for all finite subsets $S \subset I$ and all $(a_i)_{i \in S} \in K^{|S|}$, the relation $\sum_{s \in S} a_s \cdot v_s = 0$ forces $a_s = 0$ for all $s \in S$. A family $(v_i)_{i \in I}$ is called generating if any vector $v \in V$ can be written $v = \sum_{s \in S} a_s \cdot v_s$ for some finite subset $S \subset I$ and some $a_s \in K$. Finally, a basis of V is a linearly independent family which is simultaneously generating. Using Zorn's lemma, one can prove that any vector space has a basis and that any linearly independent family of vectors can be extended to a basis. In most of the following problems we will only consider finite dimensional spaces, i.e. vector spaces having a finite generating family.

The dimension of such a space is by definition the number of elements of a basis (it is not obvious, but it can be proved that this does not depend on the choice of the basis).

If V and W are vector spaces over K, a map $f : V \to W$ is called linear if $f(v_1+v_2) = f(v_1)+f(v_2)$ for all $v_1, v_2 \in V$ and $f(a \cdot v) = a \cdot f(v)$ for all $v \in V$ and $a \in K$. The kernel of f is then the subspace $\operatorname{Ker}(f) = \{v \in V | f(v) = 0\}$ of V. A fundamental result in linear algebra is the following formula

$$\dim V = \dim \operatorname{Ker}(f) + \dim \operatorname{Im}(f).$$

Traditionally, $\dim \operatorname{Im}(f)$ is called the rank of f.

Let $f : V \to V$ be a linear map, where V is a K-vector space of dimension n. If $e_1, e_2, \ldots, e_n$ is a basis of V, the matrix of f in this basis is defined by the equalities $f(e_i) = \sum_{j=1}^{n} a_{ji}e_j$. It is easy to check that the matrix associated to the composition of f and g is simply the product of the matrices associated to f and g. Conversely, any matrix can be naturally seen as the matrix of an endomorphism of V in the given basis.

If A is an $n \times n$ matrix with entries in some commutative ring R, its determinant is defined by

$$\det A = \sum_{\sigma \in S_n} \varepsilon(\sigma) a_{1\sigma(1)} a_{2\sigma(2)} \cdots a_{n\sigma(n)},$$

S_n being the group of permutations of $\{1, 2, \ldots, n\}$ and $\varepsilon(\sigma)$ being the signature of σ. The most important (and rather surprising!) property of the determinant is its multiplicativity, i.e. $\det(AB) = \det A \cdot \det B$. Another important property is that A is invertible in $M_n(R)$ if and only if $\det A$ is a unit of R. Thus, if R is a field, we can test if a matrix is invertible by testing whether its determinant is nonzero and this happens if and only if its associated endomorphism is bijective. Let $M_n(K)$ be the space of $n \times n$ matrices with entries in K. If $A \in M_n(K)$, its characteristic polynomial is $\det(X \cdot I_n - A)$, a monic polynomial of degree n with coefficients in K. Its roots in an algebraic closure $\overline{K}$ of K are called the eigenvalues of A. Hence, if λ is an eigenvalue of A, then we can find $v \in \overline{K}^n$ such that $A \cdot v = \lambda \cdot v$, i.e. $\sum_{j=1}^{n} a_{ij}v_j = \lambda \cdot v_i$ for all i. It is not difficult to check that the sum of the eigenvalues of A (counted with multiplicities) is equal to the trace of A, which is by definition the sum of the diagonal elements of A. Also, $\det A$ is the product of the eigenvalues of A (again counted with multiplicities). It is a little bit trickier to prove that the eigenvalues of $g(A)$ (where $g \in K[X]$) are $g(\lambda_1), g(\lambda_2), \dots, g(\lambda_n)$.

12.1 The determinant trick

We present in this section a few tricky combinatorial problems in which the determinant of a matrix plays an important role.

1. Let $n \geq 2$. Find the greatest p such that for all $k \in \{1, 2, \dots, p\}$ we have

$$\sum_{\sigma \in A_n} \left(\sum_{i=1}^{n} i\sigma(i) \right)^k = \sum_{\sigma \in B_n} \left(\sum_{i=1}^{n} i\sigma(i) \right)^k,$$

where A_n, B_n are the sets of all even and odd permutations of the set $\{1, 2, \dots, n\}$ respectively.

Gabriel Dospinescu

Proof. The first ingredient is the following:

Lemma 12.1. *Let $a_1, a_2, \dots, a_m$ and $b_1, b_2, \dots, b_m$ be positive integers and let N be a positive integer. Then*

$$a_1^k + a_2^k + \cdots + a_m^k = b_1^k + b_2^k + \cdots + b_m^k$$

for all $1 \leq k \leq N$ if and only if $(X-1)^{N+1}$ divides

$$X^{a_1} + X^{a_2} + \cdots + X^{a_m} - X^{b_1} - X^{b_2} - \cdots - X^{b_m}.$$

Proof. This is quite easy: if

$$f(X) = X^{a_1} + X^{a_2} + \cdots + X^{a_m} - X^{b_1} - X^{b_2} - \cdots - X^{b_m},$$

then $(X-1)^{N+1}$ divides f if and only if $f^{(j)}(1) = 0$ for all $0 \leq j \leq N$. This happens if and only if

$$\sum_{i=1}^{m} a_i(a_i - 1) \cdots (a_i - j + 1) = \sum_{i=1}^{m} b_i(b_i - 1) \cdots (b_i - j + 1)$$

for all $0 \le j \le N$. As the polynomials $X(X-1)\cdots(X-j+1)$ for $1 \le j \le N$ span the same vector space as the polynomials $X, X^2, \ldots, X^N$ (this is immediate), the previous equality holds if and only if

$$a_1^k + a_2^k + \cdots + a_m^k = b_1^k + b_2^k + \cdots + b_m^k$$

for all $1 \le k \le N$. The conclusion follows. □

Using the previous result, we deduce that $p+1$ is the multiplicity of 1 as a root of the polynomial

$$f(X) = \sum_{\sigma \in A_n} X^{S(\sigma)} - \sum_{\sigma \in B_n} X^{S(\sigma)},$$

where $S(\sigma) = \sigma(1) + 2\sigma(2) + \cdots + n\sigma(n)$. The key point is that we can write

$$f(X) = \sum_{\sigma \in S_n} \varepsilon(\sigma) X^{\sigma(1)} \cdot X^{2\sigma(2)} \cdots X^{n\sigma(n)} = \det(A),$$

where $a_{ij} = X^{ij}$, S_n is the set of all permutations of $\{1, 2, \ldots, n\}$ and $\varepsilon(\sigma)$ is the signature of σ (i.e. 1 if $\sigma \in A_n$, -1 if $\sigma \in B_n$). Finally, using properties of Vandermonde determinants, we can write

$$f(X) = X^{\frac{n(n+1)}{2}} \cdot \prod_{1 \le i < j \le n} (X^j - X^i)$$

and since the multiplicity of 1 in $X^j - X^i$ is 1 for all $i \ne j$, we deduce that $p + 1 = \binom{n}{2}$. Thus the answer to the problem is $\binom{n}{2} - 1$. □

The next problem is rather strange, both because of the statement and because of one of its proofs. We also present a more natural proof, based on the inclusion-exclusion principle.

2. For a permutation σ of $\{1, 2, \ldots, n\}$ let $\varepsilon(\sigma) = 1$ if σ is even and -1 otherwise. Let $f(\sigma)$ be the number of fixed points of σ. Prove that

$$\sum_{\sigma} \frac{\varepsilon(\sigma)}{1 + f(\sigma)} = (-1)^{n+1} \cdot \frac{n}{n+1},$$

where the sum is taken over all permutations σ of $\{1, 2, \ldots, n\}$.

Putnam 2005

Proof. We start with the exotic proof: observe that

$$\sum_{\sigma} \frac{\varepsilon(\sigma)}{1+f(\sigma)} = \sum_{\sigma} \varepsilon(\sigma) \int_0^1 x^{f(\sigma)} dx = \int_0^1 \left(\sum_{\sigma} \varepsilon(\sigma) x^{f(\sigma)} \right) dx.$$

We recognize the determinant of the matrix $A(x)_{ij} = x^{1_{i=j}}$ under the integral. But $\det(A(x))$ can also be computed using row-column operations: add all columns to the first one, take out the common factor $x + n - 1$ and then subtract the first column (which consists now only of ones) from all the others, to make the elements in the first row equal to 0, except for the first one. Next, expand the determinant using the first row. We obtain in this way that

$$\det A(x) = (x + n - 1)(x - 1)^{n-1}.$$

Thus

$$\begin{aligned} \sum_{\sigma} \frac{\varepsilon(\sigma)}{1+f(\sigma)} &= \int_0^1 (x - 1 + n)(x-1)^{n-1} dx \\ &= \int_{-1}^0 (x+n) x^{n-1} dx \\ &= -\frac{(-1)^{n+1}}{n+1} - (-1)^n \\ &= (-1)^{n+1} \frac{n}{n+1}, \end{aligned}$$

which is precisely the desired result. □

Proof. Here is a more natural proof: we will partition the permutations σ according to their fixed points. For a permutation σ, let $\mathrm{Fix}(\sigma)$ be the set of its fixed points and let $f(\sigma) = |\mathrm{Fix}(\sigma)|$. As

$$\sum_{\sigma} \frac{\varepsilon(\sigma)}{1+f(\sigma)} = \sum_{k=0}^{n} \frac{1}{k+1} \sum_{|A|=k} \left(\sum_{\mathrm{Fix}(\sigma)=A} \varepsilon(\sigma) \right),$$

we naturally try to understand $F(A) = \sum_{\mathrm{Fix}(\sigma)=A} \varepsilon(\sigma)$. We will compute $F(A)$ using the inclusion-exclusion principle, according to which

$$F(A) = \sum_{A\subset B} (-1)^{|B|-|A|} \sum_{B\subset C} F(C).$$

On the other hand,

$$\sum_{B\subset C} F(C) = \sum_{\sigma|B=1} \varepsilon(\sigma).$$

The notation $\sigma|B = 1$ means that $\sigma(b) = b$ for all $b \in B$. Now, permutations such that $\sigma|B = 1$ correspond to permutations of the complement of B, so if $n - |B| \geq 2$ we have $\sum_{\sigma|B=1} \varepsilon(\sigma) = 0$. We deduce that $\sum_{B\subset C} F(C) = 0$ unless $|B| \geq n - 1$, when it is equal to 1. Combining this with the previous formula (inclusion-exclusion principle) yields $F(A) = (-1)^{n-|A|}(|A| + 1 - n)$.

We deduce that

$$\begin{aligned}\sum_{\sigma} \frac{\varepsilon(\sigma)}{1+f(\sigma)} &= \sum_{k=0}^{n}(-1)^{n-k}\binom{n}{k}\frac{k+1-n}{k+1} \\ &= \sum_{k=0}^{n}(-1)^{n-k}\binom{n}{k} - n\sum_{k=0}^{n}\frac{1}{k+1}\binom{n}{k}(-1)^{n-k} \\ &= -n\sum_{k=0}^{n}\frac{1}{n+1}\binom{n+1}{k+1}(-1)^{n-k}\end{aligned}$$

and the result follows immediately. □

The next problem crucially uses the multiplicativity of the determinant. It is (rather amusingly) a consequence of the irrationality of $\sqrt{5}$.

3. Is there in the plane a configuration of 22 circles and 22 points on their union (the union of their circumferences) such that any circle contains at least 7 points and any point belongs to at least 7 circles?

Gabriel Dospinescu, Moldavian TST 2004

Proof. The answer is negative. First, we will use the standard trick of counting pairs $(P, \{C_1, C_2\})$, where C_1, C_2 are distinct circles among the 22 ones and P is a point among the 22 ones that belongs to $C_1 \cap C_2$. Each point belongs to at least 7 circles, so for each P there are at least $\binom{7}{2}$ sets $\{C_1, C_2\}$ living in a triple with P. So the number of triples is at least $22 \cdot \binom{7}{2}$. On the other hand, for each $\{C_1, C_2\}$, $C_1 \cap C_2$ has at most 2 elements, so there are at most $2 \cdot \binom{22}{2}$ triples. The miracle is that we actually have $2 \cdot \binom{22}{2} = 22 \cdot \binom{7}{2}$. Thus, all previous inequalities are actually equalities.

Let $P_1, P_2, \ldots, P_{22}$ be the points and let $C_1, C_2, \ldots, C_{22}$ be the circles. Consider now the matrix A whose (i, j)-entry is 1 if $P_j \in C_i$ and 0 otherwise. The previous paragraph shows that $P \cdot P^t$ is the matrix whose elements on the main diagonal are equal to 7, all other elements being equal to 2. An easy computation shows that the determinant of this matrix is $49 \cdot 5^{21}$. But this determinant is also equal to $(\det A)^2$, which is a perfect square. We deduce that 5 is a perfect square, which is a bit difficult to make happen... □

We end this section with another rather challenging problem on permutations.

4. A permutation σ of $\{1, 2, \ldots, n\}$ is called k-limited if $|\sigma(i) - i| \leq k$ for all $1 \leq i \leq n$. Prove that the number of k-limited permutations of $\{1, 2, \ldots, n\}$ is odd if and only if $n \equiv 0, 1 \pmod{2k+1}$.

Putnam 2008

Proof. Let M be the $n \times n$ matrix defined by $M_{ij} = 1_{|i-j| \leq k}$. It is clear that the number of k-limited permutations has the same parity as $\det M$. So the question is when M is invertible in $M_n(\mathbb{F}_2)$. From now on, we always work in $\mathbb{F}_2$. Let us prove that if $n \equiv 2, \ldots, 2k \pmod{2k+1}$, then M is not invertible. Pick integers $0 \leq a < b \leq k$ such that $n + a + b \equiv 0 \pmod{2k+1}$ (it is clear that they exist under the assumption that $n \equiv 2, \ldots, 2k \pmod{2k+1}$) and set $j = (n + a + b)/(2k + 1)$. If r_i is the i-th row of M, then the vector all of whose components are 1 can be written as $\sum_{i=0}^{j-1} r_{k+1-a+(2k+1)i}$ and as $\sum_{i=0}^{j-1} r_{k+1-b+(2k+1)i}$. Hence there is a nontrivial combination of the rows of M which yields the zero vector. Thus M is not invertible in this case.

Assume now that $n \equiv 0, 1 \pmod{2k+1}$ and that $a_1, a_2, \ldots, a_n$ are scalars such that $a_1r_1+a_2r_2+\cdots+a_nr_n$ is the zero vector. Put $a_i = 0$ if $i \notin \{1, \ldots, n\}$. Then $a_{m-k} + a_{m-k+1} + \cdots + a_{m+k} = 0$ for all m. By comparing the relations for m and $m+1$, we obtain $a_{m-k} = a_{m+k+1}$ for $1 \leq m < n$, so a_i repeat with period $2k+1$. Taking $m = 1, \ldots, k$ further yields the equalities $a_{k+2} = \cdots = a_{2k+1} = 0$. Taking $m = n-k, \ldots, n-1$ gives another chain of equalities $a_{n-2k} = \cdots = a_{n-1-k} = 0$. If $n \equiv 0 \pmod{2k+1}$, this can be rewritten as $a_1 = \cdots = a_k = 0$, whereas for $n \equiv 1 \pmod{2k+1}$, it can be rewritten as $a_2 = \cdots = a_{k+1} = 0$. In either case, since we also have $a_1 + \cdots + a_{2k+1} = 0$, we deduce that all of the a_i must be zero. The conclusion follows. □

12.2 Matrices over $\mathbb{F}_2$

The reduction map $\mathbb{Z} \to \mathbb{Z}/2\mathbb{Z}$ induces a natural map $M_n(\mathbb{Z}) \to M_n(\mathbb{Z}/2\mathbb{Z})$, which is easily seen to be a ring homomorphism. This map is very useful when dealing with applications of linear algebra in combinatorics, since for a lot of problems the parity gives already enough information. Also, the incidence matrices of families of sets are binary matrices and so are the adjacency matrices of graphs. To simplify notation, we let $\mathbb{F}_2 = \mathbb{Z}/2\mathbb{Z}$.

5. $2n+1$ real numbers have the property that no matter how we eliminate one of them, the rest can be divided into two groups of n numbers, the sum of the numbers in the two groups being the same. Prove that the numbers are equal.

Proof. Let $x_1, x_2, \ldots, x_{2n+1}$ be real numbers as in the problem and let X be the column vector whose coordinates are the x_i's. We can write the hypothesis in the form $AX = 0$ for some matrix $A = (a_{ij})$ with $a_{ij} \in \{-1, 0, 1\}$, $a_{ii} = 0$ and such that the sum of the numbers in each row of the matrix (a_{ij}) is zero. Of course, this linear system of equations has the trivial solution $x_1 = x_2 = \cdots = x_{2n+1}$ and the problem asks to prove that this is the only solution. Now, the dimension of the vector space of solutions is $\dim \mathrm{Ker} A$, which is also $2n + 1 - \mathrm{rk} A$. Thus, it is enough to prove that A has rank $2n$. Of course, the rank is at most $2n$, since the sum of the elements in each line is

0. But if we see A as a matrix over $\mathbb{F}_2$, A becomes the matrix with 0 on the diagonal and 1 elsewhere. Since $2n+1$ is odd, it is easy to check that this matrix has rank $2n$ (the associated system of equations over $\mathbb{F}_2$ is simply $x_2+\cdots+x_{2n+1}=0,\ldots,x_1+\cdots+x_{2n}=0$, which clearly has the only solution $x_1=x_2=\cdots=x_{2n+1}$). But then A must have rank at least $2n$, since the rank can only decrease after reduction mod 2 (this is simply saying that a nonzero element of $\mathbb{F}_2$ lifts to an odd, thus nonzero, integer). The result follows. □

Proof. The classical proof is done in three steps. Let $x_1,\ldots,x_{2n+1}$ be the numbers in the problem. In the first step we assume that all x_i are integers. The hypothesis clearly implies that all x_i have the same parity as $x_1+x_2+\cdots+x_{2n+1}$. Thus all x_i have the same parity. Writing them either as $2y_i$ or as $2y_i+1$, it is clear that $y_1,\ldots,y_{2n+1}$ satisfy the same property as the x_i's. Since the sum of the absolute values of the y_i's is smaller than that of the x_i's, after finitely many steps deduce that all x_i are equal (because if the process continues forever then all the x_i must have been 0).

The second step treats the case when the x_i's are rational numbers. Multiplying them by a suitable positive integer N to make them integers, we reduce trivially this case to the one considered above.

Finally, we consider the general case when the x_i's are real numbers. Let $e_1,e_2,\ldots,e_k$ be a basis of the $\mathbb{Q}$-vector space generated by the x_i's. Write each $x_i=a_{i1}e_1+a_{i2}e_2+\cdots+a_{ik}e_k$. Since the e_i's are linearly independent, it follows that for each fixed j, the numbers $a_{1j},a_{2j},\ldots,a_{2n+1,j}$ satisfy the same properties as the x_i's. Since these numbers are rational, it follows that they are equal by the second step. But since this holds for every j, we have $x_1=x_2=\cdots=x_{2n+1}$. □

Binary matrices are very useful when dealing with iterations of some process on a combinatorial structure. The fact that 2 kills $M_n(\mathbb{F}_2)$ implies that for any commuting matrices $A,B\in M_n(\mathbb{F}_2)$ and for any $k\geq 1$ we have $(A+B)^{2^k}=A^{2^k}+B^{2^k}$. This is very useful in computations.

6. The edges of a regular 2^n-gon are colored red and blue. A step consists in recoloring each edge whose neighbors have the same color in red and recoloring each edge whose neighbors have different colors in blue. Prove

that after 2^{n-1} steps all of the edges will be red and that this need not hold after fewer steps.

Iranian Olympiad 1998

Proof. The easy part is the second question: if there is exactly one blue edge, then clearly after $k < 2^{n-1}$ steps the k-th edge after this blue edge is blue, so it is impossible to make all edges red after less than 2^{n-1} steps.

Let us now prove that after 2^{n-1} steps all edges are red. Let X_j be the vector giving the state of the edges after the j-th step, so X_j is the column vector with 2^n coordinates, the k-th coordinate being 1 if the k-th edge is blue after j steps and 0 otherwise. By definition we have $X_{j+1} = AX_j$, where everything is taken modulo 2, $A = B + B^{-1}$ and $B_{ij} = 1$ if $j \equiv i+1 \pmod{2^n}$ and 0 otherwise. Since B and B^{-1} commute, the binomial formula and the fact that $\binom{2^{n-1}}{k}$ is even for all $1 \le k \le 2^{n-1} - 1$ show that $A^{2^{n-1}} \equiv B^{2^{n-1}} + B^{-2^{n-1}} \pmod 2$. However, it is very easy to compute the successive powers of B: a trivial induction shows that B^k is simply the matrix whose entry is 1 if $j - i \equiv k \pmod{2^n}$ and 0 otherwise. In particular, $B^{2^n} = I$ and so by the previous formula $A^{2^{n-1}} \equiv 0 \pmod 2$. But since $X_j = A^j X_0$, it is clear that after 2^{n-1} steps all edges are red. □

The following problem is a classical old result, which has a very beautiful linear-algebraic proof.

7. The map $s : \mathbb{R}^r \to \mathbb{R}^r$ is defined by

$$s(a_1, a_2, \ldots, a_r) = (|a_1 - a_2|, |a_2 - a_3|, \ldots, |a_r - a_1|).$$

Prove the equivalence of the following statements:

i) for all nonnegative integers $a_1, a_2, \ldots, a_r$, there exists n such that the n-th iterate of s evaluated at $(a_1, a_2, \ldots, a_r)$ is $(0, 0, \ldots, 0)$;

ii) r is a power of 2.

Ducci's problem

Proof. Suppose first that r is a power of 2. Observe that the maximal coordinate of $s(a_1, a_2, \ldots, a_r)$ is clearly smaller than or equal to the maximal coordinate of $(a_1, a_2, \ldots, a_r)$. Start with some nonnegative integers $(a_1, a_2, \ldots, a_r)$ and define $X_0 = (a_1, a_2, \ldots, a_r)$ and $X_{n+1} = s(X_n)$. By the previous remark, all X_n's live in the hypercube $[0, \max a_i]^r$. We will prove that the X_n's become arbitrarily divisible by 2, thus one of them will have to be equal to 0. To prove this, it is enough to prove that one of the X_n's has all coordinates even numbers, since then we can repeat the procedure to make the coordinates of another X_n multiples of 4 and so on. Now, note that if $X_n = (a_1(n), a_2(n), \ldots, a_r(n))$ then $X_{n+1} \equiv (1+T)(X_n) \pmod 2$, where $T(x_1, x_2, \ldots, x_r) = (x_2, x_3, \ldots, x_1)$. Thus $X_n \equiv (1+T)^n X_0 \pmod 2$. Now, since $\binom{r}{k}$ is even for all $1 \le k \le r-1$ (because r is a power of 2), we have

$$(1+T)^r X_0 \equiv \sum_{k=0}^{r} \binom{r}{k} T^k X_0 \equiv X_0 + T^r X_0 \pmod 2.$$

But trivially $T^r = 1$, the identity map, so $X_0 + T^r X_0 \equiv 2X_0 \equiv 0 \pmod 2$. Thus all coordinates of X_r are even and we are done.

Assume now that i) holds. We deduce that for any vector $x \in \mathbb{F}_2^r$ there exists n such that $(1+T)^n x = 0$. Indeed, any such x is the reduction mod 2 of an r-tuple of nonnegative integers $(a_1, a_2, \ldots, a_r)$ and we know that we can find n such that $s^n((a_1, a_2, \ldots, a_r)) = (0, \ldots, 0)$. But we also saw that

$$s^n((a_1, a_2, \ldots, a_r)) \equiv (1+T)^n (x_1, x_2, \ldots, x_r) \pmod 2.$$

Now, since there are finitely many $x \in \mathbb{F}_2^r$, it follows that there exists $n \ge 1$ such that $(1+T)^n x = 0$ for all $x \in \mathbb{F}_2^r$ (simply choose n_x for each x and put $n = \max n_x$). We claim that this forces r to be a power of 2. Notice that the minimal polynomial of T as an endomorphism of $\mathbb{F}_2^r$ is $X^r + 1$ (T is simply the shift, so we can easily compute $P(T)$ for any polynomial P). Thus, we must have $X^r + 1 | (X+1)^n$ in $\mathbb{F}_2[X]$. In particular, we must have $X^r + 1 = (X+1)^j$ for some $j \le n$ and so $\binom{j}{u}$ is even for all $1 \le u \le j-1$, forcing j to be a power of 2. But then $(1+X)^j = 1 + X^j$ and so $r = j$ is also a power of 2. □

Before discussing the next problems, we need some preliminaries. First, an easy but fundamental remark: suppose that $f \in \mathbb{Z}[X_1, X_2, \ldots, X_n]$ satisfies

$f(x_1, x_2, \ldots, x_n) = 0$ for all integers x_i. Then an easy induction on n shows that all coefficients of f are zero. Thus, for all fields K and all $x_1, x_2, \ldots, x_n \in K$ we have $f(x_1, x_2, \ldots, x_n) = 0 \in K$. Let us apply this observation to establish the following

Theorem 12.2. *Let K be a field and let $A \in M_n(K)$ be such that $a_{ij} + a_{ji} = 0$ for all i, j. If n is odd and if $a_{ii} = 0$ for all i, then* $\det(A) = 0$.

Proof. Consider the polynomial

$$f(a_{12}, a_{13}, \ldots, a_{n,n-1}) = \begin{vmatrix} 0 & a_{12} & a_{13} & \cdots & a_{1n} \\ -a_{12} & 0 & a_{23} & \cdots & a_{2n} \\ -a_{13} & -a_{23} & 0 & \cdots & a_{3n} \\ \vdots & \vdots & \vdots & \ddots & \vdots \\ -a_{1n} & -a_{2n} & -a_{3n} & \cdots & 0 \end{vmatrix} \in \mathbb{Z}[a_{12}, a_{13}, \ldots, a_{n,n-1}].$$

By the previous discussion, it suffices to prove that f vanishes when evaluated at any $a_{ij} \in \mathbb{Z}$. But if $X \in M_n(\mathbb{Z})$ is antisymmetric, then

$$\det(X) = \det(X^t) = \det(-X) = (-1)^n \det(X) = -\det(X),$$

the first equality being standard, the second one the hypothesis and the last one a consequence of the fact that n is odd. Thus $\det(X) = 0$ for all antisymmetric matrices $X \in M_n(\mathbb{Z})$. But this is precisely what we wanted to prove. □

We consider next two applications of the previous theorem.

8. Suppose that n teams compete in a tournament wherein each team plays against any other team exactly once. In each game, 2 points are given to the winner, 1 point for a draw, and 0 points for the loser. It is known that for any subset S of teams, one can find a team (possibly in S) whose total score in the games with teams in S is odd. Prove that n is even.

D. Karpov, Russian Olympiad 1972

Proof. Define the matrix $A = (a_{ij})$ by $a_{ij} = 1$ if there is a draw between i, j and 0 otherwise. Clearly, $a_{ii} = 0$ and $a_{ij} = a_{ji}$ for all i, j. See A as a matrix with coefficients in $\mathbb{F}_2$. We claim that A is invertible. Indeed, otherwise we can find $x \in \mathbb{F}_2^n$ nonzero such that $Ax = 0$. If $I = \{1 \le i \le n | x_i = 1\}$, then the condition $Ax = 0$ can be written $\sum_{j \in I} a_{ij} = 0$ for all $1 \le i \le n$. Since each match which is not a draw yields an even number of points, the above condition implies that for all i, the total score in the games between i and the teams in I is even, contradicting the hypothesis. Now, it is enough to use theorem 12.2 to conclude that n is even. □

The following problem is essentially the same as the previous one, but written in different language.

9. A simple graph has the property: given any nonempty set H of its vertices, there is a vertex x of the graph such that the number of edges connecting x with the points in H is odd. Prove that the graph has an even number of vertices.

Proof. If A is the adjacency matrix of the graph, seen as matrix over $\mathbb{F}_2$, then the hypothesis of the problem translates into: for all nonzero $x \in \mathbb{F}_2^n$ we have $Ax \neq 0$. Thus A is invertible over $\mathbb{F}_2$. Since A is clearly antisymmetric (because it is symmetric and we are in characteristic two) with zero diagonal, the result follows from theorem 12.2. □

The following problem is rather easy.

10. Let $A_1, A_2, \ldots, A_n$ be subsets of $A = \{1, 2, \ldots, n\}$ such that for any nonempty subset T of A, there is $i \in A$ such that $|A_i \cap T|$ is odd. Suppose that B_1, B_2 are subsets of A such that $|A_i \cap B_1| = |A_i \cap B_2| = 1$ for all i. Prove that $B_1 = B_2$.

Gabriel Dospinescu, Mathematical Reflections

Proof. Let's consider the subsets of A as vectors in $\mathbb{F}_2^n$, where the entry in the ith position is 1 if the subset contains the element, 0 otherwise. Let the vectors corresponding to A_i's form the rows of the matrix M. The first condition can be expressed as $Mv \neq 0$ for all $v \neq 0$. Let v be the vector all of whose components are 1. The second condition says that $Mb_j = v$ for all j, where b_j are the vectors representing the B_j's. So we have $M(b_1 - b_2) = 0$, whence $b_1 = b_2$ and $B_1 = B_2$. □

12.3 Applications of bilinear algebra

In problems concerning incidences, it is sometimes useful to consider properties of inner products and bilinear algebra. If K is any field, we define the inner product of two vectors $x = (x_1, x_2, \ldots, x_n) \in K^n$ and $y = (y_1, y_2, \ldots, y_n) \in K^n$ as

$$\langle x, y \rangle = x_1 y_1 + x_2 y_2 + \cdots + x_n y_n.$$

If $K = \mathbb{R}$, we let $||x|| = \sqrt{\langle x, x \rangle}$ and call it the norm of x. The Cauchy-Schwarz inequality is equivalent to the triangle inequality $||x + y|| \leq ||x|| + ||y||$.

Let us start with a rather standard problem, which is nearly optimal. As the remark following the proof of it shows, this problem is actually rather subtle. We will see some variants of it later in the chapter.

11. A handbook classifies plants by 100 attributes (each plant either has a given attribute or does not have it). Two plants are dissimilar if they differ in at least 51 attributes. Show that the handbook cannot give 51 plants all dissimilar from each other.

Tournament of the Towns 1993

Proof. More generally, assume that the handbook considers $2n$ attributes, that two plants are dissimilar if they differ in at least $n+1$ of these, and that we have $n+1$ plants all dissimilar from each other. For each i, let v_i be the vector whose j-th coordinate is 1 if plant i has attribute j and -1 otherwise. Let $\langle , \rangle$ be the standard inner product and observe that the hypothesis implies that $\langle v_i, v_j \rangle < 0$ for all $i \neq j$. It is however clear that $\langle v_i, v_j \rangle$ is an even integer, thus we actually have $\langle v_i, v_j \rangle \leq -2$ for all $i \neq j$. But then

$$||\sum_{i=1}^{n+1} v_i||^2 = \sum_{i=1}^{n+1} (2n) + 2\sum_{i<j} \langle v_i, v_j \rangle \leq 2n(n+1) - 4\binom{n+1}{2} = 0.$$

This implies that $v_1 + v_2 + \cdots + v_{n+1} = 0$. If n is odd, then we are quite stuck, but if n is even, then last equality certainly cannot hold, as each component of $v_1 + v_2 + \cdots + v_{n+1}$ is an odd number. Fortunately, $n = 50$ is even in our problem and so we are done. □

Proof. Let us assume that the book has 51 pairwise dissimilar plants, call them $a_1, a_2, \ldots, a_{51}$ and call the attributes $P_1, P_2, \ldots, P_{100}$. We will count triples (a_i, a_j, P_k) with $i < j$ so that a_i and a_j differ on attribute P_k. On the one hand, for each (a_i, a_j) there are at least 51 triples containing this pair, so the total number of triples is at least $51\binom{51}{2}$. On the other hand, if A_k is the set of plants having attribute P_k, then there are $|A_k|(51 - |A_k|) \leq 25 \cdot 26$ pairs (a_i, a_j) living in triples with attribute P_k. We deduce that

$$100 \cdot 25 \cdot 26 \geq 51 \cdot \binom{51}{2} \iff 2600 \geq 2601,$$

a contradiction. □

Remark 12.3. A Hadamard matrix is a square matrix with entries ± 1 such that the rows are pairwise orthogonal vectors. It is a famous conjecture that for all $n \in 4\mathbb{N}$ there exists an $n \times n$ Hadamard matrix. Assume that $n + 1$ is even and suppose that the previous conjecture holds. Thus we can find a Hadamard matrix of order $2n + 2$. By multiplying some rows by -1, we may assume that all entries of the first column are 1. Since the columns are orthogonal, each column except the first now has $n + 1$ positive and $n + 1$ negative terms. Now consider all the rows in which the first two terms are $+1$. There are $n + 1$ such rows. Take these rows and chop away the first two columns. The result is $n + 1$ vectors of length $2n$ such that the inner product of every pair of vectors is -2. Interpreting the rows as plants and the columns as attributes, we get the required set of $n + 1$ plants which pairwise differ in strictly more than half of their $2n$ attributes (in fact, every pair differs in exactly $n + 1$ attributes). One can actually exhibit Hadamard matrices of order 104, yielding 52 plants which pairwise differ in 52 out of 102 attributes.

The next two problems are variations on the following nice and rather classical result. We will also present purely combinatorial proofs and we will leave the reader to decide which method is more efficient.

Theorem 12.4. *Let $v_1, v_2, \ldots, v_k$ be nonzero vectors in $\mathbb{R}^n$ such that*

$$\langle v_i, v_j \rangle \leq 0$$

for all $i \neq j$. Then $k \leq 2n$.

Proof. We will induct on n. For $n = 1$, it is an obvious application of the pigeonhole principle. Now, suppose that the result holds for $n - 1$ and consider $v_1, v_2, \ldots, v_k \neq 0$ in $\mathbb{R}^n$ such that $\langle v_i, v_j \rangle \leq 0$ for all $i \neq j$. Dividing each v_i by its norm and working with the new set of vectors, we may assume that v_i have norm 1. Choose an isometry A sending v_1 to the vector e_1, whose first coordinate is 1 and all the others are 0. Working with the vectors $Av_1, Av_2, \ldots, Av_k$ instead of $v_1, v_2, \ldots, v_k$ (they satisfy the same hypothesis, as an isometry is invertible and preserves the inner product), we may assume that A is the identity and so $v_1 = e_1$. Then the first coordinate of each of the vectors $v_2, v_3, \ldots, v_k$ is non-positive by hypothesis. Consider the vectors $w_2, w_3, \ldots, w_k \in \mathbb{R}^{n-1}$ obtained by deleting the first coordinate of $v_2, v_3, \ldots, v_k$. They are the projections of $v_2, v_3, \ldots, v_k$ on the orthogonal complement to e_1. Since the first coordinate of each of $v_2, v_3, \ldots, v_k$ is non-positive and since $\langle v_i, v_j \rangle \leq 0$ for $2 \leq i \neq j \leq k$, we also have $\langle w_i, w_j \rangle \leq 0$ for all $i \neq j$. If $w_2, w_3, \ldots, w_k$ are nonzero we obtain $k \leq 2n - 1 < 2n$ by the inductive hypothesis. So, assume that some w_i's are zero. We claim that at most one of them is zero. If we manage to prove this, it will follow that $k - 2 \leq 2(n - 1)$ and the inductive step will be proved. So, assume that $w_2 = w_3 = 0$, then v_1, v_2, v_3 are multiples of e_1 and since they are norm 1, each of them is equal to e_1 or $-e_1$. But then we can find $1 \leq i < j \leq 3$ such that $v_i = v_j$. As $\langle v_i, v_j \rangle \leq 0$, it follows that $v_i = 0$, a contradiction. The result follows. □

12. An $m \times n$ matrix is filled with 0 and 1 such that any two rows differ in at least $n/2$ positions. Prove that $m \leq 2n$.

Iranian Olympiad

Proof. It is more convenient to replace all zeros in the table by -1. Of course, this does not change anything to the hypothesis or conclusion of the problem. See each row as a vector in $\mathbb{R}^n$ and let $v_1, v_2, \ldots, v_m$ be these vectors. For any

$i \neq j$, the inner product $\langle v_i, v_j \rangle$ is simply the difference between the number of positions where v_i and v_j coincide and the number of positions where the two vectors differ. Taking into account the hypothesis, we deduce that $\langle v_i, v_j \rangle \leq 0$ for all $i \neq j$. Of course, $v_i \neq 0$ for all i. The result follows from theorem 12.4. □

Proof. The crucial ingredient in the proof is the following

Lemma 12.5. *If in an $m \times n$ binary matrix every two rows differ in at least d positions and if $2d > n$, then $m \leq \frac{2d}{2d-n}$.*

Proof. Call a pair (x, y) of matrix entries good if x and y are in the same column and if $(x, y) = (0, 1)$. If d_i is the number of ones in the i-th column, then clearly there are $\sum_{i=1}^{n} d_i(m - d_i)$ good pairs. Since $d_i(m - d_i) \leq \frac{m^2}{4}$, there are at most $n.\frac{m^2}{4}$ good pairs. However, if we fix two rows, we find at least d good pairs, by hypothesis. Thus we find at least $d\binom{m}{2}$ good pairs and so $d\binom{m}{2} \leq n \cdot \frac{m^2}{4}$, which immediately yields the lemma. □

Passing now to the proof, let A_1 be the number of rows whose first element is 1 and let A_2 be the number of rows whose first element is 0. If we consider only the rows whose first element is 1 and then delete the first element in each such row, we obtain a set of rows such that any two differ in at least $\frac{n}{2}$ positions. By the previous lemma, we deduce that $A_1 \leq n$ and $A_2 \leq n$. Since $m = A_1 + A_2$, the result follows. □

13. Let n be a positive integer. If $u = (u_i)_{i=1}^{2^n}$ and $v = (v_i)_{i=1}^{2^n}$ are binary sequences of length 2^n, let the distance between them be

$$d(u, v) = \sum_{i=1}^{2^n} |u_i - v_i|.$$

Find the greatest number of binary sequences of length 2^n whose pairwise distances are at least 2^{n-1}.

China TST 2005

Proof. Replacing each coordinate 0 by -1 in each vector, we must find the maximal number of vectors of length 2^n with coordinates ± 1 and whose pairwise inner products are nonpositive. Theorem 12.4 shows that this maximal number is at most 2^{n+1}. The problem is thus solved if we can exhibit 2^{n+1} vectors with coordinates ± 1 in $\mathbb{R}^{2^n}$ and with nonpositive inner products. We will consider the vectors $\pm e_1, \pm e_2, \dots, \pm e_{2^n}$, where $e_1, e_2, \dots, e_{2^n}$ is an orthogonal basis of $\mathbb{R}^{2^n}$ and all e_i have coordinates ± 1. Such a basis can be constructed by induction: for $n = 1$ choose $e_1 = (1, 1)$ and $e_2 = (1, -1)$. Assuming that we have such a basis for n, consider the vectors $f_i = (e_i, e_i)$ for $1 \le i \le 2^n$ (where (x, y) is the vector obtained by considering the coordinates of x followed by the coordinates of y) and $f_i = (-e_{i-2^n}, e_{i-2^n})$ for $2^n + 1 \le i \le 2^{n+1}$. We can immediately check that these vectors are linearly independent and by construction they are orthogonal. The inductive step is proved. □

Remark 12.6. A more general question is the following: given n and k, what is the maximal number of binary sequences of length n such that any two differ in at least k places? For some nontrivial cases of this problem, we refer the reader to chapter 10 of [7].

Here is another very classical and nontrivial result. There are purely combinatorial proofs, but they are really not very illuminating. On the other hand, the algebraic proof is very neat.

14. Let $A_1, A_2, \dots, A_m$ be distinct subsets of a set A with $n \ge 2$ elements. Suppose that any two of these subsets have exactly one common element. Prove that $m \le n$.

Fisher's inequality

Proof. We may assume that $A = \{1, 2, \dots, n\}$. Following the usual technique for this kind of problem, consider the incidence matrix M of the sets, that is the $m \times n$ matrix defined by $m_{i,j} = 1$ if $j \in A_i$ and 0 otherwise. By hypothesis, $M \cdot M^t$ is the matrix with i-th row $(1, 1, \dots, |A_i|, 1, \dots, 1)$.

For technical reasons, we have to discuss two cases. The first case is when $|A_i| = 1$ for some i. But then by assumption all A_j have to contain A_i and so $A_j \setminus A_i$ for $j \ne i$ are pairwise disjoint nonempty subsets of a set with $n - 1$

elements, thus certainly $m-1 \le n-1$ and we are done. The hard case is when all $|A_i| \geq 2$ and the key point (to be proved below) is that in this case $M \cdot M^t$ is invertible, thus of rank m. Since the rank of $M \cdot M^t$ is at most that of M, which is at most n, this will be enough to prove the result.

In order to prove the key point, we will argue by contradiction: if $M \cdot M^t$ is not invertible, there exists a nonzero vector with real coordinates $x_1, x_2, \ldots, x_m$ such that $M \cdot M^t x = 0$. But then ${}^t x M \cdot M^t x = 0$ and so

$$\sum_{i=1}^{m} |A_i| x_i^2 + 2 \sum_{1 \le i < j \le m} x_i x_j = 0.$$

This can be also written as

$$\sum_{i=1}^{m} (|A_i| - 1) x_i^2 + \left(\sum_{i=1}^{m} x_i \right)^2 = 0.$$

Since $|A_i| \geq 2$ for all i, the last relation obviously implies that all x_i are zero, a contradiction. Thus, the claim is proved and the problem is solved. □

Remark 12.7. One can also prove the invertibility of $M \cdot M^t$ in a different way, if $|A_i| \geq 2$ for all i. Namely, if $M \cdot M^t x = 0$, then $(|A_i| - 1)x_i + S = 0$, where $S = x_1 + x_2 + \cdots + x_n$. But then $x_i = -\frac{S}{|A_i|-1}$ and so

$$S = -S \cdot \sum_{i=1}^{m} \frac{1}{|A_i| - 1},$$

which obviously implies that $S = 0$ and so all x_i are zero.

Remark 12.8. We leave as an exercise to the reader to adapt the previous proof and deduce the following more general result: let $A_1, A_2, \ldots, A_m$ be different subsets of a set with n elements X. If there is $a \in \{1, 2, \ldots, n-1\}$ such that $|A_i \cap A_j| = a$ for all $i \neq j$, then $m \leq n$.

It is also very hard to solve the following problem without linear algebra. On the other hand, even with the help of (bi)linear algebra, this problem is quite tricky.

15. Let $A_1, A_2, \ldots, A_m$ and $B_1, B_2, \ldots, B_p$ be families of distinct subsets of $\{1, 2, \ldots, n\}$ such that $A_i \cap B_j$ is an odd number for all i and j. Then $mp \leq 2^{n-1}$.

Benny Sudakov

Proof. Let v_i be the incidence vector of A_i and let w_i be the incidence vector of B_i, seen as vectors in $\mathbb{F}_2^n$. The hypothesis becomes: $\langle v_i, w_j \rangle = 1$ for all i, j, where $\langle , \rangle$ is the standard inner product on $\mathbb{F}_2^n$. Let V be the vector space spanned by the vectors w_i. Since the vectors $v_i + v_1$ are m distinct vectors each orthogonal to every w_j, thus to V, we have

$$m \leq 2^{\dim \operatorname{Span}(v_i + v_1)} \leq 2^{n - \dim \operatorname{Span}(w_j)}.$$

Moreover, since $\langle v_1, w_j \rangle = 1$ and $\langle v_1, w_k + w_1 \rangle = 0$ for all j, k, the sets

$$\{w_i | 1 \leq i \leq p\} \text{ and } \{w_i + w_1 | 1 \leq i \leq p\}$$

are disjoint subsets of $\operatorname{Span}(w_i)$. Thus $2p \leq 2^{\dim \operatorname{Span}(w_i)}$. Multiplying these two inequalities yields the desired result. □

The next three problems also require a preliminary discussion. Namely, we will recall the proof of the following classical, but nontrivial result:

Theorem 12.9. *Let $A \in M_n(\mathbb{F}_2)$ be a symmetric matrix and let d be the column vector whose coordinates are the entries on the main diagonal of A. Then there exists $x \in \mathbb{F}_2^n$ such that $Ax = d$.*

The easiest proof uses bilinear algebra: consider the standard inner product on $\mathbb{F}_2^n$ and observe that it suffices to prove that any vector which is orthogonal to $\operatorname{Im}(A)$ is also orthogonal to d. But if v is such a vector, we have

$$\sum_{i=1}^{n} \left(\sum_{j=1}^{n} a_{ij} w_j \right) v_i = 0$$

for all $w_j \in \mathbb{F}_2$ and by exchanging the two sums we deduce that

$$\sum_{i=1}^{n} a_{ij} v_i = 0$$

for all j. But then we also have

$$\sum_{j=1}^{n} v_j \left(\sum_{i=1}^{n} a_{ij} v_i \right) = 0,$$

which can be written (taking into account that $a_{ij} + a_{ji} = 0$)

$$\sum_i a_{ii} v_i^2 = 0.$$

As $v_i^2 = v_i$, we are done.

16. Light bulbs $L_1, L_2, \ldots, L_n$ are controlled by switches $S_1, S_2, \ldots, S_n$. Switch S_i changes the on/off status of light L_i and possibly the status of some other lights. Suppose that if S_i changes the status of L_j then S_j changes the status of L_i. Initially all lights are off. Is it possible to operate the switches in such a way that all the lights are on?

Uri Peled, AMM 10197

Proof. This is a very easy application of the previous result: define $a_{ij} = 1$ if S_j changes the status of L_i and 0 otherwise. By hypothesis, we have $a_{ij} = a_{ji}$ and $a_{ii} = 1$. Thus the matrix $A = (a_{ij})$, seen as matrix over $\mathbb{F}_2$, is symmetric with diagonal elements 1. By the previous theorem, we can find a nonzero vector $x \in \mathbb{F}_2^n$ such that Ax is the diagonal vector of A. Thus, $\sum_{j=1}^{n} a_{ij} x_j = 1$ for all i. If J is the set of those j with $x_j = 1$, the previous equality tells us that if we operate the switches S_j with $j \in J$, then the state of each L_i will change an odd number of times and so all lights L_i will be on. □

The following problem is a bit trickier. We also give a purely combinatorial proof, to see the difference...

17. Let G be a graph. Prove that the set of its vertices can be partitioned in two groups (possibly empty) such that each group induces a subgraph in which all vertices have even degree.

Gallai's Cycle-Cocycle partition theorem

Proof. Consider the adjacency matrix A of the graph and perform the following operations on it: for each vertex i of odd degree, add a 1 in position (i,i) of A. For all the other vertices i, leave position (i,i) as in the original matrix (i.e. equal to 0). We thus get a new symmetric binary matrix $B=(b_{ij})$. By the previous theorem, there exists $x \in \mathbb{F}_2^n$ such that

$$\sum_{i=1}^{n} b_{ij}x_j = b_{ii}$$

for all i. Let $V_1 = \{1 \le i \le n | x_i = 1\}$ and let V_2 be its complement. We claim that this partition of the vertices satisfies the desired conditions. Indeed, for $i \in V_1$ we have $\sum_{j \in V_1 - \{i\}} b_{ij} = 0$, so i has an even degree in the subgraph induced by V_1. For $i \in V_2$, we have

$$\sum_{j \in V_2 - \{i\}} b_{ij} = \sum_{j=1}^{n} b_{ij} - \sum_{j \in V_1 \cup \{i\}} b_{ij} = \sum_{j=1}^{n} b_{ij}$$

and the last quantity is 0 by construction. Thus i has even degree in V_2 and we are done. □

We continue with a nice variation on the previous result.

18. At a certain mathematical conference, every pair of mathematicians are either friends or strangers. At mealtime, every participant eats in one of two large dining rooms. Each mathematician insists upon eating in a room which contains an even number of his or her friends. Prove that the number of ways that the mathematicians may be split between the two rooms is a power of two .

USAMO 2008

Proof. Note that this problem implies the existence of at least one admissible partition (i.e. one that satisfies the conditions of the problem), which is precisely the content of the previous problem. Fortunately, the hard work was actually done. Let G be the graph whose vertices are the mathematicians, two vertices being connected if the mathematicians are friends. Suppose that

G has n vertices $1, 2, \ldots, n$ and consider the matrix B as in the solution of the previous problem. We claim that there is a natural bijection between admissible partitions and solutions of the system $Bx = d$, where d is the main diagonal of B. Indeed, let us identify admissible partitions C_0, C_1 with vectors $x \in \mathbb{F}_2^n$, where $x_i = 0$ if $i \in C_0$ and $x_i = 1$ if $i \in C_1$. The admissibility condition can be written: for all i we have

$$\sum_{b_{ij}=1, j \neq i, j \in C(i)} 1 \equiv 0 \pmod 2,$$

where $C(i)$ is the class of the partition containing i. Just as in the proof of the previous problem, this condition can also be written

$$\sum_{j=1}^{n} b_{ij} x_j = b_{ii}.$$

Thus, the number of admissible partitions is the number of vectors $x \in \mathbb{F}_2^n$ such that $Bx = d$. We know that there is at least one such vector x_0 (by the discussion preceding problem 10 or by the previous problem). But then

$$\{x | Ax = d\} = x_0 + \{x | Ax = 0\}$$

and $\{x | Ax = 0\} = Ker(A)$ is a linear subspace of $\mathbb{F}_2^n$, so it has cardinality $2^{\dim Ker(A)}$. Since this is a power of 2, the conclusion follows. □

12.4 Matrix equations

To fully appreciate the power of linear algebra in combinatorics, the reader should try to find a purely combinatorial solution to the following problem.

19. Let $G_1, G_2, \ldots, G_k$ be complete bipartite subgraphs of the complete graph K_{2n} with $2n$ vertices. Assume that every edge of K_{2n} is contained in an odd number of subgraphs $G_1, G_2, \ldots, G_k$. Prove that $k \geq n$.

Proof. Let $1, 2, \ldots, 2n$ be the vertices of K_{2n} and let L_i, R_i be disjoint subsets of $\{1, 2, \ldots, 2n\}$ such that the set of vertices of G_i is $L_i \cup R_i$ and two vertices are connected if and only if one is in L_i and the other is in R_i. If $l_i \in \mathbb{F}_2^{2n}$ is the vector whose nonzero coordinates are precisely those on positions belonging to L_i and if r_i is similarly defined, then the entry in position (u, v) of the matrix $l_i \cdot r_i^t$ is precisely $1_{u\in L_i, v\in R_i}$. We deduce that the entry in position (u, v) of

$$X = \sum_{i=1}^{k} (l_i \cdot r_i^t + r_i \cdot l_i^t) \in M_{2n}(\mathbb{F}_2)$$

is 0 if $u = v$ and it is the number of subgraphs containing the edge uv, taken mod 2, otherwise. Therefore, the hypothesis can be reformulated as an equality of matrices $X = J - I_{2n}$, where J is the matrix all of whose entries are 1. Now, matrices of the form $v_1 \cdot v_2^t$ are precisely the matrices of rank at most 1. So, we expressed X as a sum of $2k$ matrices of rank at most 1. Since the rank function is sub-additive (thought of in terms of endomorphisms, this is obvious: the image of $u + v$ is contained in the sum of the images of u and v for any endomorphisms u, v of a vector space), in order to solve the problem it is enough to prove that $J - I_{2n}$ has rank $2n$, i.e. that it is invertible. This is however trivial: if $Jx = x$ for some vector x, then $x_i = x_1 + x_2 + \cdots + x_{2n}$ for all i, so first all x_i's are equal and then $x_i = 2nx_i = 0$, since we are working in characteristic 2. The result follows. □

We also discussed the following beautiful problem in example 7, chapter 23 of [3]. We give here a very natural solution using the following classical result:

Theorem 12.10. *Let $X \in M_n(\mathbb{R})$ and $Y \in M_m(\mathbb{R})$. There exists a nonzero matrix $A \in M_{n\times m}(\mathbb{R})$ such that $XA = AY$ if and only if X, Y have a common eigenvalue.*

Proof. Assume first that there exists a nonzero matrix $A \in M_{n\times m}(\mathbb{R})$ such that $XA = AY$, but X, Y have no common eigenvalue. Thus the characteristic polynomials P, Q of X, Y are relatively prime and so there exist $R, S \in \mathbb{C}[T]$ such that $PR + QS = 1$. Now, since $XA = AY$, we have by the Cayley-Hamilton theorem $0 = P(X)A = AP(Y)$. But we also have $AQ(Y) = 0$, as $Q(Y) = 0$ (again by Cayley-Hamilton). Thus

$$A = A(PR)(Y) + A(QS)(Y) = AP(Y)R(Y) + AQ(Y)S(Y) = 0,$$

a contradiction. This settles the first part. Next, if X, Y have a common eigenvalue λ, then X and Y^t also have the common eigenvalue λ and so one can find nonzero vectors v, w such that $Xv = \lambda v$, $Y^t w = \lambda w$. It is then easy to check that $A = vw^t$ is a nonzero complex matrix such that $XA = AY$. Then one of $\mathrm{Re}(A)$ and $\mathrm{Im}(A)$ is a nonzero real matrix satisfying again $XA = AY$ and we are done. □

20. A grid divides an $n \times m$ sheet of paper into unit squares. The two sides of length n are taped together to form a cylinder. Prove that it is possible to write a real number in each square, not all zero, so that each number is the sum of the numbers in the neighboring squares, if and only if there exist integers k, l such that $n+1$ does not divide k and

$$\cos\frac{2l\pi}{m} + \cos\frac{k\pi}{n+1} = \frac{1}{2}.$$

Ciprian Manolescu, Romanian TST 1998

Proof. See the grid as an $n \times m$ matrix A and define $a_{0,i} = a_{n+1,i} = 0$ for all i. The problem becomes: there exists a nonzero matrix A such that

$$a_{ij} = a_{i-1,j} + a_{i+1,j} + a_{i,j-1} + a_{i,j+1}$$

(for all $1 \leq i \leq n$, j being taken mod m) if and only if there exist integers k, l as in the statement. The previous equality can be written very simply by introducing the matrices $B \in M_n(\mathbb{Z})$, $C \in M_m(\mathbb{Z})$ defined by $b_{ij} = 1$ if $|i-j| = 1$ and 0 otherwise, $c_{ij} = 1$ if $|i-j| = 1$ or $(i,j) \in \{(1,m),(m,1)\}$. Indeed, the previous relation is equivalent to $AC + BA = A$, or $BA = A(I - C)$. By theorem 12.10, the existence of the matrix A is equivalent to the existence of a common eigenvalue of B and $I - C$. This is equivalent to the existence of eigenvalues λ_1, λ_2 of B, respectively C such that $\lambda_1 + \lambda_2 = 1$. Let us compute the eigenvalues of C. If $Cx = \lambda x$ for a nonzero vector x, we can write

$$x_2 + x_m = \lambda x_1, \quad x_1 + x_3 = \lambda x_2, \quad \ldots, \quad x_1 + x_{m-1} = \lambda x_m$$

and so if we extend x_i by m-periodicity, we can simply write this as

$$x_{i+2} + x_i = \lambda x_{i+1}.$$

If r_1, r_2 are the roots of the equation $t^2 - \lambda t + 1 = 0$, then $x_i = C_1 r_1^i + C_2 r_2^i$ ($C_1 r_1^i + C_2 i r_1^i$ if $r_1 = r_2$). Imposing the condition of m-periodicity, we see that we should have $r_1^m = r_2^m = 1$ and (because r_1 and r_2 are reciprocal roots of unity)

$$\lambda = r_1 + r_2 = r_1 + \overline{r}_1 = 2\mathrm{Re}(r_1) = 2\cos\frac{2l\pi}{m}$$

for some $0 \le l < m$. This suggests the form of the eigenvalues of C. To make this argument completely rigorous, it suffices to reverse the computations: simply choose any $0 \le l < m$, set $\lambda = 2\cos\frac{2l\pi}{m}$, $r_1 = e^{\frac{2i\pi l}{m}}$ and choose $x_i = r_1^i$. Then x is an eigenvector with eigenvalue λ. Since we have found the correct number of eigenvectors, we are done: the eigenvalues of C are $2\cos\frac{2l\pi}{m}$ (the argument was a bit long, but we wanted to present it in this form as it is rather useful). The same kind of reasoning shows that the eigenvalues of B are $2\cos\frac{k\pi}{n+1}$ and the result follows from these computations and the previous discussion. □

We give two proofs for the following beautiful and classical result. For the first proof we follow [29], while the second proof appears in [42].

21. In a society, acquaintance is mutual and any two persons have exactly one common friend. Then there is a person who knows all the others.

The friendship theorem, Erdős, Rényi, Sós

Proof. First, we will prove that either there is a universal friend (i.e. a person knowing all the others) or any two persons have the same number of friends. If A and B are not friends, let $a_1, a_2, \ldots, a_k$ be A's friends and let $b_1, b_2, \ldots, b_l$ be B's friends. By assumption, a_i and B have a unique common friend $b_{f(i)}$ and b_j, A have a unique common friend $a_{g(j)}$. Since f and g are clearly inverses to each other we have $k = l$ and we are done. Consider the obvious graph of enemy relationships in this society and suppose that it is disconnected. Then the society can be partitioned into two sets X and Y such that everyone in set X is friends with everyone in set Y. If both X and Y have two people or more, then two people in X have at least two common friends in Y, contradiction. On the other hand, if either set contains only one person then that person is a universal friend. Thus, if the enemy graph is disconnected, there must be a universal friend. On the other hand, if the enemy graph is connected, then every person has the same number of friends, because we saw that two enemies have the same number of friends. This proves the claim and also shows that we may assume that the enemy graph is connected in what follows.

Let d be the number of friends of each person and let n be the number of people. Then, the number of pairs of people with a specific person as their common friend is $\binom{d}{2}$, so by counting the number of pairs of people in two ways we get $\binom{n}{2} = n\binom{d}{2}$ and so $n = d^2 - d + 1$. Finally, let A be the adjacency matrix of the graph. By our previous work, we can write $A^2 = (d-1)I + J$, where J is the matrix having 1 at each entry. Trivially, $(d-1)I + J$ has eigenvalues d^2 and $d-1$ (with multiplicities 1 and $n-1$ respectively). Thus the eigenvalues of A are d and $\pm\sqrt{d-1}$. Since the trace of A is zero, we deduce that $d + (a-b)\sqrt{d-1} = 0$, where a, b are the multiplicities of the eigenvalues $\pm\sqrt{d-1}$. Squaring the last relation, we deduce that $d-1$ divides d^2, so that $d = 1$ or $d = 2$. But both such cases trivially satisfy the theorem, which gives the contradiction we needed. $\square$

Proof. As in the previous solution, we prove that either there is a universal friend or the associated graph G is d-regular and connected, for some d such that $n = d^2 - d + 1$. Now comes the beautiful idea: for a positive integer p, let $c(p)$ be the number of cycles of people of length p such that each person in the cycle is friends with the next person (people can appear multiple times in such a cycle). Any such cycle can be constructed in the following way: pick a starting person, pick one of his friends, pick a friend of that friend, etc., until you are at the second to last person in the cycle. If the second to last person is different from the first person, then the last person must be their common friend, but if the second to last person is the same as the first then the last person can be any of the first person's d friends. Thus

$$c(p) = nd^{p-2} + (d-1)c(p-2).$$

If $d \neq 0$ and $d \neq 2$, pick a prime divisor p of $d-1$. Then, since $n = d^2 - d + 1$, we have $n \equiv 1 \pmod{p}$ and so $c(p) \equiv 1 \pmod{p}$. On the other hand, $c(p)$ is a multiple of p: there are no cycles that are shifts of themselves (since no person is a friend of himself or herself...), so the number of cycles of length

p must be a multiple of p. The last two results are not possible at the same time, so that we must have $d = 0$ or $d = 2$. But in this case it is clear that there is a universal friend, which finishes the proof. □

Remark 12.11. For infinitely many persons, this result does not hold anymore.

Remark 12.12. The theorem actually classifies all graphs as in the statement: all of them consist of a collection of triangles sharing a vertex.

Remark 12.13. In [42] one can also find a combination of the first two solutions, which is very elegant. Namely, first we prove that the graph is d-regular for some d such that $n = d^2 - d + 1$ and that the adjacency matrix A of the graph satisfies $A^2 = (d-1)I + J$, where J is the matrix all of whose entries are 1. Suppose that $d \geq 3$ and pick a prime p dividing $d - 1$. Working in $M_n(\mathbb{F}_p)$, we obtain $A^2 = J$ and since $AJ = dJ$, we have $AJ = J$. Then $A^p = J$ and so $\operatorname{tr}(A^p) \equiv n \pmod{p}$. On the other hand, an extension of Fermat's little theorem yields $\operatorname{tr}(A^p) \equiv (\operatorname{tr} A)^p \equiv \operatorname{tr} A \pmod{p}$ (working with the eigenvalues of A this reduces easily to theorem 9.15). But clearly $\operatorname{tr} A = 0$. We deduce that p divides n, which is clearly impossible. Thus $d \leq 2$.

We end this section with another classical result in algebraic combinatorics.

22. In a graph G with $n^2 + 1$ vertices suppose that every vertex has degree n and every cycle has length at least 5. Then $n \in \{1, 2, 3, 7, 57\}$.

Hoffman-Singleton's theorem

Proof. The combinatorial part of the theorem is contained in

Lemma 12.14. *If x, y are vertices of G, the number of common neighbors of x, y is 0 if x, y are connected and 1 otherwise.*

Proof. We will count triples $(x, (y, z))$ in which x, y, z are vertices and x is connected to both y and z. Counting them according to x, we obtain $(n^2 + 1)\binom{n}{2}$ triples (since there are $n^2 + 1$ vertices, each of degree n). On the other hand, two connected vertices cannot belong to a triple (otherwise we would obtain a 3-cycle in G) and two non-connected vertices y, z can belong to at most one triple (otherwise G would have a 4-cycle). So the number of triples is at most $\binom{n^2+1}{2}$ minus the number of edges of the graph, so at most $\binom{n^2+1}{2} - \frac{(n^2+1)n}{2} = (n^2+1)\binom{n}{2}$. Since we have already established that there are precisely $(n^2+1)\binom{n}{2}$ triples, it follows that every two non-connected vertices belong to exactly one triple and the lemma is proved. $\square$

Let A be the incidence matrix of G and let J be the matrix all of whose entries are 1. The previous lemma yields

$$A^2 = nI + J - I - A = (n-1)I - A + J.$$

Next, A is a symmetric matrix with real entries and zero diagonal, so it has zero trace and it is diagonalizable in an orthonormal basis (by the spectral theorem). In particular, all eigenvalues of A are real and the eigenvectors corresponding to different eigenvalues are orthogonal. Since the sum of entries in each row is n (because G is n-regular), we have $Av_1 = nv_1$, where v_1 is the vector all of whose coordinates are 1. So, if $Av = \lambda v$ for some $v \neq 0$ and $\lambda \neq n$, then $\langle v_1, v\rangle = 0$ and so $Jv = 0$. But this implies that $\lambda^2 = n - 1 - \lambda$. So, if r_1, r_2 are the roots of the equation $x^2 + x = n - 1$, any eigenvalue of A different from n is r_1 or r_2.

Next, we claim that n has multiplicity 1 as an eigenvalue of A. It suffices to prove that if $Av = nv$, then v is collinear to v_1. This is easy, since $Av = nv$ implies $n^2v = A^2v = (n-1)v - Av + Jv$ and so $Jv = (n^2+1)v$, hence all coordinates of v are equal and v is collinear to v_1. Let a, b be the multiplicities of r_1, r_2 as eigenvalues of A. So $a+b = n^2$ by the result we have just established and $ar_1 + br_2 + n = 0$, as A has trace 0. Since

$$r_1 = \frac{-1+\sqrt{4n-3}}{2}, \quad r_2 = \frac{-1-\sqrt{4n-3}}{2},$$

the previous relation becomes

$$-\frac{a+b}{2} + \sqrt{4n-3}\frac{a-b}{2} + n = 0.$$

Now, we have two cases: if $\sqrt{4n-3}$ is irrational, the last relation implies that $a = b$ and $n = a$. Since $a + b = n^2$, this gives $n = 2$ and we are done. If

$k = \sqrt{4n-3}$ is rational, write $a + b = n^2 = \left(\frac{k^2+3}{4}\right)^2$ and replace this in the previous relation. We obtain

$$16k(a-b) - (k^2+3)^2 + 8(k^2+3) = 0.$$

Reducing this mod k, we finally obtain $k|15$ and since $n = \frac{k^2+3}{4}$, the result follows. $\square$

Remark 12.15. The proof of the lemma actually yields the following result: if G is a graph in which any cycle has length at least 5 and any vertex has degree at least n, then G has at least n^2+1 vertices. So the theorem proved above actually classifies the extremal case. At the time of this writing, it is unknown whether or not there exists a graph as in the theorem for $n = 57$ (for all the other values of n, such graphs exist).

12.5 The linear independence trick

23. In an $m \times n$ table, real numbers are written such that for any two rows and any two columns, the sum of the numbers situated in opposite corners of the rectangle formed by them is equal to the sum of the numbers situated in the other two opposite corners. Some of the numbers are erased, but the remaining ones allow us to find the erased numbers using the above property. Prove that at least $n+m-1$ numbers remained on the table.

Russian Olympiad 1971

Proof. Consider the set X of all $m \times n$ matrices with real entries such that for any two rows and columns, the sum of the numbers situated in the opposite corners of the rectangle formed by them is equal to the sum of the numbers situated in the other two opposite corners. This set is obviously a linear subspace of $M_{m\times n}(\mathbb{R})$. If S is the set of pairs (i,j) such that the number situated at the entry (i,j) has not been erased, then the hypothesis can be translated into: the map $f : X \to \mathbb{R}^{|S|}$ obtained by sending a matrix $A \in X$ to the collection $(a_{i,j})_{(i,j)\in S}$ is injective. But then the dimension of X has to be at most the dimension of $\mathbb{R}^{|S|}$. Thus, $|S|$ is at least equal to the dimension of X as $\mathbb{R}$-vector space. Therefore it is enough to exhibit $m+n-1$ linearly independent matrices in X. This is easy: simply consider the matrices A_j having the j-th row equal to the vector $(1, 1, \ldots, 1)$ and the other rows zero (and this for all $1 \leq j \leq m$), then the matrices B_i having the i-th column consisting of ones and the remaining entries 0 (and this for all $1 \leq i \leq n-1$) It is immediate to check that they are in X and linearly independent. □

Once the necessary mathematical translations are done, the proof of the following problem is shorter than its statement...

24. In a contest consisting of n problems, the jury defines the difficulty of each problem by assigning it a positive integral number of points.[1] Any participant who answers the problem correctly receives that number of points for the problem; any other participant receives 0 points. After the participants submitted their answers in such a contest, the jury realized that given any ordering of the participants, it could have defined the problems' difficulty levels to make that ordering coincide with the participants' ranking according to their total scores.[2] Determine, in terms of n, the maximum number of participants for which such a scenario could occur.

Russian Olympiad 2001

Proof. It is clear that we may have n participants: impose that participant j solves problem j and nothing else. We will prove that we cannot have more than n participants. By associating to participant j the vector x_j consisting of the number of points he received for each problem, the hypothesis implies that $x_j \in \mathbb{R}_+^n$ and for every ordering of the x_j's, there is a vector $y \in \mathbb{R}_+^n$ which gives that ordering of $\langle x_j, y \rangle$. On the other hand, if the number of participants m is greater than n, there is a nontrivial linear combination of the x_j's that vanishes. By separating the coefficients of such a combination according to their sign, we find an equality $\sum_I \alpha_i x_i = \sum_J \beta_j x_j$ for some disjoint subsets

[1] The same number of points may be assigned to different problems.

[2] Ties are not permitted.

I, J of indices and positive numbers α_i, β_j such that $\sum_i \alpha_i \geq \sum_j \beta_j$. But it is clearly impossible to rank all x_i above all x_j. $\square$

There is no known combinatorial proof of the following result. On the other hand, the algebraic proof is rather simple.

25. Let $m > n+1$ and let $A_1, A_2, \ldots, A_m$ be subsets of $\{1, 2, \ldots, n\}$. Then there are disjoint sets I, J such that $\bigcup_{i \in I} A_i = \bigcup_{j \in J} A_j$ and $\bigcap_{i \in I} A_i = \bigcap_{j \in J} A_j$.

Lindstrom's theorem

Proof. Associate to each set A_i a vector $v_i \in \mathbb{R}^{2n}$, say

$$v_i = (x_1, x_2, \ldots, x_n, 1 - x_1, 1 - x_2, \ldots, 1 - x_n),$$

where $x_j = 1$ if $j \in A_i$ and 0 otherwise. We obtain at least $n+2$ vectors that lie in the $(n+1)$-dimensional vector subspace of $\mathbb{R}^{2n}$ defined by the equations $x_1 + y_1 = x_2 + y_2 = \cdots = x_n + y_n$. Thus, these vectors are linearly dependent and we can find $a_1, a_2, \ldots, a_m \in \mathbb{R}$ not all zero and such that $\sum_{i=1}^m a_i v_i = 0$. Define now $I = \{i | a_i > 0\}$ and $J = \{i | a_i < 0\}$. Clearly I, J are disjoint and we prove that $\bigcup_{i \in I} A_i = \bigcup_{j \in J} A_j$ and $\bigcap_{i \in I} A_i = \bigcap_{j \in J} A_j$. This is very easy: if $j \in \bigcup_{i \in I} A_i$ but $j \notin \bigcup_{j \in J} A_j$, then by definition the jth component of all v_i with $i \in J$ is 0, whereas there exists an $i \in I$ with the jth component of v_i nonzero. But then, looking at the jth component in the equality $\sum_{i=1}^m a_i v_i = 0$ yields a contradiction. By symmetry, we deduce that $\bigcup_{i \in I} A_i = \bigcup_{j \in J} A_j$. Similarly, looking at the $n+1, \ldots, 2n$th coordinate in the equality $\sum_{i=1}^m a_i v_i = 0$ we can show that $\bigcap_{i \in I} A_i = \bigcap_{j \in J} A_j$, which finishes the proof of the theorem. $\square$

In order to motivate the following result, we need some preliminaries. Suppose that F is a collection of subsets of some set X. We say that F shatters a subset S of X if $\{A \cap S | A \in F\}$ contains all subsets of S. The Vapnik-Chervonenkis dimension of F (also denoted $\mathrm{vc}(F)$) is the maximal cardinality of a set shattered by F. For instance, let $X = \mathbb{R}^n$ and let F be the collection of all affine half-spaces. Then $\mathrm{vc}(F) = n+1$, though this is not obvious. Note that F shatters the set of vertices of an $n+1$-simplex, so $\mathrm{vc}(F) \geq n+1$. The converse inequality follows from the following classical theorem of Radon: if $S \subset \mathbb{R}^n$ is a set with more than $n+1$ elements, then one can find disjoint subsets A, B of S such that the convex hulls of A and B have common points. The following beautiful result shows that large collections shatter big sets.

26. Let F be a family of subsets of $\{1, 2, \dots, n\}$ such that

$$|F| > \binom{n}{0} + \binom{n}{1} + \dots + \binom{n}{k-1}.$$

Then $\mathrm{vc}(F) \geq k$.

Sauer-Shelah's lemma

Proof. Let us assume that $\mathrm{vc}(F) \leq k-1$. We will prove that

$$|F| \leq \binom{n}{0} + \binom{n}{1} + \dots + \binom{n}{k-1}.$$

Let $\mathcal{P}_{k-1}$ be the family of all subsets A of $\{1, 2, \dots, n\}$ such that $|A| \leq k-1$. To any $X \in F$ associate the map $f_X : \mathcal{P}_{k-1} \to \mathbb{R}$ defined by $f_X(A) = 1_{A \subset X}$ (this is equal to 1 if $A \subset X$ and to 0 otherwise). We will prove that the maps f_X are linearly independent when X runs over the elements of F. This will clearly imply the desired bound on the size of F. Assume that $\sum_{X \in F} a_X f_X = 0$ for some real numbers a_X, not all equal to zero. Evaluating at elements $A \in \mathcal{P}_{k-1}$, we deduce that for any such A we have $\sum_{X \in F, A \subset X} a_X = 0$. The key step is to choose a minimal subset A such that $\sum_{X \in F, A \subset X} a_X \neq 0$. We will prove that for any $B \subset A$ we can find $X \in F$ such that $B = A \cap X$. This will contradict the property of F and the result will follow. It is of course enough to prove that $\sum_{X \in F, B = X \cap A} a_X \neq 0$. But by the inclusion-exclusion principle we have

$$\sum_{X \in F, X \cap A = B} a_X = \sum_{B \subset C \subset A} (-1)^{|C-B|} \cdot \left(\sum_{X \in F, C \subset X} a_X \right).$$

However, by minimality of A all sums $\sum_{X\in F, C\subset X} a_X$ are zero, except for $C = A$. Thus the right-hand side is nothing more than $(-1)^{|A-B|}\sum_{X\in F, A\subset X} a_X$, which is nonzero by the choice of A. This finishes the proof. $\square$

Proof. We will prove by induction on $|\mathcal{F}|$ the following statement: for any n and any family $\mathcal{F}$ of subsets of $\{1, 2, \dots, n\}$, $\mathcal{F}$ shatters at least $|\mathcal{F}|$ subsets of $\{1, 2, \dots, n\}$. This trivially implies the desired result. Now, the result clearly holds for $|\mathcal{F}| = 1$, so assume that it holds for all $|\mathcal{F}| < N$ and take a family $\mathcal{F}$ with N elements, all of them subsets of $\{1, 2, \dots, n\}$. Let $\mathcal{F}_0$ consist of the elements of $\mathcal{F}$ not containing 1 (we assume, without loss of generality, that 1 occurs in some, but not all elements of $\mathcal{F}$). By induction, $\mathcal{F}_0$ shatters at least $|\mathcal{F}_0|$ subsets of $\{2, 3, \dots, n\}$. Also by induction, $\mathcal{F}_1 = \{X\backslash\{1\} : X \in \mathcal{F}, 1 \in X\}$ shatters at least $|\mathcal{F}_1|$ subsets of $\{2, 3, \dots, n\}$. But if $A \subseteq \{2, 3, \dots, n\}$ is shattered by $\mathcal{F}_1$ then A is also shattered by $\{X \in \mathcal{F}|1 \in X\}$, and if A is shattered by both $\mathcal{F}_0$ and $\mathcal{F}_1$ then both A and $A \cup \{1\}$ are shattered by $\mathcal{F}$. We conclude that at least $|\mathcal{F}_0|+|\mathcal{F}_1| = |\mathcal{F}|$ subsets of $\{1, 2, \dots, n\}$ are shattered by $\mathcal{F}$, and the result follows. $\square$

Remark 12.16. Here is a beautiful geometrical interpretation of the previous result: let S be a subset of $\{-1, 1\}^n$ with more than $\sum_{i=0}^{d}\binom{n}{i}$ elements. Then there exists a subset $I \subset \{1, 2, \dots, n\}$ such that

$$\{(x_i)_{i\in I}|x \in S\} = \{-1, 1\}^{|I|}.$$

That is, a large subset of the vertices of the unit hypercube has a coordinate projection covering a unit hypercube in a smaller dimension!

We present two proofs of the following classical theorem of Bollobás. One of the proofs needs some preliminaries on resultants. Let K be a field and let $f = a_0 + a_1X + \dots + a_pX^p$ and $g = b_0 + b_1X + \dots + b_qX^q$ be two polynomials with coefficients in K. Let $K[X]_n$ be the $n+1$-dimensional K-vector space of polynomials of degree at most n in $K[X]$. It is easy to see that $\gcd(f, g) = 1$ if and only if the map $\varphi : K[X]_{q-1} \times K[X]_{p-1} \to K[X]_{p+q-1}$, $\varphi(U, V) = Uf + Vg$ is injective. Since the source and the target are K-vector spaces of the same finite dimension, it follows that $\gcd(f, g) = 1$ if and only if φ is an invertible linear map, i.e. if and only if the matrix of φ in the natural bases of $K[X]_{q-1} \times K[X]_{p-1}$ and $K[X]_{p+q-1}$ is invertible. This matrix has the first q columns the coordinates of $f, Xf, \dots, X^{q-1}f$ with respect to the basis $(1, X, \dots, X^{p+q-1})$ and the last p columns the coordinates of $g, Xg, \dots, X^{p-1}g$ with respect to the same basis. The determinant of this matrix is called the resultant of f and g and is denoted $\operatorname{Res}(f,g)$. The previous discussion shows that $\operatorname{Res}(f,g) = 0$ if and only if f and g have a common nontrivial divisor, which is the same as saying that they have a common root in some algebraic closure of K.

27. Let a, b be positive integers and let $A_1, A_2, \dots, A_m$ and $B_1, B_2, \dots, B_m$ be sets such that

    a) $|A_1| = |A_2| = \cdots = |A_m| = a$ and $|B_1| = |B_2| = \cdots = |B_m| = b$.
    b) $A_i \cap B_j$ is nonempty if and only if $i \neq j$.

    Then $m \leq \binom{a+b}{b}$.

Bollobás' theorem

Proof. We will only assume that $A_i \cap B_j \neq \emptyset$ for $i < j$. We may assume that A_i, B_j are subsets of $\{1, 2, \dots, n\}$ for some n. Choose $v_1, v_2, \dots, v_n \in \mathbb{R}^{a+1}$ such that any $a+1$ vectors among them are linearly independent (this is easily achievable). Let $v_k^\perp \in \mathbb{R}^{a+1}$ be a vector orthogonal to all v_u with $u \in A_k$. Such a vector exists and is unique up to a scalar multiple as by construction the vectors $(v_u)_{u \in A_k}$ span a hyperplane of $\mathbb{R}^{a+1}$.

Consider the polynomials

$$f_j(X) = \prod_{x \in B_j} \langle v_x, X \rangle,$$

where $\langle , \rangle$ is the standard inner product and $X = (X_1, X_2, \dots, X_{a+1})$. These are homogeneous polynomials of degree b in $a+1$ variables, living thus in a space of dimension $\binom{a+b}{b}$. We claim that $f_j(v_i^\perp) = 0$ for $i < j$ and $f_i(v_i^\perp) \neq 0$, which implies that the f_i are linearly independent and so $m \leq \binom{a+b}{b}$. To prove the claim, let $i < j$. Since $A_i \cap B_j \neq \emptyset$, there is $x \in B_j \cap A_i$ and so $\langle v_x, v_i^\perp \rangle = 0$. Thus $f_j(v_i^\perp) = 0$ for $i < j$. Also, since $A_i \cap B_i = \emptyset$, we have $\langle v_x, v_i^\perp \rangle \neq 0$ for all $x \in B_i$, thus $f_i(v_i^\perp) \neq 0$. □

Proof. Of course, we may assume that all A_i, B_j are subsets of $\mathbb{R}$. Consider the polynomials

$$f_i(X) = \prod_{a\in A_i}(X-a), \quad g_j(X) = \prod_{b\in B_j}(X-b).$$

By assumption, f_i and g_j have a common root if and only if $i \neq j$. Thus, if $R(f,g)$ is the resultant of two polynomials f, g, we have $R(f_i, g_j) = 0$ if and only if $i \neq j$.

Now, write

$$g(X) = g_0 + g_1 X + \cdots + g_b X^b$$

and observe that the expression

$$A_i(g_0, \ldots, g_b) = R(f_i, g)$$

is a homogeneous polynomial of degree a in the variables $g_0, \ldots, g_b$. Moreover, the previous results show that $A_i(g_{j0}, \ldots, g_{jb}) = 0$ if and only if $i \neq j$. Thus, the polynomials A_i are linearly independent and homogeneous of degree a in $b+1$ variables. Since the vector space of such polynomials has dimension $\binom{a+b}{a}$, the result follows. □

The following problem is very challenging. We present the author's proof.

28. Let $x_1, x_2, \ldots, x_n$ be distinct real numbers and suppose that the vector space spanned by $x_i - x_j$ over the rationals has dimension m. Then the vector space spanned only by those $x_i - x_j$ for which $x_i - x_j \neq x_k - x_l$ whenever $(i,j) \neq (k,l)$ also has dimension m.

Straus's theorem

Proof. By working with the numbers $x_i - x_1$, we may assume that $x_1 = 0$. Let V be the span of all differences $x_i - x_j$, so $\dim V = m$. Say a difference $x_i - x_j$ is unique if $x_i - x_j = x_k - x_l$ implies that $i = k$ and $j = l$. Let V' be the span of the unique differences, let m' be its dimension and suppose that $m' < m$. As V' is a subspace of V, there is a basis $v_1, v_2, \ldots, v_m$ of V such that $v_1, v_2, \ldots, v_{m'}$ is a basis of V'. As $x_1 = 0$, we have $x_i \in V$ for all i, so we can write

$$x_i = \sum_{l=1}^{m} x_i^{(l)} v_l$$

with $x_i^{(l)} \in \mathbb{Q}$. Note that there exists r such that $x_r^{(m)} \neq 0$, as otherwise V would be included in the span of the vectors $v_1, v_2, \ldots, v_{m-1}$, contradicting that $\dim V = m$.

Let now $x_i(t) = x_i + tx_i^{(m)}$. The crucial fact is the following

Lemma 12.17. *There exists $t \in \mathbb{R}$ such that*

1) If $x_i(t) = x_j(t)$, then $i = j$.

2) $\max x_i(t) - \min x_i(t) > \max x_i - \min x_i$.

Proof. For all $i \neq j$, the equation $x_i(t) = x_j(t)$ has at most one solution (recall that $x_i \neq x_j$), so 1) fails for at most finitely many t. On the other hand, $x_1(t) = 0$, so if r is chosen such that $x_r^{(m)} \neq 0$, then the left-hand side of 2) is at least $x_r + tx_r^{(m)}$ and this becomes arbitrarily large for $t >> 0$ or $t << 0$. The conclusion follows. □

Coming back to the proof, fix such a t and choose j, k such that $x_j(t) = \max x_i(t)$ and $x_k(t) = \min x_i(t)$. We claim that the difference $x_j - x_k$ is unique and $x_j^{(m)} \neq x_k^{(m)}$. This will yield a contradiction, as $x_j - x_k \in V'$ in this case, so its coordinate with respect to v_m has to be zero. To prove the claim, assume that $x_j - x_k = x_u - x_v$. Looking at the mth coordinate, we obtain $x_j^{(m)} - x_k^{(m)} = x_u^{(m)} - x_v^{(m)}$, so that $x_j(t) - x_k(t) = x_u(t) - x_v(t)$. But then clearly $x_j(t) = x_u(t)$ and $x_k(t) = x_v(t)$, so that by the choice of t we must have $j = u$ and $k = v$. Finally, if $x_j^{(m)} = x_k^{(m)}$, then

$$x_j(t) - x_k(t) = x_j - x_k \leq \max x_i - \min x_i,$$

a contradiction. □

12.6 Applications to geometry

In this section we discuss some rather hard problems with geometric flavor. Before passing to the first problem, we need to recall a few basic facts from convex geometry. Suppose that K is a closed convex subset of $\mathbb{R}^n$. The dual of K is

$$K^* = \{x \in \mathbb{R}^n | \langle x, v\rangle \geq 0, \forall v \in K\},$$

where $\langle\cdot\rangle$ is the standard inner product on $\mathbb{R}^n$. It is easy to see that K^* is again a closed convex set. Standard but nontrivial separation theorems in convex geometry imply the fundamental equality $K^{**} = K$ for any closed and convex subset K of $\mathbb{R}^n$. Classical examples of closed convex sets are the convex hull of finitely many points and the cone generated by finitely many vectors.[3] Applying the equality $K^{**} = K$ to the cone generated by $v_1, v_2, \ldots, v_k$, we obtain the following beautiful result, which is one of the many versions of Farkas' lemma:

Theorem 12.18. *(Farkas' lemma) Let $v, v_1, v_2, \ldots, v_k \in \mathbb{R}^n$ and suppose that $\langle w, v\rangle \geq 0$ whenever $\min\langle w, v_i\rangle \geq 0$. Then v is in the cone generated by $v_1, v_2, \ldots, v_k$.*

The following hard problem is a very nice way to re-state Farkas' lemma.

29. A figure composed of 1 by 1 squares has the property that if the squares of a fixed m by n rectangle are filled with numbers the sum of all of which is positive, the figure can be placed on the rectangle[4] so that the numbers it covers also have positive sum.[5] Prove that a number of such figures can be placed on the rectangle such that each square is covered by the same number of figures.

Russia 1998

[3]If $v_1, v_2, \ldots, v_k \in \mathbb{R}^n$, the cone generated by these vectors is the set of linear combinations $a_1v_1 + a_2v_2 + \cdots + a_kv_k$, where a_i are nonnegative real numbers. It is clear that this cone is a convex set, but it is not really obvious that it is a closed subset of $\mathbb{R}^n$.

[4]Possibly after being rotated by a multiple of $\frac{\pi}{2}$.

[5]However, the figure may not have any of its squares outside the rectangle.

Proof. The first step (and the trickiest...) is to restate the problem in a more algebraic way. Suppose that we have N ways to put the figure in the rectangle, possibly after being rotated and call these ways $1, 2, \ldots, N$. To each such way i associate its incidence vector $v_i \in \mathbb{R}^{mn}$, where the jth coordinate of v_i is 1 if the square j is covered and 0 otherwise (we number the squares of the table in some way from 1 to mn). Now, the hypothesis becomes: for any $v \in \mathbb{R}^{mn}$ such that $\langle v, 1\rangle > 0$, we also have $\max\langle v, v_i\rangle > 0$. Here $1 \in \mathbb{R}^{mn}$ is the vector all of whose coordinates are 1. The result then follows from the following

Lemma 12.19. *Let $v, v_1, v_2, \ldots, v_N \in \mathbb{Q}^n$ be vectors, with v nonzero. Suppose that for all $w \in \mathbb{R}^n$, if $\langle w, v\rangle > 0$, then $\max\langle w, v_i\rangle > 0$. Then there are rational nonnegative numbers $a_1, a_2, \ldots, a_N$ such that $v = a_1v_1 + a_2v_2 + \cdots + a_Nv_N$.*

If we accept this for a moment, we deduce the existence of nonnegative integers $a_1, a_2, \ldots, a_N$ and of a positive integer a such that $a \cdot 1 = a_1v_1 + a_2v_2 + \cdots + a_Nv_N$. So, if we put a_i copies of the ith placement of the figure, each square is covered a times and the problem is solved.

Now, let us turn to the proof of the lemma. By Farkas' lemma, we can find nonnegative real numbers $x_1, x_2, \ldots, x_N$ such that $v = x_1v_1 + x_2v_2 + \cdots + x_Nv_N$. Forgetting about those indices i such that $x_i = 0$, we find a linear system in the x_i's (obtained by writing the previous equality coordinate by coordinate), whose associated matrix has rational entries and which has positive solutions. We claim that such a system also has positive rational solutions. This follows from the following standard:

Lemma 12.20. *If a linear system of equations with rational coefficients has real solutions, then the rational solutions are dense in the set of real solutions.*

Proof. As the associated matrix has rational entries, if we put it in reduced row-echelon form (i.e. perform Gaussian elimination), we immediately deduce that the space of solutions has a basis with rational coordinates. The result follows from the density of $\mathbb{Q}$ in $\mathbb{R}$. □

We're done. □

To motivate the following problems, we need to recall a famous result of Dehn, also known as Hilbert's third problem: given two polyhedra in $\mathbb{R}^3$ with equal volume, can we always cut the first one into finitely many polyhedrons which can be reassembled to yield the second polyhedron? Dehn showed that the answer is negative in a rather complicated way, but nowadays there is a very beautiful and short proof due to Hadwiger, which we will sketch now. We claim that the regular simplex and the cube yield a negative answer to Hilbert's problem. Let $a = \arccos\frac{1}{3}$, the measure of the dihedral angles of the regular simplex. It is easy to see that $\frac{a}{\pi}$ is irrational (using the fact that if r and $\cos r\pi \in \mathbb{Q}$, then $\cos r\pi \in \{\pm\frac{1}{2}, \pm 1, 0\}$, which was explained in chapter 9, section 9.2). Let $\alpha_1, \ldots, \alpha_k$ be the dihedral angles appearing at the edges of the pieces of the dissection and let V be the $\mathbb{Q}$-vector space generated by them. As $\frac{a}{\pi}$ is irrational, there is a linear form $f : V \to \mathbb{Q}$ such that $f(\pi) = 0$ and $f(a) = 1$. If P is a polyhedron appearing in the dissection, define its Dehn invariant by $H(P) = \sum_e |e| f(\alpha)$, the summation being taken over the edges of P, $|e|$ being the length of e and α the dihedral angle at e. Simple geometry shows that this invariant is additive. Since the invariant of the cube is 0 and that of the regular simplex is not, it follows that we cannot cut the regular simplex into pieces that can be reassembled to form the unit cube. The following problem uses a similar argument, which is extremely useful in tiling problems. We need one more preliminary result: any vector space has a basis and any linearly independent set of vectors extends to a basis.

30. An $a \times b$ rectangle is divided into squares with sides parallel to that of the rectangle and with side lengths $x_1, x_2, \ldots, x_n$. Prove that $\dfrac{a}{x_i}$ and $\dfrac{b}{x_i}$ are rational numbers.

Dehn's theorem

Proof. The first ingredient is the following

Lemma 12.21. *There exists an additive map $f : \mathbb{R} \to \mathbb{R}$ such that $f(x) = 0$ if and only if x is a rational multiple of a.*

Proof. Consider a basis B of $\mathbb{R}$ as a $\mathbb{Q}$-vector space and such that $a \in B$. Choose a bijection $g : B - \{a\} \to B$ (g exists, because B is an infinite set) and define $f(a) = 0$ and $f(x) = g(x)$ for all $x \in B - \{a\}$. To define f for any real number r, express r as a finite combination $r = x_1b_1 + x_2b_2 + \cdots + x_nb_n$ of elements $b_i \in B$ with rational coefficients x_i and define

$$f(x) = x_1f(b_1) + x_2f(b_2) + \cdots + x_nf(b_n).$$

We obtain in this way an additive map f and we have $f(r) = 0$ if and only if r is a rational multiple of a. Indeed, if $f(r) = 0$ and $r = x_1b_1 + x_2b_2 + \cdots + x_nb_n$ is as above, then

$$x_1f(b_1) + x_2f(b_2) + \cdots + x_nf(b_n) = 0.$$

Now the elements $f(b_i)$ are either equal to 0 (if $b_i = x$) or equal to $g(b_i)$. Moreover, the $g(b_i)$ are distinct (because g is injective and the b_i's are distinct) and are linearly independent (since they are elements of a basis B). The only possibility to have the previous relation is $x_i = 0$ whenever $b_i \neq x$, which means precisely that r is a rational multiple of a. $\square$

Define the invariant of a rectangle R with side-lengths x, y by

$$H(R) = f(x)f(y).$$

The following lemma establishes the additivity of this invariant.

Lemma 12.22. *If a rectangle R is partitioned into finitely many rectangles $R_1, R_2, \ldots, R_k$, then $H(R) = H(R_1) + H(R_2) + \cdots + H(R_k)$.*

Proof. This is essentially obvious by additivity of f: any finite partition with rectangles can be refined to a partition which forms a grid (just extend all sides of all rectangles in the partition until they meet the sides of R), so it is enough to check the claim for a grid. This reduces then to proving that if R is partitioned into two rectangles R_1, R_2 by a line parallel to one of its sides, then $H(R) = H(R_1) + H(R_2)$. But this is clear, by definition of H and because f is additive. $\square$

Coming back to the proof and using the lemma, we obtain

$$0 = f(a)f(b) = \sum_{i=1}^{n} f(x_i)^2,$$

which yields $f(x_i) = 0$ for all i and so $\frac{x_i}{a}$ are rational numbers. Since we could have worked with b instead of a from the very beginning, it follows that $\dfrac{b}{x_i}$ are also rational numbers and the theorem is proved. □

Remark 12.23. In [51], the following similar result is proved using Dehn's original method: a rectangle R has at least one rational sidelength and is tiled by smaller rectangles, each having a rational perimeter. Then all sidelengths of R and of the tiling rectangles are rational.

The following stronger result will be used in the next problems. The proof is very similar to that of problem 30.

Theorem 12.24. *(Dehn) Let the rectangle R_0 be tiled with rectangles $R_1, R_2, \ldots, R_n$. Let r_i be the side-ratio of R_i. Then $r_0 \in \mathbb{Q}(r_1, r_2, \ldots, r_n)$.*

Proof. Suppose that $r_0 \notin L = \mathbb{Q}(r_1, r_2, \ldots, r_n)$, so we may pick a basis B of $\mathbb{R}$ as L-vector space, such that the base and height b_0, h_0 of R_0 are in B. Define the invariant of a rectangle $b \times h$ to be $c_1(b)c_2(h) - c_1(h)c_2(b)$, where $c_1(x), c_2(x)$ are the coordinates of x with respect to $b_0, h_0 \in B$. As c_i are L-linear maps, it is easy to check that R_i has invariant 0. An argument similar to the one used in lemma 12.22 shows that the invariant is additive, so the invariant of R_0 is the sum of the invariants of the R_i's, that is 0. This contradicts the fact that R_0 has invariant 1. □

We end this chapter with a truly amazing result, whose proof is taken from the beautiful paper [34]. The proof requires the following rather technical theorem, which we will take for granted, since the proof would take us quite far afield.

Theorem 12.25. *(Wall) If r is an algebraic number all of whose conjugates have positive real part, then there exist $n \geq 1$ and positive rational numbers $c_1, c_2, \ldots, c_n$ such that*

$$1 = c_1 r + \cfrac{1}{c_2 r + \cfrac{1}{c_3 r + \cdots + \cfrac{1}{c_n r}}}.$$

31. Let r be a positive real number. Prove the equivalence of the following statements:

    1) the unit square can be tiled with finitely many rectangles similar to the $1 \times r$ rectangle and having sides parallel to the sides of the square.
    2) r is algebraic and all of its conjugates have positive real part.

Laczkovich-Szekeres theorem

Proof. Prepare for a long but exciting battle. Recall that the side-ratio of a rectangle is the quotient between its horizontal side and its vertical side.

Step 0. Let us suppose that r is algebraic and all of its conjugates have positive real part. Choose $c_1, c_2, \dots$ as in Wall's theorem and cut off a rectangle of side-ratio $c_1 r$ from a unit square, by a vertical cut. From the remaining rectangle, cut off a rectangle of ratio $\frac{1}{c_2 r}$ by a horizontal cut. The remaining rectangle has side-ratio $c_3 r + \frac{1}{c_4 r + \cdots}$. Repeating the process, we tile the unit square by rectangles with side-ratios $c_1 r, \frac{1}{c_2 r}, c_3 r, \dots$. We conclude using the fact that the c_i's are rational. This proves one implication. The other implication is harder and requires the following steps.

Step 1. First, we prove that if there is a partition of a square with rectangles whose side-ratio is r or $1/r$, then r is algebraic. Let x be any real number which is transcendental over $\mathbb{Q}(r)$ and scale the vertical axis by a factor of x. Applying theorem 12.24 to this new situation, we deduce that $x \in \mathbb{Q}(rx, x/r)$, so we have an equality $xP(rx, x/r) = Q(rx, x/r)$ for $P, Q \in \mathbb{Q}[X, Y]$. As x is transcendental over $\mathbb{Q}(r)$, the previous equality yields $XP(rX, X/r) = Q(rX, X/r)$ in $\mathbb{Q}(r)[X]$. Looking at the leading coefficient, we easily obtain a nontrivial polynomial with rational coefficients killing r, thus r is algebraic.

Step 2. We claim that if $r > 0$ is algebraic and has a conjugate with real part smaller than or equal to 0, then it has a conjugate with negative real part. Indeed, otherwise r has a purely imaginary conjugate and so there is a root x of the minimal polynomial p of r such that $p(x) = p(-x) = 0$. As p is irreducible, p must divide $p(X) + p(-X)$ and so $p(r) + p(-r) = 0$. Thus $-r < 0$ is a root of p and we are done.

Step 3. Assume now that r has a conjugate with negative real part. Let p be the minimal polynomial of r and let Q be the companion matrix[6] of p. Let B be a basis of $\mathbb{R}$ as $\mathbb{Q}(r)$-vector space, chosen such that $1 \in B$. If $x \in \mathbb{R}$, its coordinate with respect to $1 \in B$ is of the form $a_0 + a_1 + \cdots + a_{n-1}r^{n-1}$ with $a_i \in \mathbb{Q}$ and we let v_x be the column vector with coordinates $a_0, a_1, \ldots, a_{n-1}$. Thus $v_{rx} = Qv_x$. The key point is the following:

Lemma 12.26. *There exists a symmetric matrix $M \in M_n(\mathbb{R})$ such that*

1) $v^t MQv \geq 0$ for all $v \in \mathbb{R}^n$.

2) There exists $s \in \mathbb{Q}(r)$ such that $v_s^t M v_s < 0$.

The proof of this lemma will be given in the next step. Let us see why this finishes the proof of the theorem. Define the invariant of a rectangle $b \times h$ to be $v_b^t M v_h$. In the usual way, we obtain that this invariant is additive. The invariant of a rectangle $b \times br$ is $v_b^t M v_{br} = v_b^t MQv_b \geq 0$ and the invariant of a rectangle $br \times b$ is equal to that of a rectangle $b \times br$, as M is symmetric. So rectangles of side-ratio r have nonnegative invariant. On the other hand, 2) ensures the existence of an $s \times s$ square with negative invariant. This square cannot be tiled by rectangles with side-ratio r and we are done.

Step 4. We prove the lemma. As p has no multiple root (because it is irreducible), Q is diagonalizable over $\mathbb{C}$. As it has real entries, it is immediate to deduce the existence of a block-diagonal matrix D and of $P \in GL_n(\mathbb{R})$ such that $Q = PDP^{-1}$. Moreover, each block of D has side 1 (and corresponds to a real eigenvalue of Q) or is of the form $\begin{pmatrix} \alpha & \beta \\ -\beta & \alpha \end{pmatrix}$ (and corresponds to complex conjugate eigenvalues $\alpha \pm i\beta$ of Q). Let T be the diagonal matrix whose entries are the real parts of the eigenvalues of Q. We choose $M = (P^{-1})^t TP^{-1}$. Then for all v we have

$$v^t MQv = (P^{-1}v)^t TD(P^{-1}v) = \sum_i w_i^2 \alpha_i \geq 0,$$

[6] Recall that this means that Q is the matrix of the multiplication by r, seen as an endomorphism of the $\mathbb{Q}$-vector space $\mathbb{Q}(r)$.

where w_i are the coordinates of $P^{-1}v$ and α_i are the squares of the real parts of the eigenvalues of Q. So the first property is satisfied. On the other hand, T has a negative entry (as r has a conjugate of negative real part by step 2 and the conjugates of r are exactly the eigenvalues of Q), which we may assume to be the first one. If e_1 is the column vector whose first coordinate is 1 and the other are 0, we deduce that $(Pe_1)^t MPe_1 < 0$. Simply choose $v \in \mathbb{Q}^n$ sufficiently close to Pe_1 to ensure that $v^t Mv < 0$ and then choose $s \in \mathbb{Q}(r)$ such that $v_s = v$. The lemma is proved. $\square$

12.7 Notes

The solutions to some of the problems in this chapter are due to the following people: Alon Amit (problem 11), Tom Belulovich (problem 4), Aart Blokhuis (problem 27), Iurie Boreico (problem 15), Zeb Brady (problem 21), Alexandru Chirvăsitu (problem 13), Darij Grinberg (problems 12, 15, 19), Omid Hatami (problem 12), Xiangyi Huang (problems 15, 20), Laszlo Lovasz (problem 27), James Merryfield (problem 2), Jorge Miranda (problem 14), Fedja Nazarov (problems 7, 24, 29), Hunter Spink (problem 8), J. Steinhardt (problem 18), Gjergji Zaimi (problems 5, 7, 9, 10, 11, 20).

Bibliography

[1] M. Ajtai, V. Chvátal, M. Newborn, and E. Szemerédi. Crossing-free subgraphs. *North-Holland Mathematics Studies*, 60:9–12, 1982.

[2] N. Alon and J. H. Spencer. *The probabilistic method*, volume 73 of *Wiley-Interscience in Discrete Mathematics and Optimization*. John Wiley & Sons, Inc., 3rd edition, 2008.

[3] T. Andreescu and G. Dospinescu. *Problems from the Book*. XYZ Press, 2nd edition, 2010.

[4] N. C. Ankeny and C. A. Rogers. A conjecture of Chowla. *The Annals of Mathematics, Second Series*, 53(3):541–550, May 1951.

[5] A. S. Besicovitch. On the linear independence of fractional powers of integers. *Journal of the London Mathematical Society*, 15(1):3–6, 1940.

[6] M. Bhargava. The factorial function and generalizations. *The American Mathematical Monthly*, 107(9):783–799, November 2000.

[7] B. Bollobás. *Combinatorics: set systems, hypergraphs, families of vectors, and combinatorial probability*. Cambridge University Press, 1986.

[8] E. Bombieri. On the large sieve. *Mathematika*, 12(2):201–225, 1965.

[9] E. Bombieri. *Le grand crible dans la théorie analytique des nombres*. Number 18 in Astérisque. Société mathématique de France, 2nd edition, 1974.

[10] I. Boreico. My favorite problem: linear independence of radicals. *The Harvard College Mathematics Review*, 2(1), 2008.

[11] J. Bourgain, N. Katz, and T. Tao. A sum-product estimate in finite fields, and applications. *Geometric and Functional Analysis*, 14(1):27–57, 2004.

[12] T. C. Brown and J. P. Buhler. A density version of a geometric Ramsey theorem. *Journal of Combinatorial Theory, Series A*, 32(1):20–34, 1982.

[13] J. W. S. Cassels. *Local fields*, volume 3 of *London Mathematical Society Student Texts*. Cambridge University Press, 1986.

[14] K. Chandrasekharan. The work of Enrico Bombieri. In R. D. James, editor, *Proceedings of the International Congress of Mathematicians*, volume 1, pages 3–10. Canadian Mathematical Congress, 1974.

[15] F. R. K. Chung. Sphere-and-point incidence relations in high dimensions with applications to unit distances and furthest-neighbor pairs. *Discrete & Computational Geometry*, 4(1):183–190, 1989.

[16] F. R. K. Chung, E. Szemerédi, and W. T. Trotter. The number of different distances determined by a set of points in the Euclidean plane. *Discrete & Computational Geometry*, 7(1):1–11, 1992.

[17] F. Clarke and C. Jones. A congruence for factorials. *Bulletin of the London Mathematical Society*, 36(4):553–558, 2004.

[18] J. H. E. Cohn. On square Fibonacci numbers. *Journal of the London Mathematical Society*, s1-39(1):537–540, 1964.

[19] A. Cojocaru and M. R. Murty. *An introduction to sieve methods and their applications*, volume 66 of *London Mathematical Society Student Texts*. Cambridge University Press, 2005.

[20] J. H. Conway and A. J. Jones. Trigonometric diophantine equations (on vanishing sums of roots of unity). *Acta Arithmetica*, 30(3):229–240, 1976.

[21] L. E. Dickson. *Introduction to the theory of numbers*. Dover Publications, Inc. New York, 1957.

[22] A. Dumitrescu, M. Sharir, and C. D. Tóth. Extremal problems on triangle areas in two and three dimensions. *Journal of Combinatorial Theory, Series A*, 116(7):1177–1198, 2009.

[23] G. Elekes. On the number of sums and products. *Acta Arithmetica*, 81(4):365–367, 1997.

[24] P. D. T. A. Elliott. The Turán-Kubilius inequality. *Proceedings of the American Mathematical Society*, 65(1):8–10, July 1977.

[25] P. D. T. A. Elliott. *Probabilistic number theory*, volume 1, 2. Springer-Verlag, 1979, 1980.

[26] P. Erdős. A theorem of Sylvester and Schur. *Journal of the London Mathematical Society*, s1-9(4):282–288, 1934.

[27] P. Erdős. On the greatest prime factor of $\prod_{k=1}^{n} f(k)$. *Journal of the London Mathematical Society*, s1-27(3):379–384, 1952.

[28] P. Erdős and M. Kac. The Gaussian law of errors in the theory of additive number theoretic functions. *American Journal of Mathematics*, 62(1):738–742, 1940.

[29] P. Erdős, A. Rényi, and V. T. Sós. On a problem of graph theory. *Studia Sci. Math. Hungar*, 1:215–235, 1966.

[30] J. H. Evertse. The number of solutions of linear equations in roots of unity. *Acta Arithmetica*, 89(1):45–51, 1999.

[31] B. Farhi. An identity involving the least common multiple of binomial coefficients and its application. *The American Mathematical Monthly*, 116(9):836–839, 2009.

[32] H. Flanders. Generalization of a theorem of Ankeny and Rogers. *The Annals of Mathematics, Second Series*, 57(2):392–400, March 1953.

[33] P. Frankl, R. L. Graham, and V. Rodl. On subsets of abelian groups with no three term arithmetic progression. *Journal of Combinatorial Theory, Series A*, 45(1):157–161, 1987.

[34] C. Freiling and D. Rinne. Tiling a square with similar rectangles. *Math. Research Letters*, 1:547–558, 1994.

[35] D. Grinberg. An inequality involving $2n$ numbers. `http://www.cip.ifi.lmu.de/~grinberg/Yugoslavia1998.pdf`.

[36] L. Guth and N. H. Katz. On the Erdős distinct distance problem in the plane. `arXiv:1011.4105`.

[37] H. Halberstam and H. E. Richert. *Sieve Methods.* Academic Press New York, 1974.

[38] D. Hanson. On the product of the primes. *Canad. Math. Bull*, 15:33–37, 1972.

[39] G. H. Hardy and S. Ramanujan. The normal number of prime factors of a number n. *Quart. Journal Math*, 48:76–92, 1917.

[40] A. Hildebrand. On Wirsing's mean value theorem for multiplicative functions. *Bulletin of the London Mathematical Society*, 18(2):147–152, 1986.

[41] C. Hooley. On the greatest prime factor of a quadratic polynomial. *Acta Mathematica*, 117(1):281–299, 1967.

[42] C. Huneke. The friendship theorem. *The American Mathematical Monthly*, 109(2):192–194, February 2002.

[43] K. F. Ireland and M. I. Rosen. *A Classical Introduction to Modern Number Theory*, volume 84 of *Graduate Texts in Mathematics.* Springer, 1990.

[44] H. Iwaniec and E. Kowalski. *Analytic Number Theory*, volume 53 of *Colloquium Publications.* American Mathematical Society, 2004.

[45] C. Lech. A note on recurring series. *Arkiv för Matematik*, 2(5):417–421, 1953.

[46] Y. V. Linnik. The large sieve. In *Dokl. Akad. Nauk SSSR*, volume 30, pages 292–294, 1941. in Russian.

[47] Y. V. Linnik. A remark on the least quadratic non-residue. In *C. R. (Doklady) Acad. Sci. URSS (N.S.)*, volume 36, pages 119–120, 1942.

[48] K. Mahler. On the fractional parts of the powers of a rational number. *Acta Arithmetica*, 3:89–93, 1938.

[49] K. Mahler. *p-adic numbers and their functions.*, volume 76 of *Cambridge Tracts in Mathematics*. Cambridge University Press, 2nd edition, 1981.

[50] H. B. Mann. On linear relations between roots of unity. *Mathematika*, 12(2):107–117, 1965.

[51] D. G. Mead and S. K. Stein. More on rectangles tiled by rectangles. *The American Mathematical Monthly*, 100(7):641–643, August-September 1993.

[52] R. Meshulam. On subsets of finite abelian groups with no 3-term arithmetic progressions. *J. Comb. Theory, Series A*, 71(1):168–172, 1995.

[53] P. Monsky. On dividing a square into triangles. *The American Mathematical Monthly*, 77(2):161–164, February 1970.

[54] P. Monsky. Simplifying the proof of Dirichlet's theorem. *The American Mathematical Monthly*, 100(9):861–862, 1993.

[55] H. L. Montgomery. *Topics in Multiplicative Number Theory*, volume 227 of *Lecture Notes in Mathematics*. Springer-Verlag, 1971.

[56] H. L. Montgomery. The analytic principle of the large sieve. *Bull. Amer. Math. Soc.*, 84:547–567, 1978.

[57] H. L. Montgomery and R. C. Vaughan. The large sieve. *Mathematika*, 20(2):119–134, 1973.

[58] L. J. Mordell. On the linear independence of algebraic numbers. *Pacific Journal of Mathematics*, 3(3):625–630, 1953.

[59] M. R. Murty. Small solutions of polynomial congruences. *Indian Journal of Pure and Applied Mathematics*, 41(1):15–23, 2010.

[60] T. Nagell. Généralisation d'un théorème de Tchebycheff. *J. Math. Pures Appl., Série 8*, 4:343–356, 1921.

[61] M. B. Nathanson. An exponential congruence of Mahler. *The American Mathematical Monthly*, 79(1):55–57, January 1972.

[62] D. J. Newman. *Analytic number theory*, volume 177 of *Graduate Texts in Mathematics*. Springer Verlag, 1998.

[63] P. J. O'Hara. Another proof of Bernstein's theorem. *The American Mathematical Monthly*, 80(6):673–674, June-July 1973.

[64] H. Pan and Z. W. Sun. A combinatorial identity with application to Catalan numbers. *Discrete Mathematics*, 306(16):1921–1940, 2006.

[65] A. Postnikov. Intransitive trees. *Journal of Combinatorial Theory, Series A*, 79(2):360–366, August 1997.

[66] V. V. Prasolov. *Polynomials*, volume 11 of *Algorithms and Computation in Mathematics*. Springer Verlag, 2009.

[67] A. Robert. *A course in p-adic analysis*, volume 198 of *Graduate Texts in Mathematics*. Springer-Verlag New York, Inc., 2000.

[68] A. Robert and M. Zuber. The Kazandzidis supercongruences. a simple proof and an application. *Rendiconti del Seminario Matematico della Università di Padova*, 94:235–243, 1995.

[69] B. E. Sagan. Proper partitions of a polygon and k-Catalan numbers. *Ars Combinatoria*, 88:109–124, 2008.

[70] J. L. Selfridge and E. G. Straus. On the determination of numbers by their sums of a fixed order. *Pacific Journal of Mathematics*, 8(4):847–856, 1958.

[71] I. E. Shparlinski. Exponential sums in coding theory, cryptology and algorithms. In H. Niederreiter, editor, *Coding Theory and Cryptology*, pages 323–383. World Scientific Publishing, Singapore, 2002.

[72] J. Solymosi. On sum-sets and product-sets of complex numbers. *Journal de Théorie des Nombres de Bordeaux*, 17(3):921–924, 2005.

[73] J. Solymosi. On the number of sums and products. *Bulletin of the London Mathematical Society*, 37(4):491–494, 2005.

[74] J. Solymosi and V. Vu. Distinct distances in high dimensional homogeneous sets. In *Towards a Theory of Geometric Graphs*, volume 342 of *Contemporary Mathematics*, pages 259–268. American Mathematical Society, 2004.

[75] J. Spencer, E. Szemerédi, and W. T. Trotter. Unit distances in the Euclidean plane. In B. Bollobás, editor, *Graph theory and Combinatorics*, pages 293–303. Academic Press, New York, 1984.

[76] R. P. Stanley. *Enumerative combinatorics, vol. II*, volume 62 of *Cambridge Studies in Advanced Mathematics*. Cambridge University Press, 1999.

[77] Z. W. Sun. On sums of binomial coefficients modulo p^2. `http://math.nju.edu.cn/~zwsun/`.

[78] Z. W. Sun and R. Tauraso. New congruences for central binomial coefficients. *Advances in Applied Mathematics*, 45(1):125–148, 2010.

[79] Z. W. Sun and R. Tauraso. On some new congruences for binomial coefficients. *Int. J. Number Theory*, 7(3):645–662, 2011.

[80] L. A. Székely. Crossing numbers and hard Erdős problems in discrete geometry. *Combinatorics, Probability and Computing*, 6(3):353–358, 1997.

[81] E. Szemerédi and W. T. Trotter. Extremal problems in discrete geometry. *Combinatorica*, 3(3):381–392, 1983.

[82] T. Tao and V. Vu. *Additive combinatorics*, volume 105 of *Cambridge Studies in Advanced Mathematics*. Cambridge University Press, 2006.

[83] R. Tauraso. An elementary proof of a Rodriguez-Villegas supercongruence. `arXiv:0911.4261`.

[84] A. Weil. Numbers of solutions of equations in finite fields. *Bulletin of the American Mathematical Society*, 55(5):497–508, 1949.

[85] E. Wirsing. Das asymptotische Verhalten von Summen über multiplikative Funktionen. *Mathematische Annalen*, 143(1):75–102, 1961.

[86] E. Wirsing. Das asymptotische Verhalten von Summen über multiplikative Funktionen II. *Acta Mathematica Hungarica*, 18(3):411–467, 1967.

Printed by "Combinatul Poligrafic"